Springer Proceedings in Mathematics

Volume 13

For further volumes:
http://www.springer.com/series/8806

Springer Proceedings in Mathematics

This book series features volumes of selected contributions from workshops and conferences in all areas of current research activity in mathematics. After an overall evaluation, at the hands of the publisher, of the interest, scientific quality, and timeliness of each proposal, every individual contribution has been refereed to standards comparable to those of leading mathematics journals. This series thus presents to the research community well-edited and authoritative reports on newest developments in the most interesting and promising areas of mathematical research today.

Marian Neamtu • Larry Schumaker
Editors

Approximation Theory XIII: San Antonio 2010

Springer

Editors
Marian Neamtu
Center for Constructive Approximation
Department of Mathematics
Vanderbilt University
Nashville, TN 37240
USA
mike.neamtu@gmail.com

Larry Schumaker
Center for Constructive Approximation
Department of Mathematics
Vanderbilt University
Nashville, TN 37240
USA
larry.schumaker@vanderbilt.edu

ISSN 2190-5614 e-ISSN 2190-5622
ISBN 978-1-4614-0771-3 e-ISBN 978-1-4614-0772-0
DOI 10.1007/978-1-4614-0772-0
Springer New York Dordrecht Heidelberg London

Mathematics Subject Classification (2010): 41Axx, 65Dxx

Printed on acid-free paper

Springer is part of Springer Science+Business Media (www.springer.com)

Preface

These proceedings were prepared in connection with the international conference *Approximation Theory XIII*, which was held during March 7–10, 2010 in San Antonio, Texas. The conference was the thirteenth in a series of meetings in Approximation Theory held at various locations in the United States, and was attended by 144 participants. Previous conferences in the series were held in Austin, Texas (1973, 1976, 1980, 1992); College Station, Texas (1983, 1986, 1989, 1995); Nashville, Tennessee (1998), St. Louis, Missouri (2001); Gatlinburg, Tennessee (2004); and San Antonio, Texas (2007).

We are particularly indebted to our plenary speakers: Albert Cohen (Paris), Oleg Davydov (Strathclyde), Gregory Fasshauer (Illinois Institute of Technology), Anne Gibert (University of Michigan), Bin Han (University of Alberta), Kirill Kopotun (University of Manitoba), and Vilmos Totik (University of South Florida), who provided inspiring talks and set a high standard of exposition in their descriptions of new directions for research. The conference also provided a forum for the awarding of the Popov Prize in Approximation Theory. The sixth Vasil A. Popov Prize was awarded to Joel A. Tropp (Cal Tech), who also presented a plenary lecture. Thanks are also due to the presenters of contributed papers, as well as everyone who attended, for making the conference a success.

We are especially grateful to the National Science Foundation for financial support, and also to the Department of Mathematics at Vanderbilt University for its logistical support.

We would also like to express our sincere gratitude to the reviewers who helped select articles for inclusion in this proceedings volume, and also for their suggestions to the authors for improving their papers.

Nashville, TN

Marian Neamtu
Larry L. Schumaker

Contents

An Asymptotic Equivalence Between Two Frame Perturbation Theorems 1
B. A. Bailey
1 The Perturbation Theorems 1
2 An Asymptotic Equivalence 3
References 7

Growth Behavior and Zero Distribution of Maximally Convergent Rational Approximants 9
Hans-Peter Blatt, René Grothmann, and Ralitza K. Kovacheva
References 15

Generalization of Polynomial Interpolation at Chebyshev Nodes 17
Debao Chen
1 Introduction 17
2 Generalization of Chebyshev Nodes 21
3 Lebesgue Functions 23
4 Properties of Pairs of Auxiliary Functions 30
5 Optimal Nodes for Lagrange Polynomial Interpolation 32
References 35

Green's Functions: Taking Another Look at Kernel Approximation, Radial Basis Functions, and Splines 37
Gregory E. Fasshauer
1 Introduction 37
2 Toward an Intuitive Interpretation of Native Spaces 39
2.1 What is the Current Situation? 39
2.2 Mercer's Theorem and Eigenvalue Problems 41
2.3 Green's Functions and Eigenfunction Expansions 42
2.4 Generalized Sobolev Spaces 45
3 Flat Limits 48
3.1 Infinitely Smooth RBFs 49

3.2 Finitely Smooth RBFs ... 50
4 Stable Computation ... 53
4.1 An Eigenfunction Expansion for Gaussians ... 53
4.2 The RBF-QR Algorithm ... 54
5 Dimension Independent Error Bounds ... 57
5.1 The Current Situation ... 57
5.2 New Results on (Minimal) Worst-Case Weighted L_2 Error ... 58
6 Summary ... 60
References ... 61

Sparse Recovery Algorithms: Sufficient Conditions in Terms of Restricted Isometry Constants ... 65
Simon Foucart
1 Introduction ... 65
1.1 Basis Pursuit ... 66
1.2 Iterative Hard Thresholding ... 66
1.3 Compressive Sampling Matching Pursuit ... 66
2 Restricted Isometry Constants ... 66
3 Basis Pursuit ... 69
4 Iterative Hard Thresholding ... 72
5 Compressive Sampling Matching Pursuit ... 74
References ... 77

Lagrange Interpolation and New Asymptotic Formulae for the Riemann Zeta Function ... 79
Michael I. Ganzburg
1 Introduction ... 79
2 Lagrange Interpolation ... 80
3 Asymptotic Behavior of the Interpolation Error ... 83
4 Asymptotic Formulae for $\zeta(\mathbf{s})$ ... 86
5 $\mathbf{L_p(-1,1)}$-Asymptotics and Criteria for $\zeta(\mathbf{s}) = \mathbf{0}$ and $\zeta(\mathbf{s}) \neq \mathbf{0}$ ... 89
6 Remarks ... 92
References ... 92

Active Geometric Wavelets ... 95
Itai Gershtansky and Shai Dekel
1 Introduction ... 95
2 Theoretical Background ... 98
2.1 A Jackson Estimate for Piecewise Polynomial Approximation Using Non-convex Domains ... 98
2.2 Adaptive Local Selection of the Weight μ ... 100
3 Overview of the AGW Algorithm ... 103
4 Experimental Results ... 108
References ... 109

Interpolating Composite Systems 111
Philipp Grohs
1 Introduction 111
2 Composite Dilation Systems 112
3 Interpolating Systems 114
4 Shearlets 118
5 Conclusion 120
References 120

Wavelets and Framelets Within the Framework of Nonhomogeneous Wavelet Systems 121
Bin Han
1 Introduction 121
2 Nonhomogeneous Wavelet Systems in $L_2(\mathbb{R}^d)$ 125
3 Frequency-Based Nonhomogeneous Dual Framelets in the Distribution Space 132
4 Wavelets and Framelets in Function Spaces 140
5 Wavelets and Framelets Derived from Filter Banks 145
6 Discrete Framelet Transform and Its Basic Properties 149
7 Directional Tight Framelets in $L_2(\mathbb{R}^d)$ and Projection Method 154
References 158

Compactly Supported Shearlets 163
Gitta Kutyniok, Jakob Lemvig, and Wang-Q Lim
1 Introduction 164
1.1 Directional Representation Systems 165
1.2 Anisotropic Features, Discrete Shearlet Systems, and Quest for Sparse Approximations 166
1.3 Continuous Shearlet Systems 167
1.4 Applications 168
1.5 Outline 170
2 2D Shearlets 171
2.1 Preliminaries 171
2.2 Classical Construction 172
2.3 Constructing Compactly Supported Shearlets 173
3 Sparse Approximations 175
3.1 Cartoon-like Image Model 175
3.2 Optimally Sparse Approximation of Cartoon-like Images 176
4 Shearlets in 3D and Beyond 177
4.1 Pyramid-Adapted Shearlet Systems 178
4.2 Sparse Approximations of 3D Data 181
5 Conclusions 184
References 184

Shearlets on Bounded Domains ... 187
Gitta Kutyniok and Wang-Q Lim
1 Introduction ... 187
1.1 Optimally Sparse Approximations of Cartoon-like Images ... 188
1.2 Shortcomings of this Cartoon-like Model Class ... 189
1.3 Our Model for Cartoon-like Images on Bounded Domains ... 189
1.4 Review of Shearlets ... 191
1.5 Surprising Result ... 191
1.6 Main Contributions ... 192
1.7 Outline ... 193
2 Compactly Supported Shearlets ... 193
2.1 Compactly Supported Shearlet Frames for $L^2(\mathbb{R}^2)$ ... 193
2.2 Compactly Supported Shearlet Frames for $L^2(\Omega)$ ... 195
3 Optimal Sparsity of Shearlets on Bounded Domains ... 196
3.1 Main Theorem 1 ... 196
3.2 Architecture of the Proof of Theorem 1 ... 197
4 Proof of Theorem 1 ... 198
4.1 Case 1: The Smooth Part ... 198
4.2 Case 2: The Non-Smooth Part ... 199
5 Discussion ... 205
References ... 206

On Christoffel Functions and Related Quantities for Compactly Supported Measures ... 207
D. S. Lubinsky
1 Introduction ... 207
2 Proof of Theorem 6 ... 214
References ... 219

Exact Solutions of Some Extremal Problems of Approximation Theory ... 221
A.L. Lukashov
1 Introduction ... 221
2 Proofs ... 223
References ... 228

A Lagrange Interpolation Method by Trivariate Cubic C^1 Splines of Low Locality ... 231
G. Nürnberger and G. Schneider
1 Introduction ... 231
2 Preliminaries ... 232
3 A Uniform Partition Consisting of Tetrahedra and Octahedra ... 235
4 A (Partial) Worsey–Farin Refinement of $\diamondsuit$ ... 239

5 Local Lagrange Interpolation by $\mathscr{S}_3^1(\widetilde{\diamondsuit})$ 241
6 Numerical Tests 247
References 248

Approximation of Besov Vectors by Paley–Wiener Vectors in Hilbert Spaces 249
Isaac Z. Pesenson and Meyer Z. Pesenson
1 Introduction 249
2 Paley–Wiener Subspaces Generated by a Self-adjoint Operator in a Hilbert Space 251
3 Direct and Inverse Approximation Theorems 253
References 261

A Subclass of the Length 12 Parameterized Wavelets 263
David W. Roach
1 Introduction 263
2 The Nonlinear Equations 264
3 Length Four 266
4 Length 12 Subclass 267
5 A Numerical Experiment 272
References 274

Geometric Properties of Inverse Polynomial Images 277
Klaus Schiefermayr
1 Introduction 277
2 The Number of (Analytic) Jordan Arcs of an Inverse Polynomial Image 278
3 The Connectedness of an Inverse Polynomial Image 285
References 287

On Symbolic Computation of Ideal Projectors and Inverse Systems 289
Boris Shekhtman
1 Introduction 289
2 Preliminaries 291
2.1 Multiplication Operators 291
2.2 Duality 291
3 The Main Result 292
4 A Couple of Examples 293
References 294

The Dimension of the Space of Smooth Splines of Degree 8 on Tetrahedral Partitions 297
Xiquan Shi, Ben Kamau, Fengshan Liu, and Baocai Yin
1 Introduction 298
2 Main Results 300
References 320

On Simultaneous Approximation in Function Spaces 321
Eyad Abu-Sirhan
1 Introduction 321
2 Preliminary Results 322
3 Main Result 326
References 328

Chalmers–Metcalf Operator and Uniqueness of Minimal Projections in ℓ_∞^n and ℓ_1^n Spaces 331
Lesław Skrzypek
1 Introduction 331
2 Chalmers–Metcalf Operator for Hyperplanes in ℓ_∞^n 334
3 Chalmers–Metcalf Operator for Hyperplanes in ℓ_n^1 337
References 344

The Polynomial Inverse Image Method 345
Vilmos Totik
1 Introduction 345
2 The Bernstein Inequality on General Sets 347
3 The Model case $[-1,1]$, Admissible Polynomial Maps, Approximation 348
4 The Model case C_1, Sharpened form of Hilbert's Lemniscate Theorem 350
5 A Critical Point in the Method 352
6 The Markoff Inequality for Several Intervals 354
7 Bernstein's Inequality on Curves 355
8 Asymptotics for Christoffel Functions 356
9 Lubinsky's Universality on General Sets 358
10 Fine Zero Spacing of Orthogonal Polynomials 359
11 Polynomial Approximation on Compact Subsets of the Real Line 360
12 Appendix: Basic Notions from Logarithmic Potential Theory 362
References 363

On Approximation of Periodic Analytic Functions by Linear Combinations of Complex Exponents 367
Vesselin Vatchev
1 Introduction 367
2 The Minimization Problem for Periodic Functions 368
3 Related Properties of Orthogonal Polynomials 370
4 Approximation with Imaginary Exponents 372
References 374

Matrix Extension with Symmetry and Its Applications ... 375
Xiaosheng Zhuang
1 Introduction and Motivation ... 375
1.1 The Matrix Extension Problems ... 376
1.2 Motivation ... 378
1.3 Prior Work and Our Contributions ... 381
1.4 Outline ... 382
2 Orthogonal Matrix Extension with Symmetry ... 382
2.1 An Algorithm for the Orthogonal Matrix Extension with Symmetry ... 383
2.2 Application to Filter Banks and Orthonormal Multiwavelets with Symmetry ... 390
3 Construction of Symmetric Complex Tight Framelets ... 394
3.1 Symmetric Complex d-Refinable Functions ... 394
3.2 Tight Framelets via Matrix Extension ... 398
4 Biorthogonal Matrix Extension with Symmetry ... 401
4.1 Proof of Theorem 4 and an Algorithm ... 402
4.2 Application to Construction of Biorthogonal Multiwavelets with Symmetry ... 408
References ... 414

List of Contributors

Eyad Abu-Sirhan
Tafila Technical University, Tafila, Jordan

B. A. Bailey
Department of Mathematics, University of Connecticut, Storrs, CT, USA

Hans-Peter Blatt
Katholische Universität Eichstätt-Ingolstadt, Mathematisch-Geographische Fakultät, Eichstätt, Germany

Debao Chen
Computer Science Department, Oklahoma State University-Tulsa, Tulsa, OK, USA

Shai Dekel
GE Healthcare and School of Mathematical Sciences, Tel Aviv University, Tel Aviv, Israel

Gregory E. Fasshauer
Illinois Institute of Technology, Chicago, IL, USA

Simon Foucart
Department of Mathematics, Drexel University, Philadelphia, PA, USA

Michael I. Ganzburg
Hampton University, Hampton, VA, USA

Itai Gershtansky
School of Mathematical Sciences, Tel Aviv University, Tel Aviv, Israel

Philipp Grohs
TU Graz, Institute of Geometry, Graz, Austria

René Grothmann
Katholische Universität Eichstätt-Ingolstadt, Mathematisch-Geographische Fakultät, Eichstätt, Germany

Bin Han
Department of Mathematical and Statistical Sciences, University of Alberta, Edmonton, Alberta, Canada

Ben Kamau
Mathematics Department, Columbus State University, Columbus, GA, USA

Ralitza K. Kovacheva
Bulgarian Academy of Sciences, Institute of Mathematics and Informatics, Sofia, Bulgaria

Gitta Kutyniok
Department of Mathematics, Technische Universitüt Berlin, Berlin

Jakob Lemvig
Department of Mathematics, Technical University of Denmark, Lyngby, Denmark

Wang-Q Lim
Department of Mathematics, Technische Universitüt Berlin, Berlin

Fengshan Liu
Department of Mathematical Sciences, Delaware State University, Dover, DE, USA

D. S. Lubinsky
School of Mathematics, Georgia Institute of Technology, Atlanta, GA, USA

A. L. Lukashov
Fatih University, Istanbul, Turkey
Saratov State University, Saratov, Russia

G. Nürnberger
University of Mannheim, Mannheim, Germany

Isaac Z. Pesenson
Department of Mathematics, Temple University, Philadelphia, PA, USA

Meyer Z. Pesenson
Department of Computing and Mathematical Sciences, California Institute of Technology, Pasadena, CA, USA

David W. Roach
Murray State University, Murray, KY, USA

Klaus Schiefermayr
University of Applied Sciences Upper Austria, School of Engineering and Environmental Sciences, Wels, Austria

G. Schneider
University of Mannheim, Mannheim, Germany

Boris Shekhtman
Department of Mathematics and Statistics, University of South Florida, Tampa, FL, USA

Xiquan Shi
Department of Mathematical Sciences, Delaware State University, Dover, DE, USA

Eyad Abu-Sirhan
Tafila Technical University, Tafila, Jordan

Lesław Skrzypek
Department of Mathematics and Statistics, University of South Florida, Tampa, FL, USA

Vilmos Totik
Bolyai Institute, Analysis Research Group of the Hungarian Academy of Sciences, University of Szeged, Szeged, Hungary

Department of Mathematics and Statistics, University of South Florida, Tampa, FL, USA

Vesselin Vatchev
University of Texas at Brownsville, Brownsville, TX, USA

Baocai Yin
Beijing Key Laboratory of Multimedia and Intelligent Software, College of Computer Science and Technology, Beijing University of Technology, Beijing, China

Xiaosheng Zhuang
Department of Mathematical and Statistical Sciences, University of Alberta, Edmonton, Alberta, Canada

An Asymptotic Equivalence Between Two Frame Perturbation Theorems

B. A. Bailey

Abstract In this paper, two stability results regarding exponential frames are compared. The theorems, (one proven herein, and the other in Sun and Zhou (J. Math. Anal. Appl. 235:159–167, 1999)), each give a constant such that if $\sup_{n\in\mathbb{Z}} \|\varepsilon_n\|_\infty < C$, and $(\mathrm{e}^{\mathrm{i}\langle\cdot,t_n\rangle})_{n\in\mathbb{Z}^d}$ is a frame for $L_2[-\pi,\pi]^d$, then $(\mathrm{e}^{\mathrm{i}\langle\cdot,t_n+\varepsilon_n\rangle})_{n\in\mathbb{Z}^d}$ is a frame for $L_2[-\pi,\pi]^d$. These two constants are shown to be asymptotically equivalent for large values of d.

1 The Perturbation Theorems

We define a frame for a separable Hilbert space H to be a sequence $(f_n)_n \subset H$ such that for some $0 < A \leq B$,

$$A^2\|f\|^2 \leq \sum_n |\langle f, f_n\rangle|^2 \leq B^2\|f\|^2, \quad f \in H.$$

The best A^2 and B^2 satisfying the inequality above are said to be the frame bounds for the frame. If $(e_n)_n$ is an orthonormal basis for H, the synthesis operator $Le_n = f_n$ is bounded, linear, and onto, iff $(f_n)_n$ is a frame. Equivalently, $(f_n)_n$ is a frame iff the operator L^* is an isomorphic embedding, (see [1]). In this case, A and B are the best constants such that

$$A\|f\| \leq \|L^*f\| \leq B\|f\|, \quad f \in H.$$

The simplest stability result regarding exponential frames for $L_2[-\pi,\pi]$ is the theorem below, which follows immediately from [2, Theorem 13, p 160].

B.A. Bailey
Department of Mathematics, University of Connecticut, Storrs, CT, 06269-3009, USA
e-mail: benjamin.bailey@uconn.edu

M. Neamtu and L. Schumaker (eds.), *Approximation Theory XIII: San Antonio 2010*, Springer Proceedings in Mathematics 13, DOI 10.1007/978-1-4614-0772-0_1,

Theorem 1. *Let $(t_n)_{n\in\mathbb{Z}} \subset \mathbb{R}$ be a sequence such that $(h_n)_{n\in\mathbb{Z}} := \left(\frac{1}{\sqrt{2\pi}}\mathrm{e}^{\mathrm{i}t_n x}\right)_{n\in\mathbb{Z}}$ is a frame for $L_2[-\pi,\pi]$ with frame bounds A^2 and B^2. If $(\tau_n)_{n\in\mathbb{Z}} \subset \mathbb{R}$ and $(f_n)_{n\in\mathbb{Z}} := \left(\frac{1}{\sqrt{2\pi}}\mathrm{e}^{\mathrm{i}\tau_n x}\right)_{n\in\mathbb{Z}}$ is a sequence such that*

$$\sup_{n\in\mathbb{Z}} |\tau_n - t_n| < \frac{1}{\pi}\ln\left(1+\frac{A}{B}\right), \tag{1}$$

then the sequence $(f_n)_{n\in\mathbb{Z}}$ is also a frame for $L_2[-\pi,\pi]$.

The following theorem is a very natural generalization of Theorem 1 to higher dimensions.

Theorem 2. *Let $(t_k)_{k\in\mathbb{N}} \subset \mathbb{R}^d$ be a sequence such that $(h_k)_{k\in\mathbb{N}} := \left(\frac{1}{(2\pi)^{d/2}}\mathrm{e}^{\langle(\cdot),t_k\rangle}\right)_{k\in\mathbb{N}}$ is a frame for $L_2[-\pi,\pi]^d$ with frame bounds A^2 and B^2. If $(\tau_k)_{k\in\mathbb{N}} \subset \mathbb{R}^d$ and $(f_k)_{k\in\mathbb{N}} := \left(\frac{1}{(2\pi)^{d/2}}\mathrm{e}^{\mathrm{i}\langle(\cdot),\tau_k\rangle}\right)_{k\in\mathbb{N}}$ is a sequence such that*

$$\sup_{k\in\mathbb{N}} \|\tau_k - t_k\|_\infty < \frac{1}{\pi d}\ln\left(1+\frac{A}{B}\right), \tag{2}$$

then the sequence $(f_k)_{k\in\mathbb{N}}$ is also a frame for $L_2[-\pi,\pi]^d$.

The proof of Theorem 2 relies on the following lemma:

Lemma 1. *Choose $(t_k)_{k\in\mathbb{N}} \subset \mathbb{R}^d$ such that $(h_k)_{k\in\mathbb{N}} := \left(\frac{1}{(2\pi)^{d/2}}\mathrm{e}^{\langle(\cdot),t_k\rangle}\right)_{k\in\mathbb{N}}$ satisfies*

$$\Big\|\sum_{k=1}^n a_k h_k\Big\|_{L_2[-\pi,\pi]^d} \le B\Big(\sum_{k=1}^n |a_k|^2\Big)^{1/2}, \quad \text{for all} \quad (a_k)_{k=1}^n \subset \mathbb{C}.$$

If $(\tau_k)_{k\in\mathbb{N}} \subset \mathbb{R}^d$, and $(f_k)_{k\in\mathbb{N}} := \left(\frac{1}{(2\pi)^{d/2}}\mathrm{e}^{\mathrm{i}\langle(\cdot),\tau_k\rangle}\right)_{k\in\mathbb{N}}$, then for all $r,s \ge 1$ and any finite sequence $(a_k)_k$, we have

$$\left\|\sum_{k=r}^s a_k(h_k - f_k)\right\|_{L_2[-\pi,\pi]^d} \le B\left(\mathrm{e}^{\pi d\left(\sup\limits_{r\le k\le s}\|\tau_k - t_k\|_\infty\right)} - 1\right)\Big(\sum_{k=r}^s |a_k|^2\Big)^{\frac{1}{2}}.$$

This lemma is a slight generalization of Lemma 5.3, proven in [3] using simple estimates. Lemma 1 is proven similarly. Now for the proof of Theorem 2.

Proof. Define $\delta = \sup_{k\in\mathbb{N}}\|\tau_k - t_k\|_\infty$. Lemma 1 shows that the map $\tilde{L}e_n = f_n$ is bounded and linear, and that

$$\|L - \tilde{L}\| \le B\left(\mathrm{e}^{\pi d\delta} - 1\right) := \beta A$$

for some $0 \le \beta < 1$. This implies

$$\|L^* f - \tilde{L}^* f\| \le \beta A, \quad \text{when} \quad \|f\| = 1. \tag{3}$$

Rearranging, we have

$$A(1-\beta) \leq \|\tilde{L}^* f\|, \quad \text{when} \quad \|f\| = 1.$$

By the previous remarks regarding frames, $(f_k)_{k\in\mathbb{N}}$ is a frame for $L_2[-\pi,\pi]^d$.

Theorem 3, proven in [4], is a more delicate frame perturbation result with a more complex proof:

Theorem 3. *Let $(t_k)_{k\in\mathbb{N}} \subset \mathbb{R}^d$ be a sequence such that $(h_k)_{k\in\mathbb{N}} := \left(\frac{1}{(2\pi)^{d/2}} e^{\langle(\cdot),t_k\rangle}\right)_{k\in\mathbb{N}}$ is a frame for $L_2[-\pi,\pi]^d$ with frame bounds A^2 and B^2. For $d \geq 1$, define*

$$D_d(x) := \left(1 - \cos\pi x + \sin\pi x + \frac{\sin\pi x}{\pi x}\right)^d - \left(\frac{\sin\pi x}{\pi x}\right)^d,$$

and let x_d be the unique number such that $0 < x_d \leq 1/4$ and $D_d(x_d) = \frac{A}{B}$. If $(\tau_k)_{k\in\mathbb{N}} \subset \mathbb{R}^d$ and $(f_k)_{k\in\mathbb{N}} := \left(\frac{1}{(2\pi)^{d/2}} e^{i\langle(\cdot),\tau_k\rangle}\right)_{k\in\mathbb{N}}$ is a sequence such that

$$\sup_{k\in\mathbb{N}} \|\tau_k - t_k\|_\infty < x_d, \tag{4}$$

then the sequence $(f_k)_{k\in\mathbb{N}}$ is also a frame for $L_2[-\pi,\pi]^d$.

2 An Asymptotic Equivalence

It is natural to ask how the constants x_d and $\frac{1}{\pi d}\ln\left(1+\frac{A}{B}\right)$ are related. Such a relationship is given in the following theorem.

Theorem 4. *If x_d is the unique number satisfying $0 < x_d < 1/4$ and $D_d(x_d) = \frac{A}{B}$, then*

$$\lim_{d\to\infty} \frac{x_d - \frac{1}{\pi d}\ln\left(1+\frac{A}{B}\right)}{\frac{\left[\ln\left(1+\frac{A}{B}\right)\right]^2}{6\pi\left(1+\frac{B}{A}\right)d^2}} = 1.$$

We prove the theorem with a sequence of propositions.

Proposition 1. *Let d be a positive integer. If*

$$f(x) := 1 - \cos(x) + \sin(x) + \mathrm{sinc}(x),$$
$$g(x) := \mathrm{sinc}(x),$$

then

$$(1) \quad f'(x) + g'(x) > 0, \quad x \in (0,\pi/4),$$
$$(2) \quad g'(x) < 0, \quad x \in (0,\pi/4),$$
$$(3) \quad f''(x) > 0, \quad x \in (0,\Delta) \quad \text{for some} \quad 0 < \Delta < 1/4.$$

The proof of Proposition 1 involves only elementary calculus and is omitted.

Proposition 2. *The following statements hold:*
(1) For $d > 0$, $D_d(x)$ and $D'_d(x)$ are positive on $(0,1/4)$.
(2) For all $d > 0$, $D''_d(x)$ is positive on $(0,\Delta)$.

Proof. Note $D_d(x) = f(\pi x)^d - g(\pi x)^d$ is positive. This expression yields

$$D'_d(x)/(d\pi) = f(\pi x)^{d-1} f'(\pi x) - g(\pi x)^{d-1} g'(\pi x) > 0 \quad \text{on} \quad (0,1/4)$$

by Proposition 1. Differentiating again, we obtain

$$\begin{aligned} D''_d(x)/(d\pi^2) &= (d-1)\left[f(\pi x)^{d-2}(f'(\pi x))^2 - g(\pi x)^{d-2}(g'(\pi x))^2\right] \\ &+ \left[f(\pi x)^{d-1} f''(\pi x) - g(\pi x)^{d-1} g''(\pi x)\right] \quad \text{on} \quad (0,1/4). \end{aligned}$$

If $g''(\pi x) \le 0$ for some $x \in (0,1/4)$, then the second bracketed term is positive. If $g''(\pi x) > 0$ for some $x \in (0,1/4)$, then the second bracketed term is positive if $f''(\pi x) - g''(\pi x) > 0$, but

$$f''(\pi x) - g''(\pi x) = \pi^2(\cos(\pi x) - \sin(\pi x))$$

is positive on $(0,1/4)$.

To show the first bracketed term is positive, it suffices to show that

$$f'(\pi x)^2 > g'(\pi x)^2 = (f'(\pi x) + g'(\pi x))(f'(\pi x) - g'(\pi x)) > 0$$

on $(0,\Delta)$. Noting $f'(\pi x) - g'(\pi x) = \pi(\cos(\pi x) + \sin(\pi x)) > 0$, it suffices to show that $f'(\pi x) + g'(\pi x) > 0$, but this is true by Proposition 1. □

Note that Proposition 2 implies x_d is unique.

Corollary 1. *We have* $\lim_{d\to\infty} x_d = 0$.

Proof. Fix $n > 0$ with $1/n < \Delta$, then $\lim_{d\to\infty} D_d(1/n) = \infty$ (since f increasing implies $0 < -\cos(\pi/n) + \sin(\pi/n) + \operatorname{sinc}(\pi/n)$). For sufficiently large d, $D_d(1/n) > \frac{A}{B}$. But $\frac{A}{B} = D_d(x_d) < D_d(1/n)$, so $x_d < 1/n$ by Proposition 2.

Proposition 3. *Define* $\omega_d = \frac{1}{\pi d}\ln\left(1+\frac{A}{B}\right)$. *We have*

$$\begin{aligned} \lim_{d\to\infty} d\left(\frac{A}{B} - D_d(\omega_d)\right) &= \frac{A}{6B}\left[\ln\left(1+\frac{A}{B}\right)\right]^2, \\ \lim_{d\to\infty} \frac{1}{d} D'_d(\omega_d) &= \pi\left(1+\frac{A}{B}\right), \\ \lim_{d\to\infty} \frac{1}{d} D'_d(x_d) &= \pi\left(1+\frac{A}{B}\right). \end{aligned}$$

Proof. (1) For the first equality, note that

$$D_d(\omega_d) = \left[(1+h(x))^{\ln(c)/x} - g(x)^{\ln(c)/x}\right]\Big|_{x=\frac{\ln(c)}{d}} \tag{5}$$

where $h(x) = -\cos(x)+\sin(x)+\mathrm{sinc}(x)$, $g(x) = \mathrm{sinc}(x)$, and $c = 1+\frac{A}{B}$. L'Hospital's rule implies that

$$\lim_{x\to 0}(1+h(x))^{\ln(c)/x} = c \quad \text{and} \quad \lim_{x\to 0} g(x)^{\ln(c)/x} = 1.$$

Looking at the first equality in the line above, another application of L'Hospital's rule yields

$$\lim_{x\to 0} \frac{(1+h(x))^{\ln(c)/x} - c}{x} = c\ln(c)\left[\frac{\frac{h'(x)}{1+h(x)} - 1}{x} - \frac{\ln(1+h(x)) - x}{x^2}\right]. \tag{6}$$

Observing that $h(x) = x + x^2/3 + O(x^3))$, we see that

$$\lim_{x\to 0} \frac{\frac{h'(x)}{1+h(x)} - 1}{x} = -\frac{1}{3}.$$

L'Hospital's rule applied to the second term on the right hand side of (6) gives

$$\lim_{x\to 0} \frac{(1+h(x))^{\ln(c)/x} - c}{x} = \frac{-c\ln(c)}{6}. \tag{7}$$

In a similar fashion,

$$\lim_{x\to 0} \frac{g(x)^{\ln(c)/x} - 1}{x} = \ln(c)\lim_{x\to 0}\left[\frac{\frac{g'(x)}{g(x)}}{x} - \frac{\ln(g(x))}{x^2}\right]. \tag{8}$$

Observing that $g(x) = 1 - x^2/6 + O(x^4)$, we see that

$$\lim_{x\to 0} \frac{\frac{g'(x)}{g(x)}}{x} = -\frac{1}{3}.$$

L'Hospital's rule applied to the second term on the right hand side of (8) gives

$$\lim_{x\to 0} \frac{g(x)^{\ln(c)/x} - 1}{x} = -\frac{\ln(c)}{6}. \tag{9}$$

Combining (5), (7), and (9), we obtain

$$\lim_{d\to\infty} d\Big(\frac{A}{B} - D_d(\omega_d)\Big) = \frac{A}{6B}\left[\ln\left(1+\frac{A}{B}\right)\right]^2.$$

(2) For the second equality we have, (after simplification),

$$\frac{1}{d}D_d'(\omega_d) = \pi\left[\frac{\left(1+h\left(\frac{\ln(c)}{d}\right)\right)^{\left(\ln(c)\right)/\left(\frac{\ln(c)}{d}\right)}}{1+h\left(\frac{\ln(c)}{d}\right)} - \frac{g\left(\frac{\ln(c)}{d}\right)^{\left(\ln(c)\right)/\left(\frac{\ln(c)}{d}\right)}}{g\left(\frac{\ln(c)}{d}\right)}g'\left(\frac{\ln(c)}{d}\right)\right].$$

In light of the previous work, this yields

$$\lim_{d\to\infty}\frac{1}{d}D_d'(\omega_d) = \pi\left(1+\frac{A}{B}\right).$$

(3) To derive the third equality, note that $(1+h(\pi x_d))^d = \frac{A}{B}+g(\pi x_d)^d$ yields

$$\frac{1}{d}D_d'(x_d) = \pi\left[\frac{\frac{A}{B}+g(\pi x_d)^d}{1+h(\pi x_d)}h'(\pi x_d) - \frac{g(\pi x_d)^d}{g(\pi x)}g'(\pi x_d)\right]. \tag{10}$$

Also, the first inequality in Proposition 3 yields that, for sufficiently large d (also large enough so that $x_d < \Delta$ and $\omega_d < \Delta$), that $D_d(\omega_d) < \frac{A}{B} = D_d(x_d)$. This implies $\omega_d < x_d$ since D_d is increasing on $(0,1/4)$. But D_d is also convex on $(0,\Delta)$, so we can conclude

$$D_d'(\omega_d) < D_d'(x_d). \tag{11}$$

Combining this with (10), we obtain

$$\left[\frac{1}{d}D_d'(\omega_d) + \frac{\pi g(\pi x_d)^d}{g(\pi x_d)}g'(\pi x_d)\right]\left(\frac{1+h(\pi x_d)}{h'(\pi x_d)}\right) < \pi\left(\frac{A}{B}+g(\pi x_d)^d\right) < \pi\left(1+\frac{A}{B}\right).$$

The limit as $d\to\infty$ of the left hand side of the above inequality is $\pi\left(1+\frac{A}{B}\right)$, so

$$\lim_{d\to\infty}\pi\left(\frac{A}{B}+g(\pi x_d)^d\right) = \pi\left(1+\frac{A}{B}\right).$$

Combining this with (10), we obtain

$$\lim_{d\to\infty}\frac{1}{d}D_d'(x_d) = \pi\left(1+\frac{A}{B}\right).$$

□

Now we complete the proof of Theorem 4. For large d, the mean value theorem implies

$$\frac{D_d(x_d)-D_d(\omega_d)}{x_d-\omega_d} = D_d'(\xi), \quad \xi\in(\omega_d,x_d),$$

so that

$$x_d - \omega_d = \frac{\frac{A}{B} - D_d(\omega_d)}{D_d'(\xi)}.$$

For large d, convexity of D_d on $(0, \Delta)$ implies

$$\frac{d\left(\frac{A}{B} - D_d(\omega_d)\right)}{\frac{1}{d}D_d'(x_d)} < d^2(x_d - \omega_d) < \frac{d\left(\frac{A}{B} - D_d(\omega_d)\right)}{\frac{1}{d}D_d'(\omega_d)}.$$

Applying Proposition 3 proves the theorem.

Acknowledgements This research was supported in part by the NSF Grant DMS0856148.

References

1. Casazza, P.G.: The art of frames. Taiwanese J. Math. **4**. No. 2 129–201 (2001)
2. Young, R.M.: An Introduction to Nonharmonic Fourier Series. Academic Press (2001)
3. Bailey, B.A.: Sampling and recovery of multidimensional bandlimited functions via frames. J. Math. Anal. Appl. **367**, Issue 2 374–388 (2010)
4. Sun, W., Zhou, X.: On the stability of multivariate trigonometric systems. J. Math. Anal. Appl. **235**, 159–167 (1999)

Growth Behavior and Zero Distribution of Maximally Convergent Rational Approximants

Hans-Peter Blatt, René Grothmann, and Ralitza K. Kovacheva

Abstract Given a compact set E in $\mathbb{C}$ and a function f holomorphic on E, we investigate the distribution of zeros of rational uniform approximants $\{r_{n,m_n}\}$ with numerator degree $\leq n$ and denominator degree $\leq m_n$, where $m_n = o(n/\log n)$ as $n \to \infty$. We obtain a Jentzsch–Szegő type result, i.e., the zero distribution converges weakly to the equilibrium distribution of the maximal Green domain $E_{\rho(f)}$ of meromorphy of f if f has a singularity of multivalued character on the boundary $\partial E_{\rho(f)}$. Further, we show that any singular point of f on the boundary $\partial E_{\rho(f)}$, that is not a pole, is a limit point of zeros of the sequence $\{r_{n,m_n}\}$.

Let $f(z) = \sum_{\nu=0}^{\infty} a_\nu z^\nu$ be a power series with radius of convergence 1; set

$$s_n(z) := \sum_{\nu=0}^{n} a_\nu z^\nu, \quad n = 0,1,2,\ldots .$$

The classical theorem of Jentzsch concerns the limiting behavior of the zeros of the partial sums s_n as $n \to \infty$. Jentzsch [9] proved that each point of the unit circle is a limit point of zeros of $s_n(z)$, $n = 1,2,\ldots$. Later, Szegő [13] showed that there is an infinite sequence $\Lambda \subset \mathbb{N}$ such that the zeros of s_n, $n \in \Lambda$, are asymptotically uniformly distributed in the sense of Weyl.

Since the radius of convergence equals 1,

$$\limsup_{n\to\infty} |a_n|^{1/n} = 1.$$

H.-P. Blatt • R. Grothmann
Mathematisch-Geographische Fakultät, Katholische Universität Eichstätt-Ingolstadt, 85071 Eichstätt, Germany e-mail: hans.blatt@ku-eichstaett.de; rene.grothmann@ku-eichstaett.de

R.K. Kovacheva
Bulgarian Academy of Sciences, Institute of Mathematics and Informatics, Acad. Bonchev Str. 8, 1113 Sofia, Bulgaria e-mail: rkovach@math.bas.bg

M. Neamtu and L. Schumaker (eds.), *Approximation Theory XIII: San Antonio 2010*, Springer Proceedings in Mathematics 13, DOI 10.1007/978-1-4614-0772-0_2,

In the proof of Szegő's theorem, it is shown that the asymptotic uniform distribution for the zeros in the sense of Weyl holds for any sequence Λ that satisfies

$$\lim_{n\in\Lambda, n\to\infty} |a_n|^{1/n} = \limsup_{n\to\infty} |a_n|^{1/n} = 1.$$

Throughout this paper, we shall consider compact sets $E \subset \mathbb{C}$ whose complements $\Omega = \overline{\mathbb{C}} \setminus E$ with respect to the extended complex plane $\overline{\mathbb{C}}$ are connected. A set E is called *regular* if there exists a Green function $G(z) = G_E(z, \infty)$ on Ω with pole at ∞, and $G(z) \to 0$ as $z \to \partial\Omega$. We define the Green domains E_ρ by

$$E_\rho := \{z \in \Omega : G(z) < \log\rho\} \cup E, \ \rho > 1. \tag{1}$$

Since Ω is regular, the *equilibrium distribution* μ_E of E exists, as well as the equilibrium distribution μ_ρ for every $\overline{E}_\rho, \rho > 1$ (cf. [12] or [14]).

Given an open set $B \subset \mathbb{C}$, $\mathscr{A}(B)$ represents the class of functions f that are *holomorphic* (analytic and single-valued) in the set B. Further, the function f is *meromorphic* in $B\,(f \in \mathscr{M}(B))$ if in any closed subdomain of B, f has no more than a finite number of poles. We say that a function f is holomorphic (meromorphic) on the compact set E if f is holomorphic (meromorphic) in some open set U containing E and write $f \in \mathscr{A}(E)$ $(f \in \mathscr{M}(E))$. Further, $f \in \mathscr{M}_m(B)$ (f is m-meromorphic in B) iff f is meromorphic with no more than m poles in B. As usual, poles are counted with respect to their multiplicities. Finally, $\mathscr{R}_{n,m}$ is the collection of all rational functions $\{p/q : \deg p \le n, \deg q \le m, q \not\equiv 0\}$.

For $f \in C(E)$, we introduce the *radius of holomorphy* $\rho_0(f)$, resp. the *radius* $\rho_m(f)$ *of m-meromorphy, $m \in \mathbb{N}_0$* as follows:

$$\rho_0(f) := \begin{cases} 1, & \text{if } f \notin \mathscr{A}(E), \\ \sup\{\rho > 1 : f \in \mathscr{A}(E_\rho)\}, & \text{otherwise} \end{cases}$$

and, respectively,

$$\rho_m(f) := \begin{cases} 1, & \text{if } f \notin \mathscr{A}(E), \\ \sup\{\rho > 1 : f \in \mathscr{M}_m(E_\rho)\}, & \text{otherwise.} \end{cases}$$

Given a pair (n,m), $n,m \in \mathbb{N}_0 := \mathbb{N} \cup \{0\}$, let $R_{n,m}$ denote a best uniform approximation to $f \in C(E)$ in the class $\mathscr{R}_{n,m}$; that is,

$$e_{n,m}(f) := \inf_{r\in\mathscr{R}_{n,m}} \|f - r\|_E = \|f - R_{n,m}\|_E \tag{2}$$

and $\|\cdot\|_E$ denotes the max-norm (uniform norm of Chebyshev) on E. In the following, we assume that

$$R_{n,m} = P_{n,m}/Q_{n,m},$$

where the polynomials $P_{n,m}$ and $Q_{n,m}$ do not have a common factor. The starting point of our considerations is the following theorem.

Theorem 1 ([6]). *Let E be a regular compact set in $\mathbb{C}$ with connected complement and $m \in \mathbb{N}_0$ be fixed. Suppose that $f \in \mathscr{A}(E^o) \cap C(E)$ and $\rho_m(f) < \infty$. Then*

$$\limsup_{n\to\infty} e_{n,m}^{1/n} = 1/\rho_m(f) \tag{3}$$

and $f \in \mathscr{M}_m(E_{\rho_m(f)})$.

As known, Mergelyan's theorem says that $\lim_{n\to\infty} e_{n,m}(f) = 0$, where $m \in \mathbb{N}_0$ is fixed. Theorem 1 reveals the relation of the rate of decay to the analyticity of f.

Concerning the distribution of the zeros of $R_{n,m}(f)$, let us denote by $\nu_n = \nu(P_{n,m})$ the *normalized zero counting measure* of $P_{n,m}(f)$, i.e.,

$$\nu_n(B) = \frac{\#\text{ zeros of } P_{n,m} \text{ in } B}{\deg P_{n,m}}, \quad B \subset \mathbb{C}.$$

The next theorem characterizes the zero distribution of the sequence $\{R_{n,m}\}$ as $n \to \infty$, where m is fixed.

Theorem 2. *Under the conditions of Theorem 1, suppose that f is not identically 0 on any component of E. Suppose also that $\rho_m(f) < \infty$. Then there exists a subsequence $\Lambda \subset \mathbb{N}$ such that the normalized zero counting measures ν_n of the numerators of $R_{n,m}$ converge weakly to the equilibrium distribution $\mu_{\rho_m(f)}$ of $\overline{E}_{\rho_m(f)}$.*

Theorem 2 follows from results of [1, 2]. In an independent way, it was proved in [10].

For our further purposes we use the concept of *convergence in m_1-measure*. Let e be a set in $\mathbb{C}$. We set

$$m_1(e) := \inf(\sum_\nu |U_\nu|),$$

where the infimum is taken over all countable coverings $\{U_\nu\}$ by disks U_ν, where $|U_\nu|$ denotes the radius of the disk U_ν.

Let D be a domain in $\mathbb{C}$ and φ a function defined in D with values in $\overline{\mathbb{C}}$. A sequence of functions $\{\varphi_n\}$, meromorphic in D, is said to converge to a function φ with respect to the *m_1-measure inside D* if for every $\varepsilon > 0$ and any compact set $K \subset D$ we have

$$m_1\{z \in K : |(\varphi - \varphi_n)(z)| \geq \varepsilon\} \to 0 \text{ as } n \to \infty.$$

The sequence $\{\varphi_n\}$ is said to converge to φ *m_1-almost locally uniformly inside D* if for any compact set $K \subset D$ and any $\varepsilon > 0$ there exists a set $K_\varepsilon \subset K$ such that $m_1(K \setminus K_\varepsilon) < \varepsilon$ and the sequence $\{\varphi_n\}$ converges uniformly to φ on K_ε. Hence, the m_1-almost local uniform convergence inside D implies m_1-convergence inside D [7].

We recall the basic properties of the convergence in m_1-measure [7]. Suppose that $\{\varphi_n\}$ converges in m_1-measure to φ inside the domain D, then:

(a) If $\{\varphi_n\} \in \mathscr{A}(D)$, then $\{\varphi_n\}$ converges locally uniformly in D. Thus, the limit function is analytic in D.

(b) If the functions φ_n are m-meromorphic in D, where $m \geq 0$ is a fixed integer, then the limit function φ is m_1-equivalent to a function which is also m-meromorphic in D. Hence, if φ has a pole of order τ at a point $a \in D$, then at least τ poles of φ_n tend to a.

(c) If $\varphi \in \mathscr{M}_m(D)$, then all functions φ_n (for n large enough) have at least m poles in D.

Let us return to Theorem 1. If $m \in \mathbb{N}$ is fixed and $\rho_m(f) > 1$, then it was shown in [6] that the best Chebyshev approximants $R_{n,m}$ converge m_1-almost locally uniformly inside the domain $E_{\rho_m(f)}$ as $n \to \infty$. Denote by $\tilde{f}$ the limit function. By (b), $\tilde{f}$ is m_1-equivalent to a function in $\mathscr{M}_m(E_{\rho_m(f)})$. Furthermore, Gončar proved a quantitative characteristic of the convergence of the sequence $\{R_{n,m}\}$ inside $E_{\rho_m(f)}$; namely, for every compact set $K \subset E_{\rho_m(f)}$ and $\varepsilon > 0$ there exists an open set Ω_ε with $m_1(\Omega_\varepsilon) < \varepsilon$ such that

$$\limsup_{n\to\infty} \|f - R_{n,m}\|_{K\setminus\Omega_\varepsilon}^{1/n} \leq \frac{\exp(\max_K G_E(z,\infty))}{\rho_m(f)}$$

([6], p. 153, Remark 1).

We consider here the question about the rate of convergence and the distribution of the zeros in the case of rational functions if the degrees of the denominators are not bounded.

For $f \in C(E)$, let us introduce the *radius of meromorphy* $\rho(f)$ of f, that is

$$\rho(f) := \sup_{m\geq 0} \rho_m(f).$$

Our new results are concerned with a sequence $\{m_n\}_{n=1}^{\infty}, m_n \in \mathbb{N}_0, n \in \mathbb{N}$, such that

$$m_n \to \infty \text{ and } m_n = o(n/\log n) \text{ as } n \to \infty. \tag{4}$$

Assume that $f \in \mathscr{A}(E)$ and $\rho(f) < \infty$. Then for sequences as in (4), Walsh's theorem ([15], p. 378, Theorem 3) implies that

$$\limsup_{n\to\infty} \|f - R_{n,m_n}\|_E^{1/n} \leq \frac{1}{\rho(f)}. \tag{5}$$

In Theorem 3, we state that the uniform geometric convergence of a sequence $\{r_{n,m_n}\}_{n\in\mathbb{N}}$ to f on E implies the geometric convergence of this sequence to a continuous function m_1-almost locally uniformly inside some domain E_τ, $\tau > 1$.

Theorem 3 ([5]). *Let E be compact in $\mathbb{C}$ with regular, connected complement $\Omega = \overline{\mathbb{C}} \setminus E$, $\{m_n\}_{n=1}^{\infty}$ a sequence in $\mathbb{N}_0$ with $m_n = o(n/\log n)$ as $n \to \infty$, $\{r_{n,m_n}\}_{n\in\mathbb{N}}$ a sequence of rational functions, $r_{n,m_n} \in \mathscr{R}_{n,m_n}$, such that for $f \in \mathscr{M}(E)$*

$$\limsup_{n\to\infty} \|f - r_{n,m_n}\|_{\partial E}^{1/n} \le \frac{1}{\tau} < 1. \tag{6}$$

Then there exists an extension $\widetilde{f}$ of f to E_τ with the following property: For any $\varepsilon > 0$ there exists a subset $\Omega(\varepsilon) \subset \mathbb{C}$ with $m_1(\Omega(\varepsilon)) < \varepsilon$ such that $\widetilde{f}$ is a continuous function on $E_\tau \setminus \Omega(\varepsilon)$ and

$$\limsup_{n\to\infty} \|\widetilde{f} - r_{n,m_n}\|_{\overline{E}_\sigma \setminus \Omega(\varepsilon)}^{1/n} \le \frac{\sigma}{\tau} \tag{7}$$

for any σ with $1 < \sigma < \tau$, and $\{r_{n,m_n}\}_{n\in\mathbb{N}}$ converges m_1-almost uniformly to $\widetilde{f}$ inside E_τ.

Remark 1. Under the conditions of Theorem 1, in particular m is fixed, the function f can be continued to $f \in \mathscr{M}_m(E_{\rho_m(f)})$. In this case, $\widetilde{f}$ of Theorem 3 is m_1-equivalent to f in $E_{\rho_m(f)}$. But under the conditions of Theorem 3, the sequence $\{m_n\}_{n=1}^{\infty}$ need not be bounded. If f can be continued to $f \in \mathscr{M}(E_\rho)$, $1 < \rho$, then it is an open problem whether the continuous extension $\widetilde{f}$ of f on $E_\rho \setminus \Omega(\varepsilon)$ is m_1-equivalent to f on $(E_\tau \cap E_\rho) \setminus \Omega(\varepsilon)$, where $\varepsilon > 0$ is arbitrarily small and $m_1(\Omega(\varepsilon)) < \varepsilon$.

For obtaining results about the distribution of the zeros of the approximants r_{n,m_n}, we will assume that the continuous extension $\widetilde{f}$ of Theorem 3 for $\tau \le \rho(f)$ coincides with the meromorphic continuation of f into $E_{\rho(f)}$.

Definition 1. Let $f \in \mathscr{M}(E)$ and $\rho(f) < \infty$. A sequence $\{r_n\}_{n\in\mathbb{N}}$ with $r_n \in \mathscr{R}_{n,n}$ is called m_1-*maximally convergent* to f on E if

$$\limsup_{n\to\infty} \|f - r_n\|_{\partial E}^{1/n} \le \frac{1}{\rho(f)} \tag{A}$$

and for any $\varepsilon > 0$ there exists a set $\Omega(\varepsilon) \subset \mathbb{C}$ with $m_1(\Omega(\varepsilon)) < \varepsilon$ such that

$$\limsup_{n\to\infty} \|f - r_n\|_{\overline{E}_\sigma \setminus \Omega(\varepsilon)}^{1/n} \le \frac{\sigma}{\rho(f)} \tag{B}$$

for all σ, $1 < \sigma < \rho(f)$.

Remark 2. As an example for such m_1-maximal convergent sequence one can take best real rational approximants of $f \in C(E)$, where $E = [-1,1]$ and f is real-valued [3]. Another example is classical Padé approximation to a function f meromorphic in a neighborhood of 0 ([11] and [7]).

In our results for the distribution of the zeros of the rational approximants, a special class of meromorphic functions will play an essential role.

Definition 2. Let $f \in \mathscr{M}(E)$ with $\rho(f) < \infty$. A point $z_0 \in \partial E_{\rho(f)}$ is called a *singularity of multivalued character* of the function f if there exists a neighborhood U of z_0 such that f can be continued to any point of $U \setminus \{z_0\}$ and f is locally holomorphic, but not single-valued.

Examples of singularities of multivalued character are branch points.

The next theorem provides for functions with a multivalued singularity an exact estimate of the rate of approximation.

Theorem 4 ([5]). *Let E be a compact, connected set in $\mathbb{C}$ with regular and connected complement. Moreover, let $\{m_n\}_{n=1}^{\infty}$ be a subsequence of $\mathbb{N}_0$ satisfying (4). Suppose that the sequence of rational functions $\{r_{n,m_n}\}_{n\in\mathbb{N}}$, $r_{n,m_n} \in \mathscr{R}_{n,m_n}$ is m_1-maximally convergent to $f \in \mathscr{M}(E)$ on E with $\rho(f) < \infty$. If there exists a singularity of multivalued character of f on $\partial E_{\rho(f)}$, then*

$$\limsup_{n\to\infty} \|f - r_{n,m_n}\|_{\partial E}^{1/n} = \frac{1}{\rho(f)}. \tag{8}$$

Theorem 4 and the results of [8] are fundamental for establishing an analogue of Szegő's theorem for the case being considered.

Theorem 5 ([5]). *Under the conditions of Theorem 4, the normalized zero counting measures ν_n of r_{n,m_n} converge weakly to the equilibrium distribution of $\overline{E}_{\rho(f)}$, at least for a subsequence $\Lambda \subset \mathbb{N}$ as $n \to \infty$ with $n \in \Lambda$.*

We are now interested how a singularity on the boundary $\partial E_{\rho(f)}$ of $E_{\rho(f)}$ impacts the behavior of the zeros of a m_1-maximally convergent sequence $\{r_{n,m_n}\}$ as $n \to \infty$. If among the singularities on $\partial E_{\rho(f)}$ there is at least one of multivalued character, then by Theorem 5 each point on $\partial E_{\rho(f)}$ is a limit point of zeros of $\{r_{n,m_n}\}$ as $n \to \infty$. Furthermore, there is a sequence Λ such that the associated zero distributions ν_n converge weakly to the equilibrium measure of $\overline{E}_{\rho(f)}$ as $n \in \Lambda$, $n \to \infty$. It is of interest to study the case where there is no singular point of multivalued character on $\partial E_{\rho(f)}$.

Given a domain B and a function $g \in \mathscr{M}(B)$, we introduce the notation $Z(g,B)$; that is the number of zeros of g in B. Analogously, $P(g,B)$ denotes the number of poles of g in B. Recall that poles and zeros are counted with their multiplicities.

Concerning the limit distribution of the zeros and poles of a m_1-maximally convergent sequence of rational function, we have the following result:

Theorem 6 ([4]). *Let E be a regular compact set and let $f \in \mathscr{M}(E)$. Suppose that $\rho(f) < \infty$ and assume that the point $z_0 \in \partial E_{\rho(f)}$ is a singularity of the function f, but neither a pole nor a removable singularity. Suppose that the sequence of* *rational functions* $\{r_n\}$, $r_n \in \mathscr{R}_{n,n}$, *is* m_1*-maximally convergent to* f *on* E. *Then for any neighborhood* U *of* z_0 *the following statements hold:*

(a1) *If* $P(r_n, U) = o(n)$ *as* $n \to \infty$, *then* $\limsup_{n\to\infty} Z(r_n, U) = \infty$.
(a2) *If* $P(r_n, U) = O(1)$ *as* $n \to \infty$, *then* $\limsup_{n\to\infty} Z(r_n, U)/n > 0$.
(b1) *If* $Z(r_n, U) = o(n)$ *as* $n \to \infty$, *then* $\limsup_{n\to\infty} P(r_n, U) = \infty$.
(b2) *If* $Z(r_n, U) = O(1)$ *as* $n \to \infty$, *then* $\limsup_{n\to\infty} P(r_n, U)/n > 0$.

In the proof of Theorem 6, we essentially use that $f \in \mathscr{M}(E)$, as well as the m_1-maximal convergence of the sequence $\{r_n\}$ to the function f. Theorem 6 applies to Padé approximants and to rational approximations to a real valued function on a bounded real interval, when the degrees of the denominators satisfy the condition (4) (see [5] and [4]).

Let $a \in \mathbb{C}$, $B \subset \mathbb{C}$, r a rational function, and let $a(r,B)$ denote the number of a-values of r in B, i.e.,

$$a(r,B) := \#\{z \in B : r(z) = a\}.$$

Corollary 1. *Let* E *and* f *be as in Theorem 6. Suppose that* $z_0 \in \partial E_{\rho(f)}$ *is a singular point of* f *which is neither a pole nor a removable singularity. Suppose that* $\{m_n\}_{n\in\mathbb{N}}$ *with* $m_n = o(n)$ *as* $n \to \infty$, *and let* $\{r_{n,m_n}\}_{n\in\mathbb{N}}$ *be* m_1*-maximally convergent to* f *on* E. *Then, for every neighborhood* U *of* z_0

$$\limsup_{n\to\infty} a(r_{n,m_n}, U) = \infty, \quad a \in \mathbb{C}.$$

Proof. Fix $a \in \mathbb{C}$. The proof follows from statement (a1) in Theorem 6, applied to the functions $r_{n,m_n} - a$ and $f - a$, after taking into account that $m_n = o(n)$ as $n \to \infty$.
□

Recalling the classical theorem of Picard concerning the behavior of a holomorphic function in a neighborhood of an essential singularity, we can summarize Corollary 1 by saying that a m_1-maximally convergent sequence $\{r_{n,m_n}\}$ with $\{m_n\}$ as in (4) has an asymptotic essential singularity at each singularity of the curve of meromorphy which is neither a pole nor a removable singularity.

Acknowledgements This work was supported by DFG-Research Grant (Germany), BL 272/10-1.

References

1. H.-P. Blatt, E. B. Saff, Behavior of zeros of polynomials of near best approximation, *J. Approx. Theory*, **46** (1986), 323 - 344.

2. H.-P. Blatt, E. B. Saff, M. Simkani, Jentzsch-Szegő type theorems for the zeros of best approximants, *J. Lond. Math. Soc.*, **38** (1988), No. 2, 307 - 316.
3. H.-P. Blatt, R. Grothmann, R. K. Kovacheva, Poles and alternation points in real rational Chebyshev approximation, *Comput. Methods Funct. Theory,* **3** (2003), No. 1 - 2, 165 - 177.
4. H.-P. Blatt, R. Grothmann, R.K. Kovechava, Regions of meromorphy and value distribution of geometrically converging rational functions, *J. Math. Anal. Appl.* **382** (2011), 66 - 76.
5. H.-P. Blatt, R. K. Kovacheva, Growth behaviour and zero distribution of rational approximants, to appear in Constr. Approx.
6. A. A. Gončar, On a theorem of Saff, *Mat. Sbornik,* **94** (136) (1974), 152 - 157, *english translation in Math. USSR Sbornik,* **23** (1974), No. 1, 149 - 154.
7. A. A. Gončar, On the convergence of generalized Padé approximants of meromorphic functions, *Mat. Sbornik,* **98** (140) (1975), 564 - 577, *english translation in Math. USSR Sbornik,* **27** (1975), No. 4, 503 - 514.
8. R. Grothmann, On the zeros of sequences of polynomials, *J. Approx. Theory,* **61** (1990), No. 3, 351 - 359.
9. R. Jentzsch, Untersuchungen zur Theorie der Folgen analytischer Funktionen, *Acta Math.,* **41** (1917), 219 - 251.
10. R. K. Kovacheva, On the behavior of Chebyshev approximants with a fixed number of poles, *Math. Balk.,* **3** (1989), 244 - 256.
11. O. Perron, Die Lehre von den Kettenbrüchen, *Teubner, Leipzig* (1929).
12. E. B. Saff, V. Totik, Logarithmic Potentials with External Fields, *Springer, Heidelberg,* 1997.
13. G. Szegő, Über die Nullstellen von Polynomen, die in einem Kreis gleichmäßig konvergieren, *Sitzungsber. Berl. Math. Ges.,* **21** (1922), 59 - 64.
14. M. Tsuji, Potential Theory in Modern Function Theory, *Maruzen Co., Tokyo* (1959).
15. J. L. Walsh, Interpolation and Approximation by Rational Functions in the complex domain, *Amer. Math. Soc. Colloq. Pub.*, New York, Vol. 20 (1969).

Generalization of Polynomial Interpolation at Chebyshev Nodes

Debao Chen

Abstract Previously, we generalized the Lagrange polynomial interpolation at Chebyshev nodes and studied the Lagrange polynomial interpolation at a special class of sets of nodes. This special class includes some well-known sets of nodes, such as zeros of the Chebyshev polynomials of first and second kinds, Chebyshev extrema, and equidistant nodes. In this paper, we view our previous work from a different perspective and further generalize and study the Lagrange polynomial interpolation at a larger class of sets of nodes. In particular, the set of optimal nodes is included in this extended class.

1 Introduction

Let $X = X^{[n]} = \{x_j = x_j^{[n]} : j = 0, 1, 2, \ldots, n\} \subset [-1, 1]$, $n = 2, 3, 4, \ldots$ be an interpolatory matrix such that

$$-1 \le x_0 < x_1 < \cdots < x_{n-1} < x_n \le 1. \tag{1}$$

The corresponding Lagrange interpolation polynomial of degree at most n is defined as

$$L_n(f,x) = L_n(f,X,x) = \sum_{k=0}^{n} f(x_k) l_k(x),$$

where, with $\omega_n(x) = \omega_n(X,x) = \prod_{i=0}^{n}(x - x_i)$,

Debao Chen
Computer Science Department, Oklahoma State University-Tulsa,
Tulsa, OK 74106, USA e-mail: debao.chen@okstate.edu

M. Neamtu and L. Schumaker (eds.), *Approximation Theory XIII: San Antonio 2010*, Springer Proceedings in Mathematics 13, DOI 10.1007/978-1-4614-0772-0_3,

$$l_k(x) = l_k^{[n]}(X,x) = \prod_{\substack{i=0\\ i\neq k}}^{n} \frac{x-x_i}{x_k-x_i} = \frac{\omega_n(x)}{\omega_n'(x_k)(x-x_k)}.$$

The corresponding Lebesgue function and Lebesgue constant are defined as

$$\Lambda(x) = \Lambda^{[n]}(x) = \Lambda^{[n]}(X,x) = \sum_{k=0}^{n} |l_k(x)| = |\omega_n(x)| \sum_{k=0}^{n} \frac{1}{|\omega_n'(x_k)||x-x_k|}$$

and

$$\lambda_n = \lambda_n(X) = ||\Lambda^{[n]}|| = \max_{x_0\le x\le x_n} \Lambda^{[n]}(x).$$

To standardize the problem, we always let the endpoints of the interval be nodes of the interpolation. Contrary to the decreasing order of the nodes we used in [5], in this paper we purposely use an increasing order for reasons explained below.

The Lebesgue function $\Lambda^{[n]}$ has one and only one relative maximum on every interval between adjacent nodes. Let the local maxima be

$$\lambda_{n,p} = \max_{x_p\le x\le x_{p+1}} \Lambda^{[n]}(x) \qquad p = 0,1,2,\ldots,n-1.$$

It is well known that the minimum norm of the interpolation operator is achieved if and only if all the local maxima are equal $[1,2,7,8]$. It is also well known that an affine transformation does not change the local maxima. Therefore, we view two sets of nodes as the same, if one of them can be obtained by an affine transformation of another one. Under such a consideration, there is one and only one set of optimal nodes.

There have been extensive investigations of Lagrange polynomial interpolation at some well-known sets of nodes, such as zeros of the Chebyshev polynomials of first and second kinds, Chebyshev extrema, and equidistant nodes. In the literature, most authors studied the Lagrange polynomial interpolation at a single set of nodes. Particularly, authors paid special attention to the polynomial interpolation at Chebyshev nodes (zeros of the Chebyshev polynomials of first kind), since the set of Chebyshev nodes is very close to the set of optimal nodes [3].

In a previous work [5], we generalized the Lagrange polynomial interpolation at the Chebyshev nodes and studied the Lagrange polynomial interpolation at a special class of sets of nodes, which is defined as follows. We take a subset of the unit semicircle that is symmetric about the vertical axis and divide it into n equal parts with $n+1$ points on this subset. Then we introduce a parameter $0 \le \alpha \le \pi/n$. Each α in this range corresponds to a set of nodes. First, we let α $(0 < \alpha \le \pi/n)$ be the difference between two adjacent angles. The angles of these points are

$$\theta_j = \theta_j(\alpha) = \pi/2 - (n/2 - j)\alpha, \quad j = 0,1,2,\ldots,n. \tag{2}$$

Then we project these $n+1$ points down on the horizontal axis and obtain a set of nodes for each α,

$$x_j = x_j(\alpha) = \cos\theta_j, \qquad j = 0,1,2\dots,n, \tag{3}$$

$$X_\alpha = X_\alpha^{[n]} = \{-1 \le x_n(\alpha) < x_{n-1}(\alpha) < \cdots < x_1(\alpha) < x_0(\alpha) \le 1\}. \tag{4}$$

We emphasize that a decreasing order of the nodes is used as we did in [5]. As we pointed out, when $\alpha = \pi/n, \pi/(n+1), \pi/(n+2)$ we obtain the Chebyshev extrema and zeros of the Chebyshev polynomial of first and second kinds, respectively. Furthermore equidistant nodes can be viewed as the limiting case as $\alpha \to 0$. In fact, we may take a linear transformation for each $0 < \alpha \le \pi/n$ by letting $\overline{x}_j = (1/x_0)x_j$, which transforms the interval $[x_n(\alpha), x_0(\alpha)]$ to the interval $[-1,1]$. When $\alpha \to 0$, we have

$$\overline{x}_j(\alpha) = \frac{\cos\theta_j}{\cos\theta_0} = \frac{\sin(n/2-j)\alpha}{\sin(n/2)\alpha} \to \overline{x}_j(0) = x_j(0) = \frac{n-2j}{n}, \qquad j = 0,1,2,\dots,n, \tag{5}$$

which is the set of equidistant nodes. Obviously, for each j, the node $x_j(\alpha)$ is not continuous at $\alpha = 0$, while the node $\overline{x}_j(\alpha)$ is continuous on $[0,\pi/n]$. Since a linear transformation does not change the values of local maxima, the corresponding local maxima as well as the Lebesgue constants are continuous functions of α on $[0,\pi/n]$. In fact, they are also infinitely differentiable functions. Therefore, it is convenient for us to study the local maxima as well as the Lebesgue constants as functions on the closed interval $[0,\pi/n]$ by using derivatives. In [5], we initiated the study of the derivative of the local maxima with respect to α.

We reemphasize that in the above class of sets of nodes, the order of the nodes is decreasing as follow

$$-1 \le x_n(\alpha) < x_{n-1}(\alpha) < \cdots < x_1(\alpha) < x_0(\alpha) \le 1. \tag{6}$$

We notice that

$$x_j = \cos\theta_j = \cos(\pi/2 - (n/2-j)\alpha) = \sin(n/2-j)\alpha, \qquad j = 0,1,2\dots,n. \tag{7}$$

It seems that most people prefer the decreasing order of the nodes when they study the polynomial interpolation at Chebyshev nodes. Perhaps people want to emphasize that the Chebyshev nodes are the zeros of the Chebyshev polynomial $T_{n+1}(x) = \cos(n+1)\theta$, with $x = \cos\theta$. We also used the decreasing order when we study the generalization of polynomial interpolation at Chebyshev nodes in [5]. However, it seems that the particular properties of the Chebyshev polynomials cannot be further generalized when one studies the polynomial interpolation. People attempted to find some particular polynomials (functions) and use the zeros of such polynomials (functions) as the nodes for polynomial interpolation. We do not think that this is a very good approach for finding the optimal nodes for polynomial interpolation. Instead, one should focus more to the structures of the nodes themselves.

In this paper, we view our previous work from a different perspective and intentionally change the order of the nodes to the increasing order. In contrast with (7) and (5), we let, for $j = 0, 1, \ldots, n$,

$$x_j(\alpha) = \sin(-n/2 + j)\alpha, \qquad 0 < \alpha \leq \pi/n, \tag{8}$$

and

$$x_j(0) = (-n + 2j)/n. \tag{9}$$

Therefore, for $0 \leq \alpha \leq \pi/n$, we have

$$-1 \leq x_0(\alpha) < x_1(\alpha) < \cdots < x_n(\alpha) \leq 1. \tag{10}$$

The nodes in (10) are identical to the nodes in (6) except for the order of the nodes. The node x_j in (10) is equal to the node x_{n-j} in (6). It seems that it is insignificant to make such a change for studying the Lagrange polynomial interpolation at the sets of nodes in this special class. However, after such a change, one can view the generalization we made in our previous paper [5] from a different perspective and make a further generalization.

First, we notice that the sine function is an infinitely differentiable, strictly increasing, and odd function with a domain $[-\pi/2, \pi/2]$ and a codomain $[-1, 1]$. Essentially, in [5] we used the sine function to define a special class of sets of nodes. We mentioned that when α is larger than and sufficiently close to $\pi/(n+1)$, which corresponds to the set of Chebyshev nodes, the corresponding set of nodes is closer to the optimal nodes than the set of Chebyshev nodes. However, the set of optimal nodes is not included in this class. We will consider to extend the above class of sets of nodes to a larger one so that it contains the set of optimal nodes.

The above observation stimulates us to use a function φ, instead of the sine function, to construct the following class of sets of nodes

$$x_j(\alpha) = \varphi((-n/2 + j)\alpha), \qquad j = 0, 1, 2, \ldots, n, \quad 0 < \alpha \leq \pi/n, \tag{11}$$

where φ is a strictly increasing and odd function with a domain $[-\pi/2, \pi/2]$ and a codomain $[-1, 1]$.

An arbitrary choice for function φ is not very helpful. Obviously, for any set of symmetric nodes $\{x_j\}$, there exist such a function φ and an α such that the set of nodes $\{\varphi((-n/2 + j)\alpha)\}$ is the same as $\{x_j\}$. Instead, since the set of Chebyshev nodes is close to the optimal one, we are thinking to use a function, which is "close" and related to the sine function. In other words, we will choose functions which have similar properties with the sine function.

We give an outline of this paper. In Sect. 2, we will define a class of functions with some special properties. Then a sequence of particular functions and its limit will be given. For each of these functions, we define a class of sets of nodes, with a parameter α ($0 \leq \alpha \leq \pi/n$), for the polynomial interpolation. These classes of sets of nodes are the natural generalization of the special class given in our previous paper [5]. In Sect. 3, we will give convenient formulas for Lebesgue functions of Lagrange polynomial interpolation at the sets of nodes defined in Sect. 2. In Sect. 4, we will give some properties of a particular pair of functions and its generalization.

These pairs of auxiliary functions are particularly important tools for our study of Lagrange polynomial interpolation. In Sect. 5, we will give the set of optimal nodes.

This is a very large project. Our main goal is to give the explicit formulas of optimal nodes and optimal Lebesgue constants for the Lagrange polynomial interpolations. Due to the nature of these problems, the complete proofs of our results are extremely long. To achieve our goals we must study the deep properties of the corresponding Lebesgue functions from several aspects, including the properties of extrama points of Lebesgue functions on each subinterval. Several dozen lemmas are needed. Our previous paper [5] as well as this paper is an introduction and an outline of our whole project. The complete proofs of all of our results will be given in several upcoming papers.

2 Generalization of Chebyshev Nodes

In the previous section, we mentioned that one may use a strictly increasing and odd function φ to construct a class of sets of nodes for the polynomial interpolation. For convenience and ease of finding the optimal nodes, we require that the function φ possess some special properties.

Definition 1. We say that a function φ has Property A if φ has the following properties.

1. $\varphi : [-\pi/2, \pi/2] \to [-1,1]$.
2. φ is an odd function.
3. φ possesses a continuous second derivative.
4. $\varphi'(x) > 0$ for $-\pi/2 < x < \pi/2$, $\varphi'(0) = 1$, and $\varphi'(\pi/2) = 0$.
5. $\varphi''(x) < 0$ for $0 < x \leq \pi/2$.

We make some remarks on the above definition. It is easy to verify that the sine function has Property A. In fact, we extracted some necessary and essential properties of the sine function to make this definition, so that it will be helpful for finding the set of optimal nodes. Since φ is an odd function, the first derivative φ' is an even function and the second derivative φ'' is an odd function. Therefore, we may obtain the corresponding properties of the functions φ, φ', and φ'' on the left half interval $[-\pi/2, 0]$. For example, we have that $\varphi''(x) > 0$ for $-\pi/2 \leq x < 0$. We also have that $\varphi(0) = \varphi''(0) = 0$.

Let

$$\Phi = \{\varphi \mid \varphi \text{ has Property A}\} \qquad \text{and} \qquad \overline{\Phi} = \{I_0\} \cup \Phi, \tag{12}$$

where I_0 is the identity function such that $I_0(x) = x$.

It is natural to consider the function $\overline{\varphi} = \sin\varphi$ for $\varphi \in \overline{\Phi}$. In fact, we have the following lemma.

Lemma 1. *If* $\varphi \in \overline{\Phi}$*, then* $\overline{\varphi} = \sin\varphi \in \Phi$.

Proof. If $\varphi = I_0$, then $\overline{\varphi}(x) = \sin I_0(x) = \sin x$, which is in Φ. If $\varphi \in \Phi$, then

$$\begin{aligned}\overline{\varphi}(x) &= \sin \varphi(x), \\ \overline{\varphi}'(x) &= \varphi'(x)\cos\varphi(x), \\ \overline{\varphi}''(x) &= \varphi''(x)\cos\varphi(x) - (\varphi'(x))^2 \sin\varphi(x).\end{aligned}$$

One can easily verify that $\overline{\varphi}$ has Property A. □

The choice of a function $\varphi \in \Phi$ is still somewhat arbitrary. Since the sine function belongs to Φ, by Lemma 1, we have that $\sin\sin \in \Phi$, $\sin\sin\sin \in \Phi$, and so on. Therefore, we make the following definition.

Definition 2. We define I_m recursively as follows:

$$\begin{aligned}I_0(x) &= x, \\ I_m(x) &= \sin I_{m-1}(x), \qquad m = 1,2,3,\ldots.\end{aligned}$$

By a linear transform, we also define

$$\rho_m(x) = I_m(x)/I_m(\pi/2), \qquad m = 1,2,3,\ldots.$$

By Lemma 1 and a simple induction we have that $I_m \in \Phi$ for $m \geq 1$. However, $\rho_m \notin \Phi$ for $m \geq 2$, since $\rho_m'(0) = 1/I_m(\pi/2) > 1$. Since the limit of the sequence of the functions I_m is identical to zero, we introduce the functions ρ_m instead. It is obvious that $\rho_m(\pi/2) = 1$ for each $m \geq 1$.

If φ is an odd and strictly increasing function, then we define a class $C(\varphi) = C^{[n]}(\varphi)$ of sets of symmetric nodes as following. When $0 < \alpha \leq \pi/n$, we let

$$x_j = x_j(\alpha) = x_{j;\varphi}(\alpha) = \varphi((-n/2+j)\alpha), \qquad j = 0,1,2,\ldots,n. \tag{13}$$

If the function φ is differentiable and $\varphi'(x) > 0$ for $-\pi/2 < x < \pi/2$, then we can define the set of nodes for $\alpha = 0$. We take a linear transformation for each $0 < \alpha \leq \pi/n$ by letting $\overline{x}_j = (1/x_n)x_j$, which transforms the interval $[x_0(\alpha), x_n(\alpha)]$ to the interval $[-1,1]$. When $\alpha \to 0$, we have

$$\begin{aligned}x_j(0) &= \overline{x}_j(0) = \lim_{\alpha\to 0} \overline{x}_j(\alpha) = \lim_{\alpha\to 0} \frac{\varphi((-n/2+j)\alpha)}{\varphi((n/2)\alpha)} \\ &= \lim_{\alpha\to 0} \frac{(-n/2+j)\varphi'((-n/2+j)\alpha)}{(n/2)\varphi'((n/2)\alpha)} = \frac{-n+2j}{n}, \qquad j = 0,1,2,\ldots,n. \end{aligned}\tag{14}$$

The nodes $\{x_j\}$ depend on a function φ and a parameter $\alpha \in [0,\pi/n]$. We denote this set of nodes as $X_{\varphi,\alpha} = X^{[n]}_{\varphi,\alpha} \in C(\varphi)$. However, for any odd function φ with positive derivative, the nodes $\{x_j(0)\}$ are the equidistant nodes. Since $I_1(x) = \sin(x)$, the class $C(I_1)$ is the special class we defined in [5].

We are particularly interested in the classes $C(I_m)$ as well as the "limit" of these classes as $m \to \infty$. Since $\lim_{m\to\infty} I_m(x) \equiv 0$, we consider $\lim_{m\to\infty} \rho_m(x)$ instead. First, we establish the existence of this limit.

Lemma 2. *The limit* $\rho(x) = \lim_{m\to\infty} \rho_m(x)$ *exists for* $-\pi/2 \le x \le \pi/2$. *Also,* ρ *is a continuous, odd, and strictly increasing function with* $\rho(\pi/2) = 1$.

Proof. Since ρ_m are odd functions, it suffices to consider $0 \le x \le \pi/2$ only. Since the function $f(x) = \sin x/x$ is a strictly decreasing function for $0 \le x \le \pi/2$, we have

$$\frac{\rho_{m+1}(x)}{\rho_m(x)} = \frac{\sin I_m(x)}{I_m(x)} \Big/ \frac{\sin I_m(\pi/2)}{I_m(\pi/2)} \ge 1.$$

The equality holds only for $x = \pi/2$. We have $\rho_m(0) = 0$ and $\rho_m(\pi/2)) = 1$ for all $m \ge 1$. For $0 < x < \pi/2$ we have

$$0 < \rho_1(x) < \rho_2(x) < \rho_3(x) \cdots < 1.$$

The sequence $\{\rho_m(x)\}$ is increasing and bounded for each $x \in [0, \pi/2]$. Therefore, the limit $\rho(x) = \lim_{m\to\infty} \rho_m(x)$ exists. Since the functions ρ_m are uniformly bounded for $0 \le x \le \pi/2$ and $m \ge 1$, the convergence is uniform. In addition, since the convergence is uniform and all functions ρ_m are continuous, odd, and strictly increasing, it is easy to prove that the limit function ρ is also continuous, odd, and strictly increasing. □

We mentioned that $\rho_m \notin \Phi$ for $m \ge 2$, since $\rho_m'(0) > 1$. It is obvious that $\rho \notin \Phi$. The function ρ is even not differentiable at $x = 0$, since $\lim_{m\to\infty} \rho_m'(0) = +\infty$. However, this fact will not affect our further investigation. Since ρ is a strictly increasing and odd function, we have a class $C(\rho)$ of sets of nodes. Since the linear transformation does not change the local maxima of the Lebesgue functions as well as the Lebesgue constants, we may use the sets of nodes in class $C(I_m)$ instead of $C(\rho_m)$ for polynomial interpolation. We may also study the limit of the corresponding Lebesgue function, local maxima, as well as the Lebesgue constants to obtain the properties of Lebesgue function, local maxima, as well as the Lebesgue constants corresponding to $C(\rho)$.

3 Lebesgue Functions

The first step in studying the Lagrange polynomial interpolation at a set of nodes is always to find the explicit and convenient formula for the corresponding Lebesgue function. In this section, we first discuss general formulas for the Lebesgue functions corresponding to arbitrary symmetric nodes. Then we give explicit formulas for Lebesgue functions corresponding to the Lagrange interpolation at $X_{\varphi,\alpha}$ for a strictly increasing and odd function φ. We also give another kind of useful formulas for Lebesgue functions corresponding to the Lagrange interpolation at $X_{\overline{\varphi},\alpha}$, where $\overline{\varphi} = \sin\varphi$ with $\varphi \in \overline{\Phi}$. These formulas are slight generalizations of the formulas given in our previous paper [5].

Let $\{x_j\}$ be a set of nodes as in (1). We always suppose that the nodes are symmetric about the origin, that is

$$x_{n-j} = -x_j, \qquad j = 0, 1, 2, \ldots, n.$$

The Lebesgue function Λ is a piecewise polynomial function with nodes $\{x_j\}$. We denote the interval between adjacent nodes as $I_j = (x_j, x_{j+1})$ and $\overline{I}_j = [x_j, x_{j+1}]$, for $0 \leq j \leq n-1$.

It is easy to see that $\Lambda(x) \geq 1$ for each $x \in [x_0, x_n]$ and $\Lambda(x) = 1$ if and only if $x = x_j$, for $0 \leq j \leq n$. When $x \neq x_j$, we denote the Lebesgue function as a product of two parts:

$$\Lambda(x) = \Lambda^{[n]}(x) = |\omega_n(x)| H(x),$$

where

$$\omega_n(x) = \omega_n(X, x) = \prod_{j=0}^{n} (x - x_j)$$

and

$$H(x) = H^{[n]}(x) = \sum_{j=0}^{n} \frac{1}{|\omega_n'(x_j)||x - x_j|}.$$

When $x \in I_p$ ($x_p < x < x_{p+1}$), we have

$$F_p(x) = F_p^{[n]}(x) = |\omega_n(x)| = (-1)^{n-p} \omega_n(x)$$

and

$$H_p(x) = H_p^{[n]}(x) = \sum_{j=0}^{p} \frac{1}{|\omega_n'(x_j)|(x - x_j)} + \sum_{j=p+1}^{n} \frac{-1}{|\omega_n'(x_j)|(x - x_j)}.$$

Let

$$\Lambda_p(x) = \Lambda_p^{[n]}(x) = F_p(x) H_p(x),$$

which is the polynomial coinciding with the Lebesgue function Λ on the interval $\overline{I}_p$.

In the literature, most of the authors used the above formulas when they studied the Lagrange polynomial interpolation. Starting from this formula, they gave the specific formula of the Lebesgue function for Lagrange interpolation at a particular set of nodes. However, we found that this formula for Lebesgue function is very inconvenient for our further study. It is well known that a set of nodes is optimal if and only if all the local maxima are equal. For finding the optimal nodes, one must compare Λ_{p+1} and Λ_p as well as the local maxima $\lambda_{n,p+1}$ and $\lambda_{n,p}$. If we write Λ_p as a product of two parts, this comparison is inconvenient in general. Brutman [3] compared Λ_{p+1} and Λ_p corresponding to $X_{I_1,\pi/(n+1)}$ (Chebyshev nodes) and $X_{I_1,\pi/n}$ (Chebyshev extrema) and obtained very interesting results. He proved that the local maxima corresponding to Chebyshev nodes are strictly decreasing from the outside towards the middle of the interval, while the local maxima corresponding to Chebyshev extrema are strictly increasing from the outside towards the middle of the interval. This comparison is possible because of the special structure of the Lebesgue functions corresponding to Chebyshev nodes and Chebyshev extrema. We anticipate that similar results are also true for the sets of nodes $X_{I_m,\pi/(n+1)}$ and $X_{I_m,\pi/n}$ with $m \geq 1$. However, the direct comparison is impossible in general when we denote the Lebesgue function as a product of two parts. Therefore, one must give other expressions for the Lebesgue functions and seek alternative ways for the comparison.

For a fixed n, we always let $N = [n/2]$ and $M = [(n-1)/2]$. When $n = 2N+1$, $M = N$. When $n = 2N$, $M = N-1$. By symmetry, we have $\lambda_{n,n-p-1} = \lambda_{n,p}$. Therefore, it suffices to consider Λ_p for $0 \leq p \leq M$ only.

We recall that

$$\Lambda(x) = \sum_{j=0}^{n} |l_j(x)|,$$

where

$$l_j(x) = \prod_{\substack{k=0 \\ k\neq j}}^{n} \frac{x - x_k}{x_j - x_k}.$$

When $0 \leq p \leq M$ and $x \in I_p$ ($x_p < x < x_{p+1}$), we define

$$\Lambda_{p,j}(x) = \begin{cases} |l_j(x)| + |l_{n-j}(x)|, & \text{if } 0 \leq j \leq M, \\ |l_N(x)|, & \text{if } n = 2N \text{ and } j = N. \end{cases} \tag{15}$$

Then we have

$$\Lambda_p(x) = \sum_{j=0}^{N} \Lambda_{p,j}(x), \tag{16}$$

which is the polynomial coinciding with the Lebesgue function Λ on the interval $\overline{I}_p$. We give a formula for Λ_p in the following proposition.

Proposition 1. *Let $0 \leq p \leq M$ and $x \in I_p$ ($x_p < x < x_{p+1}$). Let $\Lambda_{p,j}$ be defined in (15). When $0 \leq j \leq M$, we have*

$$\Lambda_{p,j}(x) = \left|\frac{x}{x_j}\right|^{\varepsilon_n + \varepsilon_{p,j}} \prod_{\substack{k=0 \\ k\neq j}}^{M} \left|\frac{x^2 - x_k^2}{x_j^2 - x_k^2}\right|, \tag{17}$$

where

$$\varepsilon_n = \begin{cases} 0, & \text{if } n = 2N+1, \\ 1, & \text{if } n = 2N, \end{cases} \quad \text{and} \quad \varepsilon_{p,j} = \begin{cases} 0, & \text{if } 0 \leq j \leq p, \\ 1, & \text{if } p+1 \leq j \leq M. \end{cases} \tag{18}$$

When $n = 2N$ and $p+1 \leq j \leq N$, we have

$$\Lambda_{p,j}(x) = \prod_{\substack{k=0 \\ k\neq j}}^{N} \left|\frac{x^2 - x_k^2}{x_j^2 - x_k^2}\right|. \tag{19}$$

Proof. When $0 \le j \le M$, we have

$$\Lambda_{p,j}(x) = |l_j(x)| + |l_{n-j}(x)| = \prod_{\substack{k=0 \\ k \ne j, n-j}}^{n} \left| \frac{x - x_k}{x_j - x_k} \right| \left(\frac{|x - x_j| + |x + x_j|}{2|x_j|} \right).$$

Since

$$\frac{|x - x_j| + |x + x_j|}{2|x_j|} = \begin{cases} 1, & \text{when } 0 \le j \le p, \\ |x/x_j|, & \text{when } p+1 \le j \le M, \end{cases}$$

we obtain (17).

When $n = 2N$ and $p+1 \le j \le N-1$, we have $\varepsilon_n + \varepsilon_{p,j} = 2$. We also have $x_N = 0$. The (19) can be obtained from (17).

When $n = 2N$ and $j = N$, we have

$$\Lambda_{p,N}(x) = |l_N(x)| = \prod_{\substack{k=0 \\ k \ne N}}^{2N} \left| \frac{x - x_k}{x_j - x_k} \right| = \prod_{k=0}^{N-1} \left| \frac{x^2 - x_k^2}{x_j^2 - x_k^2} \right|.$$

□

In the above proof, we see that $\Lambda_{p,j}$ can be expressed as either (17) or (19) when $n = 2N$ and $p+1 \le j \le N-1$. Both expressions are useful and will be used in different situations. The particular case of $n = 2N$ and $j = N$ is also included in (19). This fact is very convenient for investigation of polynomial interpolation.

Instead of expressing the Lebesgue function as a product of two parts, we express Λ_p as the sum of $\Lambda_{p,j}$ as in (16) for each p. Each term $\Lambda_{p,j}$ is expressed as a product as in either (17) or (19).

Now we can obtain the corresponding formulas of Lebesgue functions for Lagrange polynomial interpolation at $X_{\varphi,\alpha}$, where φ is an odd and strictly increasing function and $0 < \alpha \le \pi/n$. Let

$$\eta_j = \eta_j(\alpha) = (-n/2 + j)\alpha, \qquad j = 0, 1, 2, \ldots, n. \tag{20}$$

Then the nodes in $X_{\varphi,\alpha}$ are

$$x_j = x_j(\alpha) = \varphi((-n/2 + j)\alpha) = \varphi(\eta_j), \qquad j = 0, 1, 2, \ldots, n. \tag{21}$$

It is obvious that every point $\eta \in [\eta_p, \eta_{p+1}]$ for $0 \le p \le n-1$ can be uniquely expressed as $\eta = \eta_p + s\alpha$ with $0 \le s \le 1$. Therefore, every point $x \in \overline{I}_p = [x_p(\alpha), x_{p+1}(\alpha)]$ can be uniquely expressed as $\varphi(\eta_p + s\alpha)$. Here, we introduced a parameter s $(0 \le s \le 1)$ to indicate the relative position of the variable x in the interval $\overline{I}_p$. We must emphasize that this parameter s $(0 \le s \le 1)$ is the same for all functions φ (which are strictly increasing and odd), all of the values of parameter α $(0 < \alpha \le \pi/n$, or $0 \le \alpha \le \pi/n)$, and all subinterval $\overline{I_p}$ $(0 \le p \le n-1)$. This fact will be extremely convenient for our further study.

By Proposition 1, the Lebesgue function for the Lagrange polynomial interpolation at $X_{\varphi,\alpha}$ can be expressed as follows. When $0 \le p \le M$ and $x = \varphi(\eta_p + s\alpha) \in \overline{I}_p = [\varphi(\eta_p(\alpha)), \varphi(\eta_{p+1}(\alpha))]$,

$$\Lambda_{p;\varphi}(\alpha,s) = \sum_{j=0}^{N} \Lambda_{p,j;\varphi}(\alpha,s). \tag{22}$$

When $0 \le j \le M$, we have

$$\Lambda_{p,j;\varphi}(\alpha,s) = \left|\frac{\varphi(\eta_p + s\alpha)}{\varphi(\eta_j)}\right|^{\varepsilon_n+\varepsilon_{p,j}} \prod_{\substack{k=0\\k\neq j}}^{M} \left|\frac{\varphi^2(\eta_p + s\alpha) - \varphi^2(\eta_k)}{\varphi^2(\eta_j) - \varphi^2(\eta_k)}\right|, \tag{23}$$

where ε_n and $\varepsilon_{p,j}$ are defined as in (18). When $n = 2N$ and $p+1 \le j \le N$, we have

$$\Lambda_{p,j;\varphi}(\alpha,s) = \prod_{\substack{k=0\\k\neq j}}^{N} \left|\frac{\varphi^2(\eta_p + s\alpha) - \varphi^2(\eta_k)}{\varphi^2(\eta_j) - \varphi^2(\eta_k)}\right|. \tag{24}$$

In our previous paper [5], we gave another kind of formula for the Lebesgue function for the Lagrange polynomial interpolation at $X_{I_1,\alpha}$, where $I_1(x) = \sin x$. This kind of formulas can be extended for the set of nodes $X_{\overline{\varphi},\alpha}$, where $\overline{\varphi} = \sin\varphi$ with $\varphi \in \overline{\Phi}$.

Since we used a decreasing order of the nodes in our previous work [5], we used the transformation $x = \cos\theta$ there. The following three propositions give the formula for Lebesgue function for arbitrary symmetric nodes with the transformation $x = \sin\theta$ and $x_j = \sin\theta_j$. These three propositions are just slight modifications of the corresponding propositions in [5]. We omit their proofs.

Proposition 2. *Let* $0 \le p \le M$ *and* $x = \sin\theta \in I_p$. *Then*

$$\omega_n(\sin\theta) = \prod_{j=0}^{n} \sin(\theta - \theta_j), \tag{25}$$

$$F_p(\sin\theta) = |\omega_n(\sin\theta)| = (-1)^{n-p} \prod_{j=0}^{n} \sin(\theta - \theta_j). \tag{26}$$

Proposition 3. *For* $j = 0,1,2,\ldots,n$, *we have*

$$\cos\theta_j \omega_n'(\sin\theta_j) = \prod_{\substack{l=0\\l\neq j}}^{n} \sin(\theta_j - \theta_l), \tag{27}$$

$$\cos\theta_j |\omega_n'(\sin\theta_j)| = (-1)^{n-j} \prod_{\substack{l=0\\l\neq j}}^{n} \sin(\theta_j - \theta_l). \tag{28}$$

Let

$$A_j = A_j(\theta_0, \theta_1, \ldots, \theta_M) = 1/\cos\theta_j|\omega'(\sin\theta_j)|, \qquad j = 0, 1, \ldots, n. \tag{29}$$

Since the nodes are symmetric about the origin and $\theta_{n-j} = -\theta_j$, A_j is a function of $\{\theta_0, \theta_1, \ldots, \theta_M\}$. We also have $A_{n-j} = A_j$, for $0 \le j \le n$, and $\lambda_{n,n-1-p} = \lambda_{n,p}$, for $0 \le p \le n-1$. It suffices to consider $\lambda_{n,p}$ only for $0 \le p \le M$.

Proposition 4. *Let $0 \le p \le M$ and $x = \sin\theta \in I_p$. Then*

$$H_p(\sin\theta) = \sum_{j=0}^{p} A_j\left(\cot(\theta - \theta_j) - \cot(\theta + \theta_j)\right) + \sum_{j=p+1}^{n-p-1} A_j \frac{1}{\sin(\theta_j - \theta)}. \tag{30}$$

Let $\theta = \varphi(\eta) = \varphi(\eta_p + s\alpha)$ and $\theta_j = \varphi(\eta_j)$ in the above formulas. We then obtain the formulas for $F_{p;\overline{\varphi}}(\alpha, s)$, $A_{j;\overline{\varphi}}(\alpha)$, and $H_{p;\overline{\varphi}}(\alpha, s)$. In other words, we obtain the formula of Lebesgue function $\Lambda_{p;\overline{\varphi}}(\alpha, s) = F_{p;\overline{\varphi}}(\alpha, s) H_{p;\overline{\varphi}}(\alpha, s)$ for polynomial interpolation at $X_{\overline{\varphi},\alpha}$. The particular case for $\overline{\varphi} = I_1$, where $I_1(x) = \sin I_0(x) = \sin x$, was obtained in our previous paper [5].

We expect to use these formulas for the functions $\overline{\varphi} = \sin(I_{m-1}) = I_m$ with $m \ge 1$ in our future study, particularly, for $\overline{\varphi} = I_1$. We give the corresponding formulas corresponding to I_1 as follows:

$$F_{p;I_1}(\alpha, s) = (-1)^{n-p} \prod_{j=0}^{n} \sin(p + s - j)\alpha = (-1)^{n-p} \prod_{l=-(n-p)}^{p} \sin(l + s)\alpha. \tag{31}$$

$$\begin{aligned} H_{p;I_1}(\alpha, s) &= \sum_{j=0}^{p} A_{j;I_1}(\alpha)\left[\cot(p + s - j)\alpha + \cot(n - j - p - s)\alpha\right] \\ &\quad + \sum_{j=p+1}^{n-p-1} A_{j;I_1}(\alpha) \frac{1}{\sin(j - p - s)\alpha} \\ &= \sum_{j=0}^{p} A_{j;I_1}(\alpha) \frac{\sin(n - 2j)\alpha}{\sin(p + s - j)\alpha \sin(n - j - p - s)\alpha} \\ &\quad + \sum_{j=p+1}^{N} A_{j;I_1}(\alpha) \frac{2\sin(n/2 - p - s)\alpha \cos(n/2 - j)\alpha}{\sin(j - p - s)\alpha \sin(n - j - p - s)\alpha} \delta_j^{[n]}, \end{aligned} \tag{32}$$

where

$$\delta_j^{[n]} = \begin{cases} 1/2, & n = 2N \text{ and } j = N, \\ 1, & \text{otherwise.} \end{cases} \tag{33}$$

$$A_{j;I_1}(\alpha) = \frac{1}{\prod_{l=1}^{j} \sin l\alpha \prod_{l=1}^{n-j} \sin l\alpha}. \tag{34}$$

We will use induction to study the Lebesgue functions, as well as local maxima and Lebesgue constants, corresponding to I_m. The Lebesgue functions corresponding to I_1 will serve as the basis. Therefore, we need the expressions for $\Lambda_{p,j;I_1}$, which can be easily obtained by using (31), (32), and (34). When $0 \le j \le p$, we have

$$\Lambda_{p,j;I_1}(\alpha,s) = F_{p;I_1}(\alpha,s)A_{j;I_1}(\alpha)\frac{\sin(n-2j)\alpha}{\sin(p+s-j))\alpha\sin(n-j-p-s)\alpha}. \tag{35}$$

When $p+1 \le j \le N$, we have

$$\Lambda_{p,j;I_1}(\alpha,s) = F_{p;I_1}(\alpha,s)A_{j;I_1}(\alpha)\frac{2\sin(n/2-p-s)\alpha\cos(n/2-j)\alpha}{\sin(j-p-s)\alpha\sin(n-j-p-s)\alpha}\,\delta_j^{[n]}. \tag{36}$$

Equations (31), (32), and (34) were given in our previous paper [5]. These equations include some particular cases of Lebesgue functions corresponding to $X_{I_1,\alpha}$ for $\alpha = \pi/n$, $\pi/(n+1)$, $\pi/(n+2)$, 0, which correspond to Chebyshev extrema, zeros of the Chebyshev polynomials of first and second kinds, and equidistant nodes, respectively. The Lagrange polynomial interpolation for these particular cases have been extensively investigated by various authors. We are particularly interested in the cases of Chebyshev nodes ($\alpha = \pi/(n+1)$) and Chebyshev extrema ($\alpha = \pi/n$). Here, we list the corresponding formulas.

$$\begin{aligned}
\Lambda_{p;I_1}(\pi/(n+1),s) &= \frac{\sin s\pi}{n+1}\left[\sum_{j=0}^{p}\left(\cot(p+s-j)\frac{\pi}{n+1}+\cot(n-j-p-s)\frac{\pi}{n+1}\right)\right.\\
&\qquad\left.+\sum_{j=p+1}^{n-p-1}\frac{1}{\sin[(j-p-s)\pi/(n+1)]}\right]\\
&= \frac{\sin s\pi}{n+1}\left[\sum_{l=0}^{p}\cot(l+s)\frac{\pi}{n+1}-\sum_{l=p+1}^{2p+1}\cot(l+s)\frac{\pi}{n+1}\right.\\
&\qquad\left.+\sum_{l=1}^{n-2p-1}\frac{1}{\sin[(l-s)\pi/(n+1)]}\right].
\end{aligned} \tag{37}$$

$$\Lambda_{p;I_1}(\pi/n,s) = \frac{\sin s\pi}{n}\left[\sum_{l=0}^{2p}\frac{1}{\sin[(l+s)\pi/n]}-\sum_{l=2p+1}^{n-1}\cot[(l+s)\pi/n]\right]. \tag{38}$$

The above formulas were first given by Brutman [3]. We express these formulas in our terminology. Brutman investigated polynomial interpolation at the Chebyshev nodes and Chebyshev extrema separately. We introduced a parameter α ($0 \le \alpha \le \pi/n$) and included the Chebyshev nodes and Chebyshev extrema as particular cases. We also introduced the same parameter s ($0 \le s \le 1$) to indicate the relative positions of the variable in the Lebesgue functions in each subinterval.

Both $\Lambda_{p;I_1}(\pi/(n+1),s)$ and $\Lambda_{p;I_1}(\pi/n,s)$ are the products of the function $\sin s\pi$ and another function of s for all p. Therefore, it is convenient to compare $\Lambda_{p+1;I_1}$ and $\Lambda_{p;I_1}$ in these two particular cases. For all other α, the direct comparing of $\Lambda_{p+1;I_1}(\alpha,s)$ and $\Lambda_{p;I_1}(\alpha,s)$ is very difficulty. For $m \ge 2$, the direct comparing of

$\Lambda_{p+1;I_m}(\alpha,s)$ and $\Lambda_{p;I_m}(\alpha,s)$ is almost impossible. Therefore, one must seek an indirect comparing and use induction.

4 Properties of Pairs of Auxiliary Functions

In our previous paper [5], we introduced some lemmas which are very useful for Lagrange polynomial interpolation. In particular, we gave some properties of the auxiliary function $g(x) = x\cot x$ $(-\pi < x < \pi)$, which are particularly important tools for our study.

Since the function g is an even and infinitely differentiable function, it suffices to give the properties of g on the interval $[0,\pi)$. In [5], we proved that, for $0 < x < \pi$, $g'(x) < 0$, $g''(x) < 0$, and $g'''(x) < 0$. We remarked that the fourth derivative $g^{(4)}$ is not negative for some x on the interval $(0,\pi)$. However, this remark is incorrect. In fact, for any $n \geq 1$, the nth derivative $g^{(n)}$ is negative on $(0,\pi)$. In our future study for the Lagrange polynomial interpolation, we will also use the properties of $g^{(n)}$ for $n \geq 4$. We give the proof of these inequalities in this section for the future reference.

To prove this property, we must introduce another auxiliary function $h(x) = x\tan x$, which is defined for $-\pi/2 < x < \pi/2$. This function is also even and infinitely differentiable. It is easy to verify that

$$g(x) = g(x/2) - h(x/2) \tag{39}$$

for $-\pi < x < \pi$.

Lemma 3. *Let $h(x) = x\tan x$ for $0 < x < \pi/2$. Then $h^{(n)}(x) > 0$ for $n \geq 0$.*

Proof. First, we have that $\tan x > 0$ for $0 < x < \pi/2$ and $(\tan x)' = 1 + \tan^2 x$. By a simple induction, it is easy to verify that $(\tan x)^{(n)} > 0$ for $0 < x < \pi/2$ and $n \geq 0$. Therefore,

$$h^{(n)}(x) = x(\tan x)^{(n)} + n(\tan x)^{(n-1)} > 0$$

for $n > 0$. When $n = 0$, it is obvious that $h(x) = x\tan x > 0$. □

In this paper, we always let g and h denote the functions $g(x) = x\cot x$ and $h(x) = x\tan x$. We treat the functions g and h as a pair (g,h). There are many pairs of functions, which have similar properties as the pair (g,h) and are also very important for our study of polynomial interpolation. Therefore, we give the following definition.

Definition 3. We say that a pair of functions $(\overline{g},\overline{h})$ has Property B if it has the following properties.

1. $\overline{g} : (-\pi,\pi) \to (-\infty, c]$, and $\overline{g}(0) = c$.
2. $\overline{h} : (-\pi/2,\pi/2) \to [0,+\infty)$, and $\overline{h}(0) = 0$.
3. Both $\overline{g}$ and $\overline{h}$ are even and infinitely differentiable functions.
4. $\overline{g}(x) = \overline{g}(x/2) - \overline{h}(x/2)$ for $-\pi < x < \pi$.
5. $\overline{h}^{(n)}(x) > 0$ for $0 < x < \pi/2$ and $n \geq 0$.

By (39) and Lemma 3, one can easily verify that the pair of functions (g,h) has Property B. In general, the inequalities corresponding to $\overline{h}$ and its derivatives $\overline{h}^{(n)}$ are easy to verify. Therefore, we include these inequalities in Definition 3. The inequalities corresponding to $\overline{g}^{(n)}$ $(n \geq 1)$ can be verified. We have the following lemma.

Lemma 4. *Let the pair of functions* $(\overline{g},\overline{h})$ *have Property B. Then* $\overline{g}^{(n)}(x) < 0$ *for* $0 < x < \pi$ *and* $n > 0$.

Proof. By a simple induction,

$$\overline{g}(x) = \overline{g}(x/2^k) - \sum_{l=1}^{k} \overline{h}(x/2^l) \tag{40}$$

for $k \geq 1$.

Let $0 < a < \pi$. Then for any $x \in [0,a]$,

$$\begin{aligned}\sum_{l=1}^{k} \overline{h}(x/2^l) \leq \sum_{l=1}^{k} \overline{h}(a/2^l) &= \sum_{l=1}^{k} \left[\overline{h}(a/2^l) - \overline{h}(0)\right] \\ &= \sum_{l=1}^{k} (a/2^l)\overline{h}'(\xi_l) < a\overline{h}'(a/2)\sum_{l=1}^{k}(1/2^l) < a\overline{h}'(a/2).\end{aligned}$$

Therefore, the sum is uniformly bounded for $0 \leq x \leq a$ and $1 \leq k < \infty$. Since a is an arbitrary number in $(0,\pi)$, for any $0 \leq x < \pi$, we may let $k \to +\infty$ in (40) and obtain

$$\overline{g}(x) = \overline{g}(0) - \sum_{l=1}^{\infty} \overline{h}(x/2^l). \tag{41}$$

Since $\overline{h}^{(n)}$ are increasing functions for all n, we also have, for $0 \leq x \leq a < \pi$ and $n \geq 1$,

$$\sum_{l=1}^{k} \frac{\mathrm{d}^n}{\mathrm{d}x^n}\overline{h}(x/2^l) = \sum_{l=1}^{k}(1/2^{nl})\overline{h}^{(n)}(x/2^l) < \overline{h}^{(n)}(a/2)\sum_{l=1}^{k}(1/2^{nl}) < \overline{h}^{(n)}(a/2).$$

Therefore, we can take the nth derivative of (41) term by term to obtain

$$\overline{g}^{(n)}(x) = -\sum_{l=1}^{\infty}(1/2^{nl})\overline{h}^{(n)}(x/2^l) < 0$$

for $0 < x < \pi$ and $n \geq 1$. □

In particular, since the pair of functions (g,h) has Property B, we have $g^{(n)}(x) < 0$ for $0 < x < \pi$ and $n = 1,2,3,\ldots$.

From a given pair of functions with Property B, one can construct various new pairs of functions which also have Property B. In particular, we have the following lemma.

Lemma 5. *Let the pair of functions* $(\overline{g},\overline{h})$ *have Property B. Let* $\overline{g}_m(x) = x^m\overline{g}^{(m)}(x)$ *and* $\overline{h}_m(x) = x^m\overline{h}^{(m)}(x)$ *with* $m \geq 0$. *Then the pairs of functions* $(\overline{g}_m,\overline{h}_m)$ *also have Property B.*

Lemma 5 can be easily and directly verified. In particular, the pair of functions (g_m,h_m), where $g_m(x) = x^m g^{(m)}(x)$ and $h_m(x) = x^m h^{(m)}(x)$ with $m \geq 0$, has Property B. These pairs of functions will also play very important roll in our investigation for the polynomial interpolation.

We make a very important remark. If the pair of functions $(\overline{g},\overline{h})$ has Property B, then the functions $\overline{g}^{(2k)}$ and $\overline{h}^{(2k)}$ are even functions, while the functions $\overline{g}^{(2k+1)}$ and $\overline{h}^{(2k+1)}$ are odd functions. Therefore, $\overline{g}^{(2k)}(x) < 0$ for $k \geq 1$ and $-\pi < x < \pi$. But we have $\overline{g}^{(2k+1)}(x) < 0$ for $0 < x < \pi$ and $\overline{g}^{(2k+1)}(x) > 0$ for $-\pi < x < 0$. When applying Lemma 4, one must very carefully distinguish the even and odd cases.

Sometimes we need to consider the values of $\overline{g}^{(n)}(0)$ and $\overline{h}^{(n)}(0)$. We take nth derivatives of both sides of the equation $\overline{g}(x) = \overline{g}(x/2) - \overline{h}(x/2)$.

$$\overline{g}^{(n)}(x) = (1/2^n)\overline{g}^{(n)}(x/2) - (1/2^n)\overline{h}^{(n)}(x/2). \tag{42}$$

Since both $\overline{g}$ and $\overline{h}$ are even functions, we have $\overline{g}^{(2k+1)}(0) = \overline{h}^{(2k+1)}(0) = 0$. When $n = 0$, we have $\overline{g}(0) = c$ $(g(0) = 1)$ and $\overline{h}(0) = 0$. When $n = 2k > 0$, by (42), we have

$$\overline{g}^{(2k)}(0) = -\frac{1}{2^{2k}-1}\overline{h}^{(2k)}(0). \tag{43}$$

5 Optimal Nodes for Lagrange Polynomial Interpolation

In Sect. 3, we expressed $\Lambda_{p;\varphi}$, which is the polynomial coinciding with the Lebesgue function on the interval $\overline{I}_p$, as a summation in (22). Each term $\Lambda_{p,j;\varphi}$ in this summation is a product as in either (23) or (24). Our aim is to compare the local maxima of the Lebesgue function on consecutive subintervals. Therefore, we must compare $\Lambda_{p+1;\varphi}$ and $\Lambda_{p;\varphi}$ for $0 \leq p < p+1 \leq M$. Since each term $\Lambda_{p,j;\varphi}$ is a product, we may use the quotients $\Lambda_{p+1,j;\varphi}/\Lambda_{p,j;\varphi}$ to compare $\Lambda_{p+1,j;\varphi}$ and $\Lambda_{p,j;\varphi}$. By comparing $\Lambda_{p+1,j;\varphi}$ and $\Lambda_{p,j;\varphi}$ for each j, one may get much important and useful information for comparing $\Lambda_{p+1;\varphi}$ and $\Lambda_{p;\varphi}$. In order to get the set of optimal nodes, finally, one must compare $\Lambda_{p+1;\varphi}$ and $\Lambda_{p;\varphi}$, at least for certain candidate functions and for a certain range of s, directly or indirectly.

We will use induction to study the Lebesgue functions, as well as local maxima, Lebesgue constants, and so on, for the corresponding functions I_m. We first must study the base case of I_1. Brutman [3] studied polynomial interpolation at the Chebyshev nodes and Chebyshev extrema. We state his results in the following two theorems using our terminology.

Theorem 1. *Let* $0 \le p < p+1 \le M$ *and* $1/2 \le s < 1$. *Then*

$$\Lambda_{p+1;I_1}(\pi/(n+1),s) < \Lambda_{p;I_1}(\pi/(n+1),s). \tag{44}$$

Theorem 2. *Let* $0 \le p < p+1 \le M$ *and* $0 < s < 1$. *Then*

$$\Lambda_{p+1;I_1}(\pi/n,s) > \Lambda_{p;I_1}(\pi/n,s). \tag{45}$$

Let $\lambda_{n,p;\varphi}$ be the local maximum such that

$$\lambda_{n,p;\varphi} = \lambda_{n,p;\varphi}(\alpha) = \max_{0\le s\le 1} \Lambda_{p;\varphi}(\alpha,s) = \max_{0\le s\le 1} \Lambda^{[n]}_{p;\varphi}(\alpha,s). \tag{46}$$

There is one and only one $s_{p;\varphi} = s_{p;\varphi}(\alpha) = s^{[n]}_{p;\varphi}(\alpha)$ such that

$$\lambda_{n,p;\varphi} = \lambda_{n,p;\varphi}(\alpha) = \Lambda_{p;\varphi}(\alpha, s_{p;\varphi}). \tag{47}$$

If $n = 2N+1$ and $p = N$, by symmetry it is easy to see that $s_{N;\varphi}(\alpha) = 1/2$ for any odd and strictly increasing function φ and any α ($0 \le \alpha \le \pi/n$ or $0 < \alpha \le \pi/n$).

Brutman [3] also proved that $1/2 < s_{p;I_1}(\pi/(n+1)) < 1$ for $0 \le p \le N-1$. Therefore, by Theorems 1 and 2, he concluded that for $0 \le p < p+1 \le M$,

$$\lambda_{n,p+1;I_1}(\pi/(n+1)) < \lambda_{n,p;I_1}(\pi/(n+1)), \tag{48}$$

$$\lambda_{n,p+1;I_1}(\pi/n) > \lambda_{n,p;I_1}(\pi/n). \tag{49}$$

In other words, he proved that the local maxima of the Lebesgue function for polynomial interpolation at the Chebyshev nodes are strictly decreasing from outside towards the middle of the interval, while the local maxima of Lebesgue function for the polynomial interpolation at the Chebyshev extrema are strictly increasing from outside towards the middle of the interval. A similar result for equidistant nodes ($\alpha = 0$) was found by Tietze [10] in 1917, almost one century ago.

Brutman got the above results because he very much benefited from the formulas in (37) and (38). We introduced a parameter α ($0 \le \alpha \le \pi/n$) and generalized Brutman's formulas [5]. However, for $\alpha \ne 0, \pi/(n+1), \pi/n$, a direct comparison of $\Lambda_{p+1;I_1}(\alpha,s)$ and $\Lambda_{p;I_1}(\alpha,s)$ is difficult.

By using induction, we found that the inequalities (48) and (49) are also true for I_m ($m \ge 2$). In the following, we state some of our results.

Theorem 3. *Let* $0 \le p < p+1 \le M$ *and* $m \ge 1$. *Then*

$$\lambda_{n,p+1;I_m}(\pi/(n+1)) < \lambda_{n,p;I_m}(\pi/(n+1)). \tag{50}$$

Theorem 4. *Let* $0 \le p < p+1 \le M$ *and* $m \ge 1$. *Then*

$$\lambda_{n,p+1;I_m}(\pi/n) > \lambda_{n,p;I_m}(\pi/n). \tag{51}$$

In other words, the local maxima of Lebesgue function for the polynomial interpolation at the set of nodes $X_{I_m,\pi/(n+1)}$ are strictly decreasing from outside towards the middle of the interval, while the local maxima of Lebesgue function for the polynomial interpolation at the set of nodes $X_{I_m,\pi/n}$ are strictly increasing from outside towards the middle of the interval. These facts are true for any $m \ge 1$. Due to this fact, we are particularly interested in the outside one and middle one of the local maxima. We compare the outside (middle) local maxima for different m, but the same α.

Theorem 5. *Let* $0 < \alpha \le \pi/n$. *Then*

$$\lambda_{n,M;I_1}(\alpha) < \lambda_{n,M;I_2}(\alpha) < \lambda_{n,M;I_3}(\alpha) < \cdots. \tag{52}$$

Theorem 6. *Let* $0 < \alpha \le \pi/n$. *Then*

$$\lambda_{n,0;I_1}(\alpha) > \lambda_{n,0;I_2}(\alpha) > \lambda_{n,0;I_3}(\alpha) > \cdots. \tag{53}$$

By Theorems 3, 5, and 6, we have

$$\begin{aligned} 0 &< \lambda_{n,0;I_{m+1}}(\pi/(n+1)) - \lambda_{n,M;I_{m+1}}(\pi/(n+1)) \\ &< \lambda_{n,0;I_m}(\pi/(n+1)) - \lambda_{n,M;I_m}(\pi/(n+1)). \end{aligned} \tag{54}$$

Therefore, the limit of $\left[\lambda_{n,0;I_m}(\pi/(n+1)) - \lambda_{n,M;I_m}(\pi/(n+1))\right]$ as $m \to \infty$ exists. In fact, this limit is zero.

Theorem 7. *We have*

$$\lim_{m\to\infty}\left[\lambda_{n,0;I_m}(\pi/(n+1)) - \lambda_{n,M;I_m}(\pi/(n+1))\right] = 0. \tag{55}$$

By Theorem 7 together with Theorem 3, we immediately obtain the following theorem.

Theorem 8. *Let* ρ *be the function defined in Lemma 2. Then*

$$\lambda_{n,0;\rho}(\pi/(n+1)) = \lambda_{n,1;\rho}(\pi/(n+1)) = \cdots = \lambda_{n,M;\rho}(\pi/(n+1)). \tag{56}$$

Therefore, the set of nodes $X_{\rho,\pi/(n+1)}$ is the set of optimal nodes for polynomial interpolation.

We are also interested in the optimal Lebesgue constants. For a brief history about the estimation of the optimal Lebesgue constants, one may refer to [4, 6, 9]. Using the optimal nodes $X_{\rho,\pi/(n+1)}$, we get the exact optimal Lebesgue constants.

Since the function ρ is defined as a limiting function of the functions ρ_m, we may not be completely satisfied with the above optimal nodes. Although the set of optimal nodes is unique, it can be given by different functions. The finite version (without limit) of optimal nodes are also given. For any sufficient large m, for example $m \geq n$, there is an α_m $(\pi/(n+1) < \alpha_m < \pi/n)$ such that the set of nodes X_{I_m,α_m} is the optimal one. We also have $\lim_{m\to\infty} \alpha_m = \pi/(n+1)$.

To complete the proofs of the above theorems and related results, we need several dozen lemmas. In particular, we have several important sets of lemmas about the following topics.

1. The partial derivatives, with respect to α and s, of the functions $\Lambda_{p,j;\varphi}$, $\Lambda_{p+1,j;\varphi}/\Lambda_{p,j;\varphi}$, etc., particularly, for $\varphi = I_1$.
2. The properties of $s_{n,p;\varphi} = s_{n,p;\varphi}(\alpha)$, particularly, for $\varphi = I_1$.
3. The relation of various functions corresponding to φ and $\overline{\varphi} = \sin\varphi$. This is the foundation for induction.

We will publish these lemmas and the complete proofs of the above theorems and related results in our upcoming papers.

Acknowledgements Professor Ward Cheney introduced me to the problem of optimal Lagrange polynomial interpolation. We have been working on this problem for many years. His advice and insights have been vital to the success of this project. I am very grateful to him for carefully reading different versions of this paper and making valuable suggestions. My appreciation is extended to Professor Larry Schumaker whose advice makes this paper much better.

References

1. S. N. Bernstein, Sur la limitation des valeurs d'un polynôme $P_n(x)$ de degré n sur tout un segment par ses valeurs en $n+1$ points du segment, Bull Acad. Sci. URSS, 1025-1050, (1931).
2. C. de Boor and A. Pinkus, Proof of the conjectures of Berstein and Erdős concerning the optimal nodes for polynomial interpolation, J. Approx. Theory **24**, 289-303, (1978)
3. L. Brutman, On the Lebesgue functions for polynomial interpolation, SIAM J. Numerical Analysis **15**, 694-704, (1978)
4. L. Brutman, Lebesgue functions for polynomial interpolation: a survey, Ann. Numer. Math. **4**, 111-127, (1997)
5. D. Chen and E. W. Cheney, Lagrange polynomial interpolation, In: Approximation Theory XII, Int. Conf., San Antonio 2007, M. Neamtu and L. Schumaker (eds.), Nashboro, Brentwood, 60-76, (2008)
6. E. W. Cheney and W. A. Light, A Course in Approximation Theory, Brooks/Cole, New York (2000)
7. T. A. Kilgore, Optimization of the norm of the Lagrange interpolation operator, Bull. Amer. Math. Soc. **83**, 1069-1071, (1977)
8. T. A. Kilgore, A characterization of the Lagrange interpolating projection with minimal Tchebycheff norm, J. Approx. Theory **24**, 273-288, (1978)
9. T. J. Rivlin, The Lebesgue constants for polynomial interpolation, In: Functional Analysis and Its Applications, Int. Conf., Madras, 1973, H. C. Carnier et al., (eds.), Springer-Verlag, Berlin, 422-437, (1974)
10. H. Tietze, Eine Bemerkung zur Interpolation, Z. Angew. Math. and Phys. **64**, 74-90, (1917)

Green's Functions: Taking Another Look at Kernel Approximation, Radial Basis Functions, and Splines

Gregory E. Fasshauer

Abstract The theories for radial basis functions (RBFs) as well as piecewise polynomial splines have reached a stage of relative maturity as is demonstrated by the recent publication of a number of monographs in either field. However, there remain a number of issues that deserve to be investigated further. For instance, it is well known that both splines and radial basis functions yield "optimal" interpolants, which in the case of radial basis functions are discussed within the so-called native space setting. It is also known that the theory of reproducing kernels provides a common framework for the interpretation of both RBFs and splines. However, the associated reproducing kernel Hilbert spaces (or native spaces) are often not that well understood — especially in the case of radial basis functions. By linking (conditionally) positive definite kernels to Green's functions of differential operators we obtain new insights that enable us to better understand the nature of the native space as a generalized Sobolev space. An additional feature that appears when viewing things from this perspective is the notion of scale built into the definition of these function spaces. Furthermore, the eigenfunction expansion of a positive definite kernel via Mercer's theorem provides a tool for making progress on such important questions as stable computation with flat radial basis functions and dimension independent error bounds.

1 Introduction

A number of monographs and survey papers dealing with splines, radial basis functions and, more generally, reproducing kernels, have appeared in recent years. The following list is representative, but certainly far from complete: [1, 3, 8, 18, 20, 29, 34, 39–41, 43]. Even though (or precisely because) there is regrettably little

Gregory E. Fasshauer
Illinois Institute of Technology, Chicago, IL 60616, USA
e-mail: fasshauer@iit.edu

M. Neamtu and L. Schumaker (eds.), *Approximation Theory XIII: San Antonio 2010*, Springer Proceedings in Mathematics 13, DOI 10.1007/978-1-4614-0772-0_4,

interaction between different mathematical communities, we have included references from approximation theory as well as probability/statistics and machine learning since reproducing kernels play a central — and perhaps increasing — role in all of these communities.

In this paper we will address several questions that — to our knowledge — have not been addressed sufficiently in the existing literature. The first few questions center around the notion of an RBF *native space* (to be defined below). We will recall existing interpretations and claim that most of them are not very "intuitive". This has, in fact, been a point of criticism of RBF methods. What are these native spaces, and what kind of functions do they contain? How do they relate to classical function spaces such as Sobolev spaces? We try to shed some light on this topic by discussing recent work of [12] in Section 2.

Another set of questions is related to the role of *scale*. RBF practitioners have known for a long time that the proper scaling of the basis functions plays a very important role. It might affect the accuracy of an approximation, its numerical stability and its efficiency. Should a notion of scale be included in the definition of the native space? Our framework of Section 2 does indeed provide a natural way of doing this.

An appropriate scaling of the kernel has been used to establish a connection between infinitely smooth RBFs and polynomial interpolants in the literature (see Section 3 and the references listed there). If the kernels are "flat", we get convergence of RBF interpolants to polynomial interpolants. We will report on a recent investigation [38] that reveals a similar connection between RBFs of limited smoothness and piecewise polynomial splines.

Even though researchers have struggled for many years with the ill-conditioning of RBF systems, relatively little progress has been made in this direction. For univariate piecewise polynomial splines it is well known that moving from the basis of truncated power functions to the B-spline basis provides well-conditioned, and even banded, matrices. Aside from some scattered work on preconditioning of RBF systems, only Bengt Fornberg together with his co-workers has tackled this problem with some success. We are especially motivated by their RBF-QR idea [14, 15] and will provide some of our own thoughts on this approach in Section 4.

Finally, many papers on rates of convergence of the RBF approximation method exist in the literature. However, none of these papers address the question of dimension-dependence of these bounds. In fact, it is quite obvious that all of the existing bounds suffer from the *curse of dimensionality*. In Section 5 we review recent work [9] on *dimension-independent* error bounds for RBF approximation methods.

It turns out that a unifying theme underlying all of these questions is the notion of *Green's functions* and *eigenfunction expansions*. Therefore, these topics will be reviewed in the next section. Connections between either splines and Green's functions or radial basis functions and Green's functions have repeatedly been made over the past decades (see, e.g., [4, 7, 19, 25, 27, 28, 37, 41]). However, many of the connections presented in the following seem to go beyond the discussion in the existing literature. Throughout the paper we will use (simple) examples to illustrate the various topics.

2 Toward an Intuitive Interpretation of Native Spaces

2.1 What is the Current Situation?

Even though piecewise polynomial splines and radial basis functions are conceptually very similar (some people do not even distinguish between the two and use the term *spline* to refer to either method), there are relatively few intersections in the literature on these two approximation methods. Perhaps the most prominent common feature of the two methods is given by the fact that they both yield *minimum norm interpolants* (see, e.g., [2, 8, 37, 43]). In fact, it is precisely this property that led Schoenberg to refer to piecewise polynomial univariate approximating functions as *splines* [36].

To begin with a specific example, we recall that the natural spline $s_{f,2m}$ of order $2m$ provides the smoothest interpolant to data sampled from any function f in the Sobolev space $H^m(a,b)$ of functions whose m^{th} derivative is square integrable on $[a,b]$ and whose derivatives of orders m through $2m-2$ vanish at the endpoints of the interval $[a,b]$, i.e.,

$$s_{f,2m} = \underset{f\in H^m(a,b)}{\operatorname{argmin}} \left\{ \int_a^b \left[f^{(m)}(x)\right]^2 \mathrm{d}x \mid f(x_i) = y_i,\ i=1,\dots,N, \right.$$
$$\left. f^{(\ell)}(a) = f^{(\ell)}(b) = 0,\ \ell = m,\dots,2m-2 \right\}. \tag{1}$$

Now let us consider the corresponding minimum norm property as it is commonly found for radial basis functions, or more generally reproducing kernel interpolants (see, e.g., [43]). The reproducing kernel interpolant $s_{f,K}$ is optimal in the sense that it is the minimum norm interpolant to data sampled from any function f in $\mathscr{H}(K,\Omega)$, the reproducing kernel Hilbert space (or *native space*) associated with K. This can be stated as

$$s_{f,K} = \underset{f\in\mathscr{H}(K,\Omega)}{\operatorname{argmin}} \left\{ \|f\|_{\mathscr{H}(K,\Omega)} \mid s_{f,K}(\mathbf{x}_i) = f(\mathbf{x}_i),\ i=1,\dots,N \right\}. \tag{2}$$

While the function space $H^m(a,b)$ that appears in (1) can be rather easily understood in terms of the smoothness and boundary conditions imposed, the native space $\mathscr{H}(K,\Omega)$ in (2) looks a bit more cryptic. What is this mysterious native space and how is its norm defined?

For a general positive definite kernel K and domain $\Omega \subseteq \mathbb{R}^d$ the native space is commonly defined as

$$\mathscr{H}(K,\Omega) = \operatorname{span}\{K(\cdot,\mathbf{z}) \mid \mathbf{z}\in\Omega\},$$

i.e., the native space is given by all linear combinations of — often infinitely many — "shifts" of the kernel K. This is certainly a valid definition, but what sort of functions does $\mathscr{H}(K,\Omega)$ contain? The literature is more specific for the case in which we use translation invariant (in the statistics literature also referred to as *stationary*) kernels on $\Omega = \mathbb{R}^d$, i.e., if the kernel is really a function of one variable, namely the difference of two points, or $\widetilde{K}(\mathbf{x}-\mathbf{z}) = K(\mathbf{x},\mathbf{z})$. In this case, if $\widetilde{K} \in C(\mathbb{R}^d) \cap L_1(\mathbb{R}^d)$, then

$$\mathscr{H}(\widetilde{K},\mathbb{R}^d) = \left\{ f \in L_2(\mathbb{R}^d) \cap C(\mathbb{R}^d) \mid \frac{\mathscr{F}f}{\sqrt{\mathscr{F}\widetilde{K}}} \in L_2(\mathbb{R}^d) \right\},$$

i.e., a function f belongs to the native space $\mathscr{H}(\widetilde{K},\mathbb{R}^d)$ of the kernel $\widetilde{K}$ if the decay of its Fourier transform $\mathscr{F}f$ relative to that of the Fourier transform $\mathscr{F}\widetilde{K}$ of the kernel is rapid enough. This characterization certainly encodes some kind of smoothness information, but it is not very intuitive. The previous material is covered in much more detail in [43].

As mentioned above, we are not only interested in understanding the type of functions contained in the native space, but also the norm this space is equipped with. Since both the spline and kernel spaces are Hilbert spaces it is natural to look at their inner products. In the natural spline case this is the standard Sobolev inner product whose induced norm appears in (1). What does the native space inner product look like?

For a general positive definite kernel K on a general domain Ω we take functions $f,g \in \mathscr{H}(K,\Omega)$ and use the notation $N_K = \dim(\mathscr{H}(K,\Omega))$ for the dimension of the native space (note that $N_K = \infty$ is common). Then

$$\langle f,g\rangle_{\mathscr{H}(K,\Omega)} = \langle \sum_{j=1}^{N_K} c_j K(\cdot,\mathbf{x}_j), \sum_{k=1}^{N_K} d_k K(\cdot,\mathbf{z}_k)\rangle_{\mathscr{H}(K,\Omega)} = \sum_{j=1}^{N_K}\sum_{k=1}^{N_K} c_j d_k K(\mathbf{x}_j,\mathbf{z}_k).$$

Once again, one might wonder how to interpret this. As before, for translation invariant kernels on $\Omega = \mathbb{R}^d$, i.e., $\widetilde{K}(\mathbf{x}-\mathbf{z}) = K(\mathbf{x},\mathbf{z})$, we can employ Fourier transforms. Then we have

$$\langle f,g\rangle_{\mathscr{H}(\widetilde{K},\mathbb{R}^d)} = \frac{1}{\sqrt{(2\pi)^d}} \langle \frac{\mathscr{F}f}{\sqrt{\mathscr{F}\widetilde{K}}}, \frac{\mathscr{F}g}{\sqrt{\mathscr{F}\widetilde{K}}} \rangle_{L_2(\mathbb{R}^d)}$$

provided $\widetilde{K} \in C(\mathbb{R}^d) \cap L_1(\mathbb{R}^d)$ and $f,g \in \mathscr{H}(\widetilde{K},\mathbb{R}^d)$.

Before we begin our discussion relating kernel methods to Green's functions — and thereby providing an interpretation of native spaces as *generalized Sobolev spaces* — we mention a few examples of kernels whose native spaces already are known to be Sobolev spaces. Since all of these kernels are radial (or *isotropic*) kernels we introduce the notation $\kappa(\|\mathbf{x}-\mathbf{z}\|) = K(\mathbf{x},\mathbf{z})$. This also helps us avoid confusion between a kernel K and the modified Bessel function of the second kind $K_{m-d/2}$ that appears below.

Matérn kernels (sometimes also called *Sobolev splines*, see, e.g., [8]) are of the form

$$\kappa(r) \doteq K_{m-d/2}(r) r^{m-d/2}, \quad m > \frac{d}{2},$$

where we have used the notation $\doteq$ to indicate that equality holds up to a multiplicative constant. It is quite natural to use the term Sobolev splines to refer to these functions since their native space is given by a classical Sobolev space, i.e., $\mathscr{H}(\kappa,\mathbb{R}^d) = H^m(\mathbb{R}^d)$.

A second example is given by the entire family of *Wendland's compactly supported radial basis functions* (see, e.g., [8, 43]). A popular member of this family is of the form

$$\kappa(r) \doteq (1-r)_+^4(4r+1),$$

and its native space $\mathscr{H}(\kappa,\mathbb{R}^3)$ is norm-equivalent to the classical Sobolev space $H^3(\mathbb{R}^3)$ (see [42]).

The family of *polyharmonic splines* is another famous (albeit only conditionally positive definite) example that fits this list. These functions are of the form

$$\kappa(r) \doteq \begin{cases} r^{2m-d}, & d \text{ odd}, \\ r^{2m-d}\log r, & d \text{ even}, \end{cases}$$

and the native space $\mathscr{H}(\kappa,\mathbb{R}^d)$ is a *Beppo-Levi space of order m*

$$\mathrm{BL}_m(\mathbb{R}^d) = \left\{ f \in C(\mathbb{R}^d) \mid D^\alpha f \in L_2(\mathbb{R}^d) \text{ for all } |\alpha| = m \right\}.$$

We may consider this space as a homogeneous Sobolev space of order m (see [8]).

This latter example is also featured in the recent paper [4], and spherical versions of the latter two examples are discussed in [25]. We became aware of both of these papers only after the initial submission of our own paper and it is interesting to note that they both use the connection between reproducing kernels and Green's functions as an essential ingredient to obtain their L_p approximation results.

2.2 Mercer's Theorem and Eigenvalue Problems

We will limit most of our discussion to positive definite kernels. A perspective on positive definite kernels that appears quite frequently in the literature on statistical learning (but not so much in approximation theory) is their characterization via an eigenfunction expansion. This fact goes back many years to the early work of James Mercer [24]. We quote here a version of this result from [29].

Theorem 1 (Mercer's theorem). *Let (Ω,μ) be a finite measure space and $K \in L_\infty(\Omega^2,\mu^2)$ be a kernel such that the integral operator $T_K : L_2(\Omega,\mu) \to L_2(\Omega,\mu)$ defined by*

$$(T_K f)(\mathbf{x}) = \int_\Omega K(\mathbf{x},\mathbf{z}) f(\mathbf{z}) \mathrm{d}\mu(\mathbf{z})$$

is positive definite. Let $\varphi_n \in L_2(\Omega,\mu)$ be the normalized eigenfunctions of T_K associated with the eigenvalues $\lambda_n > 0$. Then

1. the eigenvalues $\{\lambda_n\}_{n=1}^{\infty}$ *are absolutely summable,*

2. $K(\mathbf{x},\mathbf{z}) = \sum_{n=1}^{\infty} \lambda_n \varphi_n(\mathbf{x})\varphi_n(\mathbf{z})$ *holds* μ^2 *almost everywhere, and the series converges absolutely and uniformly* μ^2 *almost everywhere.*

More generally, Hilbert-Schmidt theory ensures the existence of L_2-convergent eigenfunction expansions of compact, self-adjoint operators.

We now consider a kernel $K : \Omega \times \Omega \to \mathbb{R}$ on a general domain Ω and define an inner product with positive weight function σ (instead of using the measure theoretic notation of the theorem) as

$$\langle f, g \rangle = \int_{\Omega} f(\mathbf{x}) g(\mathbf{x}) \sigma(\mathbf{x}) \mathrm{d}\mathbf{x}.$$

The eigenvalue problem for the integral operator $T_K : f \mapsto \int_{\Omega} K(\cdot,\mathbf{z}) f(\mathbf{z}) \sigma(\mathbf{z}) \mathrm{d}\mathbf{z}$ consists of finding solutions λ and φ of

$$\int_{\Omega} K(\mathbf{x},\mathbf{z}) \varphi(\mathbf{z}) \sigma(\mathbf{z}) \mathrm{d}\mathbf{z} = \lambda \varphi(\mathbf{x}). \tag{3}$$

This represents a homogeneous Fredholm integral equation of the 2nd kind and it is therefore not obvious how we should go about finding the eigenvalues and eigenfunctions of T_K. The idea we will pursue here is to relate the integral equation to a differential equation which may be easier to solve.

2.3 Green's Functions and Eigenfunction Expansions

Green's functions (see, e.g., [6]) play a central role in the solution of differential equations. We now consider the nonhomogeneous linear (ordinary or partial) differential equation

$$(Lu)(\mathbf{x}) = f(\mathbf{x}) \quad \text{on } \Omega \subset \mathbb{R}^d$$

with a linear and elliptic operator L and some appropriate homogeneous boundary conditions. We will be more specific about the boundary conditions later.

The solution of this differential equation can be written in terms of a *Green's function* G as

$$u(\mathbf{x}) = \int_{\Omega} f(\mathbf{z}) G(\mathbf{x},\mathbf{z}) \mathrm{d}\mathbf{z}, \tag{4}$$

where the Green's function satisfies the differential equation

$$(LG)(\mathbf{x},\mathbf{z}) = \delta(\mathbf{x} - \mathbf{z}).$$

Here δ denotes the standard delta function(al), and the point $\mathbf{z}$ denotes a fixed (and arbitrary) "source". The boundary conditions are the same as above.

We now establish a connection between the integral operator eigenvalue problem that is needed for Mercer's series representation of a positive definite kernel discussed above and a related eigenvalue problem for a differential operator. For simplicity we assume that K is a free space Green's function for the differential operator L, i.e., $(LK)(\mathbf{x},\mathbf{z}) = \delta(\mathbf{x}-\mathbf{z})$[1].

We apply the differential operator L to the integral equation (3), interchange integration and differentiation and use the definition of the Green's function to obtain

$$L\int_\Omega K(\mathbf{x},\mathbf{z})\varphi(\mathbf{z})\sigma(\mathbf{z})\mathrm{d}\mathbf{z} = L\lambda\varphi(\mathbf{x}) \iff \int_\Omega \delta(\mathbf{x}-\mathbf{z})\varphi(\mathbf{z})\sigma(\mathbf{z})\mathrm{d}\mathbf{z} = \lambda L\varphi(\mathbf{x}).$$

Using the definition of the delta function this gives us

$$L\varphi(\mathbf{x}) = \frac{1}{\lambda}\sigma(\mathbf{x})\varphi(\mathbf{x}),$$

which shows that the eigenvalues of the integral operator correspond to reciprocals of eigenvalues of the differential operator, while the corresponding eigenfunctions are the same.

We now present a simple and well-known example (on a bounded interval). The kernel in this example is sometimes referred to as the *Brownian bridge kernel* (see, e.g., [1]) since it is the covariance kernel of a Brownian motion with zero boundary conditions at both ends of the interval, also known as a Brownian bridge.

Example 1 (Brownian bridge kernel). Consider the domain $\Omega = [0,1]$, and let

$$K(x,z) = \min(x,z) - xz = \begin{cases} x(1-z), & x \le z, \\ z(1-x), & x > z. \end{cases}$$

This kernel may be obtained by integrating

$$-\frac{\mathrm{d}^2}{\mathrm{d}x^2}K(x,z) = \frac{1}{2}\delta(x-z)$$

twice using the boundary conditions

$$K(0,z) = K(1,z) = 0.$$

In other words, K is the Green's function (up to a factor 2) of the differential operator $L = -\frac{\mathrm{d}^2}{\mathrm{d}x^2}$ with corresponding boundary conditions.

We now consider the integral operator eigenvalue problem

$$\int_\Omega K(x,z)\varphi(z)\sigma(z)\mathrm{d}z = \lambda\varphi(x)$$

---

[1] The problem is considerably more difficult on bounded domains, i.e., for differential equations including boundary conditions, and we do not discuss that case here.

with $\sigma(x) \equiv 1$ and K and Ω as above. Using the piecewise definition of K this corresponds to

$$\int_0^x z\varphi(z)\mathrm{d}z + \int_x^1 x\varphi(z)\mathrm{d}z - \int_0^1 xz\varphi(z)\mathrm{d}z = \lambda\varphi(x).$$

If we apply the differential operator $L = -\frac{\mathrm{d}^2}{\mathrm{d}x^2}$ to this integral equation and use two elementary differentiation steps we obtain

$$\frac{\mathrm{d}}{\mathrm{d}x}\left\{x\varphi(x) - \int_1^x \varphi(z)\mathrm{d}z - x\varphi(x) - \int_0^1 z\varphi(z)\mathrm{d}z\right\} = \lambda\varphi''(x)$$
$$\iff \quad -\varphi''(x) = \tfrac{1}{\lambda}\varphi(x),$$

which again illustrates that the eigenvalues of the integral operator are the reciprocals of the eigenvalues of the differential operator. We will continue this example below.

The second piece of the puzzle is to express the Green's function of the differential equation in terms of the eigenvalues and eigenfunctions of a related Sturm-Liouville eigenvalue problem. To this end we start with the generic ordinary differential equation

$$(LG)(x,z) = \delta(x-z)$$

with regular Sturm-Liouville boundary conditions. The so-called *Sturm-Liouville eigenvalue problem* is then given by

$$(L\varphi)(x) = \frac{1}{\lambda}\sigma(x)\varphi(x), \tag{5}$$

where we need to add the same set of regular Sturm-Liouville boundary conditions. Here σ is a weight function whose choice is basically free, but of course determines the specific form of the eigenfunctions and eigenvalues by defining different inner products. For a fixed choice of σ we can represent the Green's function G via an eigenfunction expansion of the form

$$G(x,z) = \sum_{n=1}^{\infty} c_n(z)\varphi_n(x). \tag{6}$$

We again consider z as an arbitrary, but fixed, source point and therefore the expansion coefficients (generalized Fourier coefficients) c_n will depend on z. In order to determine these coefficients we apply the differential operator L to (6) and use linearity along with the definitions of the Green's function and the Sturm-Liouville eigenvalue problem to arrive at

$$\delta(x-z) = (LG)(x,z) = \sum_{n=1}^{\infty} c_n(z)(L\varphi_n)(x) = \sum_{n=1}^{\infty} \frac{c_n(z)\sigma(x)\varphi_n(x)}{\lambda_n}.$$

Multiplication of this identity by $\varphi_m(x)$ and integration from a to b yields

$$c_n(z) = \frac{\lambda_n \varphi_n(z)}{\int_a^b \varphi_n^2(x)\sigma(x)\mathrm{d}x},$$

where we have used the orthogonality of the eigenfunctions φ_n. If we identify the Green's function G with the kernel K in (3) then we see that the coefficients $c_n(z)$ are nothing but the generalized Fourier coefficients of G, i.e., the appropriately normalized inner product of $G(\cdot,z)$ with φ_n. In particular, if the eigenfunctions are orthonormal with respect to σ then

$$G(x,z) = \sum_{n=1}^{\infty} \lambda_n \varphi_n(x)\varphi_n(z).$$

This argument works analogously in higher space dimensions.

Example 2 (More Brownian bridge). A simple exercise in standard Sturm-Liouville theory tells us that the boundary value problem

$$-\varphi''(x) = \frac{1}{\lambda}\varphi(x), \qquad \varphi(0) = \varphi(1) = 0,$$

has eigenvalues and eigenfunctions

$$\lambda_n = \frac{1}{(n\pi)^2}, \quad \varphi_n(x) = \sin n\pi x, \qquad n = 1,2,3,\ldots,$$

and we can verify

$$G(x,z) = \min(x,z) - xz = \sum_{n=1}^{\infty} c_n(z)\sin n\pi x$$

with

$$c_n(z) = \int_0^1 (\min(x,z) - xz)\sin n\pi x \mathrm{d}x = \frac{\sin n\pi z}{(n\pi)^2} = \lambda_n \varphi_n(z).$$

2.4 Generalized Sobolev Spaces

We now briefly discuss how we interpret the native space of a kernel K in terms of an associated differential operator L. Many more details are given in [12]. A rigorous theoretical framework supporting this interpretation for general vector distributional operators is provided in [12]. While this paper contains the theory for generalized Sobolev spaces on the unbounded domain $\mathbb{R}^d$, we will illustrate the framework here with some examples on bounded domains. A theoretical framework for that (more complicated) case is the subject of [13].

Our approach depends on the ability to identify the differential operator L that corresponds to a given Green's kernel K (or vice versa). Given such an L, we then decompose it into

$$L = \mathbf{P}^{*T}\mathbf{P} = \sum_{j=1}^{J} P_j^* P_j,$$

where the P_j are themselves differential operators and P_j^* is an appropriately defined adjoint. For example, for the Brownian bridge kernel $K(x,z) = \min(x,z) - xz$ discussed in previous examples we have $L = -\frac{\mathrm{d}}{\mathrm{d}x^2}$, $\mathbf{P} = P = \frac{\mathrm{d}}{\mathrm{d}x}$, and $\mathbf{P}^* = P^* = -\frac{\mathrm{d}}{\mathrm{d}x}$, i.e., $J = 1$. We point out that the theory in [12] is not limited to finite J. In particular, the vector differential operator $\mathbf{P}$ corresponding to the Gaussian kernel is infinite-dimensional.

We then define the *generalized Sobolev space* $H_{\mathbf{P}}(\mathbb{R}^d)$ as the set of slowly increasing locally integrable functions f for which $\mathbf{P}f \in L_2(\mathbb{R}^d)$, i.e.,

$$H_{\mathbf{P}}(\mathbb{R}^d) = \left\{ f \in L_1^{loc}(\mathbb{R}^d) \cap SI \mid P_j f \in L_2(\mathbb{R}^d),\ j = 1,\ldots,J \right\}.$$

The (semi-)inner product for this space is also defined in terms of $\mathbf{P}$. Namely,

$$\langle f,g \rangle_{H_{\mathbf{P}}(\mathbb{R}^d)} = \sum_{j=1}^{J} \int_{\mathbb{R}^d} P_j f(\mathbf{x}) \overline{P_j g(\mathbf{x})} \mathrm{d}\mathbf{x}.$$

For our running Brownian bridge example the reproducing kernel Hilbert space is the standard Sobolev space

$$H_{0,1}^1(0,1) = H_{\mathbf{P}}(0,1) = \left\{ Pf = f' \in L_2(0,1) : f(0) = f(1) = 0 \right\}$$

with inner product

$$\langle f,g \rangle_{H_{\mathbf{P}}(0,1)} = \int_0^1 Pf(x)Pg(x)\mathrm{d}x = \int_0^1 f'(x)g'(x)\mathrm{d}x.$$

We then have that $K(x,z) = \min(x,z) - xz$ is the reproducing kernel of $H_{\mathbf{P}}(0,1)$.

The left graph in Figure 1 shows multiple copies of the piecewise linear Brownian bridge kernel centered at equally spaced points in the interval $[0,1]$. Note how this kernel is neither radial (isotropic), nor translation invariant (stationary).

We provide two more examples that are obtained from the previous one by a simple shift of the eigenvalues.

Example 3 (Tension spline kernel). We begin with the Sturm-Liouville ODE eigenvalue problem

$$\varphi''(x) + (\lambda - \varepsilon^2)\varphi(x) = 0, \qquad \varphi(0) = \varphi(1) = 0,$$

where ε is an additional parameter, often referred to as *shape parameter* or *tension parameter*. This means that $L = -\frac{\mathrm{d}}{\mathrm{d}x^2} + \varepsilon^2 I$ and $\mathbf{P} = \left[\frac{\mathrm{d}}{\mathrm{d}x}, \varepsilon I\right]^T$. Note that we now

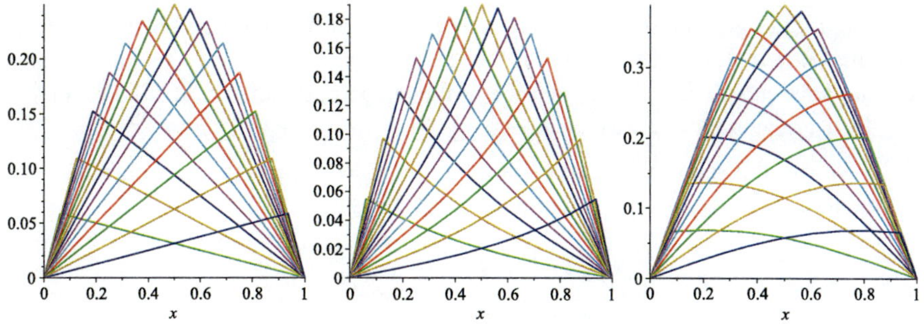

Fig. 1: Copies of the Brownian bridge (left), tension spline (middle) and relaxation spline (right) kernels $K(x,z)$ for 15 equally spaced values of z in $[0,1]$.

indeed have a vector differential operator $\mathbf{P}$. The eigenvalues and eigenfunctions can easily be found to be

$$\lambda_n = n^2\pi^2 + \varepsilon^2, \quad \varphi_n(x) = \sin n\pi x, \quad n = 1,2,\ldots.$$

The kernel (i.e., Green's function) is given by

$$K(x,z) = \begin{cases} \frac{\sinh(\varepsilon x)\sinh\varepsilon(1-z)}{\varepsilon \sinh \varepsilon}, & x < z, \\ \frac{\sinh(\varepsilon z)\sinh\varepsilon(1-x)}{\varepsilon \sinh \varepsilon}, & x > z. \end{cases}$$

For this example, the reproducing kernel Hilbert space is again a standard Sobolev space, namely

$$H_{\mathbf{P}}(0,1) = \left\{f, f' \in L_2(0,1) : f(0) = f(1) = 0\right\}.$$

However, the inner product is now given by

$$\langle f,g\rangle_{H_{\mathbf{P}}(0,1)} = \sum_{j=1}^{2}\int_0^1 P_j f(x)\overline{P_j g(x)}\mathrm{d}x = \int_0^1 f'(x)g'(x)\mathrm{d}x + \varepsilon\int_0^1 f(x)g(x)\mathrm{d}x.$$

One of the most notable points here is that the so-called *shape parameter* ε of the kernel is intimately related to the inner-product and therefore the norm of the function space. Through this feature, which is usually completely ignored in the discussion of function spaces used in approximation theory, we are able to introduce a more refined notion of a function space for a certain approximation problem at hand by our ability to capture a certain length scale represented in the data. This length scale is defined by the relative importance of function values and derivatives. Therefore, we might want to denote this Sobolev space by $H^1_\varepsilon(0,1)$. As a consequence, the definition of a generalized Sobolev space encodes both smoothness and "peakiness" information of the functions it contains.

The middle graph in Figure 1 shows multiple copies of the tension spline kernel centered at equally spaced points in the interval $[0,1]$. Note how this kernel has a certain tension specified by the choice of the shape parameter ε. For the graphs in Figure 1 the value $\varepsilon = 2$ was used.

If we use the same differential operator, i.e., $L = -\frac{\mathrm{d}}{\mathrm{d}x^2} + \varepsilon^2 I$, but eliminate the boundary conditions, then we obtain the related radial kernel

$$\kappa(r) = \frac{1}{2\varepsilon}\mathrm{e}^{-\varepsilon r}, \qquad r = |x - z|,$$

(see [12]). This kernel is well-known in the literature as one of the members of the *Matérn* family, or as the *Ornstein-Uhlenbeck* kernel. Its reproducing kernel Hilbert space is the classical Sobolev space $H^1(\mathbb{R})$ (see also [1]).

Example 4 (Relaxation spline kernel). By adding the shift ε^2 in the Sturm-Liouville equation of the previous example instead of subtracting we obtain a different set of eigenvalues and eigenfunctions, namely

$$\lambda_n = n^2\pi^2 - \varepsilon^2, \quad \varphi_n(x) = \sin n\pi x, \quad n = 1,2,\ldots.$$

The kernel in this example is given by

$$K(x,z) = \begin{cases} \frac{\sin(\varepsilon x)\sin\varepsilon(1-z)}{\varepsilon\sin\varepsilon}, & x < z, \\ \frac{\sin(\varepsilon z)\sin\varepsilon(1-x)}{\varepsilon\sin\varepsilon}, & x > z, \end{cases}$$

and the generalized Sobolev space and inner product are defined analogously. The right graph in Figure 1 shows different copies of this kernel. Since the effects of the shape parameter here amount to a relaxation instead of a tension we chose to call this kernel a *relaxation spline* kernel.

More examples of reproducing kernels, their associated differential operators, as well as eigenvalues and eigenfunctions — also in higher space dimensions — are presented below. It should also be noted that a connection between piecewise polynomial splines and Green's functions has been mentioned in the literature before (see, e.g., [37, 41]). However, in both instances the splines were related to Green's functions of *initial value problems* whereas our framework uses Green's functions of boundary value problems.

3 Flat Limits

In this section we will take a closer look at the effect of the shape parameter ε present in the definition of some of our kernels. In particular, we are interested in understanding the behavior of *radial* kernel interpolants for the limiting case of $\varepsilon \to 0$, i.e., flat kernels. A radial kernel is of the form $\kappa(\|\mathbf{x}-\mathbf{z}\|) = K(\mathbf{x},\mathbf{z})$, i.e., it is invariant under both translation and rotation. In the statistics literature such a kernel is called stationary and isotropic.

The results in this section specifically address the scattered data interpolation problem. In other words, we are given data sites $\mathscr{X} = \{\mathbf{x}_1, \ldots, \mathbf{x}_N\} \subset \mathbb{R}^d$ with associated data values $\{f(\mathbf{x}_1), \ldots, f(\mathbf{x}_N)\}$ sampled from some function f and wish to reconstruct f by a function of the form

$$s_{f,\kappa}^{\varepsilon}(\mathbf{x}) = \sum_{j=1}^{N} c_j \kappa(\varepsilon \|\mathbf{x} - \mathbf{x}_j\|), \qquad \mathbf{x} \in \mathbb{R}^d,$$

where the coefficients c_j are determined by satisfying the interpolation conditions

$$s_{f,\kappa}^{\varepsilon}(\mathbf{x}_i) = f(\mathbf{x}_i), \qquad i = 1, \ldots, N.$$

3.1 Infinitely Smooth RBFs

In recent years so-called *flat radial basis functions* (RBFs) have received much attention in the case when the kernels are infinitely smooth (see, e.g., [5, 16, 21–23, 32, 33]). We begin by summarizing the essential insight gained in these papers, and then present some recent results from [38] that deal with radial kernels of finite smoothness in the next subsection.

Theorem 2. *Assume the positive definite radial kernel κ has an expansion of the form*

$$\kappa(r) = \sum_{n=0}^{\infty} a_n r^{2n}$$

into even powers of r (i.e., κ is infinitely smooth), and that the data $\mathscr{X}$ are unisolvent with respect to any set of N linearly independent polynomials of degree at most m. Then

$$\lim_{\varepsilon \to 0} s_{f,\kappa}^{\varepsilon}(\mathbf{x}) = p_{m,f}(\mathbf{x}), \quad \mathbf{x} \in \mathbb{R}^d,$$

where $p_{m,f}$ is determined as follows:

- *If interpolation with polynomials of degree at most m is unique, then $p_{m,f}$ is that unique polynomial interpolant.*
- *If interpolation with polynomials of degree at most m is not unique, then $p_{m,f}$ is a polynomial interpolant whose form depends on the choice of RBF.*

This theorem applies to kernels such as

$$\kappa(\varepsilon r) = \frac{1}{1+\varepsilon^2 r^2} = 1 - (\varepsilon r)^2 + (\varepsilon r)^4 - (\varepsilon r)^6 + (\varepsilon r)^8 + \cdots \text{ (IQ)},$$

$$\kappa(\varepsilon r) = \mathrm{e}^{-\varepsilon^2 r^2} = 1 - (\varepsilon r)^2 + \frac{1}{2}(\varepsilon r)^4 - \frac{1}{6}(\varepsilon r)^6 + \frac{1}{24}(\varepsilon r)^8 + \cdots \text{ (Gaussian)},$$

$$\kappa(\varepsilon r) = \frac{1}{\sqrt{1+\varepsilon^2 r^2}} = 1 - \frac{1}{2}(\varepsilon r)^2 + \frac{3}{8}(\varepsilon r)^4 - \frac{5}{16}(\varepsilon r)^6 + + \frac{35}{128}(\varepsilon r)^8 + \cdots \text{ (IMQ)}.$$

The implications of this theorem are quite deep since it essentially establishes radial basis functions as generalizations of polynomial spectral methods. As a consequence, this opens the door to the design of algorithms for function approximation as well as the numerical solution of partial differential equations that are more accurate than the standard polynomial spectral methods. Moreover, the scattered data setting in which radial basis functions are used allows for more flexibility with respect to geometry and adaptivity.

We will come back to the Gaussian kernel in the next two sections of this paper where we address two important issues: computational stability and rates of convergence.

3.2 Finitely Smooth RBFs

To our knowledge, the flat limit of RBFs with finite smoothness was not studied until the recent paper [38] in which interpolation on $\mathbb{R}^d$ was investigated.

Before we explain the results obtained in [38], we look at a few finitely smooth radial kernels as full space Green's functions as discussed in the earlier sections.

Example 5 (Radial kernels with finite smoothness).

1. We have already mentioned the univariate C^0 Matérn kernel $K(x,z) \doteq \mathrm{e}^{-\varepsilon|x-z|}$. For this first example we remember that the differential operator L associated with this full-space Green's function was given by
$$L = -\frac{\mathrm{d}^2}{\mathrm{d}x^2} + \varepsilon^2 I.$$
On the other hand, it is well-known that univariate C^0 piecewise linear splines may be expressed in terms of kernels of the form $K(x,z) \doteq |x-z|$. The corresponding differential operator in this case is
$$L = -\frac{\mathrm{d}^2}{\mathrm{d}x^2}.$$
Note that the differential operator associated with the Matérn kernel "converges" to that of the piecewise linear splines as $\varepsilon \to 0$. We also remark that the piecewise linear Brownian bridge kernel does not fit into this discussion since it is associated with a boundary value problem, i.e., it is not a full-space Green's function.
2. The univariate C^2 tension spline kernel [30] $K(x,z) \doteq \mathrm{e}^{-\varepsilon|x-z|} + \varepsilon|x-z|$ is the Green's kernel of
$$L = -\frac{\mathrm{d}^4}{\mathrm{d}x^4} + \varepsilon^2\frac{\mathrm{d}^2}{\mathrm{d}x^2},$$

while the univariate C^2 cubic spline kernel $K(x,z) \doteq |x-z|^3$ corresponds to

$$L = -\frac{\mathrm{d}^4}{\mathrm{d}x^4}.$$

Again, the differential operator associated with the tension spline "converges" to that of the cubic spline as $\varepsilon \to 0$.

3. In [1] we find a so-called univariate *Sobolev kernel* of the form $K(x,z) \doteq \mathrm{e}^{-\varepsilon|x-z|}\sin\left(\varepsilon|x-z| + \frac{\pi}{4}\right)$ which is associated with

$$L = -\frac{\mathrm{d}^4}{\mathrm{d}x^4} - \varepsilon^2 I.$$

The operator for this kernel also "converges" to the cubic spline kernel, but the effect of the scale parameter is analogous to that of the relaxation spline of Example 4.

4. The general *multivariate Matérn kernels* are of the form

$$K(\mathbf{x},\mathbf{z}) \doteq K_{m-d/2}(\varepsilon\|\mathbf{x}-\mathbf{z}\|)(\varepsilon\|\mathbf{x}-\mathbf{z}\|)^{m-d/2}, \qquad m > \frac{d}{2},$$

and can be obtained as Green's kernels of (see [12])

$$L = \left(\varepsilon^2 I - \Delta\right)^m, \qquad m > \frac{d}{2}.$$

We contrast this with the *polyharmonic spline kernels*

$$K(\mathbf{x},\mathbf{z}) \doteq \begin{cases} \|\mathbf{x}-\mathbf{z}\|^{2m-d}, & d \text{ odd}, \\ \|\mathbf{x}-\mathbf{z}\|^{2m-d}\log\|\mathbf{x}-\mathbf{z}\|, & d \text{ even}, \end{cases}$$

and

$$L = (-1)^m \Delta^m, \qquad m > \frac{d}{2}.$$

In summary, all of these examples show that the differential operators associated with finitely smooth RBF kernels "converge" to those of a piecewise polynomial or polyharmonic spline kernel as $\varepsilon \to 0$. This motivates us to ask whether RBF interpolants based on finitely smooth kernels converge to (polyharmonic) spline interpolants for $\varepsilon \to 0$ mimicking the relation between infinitely smooth radial kernels and polynomials. As the following theorem shows, this is indeed true.

As mentioned in Theorem 2, infinitely smooth radial kernels can be expanded into an infinite series of even powers of r. Finitely smooth radial kernels can also be expanded into an infinite series of powers of r. However, in this case there always exists some minimal odd power of r with nonzero coefficient indicating the smoothness of the kernel. For example, for univariate C^0, C^2 and C^4 Matérn kernels, respectively, we have

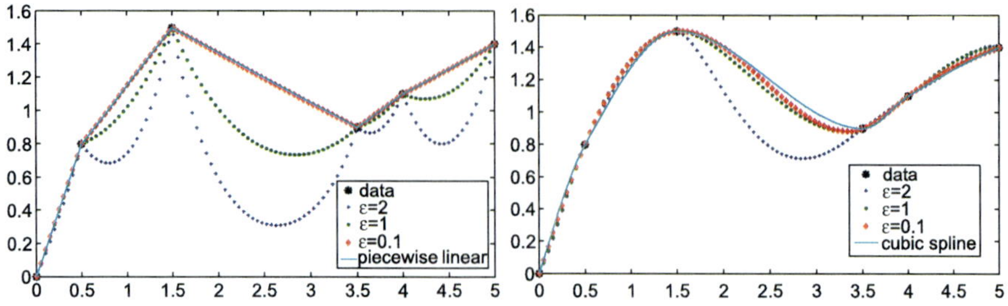

Fig. 2: Convergence of C^0 (left) and C^2 (right) Matérn interpolants to piecewise linear (left) and cubic (right) spline interpolants.

$$\kappa(\varepsilon r) \doteq \mathrm{e}^{-\varepsilon r} = 1 - \varepsilon r + \frac{1}{2}(\varepsilon r)^2 - \frac{1}{6}(\varepsilon r)^3 + \cdots,$$
$$\kappa(\varepsilon r) \doteq (1+\varepsilon r)\mathrm{e}^{-\varepsilon r} = 1 - \frac{1}{2}(\varepsilon r)^2 + \frac{1}{3}(\varepsilon r)^3 - \frac{1}{8}(\varepsilon r)^4 + \cdots,$$
$$\kappa(\varepsilon r) \doteq \left(3+3\varepsilon r+(\varepsilon r)^2\right)\mathrm{e}^{-\varepsilon r} = 3 - \frac{1}{2}(\varepsilon r)^2 + \frac{1}{8}(\varepsilon r)^4 - \frac{1}{15}(\varepsilon r)^5 + \frac{1}{48}(\varepsilon r)^6 + \cdots.$$

Theorem 3 ([38]). *Suppose κ is conditionally positive definite of order $m \le n$ with an expansion of the form*

$$\kappa(r) = a_0 + a_2 r^2 + \ldots + a_{2n} r^{2n} + a_{2n+1} r^{2n+1} + a_{2n+2} r^{2n+2} + \ldots,$$

where $2n+1$ denotes the smallest odd power of r present in the expansion (i.e., κ is finitely smooth). Also assume that the data $\mathscr{X}$ contain a unisolvent set with respect to the space $\pi_{2n}(\mathbb{R}^d)$ of d-variate polynomials of degree less than $2n$. Then

$$\lim_{\varepsilon\to 0} s^{\varepsilon}_{f,\kappa}(\mathbf{x}) = \sum_{j=1}^{N} c_j \|\mathbf{x}-\mathbf{x}_j\|^{2n+1} + \sum_{k=1}^{M} d_k p_k(\mathbf{x}), \quad \mathbf{x} \in \mathbb{R}^d,$$

where $\{p_k \mid k = 1,\ldots,M\}$ denotes a basis of $\pi_n(\mathbb{R}^d)$.

In other words, the "flat" limit of a piecewise smooth RBF interpolant is nothing but a polyharmonic spline interpolant. Therefore, just as infinitely smooth RBFs can be interpreted as generalizations of polynomials, we can view finitely smooth RBFs as generalizations of piecewise polynomial (or more generally polyharmonic) splines.

We point out that Theorem 3 does not cover Matérn kernels with odd-order smoothness. However, all other examples listed above are covered by the theorem.

Figure 2 illustrates the convergence of univariate C^0 and C^2 Matérn interpolants to piecewise linear and piecewise cubic spline interpolants, respectively.

4 Stable Computation

We now look at some practical consequences of working with "flat" RBF kernels. It is well-known that interpolation with "flat" Gaussian kernels leads to a notoriously ill-conditioned interpolation matrix K. This is due to the fact that the standard set of basis functions $\{e^{-\varepsilon^2(x-x_j)^2)}, j=1,\dots,N\}$ becomes numerically linearly dependent. It needs to be emphasized that the resulting numerical instabilities are due only to this "bad" choice of basis and not to the choice of function space itself. In fact, we will discuss in the next section how well one can approximate with linear combinations of Gaussians.

Even though Gaussian kernels are rather popular — especially in the machine learning community, it has been widely accepted that working with Gaussians is an ill-conditioned problem. As a result, the literature contains many references to a so-called *uncertainty* or *trade-off principle* (see, e.g., [31] or the more recent paper [35]). This uncertainty principle, however, is tied directly to the use of the standard ("bad") basis, and we demonstrate below how it can be circumvented by choosing a better — orthonormal — basis. The following discussion is motivated by the recent work of Bengt Fornberg and his collaborators [14, 15] in which they have proposed a so-called RBF-QR algorithm which allows for stable RBF computations. In addition to this QR-based approach they have also proposed other stable algorithms such as the Contour-Padé algorithm [16]. The guiding principle in this work is always the fact that the RBF-direct algorithm (based on the use of the "bad" standard basis) is ill-conditioned, but the RBF interpolation problem itself is not.

4.1 An Eigenfunction Expansion for Gaussians

In [29] (and already [44], albeit with incorrect normalization) one can find the following general eigenfunction expansion

$$e^{-b(x-z)^2} = \sum_{n=1}^{\infty} \lambda_n \varphi_n(x)\varphi_n(z), \tag{7}$$

where the eigenfunctions φ_n are orthonormal in $L_2(\mathbb{R},\rho)$ with weight function

$$\rho(x) = \sqrt{\frac{2a}{\pi}} e^{-2ax^2}.$$

Here a and b are arbitrary positive numbers. If we let $c=\sqrt{a^2+2ab}$, then the eigenfunctions φ_n turn out to be

$$\varphi_n(x) = \frac{1}{\sqrt{2^{n-1}(n-1)!\sqrt{\frac{a}{c}}}} e^{-(c-a)x^2} H_{n-1}(\sqrt{2c}x), \qquad n=1,2,\dots,$$

with H_n the classical Hermite polynomials of degree n, i.e.,

$$H_n(x) = (-1)^n \mathrm{e}^{x^2} \frac{\mathrm{d}^n}{\mathrm{d}x^n} \mathrm{e}^{-x^2} \quad \text{for all} \quad x \in \mathbb{R}, \quad n = 0, 1, 2, \ldots$$

so that

$$\int_{\mathbb{R}} H_n^2(x) \mathrm{e}^{-x^2} \, \mathrm{d}x = \sqrt{\pi} 2^n n! \qquad \text{for } n = 0, 1, 2, \ldots.$$

The corresponding eigenvalues are

$$\lambda_n = \sqrt{\frac{2a}{a+b+c}} \left(\frac{b}{a+b+c} \right)^{n-1}, \qquad n = 1, 2, \ldots.$$

In particular, we will want to use the Gaussian kernel in its usual form with shape parameter ε as

$$K(x,z) = \mathrm{e}^{-\varepsilon^2 (x-z)^2}$$

so that $b = \varepsilon^2$. Moreover, we take $a = \frac{1}{2}$ and therefore $c = \frac{1}{2}\sqrt{1+4\varepsilon^2}$ (see also [9]). This leads to eigenvalues

$$\begin{aligned} \lambda_n &= \frac{1}{\sqrt{\frac{1}{2}(1+\sqrt{1+4\varepsilon^2}) + \varepsilon^2}} \left(\frac{\varepsilon^2}{\frac{1}{2}(1+\sqrt{1+4\varepsilon^2}) + \varepsilon^2} \right)^{n-1} \\ &= \frac{\varepsilon^{2(n-1)}}{\left(\frac{1}{2}(1+\sqrt{1+4\varepsilon^2}) + \varepsilon^2 \right)^{n-\frac{1}{2}}}, \qquad n = 1, 2, \ldots \end{aligned} \tag{8}$$

and eigenfunctions

$$\varphi_n(x) = \sqrt{\frac{(1+4\varepsilon^2)^{1/4}}{2^{n-1}(n-1)!}} \exp\left(-\frac{\varepsilon^2 x^2}{\frac{1}{2}(1+\sqrt{1+4\varepsilon^2})} \right) H_{n-1}\left((1+4\varepsilon^2)^{1/4} x \right). \tag{9}$$

4.2 The RBF-QR Algorithm

The starting point for the bivariate Gaussian RBF-QR algorithm in [14] was an expansion of the form

$$\mathrm{e}^{-\varepsilon^2 (x-z)^2} = \sum_{n=0}^{\infty} \frac{(2\varepsilon^2)^n}{n!} x^n \mathrm{e}^{-\varepsilon^2 x^2} z^n \mathrm{e}^{-\varepsilon^2 z^2}, \qquad x, z \in \mathbb{R}. \tag{10}$$

However, the authors claimed that this series is not ideal since, coupled with the RBF-QR strategy described below, it does not provide an effective reduction of the conditioning of the RBF interpolation algorithm. Most likely, the poor conditioning of the new basis that results from this expansion is due to the fact that the functions $x \mapsto x^n e^{-\varepsilon^2 x^2}$ are not orthogonal in $L_2(\mathbb{R})$. Indeed, for $\varepsilon \to 0$ these functions converge to the standard monomial basis which is known to be ill-conditioned (cf. Vandermonde matrices). Therefore, the authors followed up their initial expansion with a transformation to polar coordinates and an expansion in terms of Chebyshev polynomials. This leads to an RBF-QR algorithm for Gaussians that is indeed stable, but limited to problems in $\mathbb{R}^2$.

The following discussion based on the eigenfunction expansion (7) of the Gaussian kernel will be applicable in any space dimension. Due to the product nature of the kernel we describe only the 1D version here. A comment at the end of this section indicates how to approach the general multivariate setting.

We will now show that if we use an expansion of the kernel in terms of orthonormal (eigen-)functions, then the source of ill-conditioning of the Gaussian basis is moved entirely into its eigenvalues. Since the eigenvalues of the Gaussian kernel decay very quickly we are now able to directly follow the QR-based strategy suggested in [14] — without the need for any additional transformation to Chebyshev polynomials.

In particular, we use the eigenvalues (8) and eigenfunctions (9) of the Gaussian kernel as discussed above.

The QR-based algorithm of [14] corresponds to the following. Starting with an expansion of the basis functions centered at x_j, $j = 1, \dots, N$, of the form

$$K(x, x_j) = \sum_{n=1}^{\infty} \varepsilon^{2(n-1)} b_n(x_j) \varphi_n(x), \qquad b_n(x_j) := \varepsilon^{-2(n-1)} \lambda_n \varphi_n(x_j),$$

i.e., a generalized Fourier expansion with x_j-dependent Fourier coefficients, we obtain

$$\begin{bmatrix} K(x,x_1) \\ K(x,x_2) \\ \vdots \\ \vdots \\ K(x,x_N) \end{bmatrix} = \begin{bmatrix} \cdots\cdots \\ \cdots\cdots \\ \cdots \mathsf{B} \cdots \\ \cdots\cdots \\ \cdots\cdots \end{bmatrix} \begin{bmatrix} \varepsilon^0 & & & & \\ & \varepsilon^2 & & & \\ & & \ddots & & \\ & & & \varepsilon^{2n} & \\ & & & & \ddots \end{bmatrix} \begin{bmatrix} \varphi_1(x) \\ \varphi_2(x) \\ \vdots \\ \varphi_n(x) \\ \vdots \end{bmatrix}.$$

Using more compact matrix-vector notation we can denote this by

$$\mathbf{k}(x) = \mathsf{B}\mathsf{E}\phi(x), \tag{11}$$

where $\mathbf{k}(x) = (K(x,x_j))_{j=1}^N$ and $\phi(x) = (\varphi_n(x))_{n=1}^{\infty}$ are the vectors of standard basis functions and eigenfunctions, respectively, evaluated at x, $\mathsf{B} = (b_n(x_j))_{j=1,n=1}^{N,\infty}$, and $\mathsf{E} = \operatorname{diag}\left(\varepsilon^0, \varepsilon^2, \dots, \varepsilon^{2n}, \dots\right)$ is the diagonal matrix of increasing even powers of ε. Note that E and ϕ are infinite and need to be appropriately truncated for practical applications. The matrix B has N rows, but infinitely many columns. However, since we are working with an eigenfunction expansion, truncating the representation at M terms will provide the best (in the L_2-sense) M-term approximation to the full series. Note that since

$$\begin{aligned} \mathsf{B}_{jn} &= \left(\frac{2}{1+\sqrt{1+4\varepsilon^2}+2\varepsilon^2}\right)^{n-\frac{1}{2}} \sqrt{\frac{(1+4\varepsilon^2)^{1/4}}{2^{n-1}(n-1)!}} \mathrm{e}^{-\frac{2\varepsilon^2 x_j^2}{1+\sqrt{1+4\varepsilon^2}}} H_{n-1}\left((1+4\varepsilon^2)^{1/4} x_j\right) \\ &\to \frac{1}{\sqrt{2^{n-1}(n-1)!}} H_{n-1}(x_j) \quad \text{as } \varepsilon \to 0 \end{aligned}$$

the matrix B remains "nice" as $\varepsilon \to 0$. Moreover, this limiting relation is another indication of the polynomial limit of Gaussian kernel interpolation as discussed in the previous section.

The QR idea now consists in first computing the QR-decomposition of B, i.e.,

$$\mathsf{B} = \mathsf{QR}$$

with unitary matrix Q and upper triangular matrix R. Next, we multiply the relation $\mathbf{k}(x) = \mathsf{B}\mathsf{E}\phi(x)$ on both sides by the non-singular matrix $\mathsf{E}^{-1}\mathsf{Q}^*$. The crucial observation here is that this does not change the function space spanned by the (poorly conditioned) standard basis functions $K(\cdot,x_1),\ldots,K(\cdot,x_N)$.

As a result, using (11), the QR-decomposition of B and the fact that Q is unitary we obtain a new basis for the Gaussian approximation space, namely

$$\psi(x) = \mathsf{E}^{-1}\mathsf{Q}^*\mathbf{k}(x) = \mathsf{E}^{-1}\mathsf{Q}^*\mathsf{QRE}\phi(x) = \mathsf{E}^{-1}\mathsf{RE}\phi(x),$$

where $\psi(x) = (\psi_n(x))_{n=1}^{\infty}$. Note that the matrix $\mathsf{E}^{-1}\mathsf{RE}$ is upper triangular and due to the scaling from the left and right should be relatively well-conditioned.

We are currently in the process of implementing this algorithm [10], and preliminary tests indicate that it is now possible to compute Gaussian RBF interpolants with this new eigenfunction basis stably also in the "flat" limit as $\varepsilon \to 0$. Incidentally, this is precisely the approach taken in [15] for stable radial basis function approximation on the sphere. It is interesting to note that traditionally there has been a much closer connection between (zonal) kernels used on the sphere and spherical harmonics, i.e., the eigenfunctions of the Laplace-Beltrami operator on the sphere (see, e.g., [11]). Furthermore, the RBF-QR approach should be successfully applicable whenever an eigenfunction expansion of the kernel is available.

As mentioned above, for the d-variate case we can use the fact that the Gaussian is a tensor product kernel:

$$K(\mathbf{x},\mathbf{z}) = \mathrm{e}^{-\varepsilon_1^2(x_1-z_1)^2-\ldots-\varepsilon_d^2(x_d-z_d)^2} = \sum_{\mathbf{n}\in\mathbb{N}^d} \lambda_{\mathbf{n}}\varphi_{\mathbf{n}}(\mathbf{x})\varphi_{\mathbf{n}}(\mathbf{z}),$$

so that the multivariate eigenvalues and eigenvectors are simply the products of the one-dimensional ones, i.e.,

$$\lambda_{\mathbf{n}} = \prod_{\ell=1}^{d} \lambda_{n_\ell} \quad \text{and} \quad \varphi_{\mathbf{n}}(\mathbf{x}) = \prod_{\ell=1}^{d} \varphi_{n_\ell}(x_\ell).$$

One of the advantages — both practical and theoretical — of this product approach is that we can take different shape parameters ε_ℓ for different dimensions, i.e., we can employ an *anisotropic* kernel K. Of course, the isotropic (or radial) case can still be recovered if we choose $\varepsilon_\ell = \varepsilon$, $\ell = 1, \ldots, d$. We will exploit this ability to generalize to anisotropic Gaussian kernels in the next section on convergence rates.

5 Dimension Independent Error Bounds

In the last section of this paper we mention some new results (see [9] for much more details) on the rates of convergence of Gaussian kernel approximation. To be more specific, we will address weighted L_2 approximation when the data is specified either by function values of an unknown function f (from the native space of the kernel) or with the help of arbitrary linear functionals. Our convergence results pay special attention to the dependence of the estimates on the space dimension d. We will see that the use of anisotropic Gaussian kernels instead of isotropic ones provides improved convergence rates. It should also be mentioned that the work in [9] deals with linear approximation algorithms, while the recent paper [17] addresses nonlinear Gaussian approximation.

5.1 The Current Situation

A good resource for standard RBF scattered data approximation results up to the year 2005 is [43]. There we can find two different L_∞ error bounds for isotropic Gaussian interpolation to data sampled from a function f in the native space $\mathcal{H}(K,\Omega)$ of the Gaussian. Both of these results are formulated in terms of the *fill distance*

$$h_{\mathcal{X},\Omega} = \sup_{\mathbf{x}\in\Omega} \min_{1\leq j\leq N} \|\mathbf{x} - \mathbf{x}_j\|,$$

where $\mathcal{X} = \{\mathbf{x}_1, \ldots, \mathbf{x}_N\}$ denotes the set of data sites as before. Since the results we mention below are in terms of N, the number of data, we will restate the error bounds from [43] also in terms of N using the fact that for quasi-uniformly distributed data sites we have $h_{\mathcal{X},\Omega} = \mathcal{O}(N^{-1/d})$.

If f has derivatives up to total order p and $s_{f,K}$ is the interpolant based on the Gaussian kernel $K(\mathbf{x},\mathbf{z}) = \mathrm{e}^{-\varepsilon^2\|\mathbf{x}-\mathbf{z}\|^2}$, i.e.,

$$s_{f,K}(\mathbf{x}) = \sum_{j=1}^{N} c_j K(\mathbf{x},\mathbf{x}_j) \quad \text{such that} \quad s_{f,K}(\mathbf{x}_i) = f(\mathbf{x}_i), \quad i = 1,\ldots,N,$$

then the first error bound is of the form

$$\|f - s_{f,K}\|_\infty \le C_d N^{-p/d} \|f\|_{\mathscr{H}(K,\Omega)}$$

with some possibly dimension-dependent constant C_d. Therefore, infinitely smooth functions can be approximated with order $p = \infty$. With some extra effort one can also obtain the spectral estimate

$$\|f - s_{f,K}\|_\infty \le \mathrm{e}^{-\frac{c}{d}N^{1/d}\log N} \|f\|_{\mathscr{H}(K,\Omega)}.$$

It is apparent from both of these bounds that the rate of convergence deteriorates as d increases. Moreover, the dependence of the constants on d is not clear. Therefore, these kinds of error bounds — and in fact almost all error bounds in the RBF literature — suffer from the *curse of dimensionality*. We will now present some results from [9] on *dimension-independent* convergence rates for Gaussian kernel approximation.

5.2 New Results on (Minimal) Worst-Case Weighted L_2 Error

As already indicated above, we will make several assumptions in order to be able to obtain dimension-independent error bounds.

We define the worst-case weighted $L_{2,\rho}$ error as

$$\mathrm{err}_{2,\rho}^{wc} = \sup_{\|f\|_{\mathscr{H}(K,\mathbb{R}^d)} \le 1} \|f - s_{f,K}\|_{2,\rho},$$

where $s_{f,K}$ is our kernel (minimum norm) approximation calculated in the usual way. Therefore

$$\|f - s_{f,K}\|_{2,\rho} \le \mathrm{err}_{2,\rho}^{wc} \|f\|_{\mathscr{H}(K,\mathbb{R}^d)} \quad \text{for all } f \in \mathscr{H}(K,\mathbb{R}^d).$$

The Nth *minimal worst case error* $\mathrm{err}_{2,\rho}^{wc}(N)$ refers to the worst case error that can be achieved with an optimal design, i.e., data generated by N optimally chosen linear functionals. For function approximation this means that the data sites have to be chosen in an optimal way. The results in [9] are non-constructive, i.e., no such optimal design is specified. However, a Smolyak or sparse grid algorithm is a natural candidate for such a design. If we are allowed to choose arbitrary linear functionals, then the optimal choice for weighted L_2 approximation is known. In this case we use generalized Fourier coefficients, i.e., the optimal linear functionals are $L_j = \langle \cdot, \varphi_j \rangle_{\mathscr{H}(K,\mathbb{R}^d)}$ and we obtain the truncated generalized Fourier series approximation

$$s_{f,K}(\mathbf{x}) = \sum_{n=1}^{N} \langle f, \varphi_n \rangle_{\mathscr{H}(K,\mathbb{R}^d)} \varphi_n(\mathbf{x}) \qquad \text{for all } f \in \mathscr{H}(K,\mathbb{R}^d),$$

where

$$K(\mathbf{x},\mathbf{z}) = \sum_{n=1}^{\infty} \lambda_n \varphi_n(\mathbf{x}) \varphi_n(\mathbf{z}), \qquad \int_{\Omega} K(\mathbf{x},\mathbf{z}) \varphi_n(\mathbf{z}) \rho(\mathbf{z}) \mathrm{d}\mathbf{z} = \lambda_n \varphi_n(\mathbf{x}).$$

It is then known [26] that

$$\mathrm{err}_{2,\rho}^{wc}(N) = \sqrt{\lambda_{N+1}},$$

the $(N+1)^{\mathrm{st}}$ largest eigenvalue, which is easy to identify in the univariate case, but takes some care to specify in the multivariate setting.

In [9] it is then proved that in the isotropic case, i.e., with a truly radial Gaussian kernel of the form

$$K(\mathbf{x},\mathbf{z}) = \mathrm{e}^{-\varepsilon^2 \|\mathbf{x}-\mathbf{z}\|^2}$$

one can approximate

- function data with an Nth minimal error of the order $\mathscr{O}(N^{-1/4+\delta})$, and
- Fourier data (i.e., arbitrary linear functional data) with an Nth minimal error of the order $\mathscr{O}(N^{-1/2+\delta})$.

Here the constants in the $\mathscr{O}$-notation do not depend on the dimension d and δ is arbitrarily small.

With anisotropic kernels, i.e.,

$$K(\mathbf{x},\mathbf{z}) = \mathrm{e}^{-\varepsilon_1^2(x_1-z_1)^2 - \ldots - \varepsilon_d^2(x_d-z_d)^2}$$

one can do much better. In this case, if the shape parameters decay like $\varepsilon_\ell = \ell^{-\alpha}$, then one can approximate

- function data with an Nth minimal error of the order $\mathscr{O}(N^{-\max(\frac{\alpha^2}{2+\alpha},1/4)+\delta})$, and
- Fourier data (i.e., arbitrary linear functional data) with an Nth minimal error of the order $\mathscr{O}(N^{-\max(\alpha,1/2)+\delta})$.

Again, the constants in the $\mathscr{O}$-notation do not depend on the dimension d.

In order to prove the above results it was essential to have the eigenvalues (cf. (8))

$$\lambda_n = \frac{\varepsilon^{2(n-1)}}{\left(\frac{1}{2}(1+\sqrt{1+4\varepsilon^2}) + \varepsilon^2\right)^{n-\frac{1}{2}}}, \qquad n = 1,2,\ldots,$$

and eigenfunctions (cf. (9))

$$\varphi_n(x) = \sqrt{\frac{(1+4\varepsilon^2)^{1/4}}{2^{n-1}(n-1)!}} \exp\left(-\frac{\varepsilon^2 x^2}{\frac{1}{2}(1+\sqrt{1+4\varepsilon^2})}\right) H_{n-1}\left((1+4\varepsilon^2)^{1/4}x\right)$$

of the univariate Gaussian kernel $K(x,z) = \mathrm{e}^{-\varepsilon^2(x-z)^2}$. As mentioned in the previous section, the multivariate (and anisotropic) case can be handled using products of univariate eigenvalues and eigenfunctions.

Even if we do not have an eigenfunction expansion of a specific kernel available, the work of [9] shows that for any radial (isotropic) kernel one has a dimension-independent Monte-Carlo type convergence rate of $\mathcal{O}(N^{-1/2+\delta})$ provided arbitrary linear functionals are allowed to generate the data. For translation-invariant (stationary) kernels the situation is similar. However, the constant in the $\mathcal{O}$-notation depends — in any case — on the sum of the eigenvalues of the kernel. For the radial case this sum is simply $\kappa(0)$ (independent of d), while for general translation invariant kernels it is $\widetilde{K}(\mathbf{0})$, which may depend on d.

These results show that — even though RBF methods are often advertised as being "dimension-blind" — their rates of convergence are only excellent (i.e., spectral for infinitely smooth kernels) if the dimension d is small. For large dimensions the constants in the $\mathcal{O}$-notation take over. If one, however, permits an anisotropic scaling of the kernel (i.e., elliptical symmetry instead of strict radial symmetry) and if those scale parameters decay rapidly with increasing dimension, then excellent convergence rates for approximation of smooth functions can be maintained independent of d.

6 Summary

In this paper we have attempted to shed some new light on the connections between piecewise polynomial splines and approximation methods based on reproducing kernels and radial basis functions in particular. Using Mercer's theorem and the resulting eigenfunction expansions of positive definite kernels along with an interpretation of these kernels as Green's functions of appropriate differential operators we provided a new interpretation of RBF native spaces as generalized Sobolev spaces (cf. [12, 13]). As a result we have a more intuitive interpretation of RBF native spaces in terms of the smoothness of the functions they contain. Moreover, special attention is paid to the native space norm and how it encodes information of the inherent scale of the functions it contains.

Extreme scaling of kernels, i.e., "flat" limits are investigated and they provide a new connection between finitely smooth RBF kernels and piecewise polynomial or polyharmonic splines (see [38]). We also use the eigenfunction expansions to move Fornberg's RBF-QR algorithm onto a more standard theoretical foundation which provides at the same time an algorithm for Gaussians that is applicable in any space dimension.

Finally, we discussed some of the results of [9] on dimension-independent convergence rates for Gaussians. The main insight obtained from these results is that one needs to allow for the use of an anisotropic scaling of the kernel with rapidly decaying scale parameters in order to be able to guarantee high rates of convergence in high space dimensions.

There is still much work to be done in the future. The theoretical framework for Green's functions on bounded domains needs to be completed, the new RBF-QR algorithm for Gaussians needs to be implemented, and the hunt for kernels with readily available or relatively easily computable eigenfunction expansions is on. Any such kernel benefits from the ideas laid out for stable computation and dimension-independent error bounds. There is also room to generalize the results on flat limits of piecewise smooth RBF kernels. Finally, it is expected that the eigenfunction expansions discussed here can be exploited to obtain fast multipole-type algorithms.

Acknowledgements The work presented in this paper was supported by NSF Grant DMS–0713848. The author is also indebted to Fred Hickernell, Mike McCourt, John Riddle, Guohui Song, Henryk Woźniakowski, and Qi Ye, all of whom played important roles in various parts of this paper.

References

1. Berlinet, A., Thomas-Agnan, C.: Reproducing Kernel Hilbert Spaces in Probability and Statistics. Kluwer, Dordrecht (2004)
2. de Boor, C.: A Practical Guide to Splines. Springer, New York (1978, revised edition 2001)
3. Buhmann, M.D.: Radial Basis Functions: Theory and Implementations. Cambridge University Press, Cambridge (2003)
4. DeVore, R., Ron, A.: Approximation using scattered shifts of a multivariate function. Trans. Amer. Math. Soc., DOI 10.1090/S0002-9947-2010-05070-6 (2010)
5. Driscoll, T.A., Fornberg, B.: Interpolation in the limit of increasingly flat radial basis functions. Comput. Math. Appl. **43**, 413–422 (2002)
6. Duffy, D.G.: Green's Funtions with Applications. Chapman & Hall/CRC, Boca Raton (2001)
7. Dyn, N., Levin, D., Rippa, S.: Numerical procedures for surface fitting of scattered data by radial functions. SIAM J. Sci. Statist. Comput. **7**, 639–659 (1986)
8. Fasshauer, G.E.: Meshfree Approximation Methods with MATLAB. World Scientific Publishers, Singapore (2007)
9. Fasshauer, G.E., Hickernell, F.J., Woźniakowski, H.: Rate of convergence and tractability of the radial function approximation problem. Under revision
10. Fasshauer, G.E., McCourt, M.J.: Stable evaluation of Gaussian RBF interpolants. Under revision
11. Fasshauer, G.E., Schumaker, L.L.: Scattered data fitting on the sphere. In: Dæhlen, M., Lyche, T., Schumaker, L.L. (eds.) Mathematical Methods for Curves and Surfaces II, pp. 117–166. Vanderbilt University Press, Nashville (1998)
12. Fasshauer, G.E., Ye, Q.: Reproducing kernels of generalized Sobolev spaces via a Green function approach with distributional operators. Numerische Mathematik DOI: 10.1007/s00211-011-0391-2
13. Fasshauer, G.E., Ye, Q.: Reproducing kernels of Sobolev spaces in bounded domains via a Green's kernel approach. Submitted

14. Fornberg, B., Larsson, E., Flyer, N.: Stable computations with Gaussian radial basis functions in 2-D. Technical Report 2009-020, Uppsala University, Department of Information Technology.
15. Fornberg, B., Piret, C.: A stable algorithm for flat radial basis functions on a sphere. SIAM J. Sci. Comp. **30**, 60–80 (2007)
16. Fornberg, B., Wright, G.: Stable computation of multiquadric interpolants for all values of the shape parameter. Comput. Math. Appl. **47**, 497–523 (2004)
17. Hangelbroek, T., Ron, A.: Nonlinear approximation using Gaussian kernels. J. Funct. Anal. **259** no. 1, 203–219 (2010)
18. Iske, A.: Multiresolution Methods in Scattered Data Modelling. Lecture Notes in Computational Science and Engineering 37, Springer Verlag, Berlin (2004)
19. Kybic, J., Blu, T., Unser, M.: Generalized sampling: A variational approach — Part I: Theory. IEEE Trans. Signal Proc. **50**, 1965–1976 (2002)
20. Lai, M.J., Schumaker, L.L.: Spline Functions on Triangulations. Cambridge University Press, Cambridge (2007)
21. Larsson, E., Fornberg, B.: A numerical study of some radial basis function based solution methods for elliptic PDEs. Comput. Math. Appl. **46**, 891–902 (2003)
22. Larsson, E., Fornberg, B.: Theoretical and computational aspects of multivariate interpolation with increasingly flat radial basis functions. Comput. Math. Appl. **49**, 103–130 (2005)
23. Lee, Y.J., Yoon, G.J., Yoon, J.: Convergence of increasingly flat radia basis interpolants to polynomial interpolants. SIAM J. Math. Anal. **39**, 537–553 (2007)
24. Mercer, J.: Functions of positive and negative type, and their connection with the theory of integral equations. Phil. Trans. Royal Soc. London Series A **209**, 415–446 (1909)
25. Mhaskar, H.N., Narcowich, F.J., Prestin, J., Ward, J.D.: L^p Bernstein estimates and approximation by spherical basis functions. Math. Comp. **79**, 1647–1679 (2010)
26. Novak, E., Woźniakowski, H.: Tractability of Multivariate Problems, Volume 1: Linear Information. EMS Tracts in Mathematics, no. 6, European Mathematical Society (2008)
27. Parzen, E.: Statistical inference on time series by RKHS methods. In: Pyke, R. (ed.) Proc. 12th Biennial Seminar, pp. 1–37. Canadian Mathematical Congress, Montreal, Canada (1970)
28. Pesenson, I.: Variational splines on Riemannian manifolds with applications to integral geometry. Adv. Appl. Math. **33**, 548–572 (2004)
29. Rasmussen, C.E., Williams, C.: Gaussian Processes for Machine Learning. MIT Press, 2006 (online version at `http://www.gaussianprocess.org/gpml/`)
30. Renka, R.J.: Interpolatory tension splines with automatic selection of tension factors. SIAM J. Sci. Stat. Comput. **8**, 393–415 (1987)
31. Schaback, R.: Error estimates and condition numbers for radial basis function interpolation. Adv. in Comput. Math. **3**, 251–264 (1995)
32. Schaback, R.: Multivariate interpolation by polynomials and radial basis functions. Constr. Approx. **21** 293–317 (2005)
33. Schaback, R.: Limit problems for interpolation by analytic radial basis functions. J. Comp. Appl. Math. **212** 127–149 (2008)
34. Schaback, R., Wendland, H.: Kernel techniques: From machine learning to meshless methods. Acta Numerica **15** 543–639 (2006)
35. Schmid, D.: A trade-off principle in connection with the approximation by positive definite kernels. In: Neamtu, M., Schumaker, L.L (eds.) Approximation Theory XII: San Antonio 2007, pp. 348–359. Nashboro Press, Brentwood, TN (2008)
36. Schoenberg, I.J.: Contributions to the problem of approximation of equidistant data by analytic functions, Parts A & B. Quart. Appl. Math. **4**, 45-99 & 112–141 (1946)
37. Schumaker, L.L.: Spline Functions: Basic Theory. John Wiley & Sons, New York (1981, reprinted by Krieger Publishing 1993)
38. Song, G., Riddle, J., Fasshauer, G.E., Hickernell, F.J.: Multivariate interpolation with increasingly flat radial basis functions of finite smoothness. Adv. in Comput. Math. To appear
39. Stein, M.L.: Interpolation of Spatial Data. Some theory for Kriging. Advances in Computational Mathematics. To appear Springer Series in Statistics, Springer-Verlag, New York (1999)

40. Steinwart, I., Christmann, A.: Support Vector Machines. Springer Verlag, Berlin (2008)
41. Wahba, G.: Spline Models for Observational Data. CBMS-NSF Regional Conference Series in Applied Mathematics 59, SIAM, Philadelphia (1990)
42. Wendland, H.: Sobolev-type error estimates for interpolation by radial basis functions. In: Le Méhauté, A., Rabut, C., Schumaker, L.L. (eds.) Surface Fitting and Multiresolution Methods, pp. 337–344. Vanderbilt University Press, Nashville (1997)
43. Wendland, H.: Scattered Data Approximation. Cambridge University Press, Cambridge (2005)
44. Zhu, H., Williams, C.K., Rohwer, R.J., Morciniec, M.: Gaussian regression and optimal finite dimensional linear models. In: Bishop, C.M. (ed.) Neural Networks and Machine Learning. Springer, Berlin (1998)

Sparse Recovery Algorithms: Sufficient Conditions in Terms of Restricted Isometry Constants

Simon Foucart

Abstract We review three recovery algorithms used in Compressive Sensing for the reconstruction s-sparse vectors $\mathbf{x} \in \mathbb{C}^N$ from the mere knowledge of linear measurements $\mathbf{y} = A\mathbf{x} \in \mathbb{C}^m$, $m < N$. For each of the algorithms, we derive improved conditions on the restricted isometry constants of the measurement matrix A that guarantee the success of the reconstruction. These conditions are $\delta_{2s} < 0.4652$ for basis pursuit, $\delta_{3s} < 0.5$ and $\delta_{2s} < 0.25$ for iterative hard thresholding, and $\delta_{4s} < 0.3843$ for compressive sampling matching pursuit. The arguments also applies to almost sparse vectors and corrupted measurements. The analysis of iterative hard thresholding is surprisingly simple. The analysis of basis pursuit features a new inequality that encompasses several inequalities encountered in Compressive Sensing.

1 Introduction

In this paper, we address the Compressive Sensing problem that consists in reconstructing an s-sparse vector $\mathbf{x} \in \mathbb{C}^N$ from the mere knowledge of the measurement vector $\mathbf{y} = A\mathbf{x} \in \mathbb{C}^m$ when $m \ll N$. We do not focus on the design of suitable measurement matrices $A \in \mathbb{C}^{m \times N}$, since we take for granted the existence of matrices having small restricted isometry constants (see Sect. 2 for the definition of these constants). Instead, we focus on three popular reconstruction algorithms that allow sparse recovery in a stable and robust fashion. For each algorithm, we present some sufficient conditions in terms of restricted isometry constants that improve on the ones currently found in the literature. The algorithms under consideration are as follows:

Simon Foucart
Department of Mathematics, Drexel University, Philadelphia, PA 19104, USA
e-mail: foucart@math.drexel.edu

M. Neamtu and L. Schumaker (eds.), *Approximation Theory XIII: San Antonio 2010*, Springer Proceedings in Mathematics 13, DOI 10.1007/978-1-4614-0772-0_5,

1.1 Basis Pursuit

Solve the convex optimization problem

$$\underset{\mathbf{z}\in\mathbb{C}^N}{\text{minimize}}\ \|\mathbf{z}\|_1 \quad \text{subject to } A\mathbf{z}=\mathbf{y}. \tag{BP}$$

1.2 Iterative Hard Thresholding

From an s-sparse vector $\mathbf{x}^0 \in \mathbb{C}^N$, iterate the single step

$$\mathbf{x}^{n+1} = H_s(\mathbf{x}^n + A^*(\mathbf{y} - A\mathbf{x}^n)), \tag{IHT}$$

where the nonlinear operator H_s keeps s largest (in modulus) entries of a vector and sets the other ones to zero, so that $H_s(\mathbf{z})$ is a – not necessarily unique – best s-term approximation to $\mathbf{z} \in \mathbb{C}^N$ in ℓ_p-norm for any $p \geq 1$.

1.3 Compressive Sampling Matching Pursuit

From an s-sparse vector $\mathbf{x}^0 \in \mathbb{C}^N$, iterate the following four steps

$$T^n := \big\{\text{indices of } 2s \text{ largest (in modulus) entries of } A^*(\mathbf{y}-A\mathbf{x}^n)\big\}, \tag{CSMP$_1$}$$
$$U^n := T^n \cup S^n, \qquad \text{where } S^n := \text{supp}(\mathbf{x}^n), \tag{CSMP$_2$}$$
$$\mathbf{u}^n := \text{argmin}\big\{\|\mathbf{y}-A\mathbf{z}\|_2, \text{supp}(\mathbf{z}) \subseteq U^n\big\}, \tag{CSMP$_3$}$$
$$\mathbf{x}^{n+1} := H_s(\mathbf{u}^n). \tag{CSMP$_4$}$$

2 Restricted Isometry Constants

We recall the definition of restricted isometry constants introduced in [3].

Definition 1. The s-th order restricted isometry constant $\delta_s = \delta_s(A)$ of a matrix $A \in \mathbb{C}^{m\times N}$ is the smallest $\delta \geq 0$ such that

$$(1-\delta)\|\mathbf{x}\|_2^2 \leq \|A\mathbf{x}\|_2^2 \leq (1+\delta)\|\mathbf{x}\|_2^2 \qquad \text{for all } s\text{-sparse vectors } \mathbf{x} \in \mathbb{C}^N. \tag{1}$$

Let us draw attention to a less common, though sometimes preferable, characterization of restricted isometry constants. This reads

$$\delta_s = \max_{S\subseteq\{1,\dots,N\}, |S|\leq s} \|A_S^* A_S - \text{Id}\|_{2\to 2}. \tag{2}$$

To justify the equivalence between the two characterizations, we start by noticing that (1) is equivalent to

$$\big|\|A_S\mathbf{x}\|_2^2 - \|\mathbf{x}\|_2^2\big| \le \delta\|\mathbf{x}\|_2^2 \quad \text{for all } S \subseteq \{1,\dots,N\},\ |S| \le s, \text{ and all } \mathbf{x} \in \mathbb{C}^{|S|}.$$

We then observe that

$$\|A_S\mathbf{x}\|_2^2 - \|\mathbf{x}\|_2^2 = \langle A_S\mathbf{x}, A_S\mathbf{x}\rangle - \langle \mathbf{x}, \mathbf{x}\rangle = \langle (A_S^*A_S - \mathrm{Id})\mathbf{x}, \mathbf{x}\rangle.$$

Now, since the matrix $(A_S^*A_S - \mathrm{Id})$ is hermitian, we have

$$\max_{\mathbf{x}\in\mathbb{C}^{|S|}\setminus\{0\}} \frac{\langle (A_S^*A_S - \mathrm{Id})\mathbf{x}, \mathbf{x}\rangle}{\|\mathbf{x}\|_2} = \|A_S^*A_S - \mathrm{Id}\|_{2\to2},$$

so that (1) is equivalent to

$$\max_{S\subseteq\{1,\dots,N\},|S|\le s} \|A_S^*A_S - \mathrm{Id}\|_{2\to2} \le \delta.$$

This establishes the identity (2), because δ_s is the smallest such δ. The expression (2) gives, for instance, an easy explanation of

$$|\langle A\mathbf{u}, A\mathbf{v}\rangle| \le \delta_{\mathrm{supp}(\mathbf{u})\cup\mathrm{supp}(\mathbf{v})}\|\mathbf{u}\|_2\|\mathbf{v}\|_2 \qquad \text{if } \mathbf{u} \text{ and } \mathbf{v} \text{ are disjointly supported},$$

a statement that can be derived in the real setting using a polarization formula, see e.g. [2]. Indeed, with $S := \mathrm{supp}(\mathbf{u}) \cup \mathrm{supp}(\mathbf{v})$, we just have to write (with slightly abusive notations)

$$\begin{aligned} |\langle A\mathbf{u}, A\mathbf{v}\rangle| &= |\langle A_S\mathbf{u}, A_S\mathbf{v}\rangle - \langle \mathbf{u}, \mathbf{v}\rangle| = |\langle (A_S^*A_S - \mathrm{Id})\mathbf{u}, \mathbf{v}\rangle| \le \|(A_S^*A_S - \mathrm{Id})\mathbf{u}\|_2\|\mathbf{v}\|_2 \\ &\le \|A_S^*A_S - \mathrm{Id}\|_{2\to2}\|\mathbf{u}\|_2\|\mathbf{v}\|_2 \le \delta_{\mathrm{supp}(\mathbf{u})\cup\mathrm{supp}(\mathbf{v})}\|\mathbf{u}\|_2\|\mathbf{v}\|_2. \end{aligned}$$

The concept of restricted isometry constant offers an elegant way to formulate sufficient conditions for the success of all the algorithms under consideration. Informally, if the restricted isometry constants are small, then all three algorithms are guaranteed to succeed in reconstructing sparse vectors. Slightly more precisely, if δ_t is small enough for some t related to s, then any s-sparse vector $\mathbf{x} \in \mathbb{C}^N$ is recovered as the output of the algorithms. The object of what follows is to quantify this statement. We note that a sufficient condition in terms of some δ_t can always be imposed by a sufficient condition in terms of some other $\delta_{t'}$, according to the comparison result given in Proposition 1 below. For instance, in view of $\delta_{3s} \le 3\delta_{2s}$, the sufficient condition $\delta_{3s} < 1/2$ obtained in Theorem 3 for iterative hard thresholding can be imposed by the condition $\delta_{2s} < 1/6$ – which will actually be improved to $\delta_{2s} < 1/4$. A heuristic way to compare such sufficient conditions is to recall that, given a prescribed $\delta > 0$, it is typical to have (c denoting an absolute constant)

$$\delta_t \le \delta \quad \text{provided} \quad m \ge c\frac{t}{\delta^2}\ln(eN/t)$$

for random measurement matrices. Therefore, it is desirable to make the ratio t/δ^2 as small as possible in order to minimize the necessary number of measurements. In this sense, the sufficient condition $\delta_{3s} < 1/2$ is heuristically better than the condition $\delta_{2s} < 1/4$, as $3s/(1/2)^2 < 2s/(1/4)^2$. Let us now state the aforementioned comparison result, which is just an extension of [7, Corollary 3.4] to the case where t is not a multiple of s.

Proposition 1. *For integers $t \geq s \geq 1$,*

$$\delta_s \leq \delta_t \leq \frac{t-d}{s}\,\delta_{2s} + \frac{d}{s}\,\delta_s, \qquad \text{where } d := \gcd(s,t).$$

Proof. The first inequality is clear. As for the second one, if d denotes a common divisor of s and t, we introduce the integers k,n such that

$$s = kd, \qquad t = nd.$$

Given a t-sparse vector $\mathbf{u} \in \mathbb{C}^N$, we need to show that

$$\big|\|A\mathbf{u}\|_2^2 - \|\mathbf{u}\|_2^2\big| \leq \Big(\frac{t-d}{s}\,\delta_{2s} + \frac{d}{s}\,\delta_s\Big)\,\|\mathbf{u}\|_2^2. \tag{3}$$

Let $T =: \{j_1, j_2, \ldots, j_t\}$ denote the support of $\mathbf{u}$. We define n subsets $S_1, S_2, \ldots, S_n$ of T, each of size s, by (the indices are meant modulo t)

$$S_i = \{j_{(i-1)d+1}, j_{(i-1)d+2}, \ldots, j_{(i-1)d+s}\}.$$

In this way, each $j \in T$ belongs to exactly $s/d = k$ sets S_i, so that

$$\mathbf{u} = \frac{1}{k}\sum_{1\leq i\leq n} \mathbf{u}_{S_i} \qquad \text{and} \qquad \|\mathbf{u}\|_2^2 = \frac{1}{k}\sum_{1\leq i\leq n} \|\mathbf{u}_{S_i}\|_2^2.$$

Inequality (3) then follows from

$$\begin{aligned}
\big|\|A\mathbf{u}\|_2^2 - \|\mathbf{u}\|_2^2\big| &= \big|\langle (A^*A - \mathrm{Id})\mathbf{u}, \mathbf{u}\rangle\big| \leq \frac{1}{k^2}\sum_{1\leq i\leq n}\sum_{1\leq j\leq n} \big|\langle (A^*A - \mathrm{Id})\mathbf{u}_{S_i}, \mathbf{u}_{S_j}\rangle\big| \\
&= \frac{1}{k^2}\Big(\sum_{1\leq i\neq j\leq n} \big|\langle (A^*_{S_i\cup S_j}A_{S_i\cup S_j} - \mathrm{Id})\mathbf{u}_{S_i}, \mathbf{u}_{S_j}\rangle\big| + \sum_{1\leq i\leq n} \big|\langle (A^*_{S_i}A_{S_i} - \mathrm{Id})\mathbf{u}_{S_i}, \mathbf{u}_{S_i}\rangle\big|\Big) \\
&\leq \frac{1}{k^2}\Big(\sum_{1\leq i\neq j\leq n} \delta_{2s}\|\mathbf{u}_{S_i}\|_2\|\mathbf{u}_{S_j}\|_2 + \sum_{1\leq i\leq n} \delta_s\|\mathbf{u}_{S_i}\|_2^2\Big) \\
&= \frac{\delta_{2s}}{k^2}\Big(\sum_{1\leq i\leq n} \|\mathbf{u}_{S_i}\|_2\Big)^2 - \frac{\delta_{2s}}{k^2}\sum_{1\leq i\leq n} \|\mathbf{u}_{S_i}\|_2^2 + \frac{\delta_s}{k^2}\sum_{1\leq i\leq n} \|\mathbf{u}_{S_i}\|_2^2
\end{aligned}$$

$$\leq \frac{\delta_{2s} n}{k^2} \sum_{1 \leq i \leq n} \|\mathbf{u}_{S_i}\|_2^2 - \frac{\delta_{2s} - \delta_s}{k^2} \sum_{1 \leq i \leq n} \|\mathbf{u}_{S_i}\|_2^2 = \Big(\frac{\delta_{2s} n}{k^2} - \frac{\delta_{2s} - \delta_s}{k^2}\Big) \sum_{1 \leq i \leq n} \|\mathbf{u}_{S_i}\|_2^2$$
$$= \Big(\frac{n}{k}\delta_{2s} - \frac{1}{k}(\delta_{2s} - \delta_s)\Big) \|\mathbf{u}\|_2^2 = \Big(\frac{t}{s}\delta_{2s} - \frac{1}{k}(\delta_{2s} - \delta_s)\Big) \|\mathbf{u}\|_2^2.$$

In order to make the latter as small as possible, we need to take k as small as possible, i.e., to take d as large as possible, hence the choice $d := \gcd(s,t)$. This finishes the proof. □

3 Basis Pursuit

In this section, we recall that s-sparse recovery via basis pursuit succeeds as soon as $\delta_{2s} < 0.46515$. Contrary to the other sections, we do not give a full proof of this statement, as this was done in [4].

Theorem 1. *Suppose that the* $2s$*-th order restricted isometry constant of the matrix* $A \in \mathbb{C}^{m \times N}$ *satisfies*

$$\delta_{2s} < \frac{3}{4+\sqrt{6}} \approx 0.46515.$$

If $\mathbf{x} \in \mathbb{C}^N$ *is an* s*-sparse vector, then it is recovered as a solution of* (BP) *with* $\mathbf{y} = A\mathbf{x}$*. More generally, if* S *denotes an index set of* s *largest (in modulus) entries of a vector* $\mathbf{x} \in \mathbb{C}^N$ *and if* $\mathbf{y} = A\mathbf{x} + \mathbf{e}$ *for some error term* $\mathbf{e} \in \mathbb{C}^m$ *satisfying* $\|\mathbf{e}\|_2 \leq \eta$*, then a minimizer* $\mathbf{x}^\star$ *of* $\|\mathbf{z}\|_1$ *subject to* $\|A\mathbf{z} - \mathbf{y}\|_2 \leq \eta$ *approximates the vector* $\mathbf{x}$ *with error*

$$\|\mathbf{x} - \mathbf{x}^\star\|_p \leq \frac{C}{s^{1-1/p}} \|\mathbf{x}_{\overline{S}}\|_1 + D s^{1/p-1/2} \eta, \qquad \text{all } p \in [1,2],$$

where the constants C *and* D *depend only on* δ_{2s}.

Classical arguments leading to more demanding sufficient conditions, such as the condition $\delta_{2s} < \sqrt{2} - 1 \approx 0.4142$ from [2], make use of the key inequality

$$\|\mathbf{v}_{S_k}\|_2 \leq \frac{1}{\sqrt{s}} \|\mathbf{v}_{S_{k-1}}\|_1, \qquad k \geq 1, \tag{4}$$

where the sets $S_1, S_2, \ldots$ of s indices are ordered by nonincreasing moduli of entries of $\mathbf{v} \in \mathbb{C}^N$. This step was refined in [1] with the introduction of the *shifting inequality* due to Cai, Wang, and Xu. This inequality was also used in [4] to obtain the sufficient condition of Theorem 1, and can more generally be used to obtain other sufficient conditions in terms of δ_{3s} or δ_{4s}, say, as was done in [1]. Instead of comparing the ℓ_2-norm of the subvector $\mathbf{v}_{S_k}$ with the ℓ_1-norm of the shifted subvector $\mathbf{v}_{S_{k-1}}$ in (4), the shifting inequality consists in reducing the size of the shift from s to roughly $s/4$. Precisely, it states that *for* $k \geq s/4$, *if* $\underbrace{a_1 \geq \cdots \geq \overbrace{a_{k+1} \geq \cdots \geq a_s}^{\mathbf{u}}}_{\mathbf{u}} \geq \cdots \geq a_{k+s} \geq 0$, *then*

$$\|\mathbf{v}\|_2 \leq \frac{1}{\sqrt{s}} \|\mathbf{u}\|_1.$$

This is in fact the particular case $p = 1$, $q = 2$, $t = s$, of the following result, which generalizes the shifting inequality to other norms and to other vector sizes.

Theorem 2. *If* $0 < p < q$ *and* $\underbrace{a_1 \geq \cdots \geq \overbrace{a_{k+1} \geq \cdots \geq a_s}_{\mathbf{u}} \geq \cdots \geq a_{k+t}}^{\mathbf{v}} \geq 0$, *then*

$$\|\mathbf{v}\|_q \leq C_{p,q}(k,s,t) \|\mathbf{u}\|_p,$$

where

$$C_{p,q}(k,s,t) = \max \left\{ \frac{t^{p/q}}{s}, \left(\frac{p}{q}\right)^{p/q} \left(1 - \frac{p}{q}\right)^{1-p/q} \frac{1}{k^{1-p/q}} \right\}^{1/p}. \tag{5}$$

When **u** *and* **v** *do not overlap much, the second term can be discarded, i.e.,*

$$C_{p,q}(k,s,t) = \frac{t^{1/q}}{s^{1/p}} \qquad \text{provided } s - k \leq \left(\frac{p}{q}\right) s. \tag{6}$$

Proof. The constant $C_{p,q}(k,s,t)^q$ is a solution of the maximization problem

$$\text{maximize } a_{k+1}^q + \cdots + a_{k+t}^q \qquad \text{subject to } a_1^p + \cdots + a_s^p \leq 1,\ a_1 \geq \cdots \geq a_{k+t} \geq 0.$$

Thus, setting $r := q/p > 1$, we aim at maximizing the convex function

$$f(x_1, \ldots, x_{k+t}) := x_{k+1}^r + \cdots + x_{k+t}^r$$

over the convex polytope

$$\mathscr{C} := \left\{ \mathbf{x} \in \mathbb{R}^{k+t} : x_1 + \cdots + x_s \leq 1, x_1 \geq \cdots \geq x_{k+t} \geq 0 \right\}.$$

The maximum is attained at one of the vertices of the convex polytope $\mathscr{C}$. These vertices are obtained by turning into equalities $k+t$ of the $k+t+1$ inequalities defining $\mathscr{C}$. We have to separate several cases:

- If $x_1 = \cdots = x_{k+t} = 0$, then

$$f(x_1, \ldots, x_{k+t}) = 0;$$

- If $x_1 + \cdots + x_s = 1$ and $x_1 = \cdots = x_h > x_{h+1} = x_{k+t} = 0$ for some $1 \leq h \leq k$, then

$$f(x_1, \ldots, x_{k+t}) = 0;$$

- If $x_1 + \cdots + x_s = 1$ and $x_1 = \cdots = x_h > x_{h+1} = x_{k+t} = 0$ for some $k \leq h \leq s$, then $x_1 = \cdots = x_h = 1/h$, and

$$f(x_1, \ldots, x_{k+t}) = \frac{h-k}{h^r};$$

- If $x_1 + \cdots + x_s = 1$ and $x_1 = \cdots = x_h > x_{h+1} = x_{k+t} = 0$ for some $s \leq h \leq k+t$, then $x_1 = \cdots = x_h = 1/s$, and

$$f(x_1, \ldots, x_{k+t}) = \frac{h-k}{s^r}.$$

It follows that the desired constant is

$$C_{p,q}(k,s,t)^q = \max\left\{ \max_{k \leq h \leq s} \frac{h-k}{h^r}, \max_{s \leq h \leq k+t} \frac{h-k}{s^r} \right\}.$$

Considering h as a continuous variable, we observe that the function $g(h) := (h-k)/h^r$ is increasing until the critical point $h^* := \big(r/(r-1)\big)k$ and decreasing thereafter, so that the first maximum is no larger than $g(h^*) = \big((r-1)^{r-1}/r^r\big)/k^{r-1}$, or than $g(s) = (s-k)/s^r$ if $h^* \geq s$. Now taking into account that $(h-k)/s^r$ increases with h on $[s, k+t]$, we deduce

$$C_{p,q}(k,s,t)^q \begin{cases} \leq \max\left\{ \dfrac{(r-1)^{r-1}}{r^r} \dfrac{1}{k^{r-1}}, \dfrac{t}{s^r} \right\}, & \\ = \dfrac{t}{s^r} & \text{if } \dfrac{r}{r-1}k \geq s. \end{cases}$$

We simply obtain (5) and (6) by rearranging the latter. It is worth noting that the constants appearing in (5) and (6) cannot be improved. □

It is interesting to point out that Theorem 2 contains two inequalities that are classical in Approximation Theory and that are often used in Compressive Sensing. These inequalities are, for $0 < p < q$ and $\mathbf{x} \in \mathbb{R}^n$,

$$\sigma_k(\mathbf{x})_q \leq \begin{cases} \dfrac{1}{k^{1/p-1/q}} \|\mathbf{x}\|_p, \\ \dfrac{D_{p,q}}{k^{1/p-1/q}} \|\mathbf{x}\|_p, \quad D_{p,q} := \dfrac{1}{(q/p-1)^{1/q}} = \left(\dfrac{(p/q)^{p/q}}{(1-p/q)^{p/q}} \right)^{1/p}. \end{cases}$$

This corresponds to the case $s = n$, $t = n-k$, for which we indeed have

$$k^{1-p/q} C_{p,q}(k,n,n-k)^p \leq \min\{1, D_{p,q}^p\},$$

since the left-hand side reduces to

$$\max\left\{ \left(1-\frac{k}{n}\right)^{p/q} \left(\frac{k}{n}\right)^{1-p/q}, \left(\frac{p}{q}\right)^{p/q} \left(1-\frac{p}{q}\right)^{1-p/q} \right\} = \left(\frac{p}{q}\right)^{p/q} \left(1-\frac{p}{q}\right)^{1-p/q},$$

and the latter is readily seen to be bounded by $\min\{1, D_{p,q}^p\}$.

4 Iterative Hard Thresholding

In this section, we study a first alternative to the basis pursuit algorithm, namely the iterative hard thresholding algorithm. Importantly, this is an easy-to-implement algorithm which requires only a few computational operations. We give an elegant and surprisingly simple justification of the success of s-sparse recovery using this algorithm as soon as $\delta_{3s} < 1/2$. This improves the result of [4], where the sufficient condition $\delta_{3s} < 1/\sqrt{8}$ was obtained – although the main theorem was stated for $\delta_{3s} < 1/\sqrt{32}$ in order to achieve a rate of convergence equal to $\rho = 1/2$.

Theorem 3. *Suppose that the* $3s$*-th order restricted isometry constant of the matrix* $A \in \mathbb{C}^{m \times N}$ *satisfies*

$$\delta_{3s} < \frac{1}{2}.$$

If $\mathbf{x} \in \mathbb{C}^N$ *is an* s*-sparse vector, then the sequence* $(\mathbf{x}^n)$ *defined by* (IHT) *with* $\mathbf{y} = A\mathbf{x}$ *converges to the vector* $\mathbf{x}$.
More generally, if S *denotes an index set of* s *largest (in modulus) entries of a vector* $\mathbf{x} \in \mathbb{C}^N$ *and if* $\mathbf{y} = A\mathbf{x} + \mathbf{e}$ *for some error term* $\mathbf{e} \in \mathbb{C}^m$*, then*

$$\|\mathbf{x}^n - \mathbf{x}_S\|_2 \le \rho^n \, \|\mathbf{x}^0 - \mathbf{x}_S\|_2 + \tau \, \|A\mathbf{x}_{\overline{S}} + \mathbf{e}\|_2, \qquad \text{all } n \ge 0, \tag{7}$$

where

$$\rho := 2\delta_{3s} < 1 \qquad \text{and} \qquad \tau := \frac{2\sqrt{1+\delta_{2s}}}{1 - 2\delta_{3s}}.$$

Remark 1. The value $\tau = 6$ was obtained in [4] for $\rho \le 1/2$, which was ensured by $\delta_{3s} \le 1/\sqrt{32}$. Theorem 3 gives the value $\tau \approx 4.4721$ for $\rho \le 1/2$, i.e., for $\delta_{3s} \le 1/4$, and the value $\tau \approx 3.3562$ for $\delta_{3s} \le 1/\sqrt{32}$.

Proof. We simply use the fact that the s-sparse vector $\mathbf{x}^{n+1}$ is a better s-term approximation to

$$\mathbf{v}^n := \mathbf{x}^n + A^*(\mathbf{y} - A\mathbf{x}^n) = \mathbf{x}^n + A^*A(\mathbf{x}_S - \mathbf{x}^n) + A^*(A\mathbf{x}_{\overline{S}} + \mathbf{e})$$

than the s-sparse vector $\mathbf{x}_S$ to write

$$\|(\mathbf{v}^n - \mathbf{x}_S) + (\mathbf{x}_S - \mathbf{x}^{n+1})\|_2^2 \le \|\mathbf{v}^n - \mathbf{x}_S\|_2^2.$$

Expanding the left-hand side and eliminating $\|\mathbf{v}^n - \mathbf{x}_S\|_2^2$ lead to, with $\mathbf{e}' := A\mathbf{x}_{\overline{S}} + \mathbf{e}$ and $V := \mathrm{supp}(\mathbf{x}) \cup \mathrm{supp}(\mathbf{x}^n) \cup \mathrm{supp}(\mathbf{x}^{n+1})$,

$$\begin{aligned}
\|\mathbf{x}^{n+1} - \mathbf{x}_S\|_2^2 &\le 2\Re\langle \mathbf{v}^n - \mathbf{x}_S, \mathbf{x}^{n+1} - \mathbf{x}_S\rangle \\
&= 2\Re\langle (\mathrm{Id} - A^*A)(\mathbf{x}^n - \mathbf{x}_S) + A^*\mathbf{e}', \mathbf{x}^{n+1} - \mathbf{x}_S\rangle \\
&\le 2\Re\langle (\mathrm{Id} - A_V^*A_V)(\mathbf{x}^n - \mathbf{x}_S), \mathbf{x}^{n+1} - \mathbf{x}_S\rangle + 2\Re\langle \mathbf{e}', A(\mathbf{x}^{n+1} - \mathbf{x}_S)\rangle \\
&\le 2\|\mathrm{Id} - A_V^*A_V\|_{2\to 2}\|\mathbf{x}^n - \mathbf{x}_S\|_2\|\mathbf{x}^{n+1} - \mathbf{x}_S\|_2 + 2\|\mathbf{e}'\|_2\|A(\mathbf{x}^{n+1} - \mathbf{x}_S)\|_2 \\
&\le 2\delta_{3s}\|\mathbf{x}^n - \mathbf{x}_S\|_2\|\mathbf{x}^{n+1} - \mathbf{x}_S\|_2 + 2\|\mathbf{e}'\|_2\sqrt{1+\delta_{2s}}\,\|\mathbf{x}^{n+1} - \mathbf{x}_S\|_2.
\end{aligned}$$

Simplifying by $\|\mathbf{x}^{n+1}-\mathbf{x}_S\|_2$, we derive

$$\|\mathbf{x}^{n+1}-\mathbf{x}_S\|_2 \le 2\delta_{3s}\|\mathbf{x}^n-\mathbf{x}_S\|_2 + 2\sqrt{1+\delta_{2s}}\,\|\mathbf{e}'\|_2.$$

This easily implies the estimate (7). In particular, if $\mathbf{x}$ is an s-sparse vector ($\mathbf{x}_{\overline{S}}=0$) and if the measurements are accurate ($\mathbf{e}=0$), then

$$\|\mathbf{x}^n-\mathbf{x}\|_2 \le \rho^n\|\mathbf{x}^0-\mathbf{x}\|,$$

so the sequence $(\mathbf{x}^n)$ converges to $\mathbf{x}$ as soon as $\rho<1$, i.e., $\delta_{3s}<1/2$. □

As already mentioned, we could call upon Proposition 1 with $t=3s$ to derive a sufficient condition in terms of δ_{2s} for the success of s-sparse recovery via iterative hard thresholding, namely $\delta_{2s}<1/6$. This condition can actually be improved using the previous technique. For simplicity, we only state the result in the case of exactly sparse vectors measured with perfect accuracy.

Theorem 4. *Suppose that the* $2s$*-th order restricted isometry constant of the matrix* $A\in\mathbb{C}^{m\times N}$ *satisfies*

$$\delta_{2s}<\frac{1}{4}.$$

If $\mathbf{x}\in\mathbb{C}^N$ *is an* s*-sparse vector, then the sequence* $(\mathbf{x}^n)$ *defined by* (IHT) *with* $\mathbf{y}=A\mathbf{x}$ *converges to the vector* $\mathbf{x}$.

Proof. We use what has been done in the proof of Theorem 3, specified to the case $\mathbf{e}'=0$, to write

$$\|\mathbf{x}^{n+1}-\mathbf{x}\|_2^2 \le 2\Re\langle(\mathrm{Id}-A^*A)(\mathbf{x}^n-\mathbf{x}),\mathbf{x}^{n+1}-\mathbf{x}\rangle.$$

Let us decompose $\mathrm{supp}(\mathbf{x})\cup\mathrm{supp}(\mathbf{x}^n)\cup\mathrm{supp}(\mathbf{x}^{n+1})$ into the three disjoint sets

$$\begin{aligned}
V_1 &:= (\mathrm{supp}(\mathbf{x})\cup\mathrm{supp}(\mathbf{x}^n)) \cap (\mathrm{supp}(\mathbf{x})\cup\mathrm{supp}(\mathbf{x}^{n+1})),\\
V_2 &:= (\mathrm{supp}(\mathbf{x})\cup\mathrm{supp}(\mathbf{x}^n)) \setminus (\mathrm{supp}(\mathbf{x})\cup\mathrm{supp}(\mathbf{x}^{n+1})),\\
V_3 &:= (\mathrm{supp}(\mathbf{x})\cup\mathrm{supp}(\mathbf{x}^{n+1})) \setminus (\mathrm{supp}(\mathbf{x})\cup\mathrm{supp}(\mathbf{x}^n)).
\end{aligned}$$

Since $V_1\cup V_2$, $V_2\cup V_3$, and $V_2\cup V_3$ all have size at most $2s$, we have

$$\begin{aligned}
\|\mathbf{x}^{n+1}-\mathbf{x}\|_2^2 &= 2\Re\langle(\mathrm{Id}-A^*A)\big((\mathbf{x}^n-\mathbf{x})_{V_1}+(\mathbf{x}^n-\mathbf{x})_{V_2}\big),(\mathbf{x}^{n+1}-\mathbf{x})_{V_1}+(\mathbf{x}^{n+1}-\mathbf{x})_{V_3}\rangle\\
&= 2\Re\langle(\mathrm{Id}-A^*A)\big((\mathbf{x}^n-\mathbf{x})_{V_1}\big),(\mathbf{x}^{n+1}-\mathbf{x})_{V_1}\rangle\\
&+ 2\Re\langle(\mathrm{Id}-A^*A)\big((\mathbf{x}^n-\mathbf{x})_{V_1}\big),(\mathbf{x}^{n+1}-\mathbf{x})_{V_3}\rangle\\
&+ 2\Re\langle(\mathrm{Id}-A^*A)\big((\mathbf{x}^n-\mathbf{x})_{V_2}\big),(\mathbf{x}^{n+1}-\mathbf{x})_{V_1}\rangle\\
&+ 2\Re\langle(\mathrm{Id}-A^*A)\big((\mathbf{x}^n-\mathbf{x})_{V_2}\big),(\mathbf{x}^{n+1}-\mathbf{x})_{V_3}\rangle\\
&\le 2\delta_{2s}\big(\|(\mathbf{x}^n-\mathbf{x})_{V_1}\|_2\|(\mathbf{x}^{n+1}-\mathbf{x})_{V_1}\|_2+\|(\mathbf{x}^n-\mathbf{x})_{V_1}\|_2\|(\mathbf{x}^{n+1}-\mathbf{x})_{V_3}\|_2\\
&\qquad+\|(\mathbf{x}^n-\mathbf{x})_{V_2}\|_2\|(\mathbf{x}^{n+1}-\mathbf{x})_{V_1}\|_2+\|(\mathbf{x}^n-\mathbf{x})_{V_2}\|_2\|(\mathbf{x}^{n+1}-\mathbf{x})_{V_3}\|_2\big)\\
&\le 2\delta_{2s}\big(\|(\mathbf{x}^n-\mathbf{x})_{V_1}\|_2^2+\|(\mathbf{x}^n-\mathbf{x})_{V_1}\|_2^2+\|(\mathbf{x}^n-\mathbf{x})_{V_2}\|_2^2+\|(\mathbf{x}^n-\mathbf{x})_{V_2}\|_2^2\big)^{1/2}
\end{aligned}$$

$$\times \left(\|(\mathbf{x}^{n+1}-\mathbf{x})_{V_1}\|_2^2+\|(\mathbf{x}^{n+1}-\mathbf{x})_{V_3}\|_2^2+\|(\mathbf{x}^{n+1}-\mathbf{x})_{V_1}\|_2^2+\|(\mathbf{x}^{n+1}-\mathbf{x})_{V_3}\|_2^2\right)^{1/2}$$
$$= 2\delta_{2s}\left(2\|\mathbf{x}^n-\mathbf{x}\|_2^2\right)^{1/2}\left(2\|\mathbf{x}^{n+1}-\mathbf{x}\|_2^2\right)^{1/2} = 4\delta_{2s}\|\mathbf{x}^n-\mathbf{x}\|_2\|\mathbf{x}^{n+1}-\mathbf{x}\|_2.$$

This yields, after simplification by $\|\mathbf{x}^{n+1}-\mathbf{x}\|_2$,

$$\|\mathbf{x}^{n+1}-\mathbf{x}\|_2 \le \rho\,\|\mathbf{x}^n-\mathbf{x}\|_2, \qquad \rho := 4\delta_{2s}.$$

Convergence of the sequence $(\mathbf{x}_n)$ towards $\mathbf{x}$ is therefore guaranteed as soon as $\rho < 1$, i.e., $\delta_{2s} < 1/4$. □

Remark 2. The better sufficient condition $\delta_{2s} < 1/3$ was obtained in [6] with a slight modification of the iterative hard thresholding algorithm, namely the iteration

$$\mathbf{x}^{n+1} = H_s\left(\mathbf{x}^n + \frac{3}{4}A^*(\mathbf{y}-A\mathbf{x}^n)\right). \qquad (\mathrm{IHT}_{3/4})$$

Note, however, that the condition $\delta_{2s} < 1/3$ is not heuristically better than the condition $\delta_{3s} < 1/2$, since $(2s)/(1/3)^2 > (3s)/(1/2)^2$.

5 Compressive Sampling Matching Pursuit

In this section, we study a second alternative to the basis pursuit algorithm, namely the compressive sampling matching pursuit algorithm. We give a proof of the success of s-sparse recovery using this algorithm as soon as $\delta_{4s} < 0.38427$. This improves the original condition of [7]. There, the authors targeted a rate of convergence equal to $\rho = 1/2$, so that they gave the sufficient condition $\delta_{4s} \le 0.1$, but their arguments actually yield $\rho < 1$ as soon as $\delta_{4s} < 0.17157$.

Theorem 5. *Suppose that the* $4s$*-th order restricted isometry constant of the matrix* $A \in \mathbb{C}^{m\times N}$ *satisfies*

$$\delta_{4s} < \sqrt{\frac{2}{5+\sqrt{73}}} \approx 0.38427.$$

If $\mathbf{x} \in \mathbb{C}^N$ *is an s-sparse vector, then the sequence* $(\mathbf{x}^n)$ *defined by* (CSMP_{1-4}) *with* $\mathbf{y} = A\mathbf{x}$ *converges to the vector* $\mathbf{x}$.
More generally, if S denotes an index set of s largest (in modulus) entries of a vector $\mathbf{x} \in \mathbb{C}^N$ *and if* $\mathbf{y} = A\mathbf{x}+\mathbf{e}$ *for some error term* $\mathbf{e} \in \mathbb{C}^m$*, then*

$$\|\mathbf{x}^n - \mathbf{x}_S\|_2 \le \rho^n\,\|\mathbf{x}^0-\mathbf{x}_S\|_2 + \tau\,\|A\mathbf{x}_{\overline{S}}+\mathbf{e}\|_2, \qquad \text{all } n \ge 0, \qquad (8)$$

where the positive constants $\rho < 1$ *and* τ *depend only on* δ_{4s}.

Remark 3. The explicit expressions for ρ and τ are given at the end of the proof (the constant τ is made dependent only on δ_{4s} by using $\delta_{3s} \le \delta_{4s}$). Note that the value $\tau = 15$ was obtained in [7] for $\rho \leq 1/2$, which was ensured by $\delta_{4s} \leq 0.1$. Theorem 5 gives the value $\tau \approx 10.369$ for $\rho \leq 1/2$, i.e., for $\delta_{4s} \leq 0.22665$, and the value $\tau \approx 5.6686$ for $\delta_{4s} \leq 0.1$.

Proof. Step (CSMP$_3$) says that $A\mathbf{u}^n$ is the best ℓ_2-approximation to $\mathbf{y}$ from the space $\{A\mathbf{z}, \text{supp}(\mathbf{z}) \subseteq U^n\}$; hence, it is characterized by

$$\langle A\mathbf{u}^n - \mathbf{y}, A\mathbf{z}\rangle = 0 \qquad \text{whenever } \text{supp}(\mathbf{z}) \subseteq U^n. \tag{9}$$

Setting $\mathbf{e}' := A\mathbf{x}_{\overline{S}} + \mathbf{e}$ to have $\mathbf{y} = A\mathbf{x}_S + \mathbf{e}'$, this can be rewritten as

$$\langle \mathbf{u}^n - \mathbf{x}_S, A^*A\mathbf{z}\rangle = \langle \mathbf{e}', A\mathbf{z}\rangle \qquad \text{whenever } \text{supp}(\mathbf{z}) \subseteq U^n. \tag{10}$$

This yields in particular

$$\begin{aligned}
\|(\mathbf{u}^n - \mathbf{x}_S)_{U^n}\|_2^2 &= \langle \mathbf{u}^n - \mathbf{x}_S, (\mathbf{u}^n - \mathbf{x}_S)_{U^n}\rangle \\
&= \langle \mathbf{u}^n - \mathbf{x}_S, (\mathrm{Id} - A^*A)\big((\mathbf{u}^n - \mathbf{x}_S)_{U^n}\big)\rangle + \langle \mathbf{e}', A\big((\mathbf{u}^n - \mathbf{x}_S)_{U^n}\big)\rangle \\
&\leq \delta_{4s}\|\mathbf{u}^n - \mathbf{x}_S\|_2 \|(\mathbf{u}^n - \mathbf{x}_S)_{U^n}\|_2 + \|\mathbf{e}'\|_2\sqrt{1+\delta_{3s}}\,\|(\mathbf{u}^n - \mathbf{x}_S)_{U^n}\|_2,
\end{aligned}$$

which gives, after simplification by $\|(\mathbf{u}^n - \mathbf{x}_S)_{U^n}\|_2$,

$$\|(\mathbf{u}^n - \mathbf{x}_S)_{U^n}\|_2 \leq \delta_{4s}\|\mathbf{u}^n - \mathbf{x}_S\|_2 + \sqrt{1+\delta_{3s}}\,\|\mathbf{e}'\|_2. \tag{11}$$

It follows that

$$\begin{aligned}
\|\mathbf{u}^n - \mathbf{x}_S\|_2^2 &= \|(\mathbf{u}^n - \mathbf{x}_S)_{\overline{U^n}}\|_2^2 + \|(\mathbf{u}^n - \mathbf{x}_S)_{U^n}\|_2^2 \\
&\leq \|(\mathbf{u}^n - \mathbf{x}_S)_{\overline{U^n}}\|_2^2 + \big(\delta_{4s}\|\mathbf{u}^n - \mathbf{x}_S\|_2 + \sqrt{1+\delta_{3s}}\,\|\mathbf{e}'\|_2\big)^2.
\end{aligned}$$

This reads $p(\|\mathbf{u}^n - \mathbf{x}_S\|_2) \leq 0$ for the quadratic polynomial defined by

$$p(t) := (1-\delta_{4s}^2)\,t^2 - (2\delta_{4s}\sqrt{1+\delta_{3s}}\,\|\mathbf{e}'\|_2)\,t - (\|(\mathbf{u}^n - \mathbf{x}_S)_{\overline{U^n}}\|_2^2 + (1+\delta_{3s})\,\|\mathbf{e}'\|_2^2).$$

This proves that $\|\mathbf{u}^n - \mathbf{x}_S\|_2$ is bounded by the largest root of p, i.e.,

$$\begin{aligned}
\|\mathbf{u}^n - \mathbf{x}_S\|_2 &\leq \frac{\delta_{4s}\sqrt{1+\delta_{3s}}\|\mathbf{e}'\|_2 + \sqrt{(1-\delta_{4s}^2)\|(\mathbf{u}^n - \mathbf{x}_S)_{\overline{U^n}}\|_2^2 + (1+\delta_{3s})\,\|\mathbf{e}'\|_2^2}}{1-\delta_{4s}^2} \\
&\leq \frac{1}{\sqrt{1-\delta_{4s}^2}}\|(\mathbf{u}^n - \mathbf{x}_S)_{\overline{U^n}}\|_2 + \frac{\sqrt{1+\delta_{3s}}}{1-\delta_{4s}}\|\mathbf{e}'\|_2.
\end{aligned} \tag{12}$$

We now turn to the estimate for $\|\mathbf{x}^{n+1} - \mathbf{x}_S\|_2$. We start by writing

$$\begin{aligned}
\|\mathbf{x}^{n+1} - \mathbf{x}_S\|_2^2 &= \|(\mathbf{u}^n - \mathbf{x}_S) - (\mathbf{u}^n - \mathbf{x}^{n+1})\|_2^2 \\
&= \|\mathbf{u}^n - \mathbf{x}_S\|_2^2 + \|\mathbf{u}^n - \mathbf{x}^{n+1}\|_2^2 - 2\Re\,\langle \mathbf{u}^n - \mathbf{x}_S, \mathbf{u}^n - \mathbf{x}^{n+1}\rangle.
\end{aligned} \tag{13}$$

Step (CSMP$_4$) implies that $\mathbf{x}^{n+1}$ is a better s-term approximation to $\mathbf{u}^n$ than $\mathbf{x}_{S\cap U^n}$, so that

$$\|\mathbf{u}^n - \mathbf{x}^{n+1}\|_2 \leq \|(\mathbf{u}^n - \mathbf{x}_S)_{U^n}\|_2. \tag{14}$$

We also note, in view of (10) and of $\mathrm{supp}(\mathbf{u}^n - \mathbf{x}^{n+1}) \subseteq U^n$, that

$$\begin{aligned} |\langle \mathbf{u}^n - \mathbf{x}_S, \mathbf{u}^n - \mathbf{x}^{n+1}\rangle| &= |\langle \mathbf{u}^n - \mathbf{x}_S, (\mathrm{Id} - A^*A)(\mathbf{u}^n - \mathbf{x}^{n+1})\rangle + \langle \mathbf{e}', A(\mathbf{u}^n - \mathbf{x}^{n+1})\rangle| \\ &\leq \delta_{4s}\|\mathbf{u}^n - \mathbf{x}_S\|_2\|\mathbf{u}^n - \mathbf{x}^{n+1}\|_2 + \|\mathbf{e}'\|_2\sqrt{1+\delta_{3s}}\,\|\mathbf{u}^n - \mathbf{x}^{n+1}\|_2. \end{aligned} \tag{15}$$

Substituting (14) and (15) into (13), then using (11), we obtain

$$\begin{aligned} \|\mathbf{x}^{n+1} - \mathbf{x}_S\|_2^2 &\leq \|\mathbf{u}^n - \mathbf{x}_S\|_2^2 + \|(\mathbf{u}^n - \mathbf{x}_S)_{U^n}\|_2^2 + 2\delta_{4s}\|\mathbf{u}^n - \mathbf{x}_S\|_2\|(\mathbf{u}^n - \mathbf{x}_S)_{U^n}\|_2 \\ &\qquad + 2\sqrt{1+\delta_{3s}}\,\|\mathbf{e}'\|_2\|(\mathbf{u}^n - \mathbf{x}_S)_{U^n}\|_2 \\ &= (1+3\delta_{4s}^2)\|\mathbf{u}^n - \mathbf{x}_S\|_2^2 + 6\delta_{4s}\sqrt{1+\delta_{3s}}\|\mathbf{u}^n - \mathbf{x}_S\|_2\|\mathbf{e}'\|_2 + 3(1+\delta_{3s})\|\mathbf{e}'\|_2^2 \\ &\leq (1+3\delta_{4s}^2)\left(\|\mathbf{u}^n - \mathbf{x}_S\|_2 + \sqrt{\frac{3(1+\delta_{3s})}{1+3\delta_{4s}^2}}\|\mathbf{e}'\|_2\right)^2. \end{aligned}$$

Combining the latter with (12), we deduce

$$\|\mathbf{x}^{n+1} - \mathbf{x}_S\|_2 \leq \sqrt{\frac{1+3\delta_{4s}^2}{1-\delta_{4s}^2}}\|(\mathbf{u}^n - \mathbf{x}_S)_{\overline{U^n}}\|_2 + \left(\frac{\sqrt{1+3\delta_{4s}^2}}{1-\delta_{4s}} + \sqrt{3}\right)\sqrt{1+\delta_{3s}}\|\mathbf{e}'\|_2. \tag{16}$$

It remains to bound $\|(\mathbf{u}^n - \mathbf{x}_S)_{\overline{U^n}}\|_2$ in terms of $\|\mathbf{x}^n - \mathbf{x}_S\|_2$. For this, we notice that ${\mathbf{u}^n}_{\overline{U^n}} = 0 = {\mathbf{x}^n}_{\overline{U^n}}$, so that

$$\|(\mathbf{u}^n - \mathbf{x}_S)_{\overline{U^n}}\|_2 = \|(\mathbf{x}^n - \mathbf{x}_S)_{\overline{U^n}}\|_2 \leq \|(\mathbf{x}^n - \mathbf{x}_S)_{\overline{T^n}}\|_2 = \|(\mathbf{x}^n - \mathbf{x}_S)_{(S\cup S^n)\setminus T^n}\|_2. \tag{17}$$

Step (CSMP$_1$) means that $A^*(\mathbf{y} - A\mathbf{x}^n)_{T^n}$ is a best $2s$-term approximation to $A^*(\mathbf{y} - A\mathbf{x}^n)_{S\cup S^n\cup T^n}$ among all vectors supported on $S \cup S^n \cup T^n$. In particular,

$$\begin{aligned} \|A^*(\mathbf{y} - A\mathbf{x}^n)_{(S\cup S^n)\setminus T^n}\|_2 &\leq \|A^*(\mathbf{y} - A\mathbf{x}^n)_{T^n\setminus(S\cup S^n)}\|_2 \\ &\leq \|A^*A(\mathbf{x}_S - \mathbf{x}^n)_{T^n\setminus(S\cup S^n)}\|_2 + \|(A^*\mathbf{e}')_{T^n\setminus(S\cup S^n)}\|_2 \\ &= \|((A^*A - \mathrm{Id})(\mathbf{x}_S - \mathbf{x}^n))_{T^n\setminus(S\cup S^n)}\|_2 + \|(A^*\mathbf{e}')_{T^n\setminus(S\cup S^n)}\|_2 \\ &\leq \delta_{4s}\|\mathbf{x}_S - \mathbf{x}^n\|_2 + \|(A^*\mathbf{e}')_{T^n\setminus(S\cup S^n)}\|_2. \end{aligned} \tag{18}$$

On the other hand, we have

$$\begin{aligned} &\|A^*(\mathbf{y} - A\mathbf{x}^n)_{(S\cup S^n)\setminus T^n}\|_2 \geq \|A^*A(\mathbf{x}_S - \mathbf{x}^n)_{(S\cup S^n)\setminus T^n}\|_2 - \|(A^*\mathbf{e}')_{(S\cup S^n)\setminus T^n}\|_2 \\ &\geq \|(\mathbf{x}_S - \mathbf{x}^n)_{(S\cup S^n)\setminus T^n}\|_2 - \|((A^*A - \mathrm{Id})(\mathbf{x}_S - \mathbf{x}^n))_{(S\cup S^n)\setminus T^n}\|_2 - \|(A^*\mathbf{e}')_{(S\cup S^n)\setminus T^n}\|_2 \\ &\geq \|(\mathbf{x}_S - \mathbf{x}^n)_{(S\cup S^n)\setminus T^n}\|_2 - \delta_{2s}\|\mathbf{x}_S - \mathbf{x}^n\|_2 - \|(A^*\mathbf{e}')_{(S\cup S^n)\setminus T^n}\|_2. \end{aligned} \tag{19}$$

From (18), (19), and (2), we derive that

$$\begin{aligned}\|(\mathbf{u}^n - \mathbf{x}_S)_{\overline{U^n}}\|_2 &\le (\delta_{2s} + \delta_{4s})\|\mathbf{x}_S - \mathbf{x}^n\|_2 + \|(A^*\mathbf{e}')_{T^n \setminus (S \cup S^n)}\|_2 + \|(A^*\mathbf{e}')_{(S \cup S^n) \setminus T^n}\|_2 \\ &\le 2\delta_{4s}\|\mathbf{x}_S - \mathbf{x}^n\|_2 + \sqrt{2}\,\|(A^*\mathbf{e}')_{(S \cup S^n) \Delta T^n}\|_2 \\ &\le 2\delta_{4s}\|\mathbf{x}_S - \mathbf{x}^n\|_2 + \sqrt{2(1+\delta_{4s})}\,\|\mathbf{e}'\|_2. \end{aligned} \tag{20}$$

Putting (16) and (20) together, we finally conclude that

$$\|\mathbf{x}^{n+1} - \mathbf{x}_S\|_2 \le \rho\, \|\mathbf{x}^n - \mathbf{x}_S\|_2 + (1-\rho)\tau\, \|\mathbf{e}'\|_2. \tag{21}$$

where

$$\rho := \sqrt{\frac{4\delta_{4s}^2(1+3\delta_{4s}^2)}{1-\delta_{4s}^2}},$$

$$(1-\rho)\tau := \sqrt{\frac{2(1+3\delta_{4s}^2)}{1-\delta_{4s}}} + \frac{\sqrt{(1+3\delta_{4s}^2)(1+\delta_{3s})}}{1-\delta_{4s}} + \sqrt{3(1+\delta_{3s})}.$$

To finish, we point out that the constant ρ is less than one when

$$12\delta_{4s}^4 + 5\delta_{4s}^2 - 1 < 0, \qquad \text{i.e.,} \qquad \delta_{4s} < \sqrt{\frac{2}{5+\sqrt{73}}} \approx 0.38427,$$

and that (21) readily implies (8). $\square$

Acknowledgements The author thanks the meeting organizers, Mike Neamtu and Larry Schumaker, for welcoming a minisymposium on Compressive Sensing at the Approximation Theory conference. He also acknowledges support from the French National Research Agency (ANR) through project ECHANGE (ANR-08-EMER-006).

References

1. Cai, T.T., Wang, L., Xu, G.: Shifting inequality and recovery of sparse signals. IEEE Transactions on Signal Processing **58**, 1300–1308 (2010).
2. Candès, E.J.: The restricted isometry property and its implications for compressed sensing. Comptes Rendus de l'Académie des Sciences, Série I, **346**, 589–592 (2008).
3. Candès, E., Tao. T.: Decoding by linear programing. IEEE Trans. Inf. Theory **51**, 4203–4215 (2005).
4. Davies, M.E., Blumensath, T.: Iterative hard thresholding for compressed sensing. Appl. Comput. Harmon. Anal. **27**, 265–274 (2009).
5. Foucart, S.: A note on guaranteed sparse recovery via ℓ_1-minimization. Applied and Comput. Harmonic Analysis, To appear. Appl. Comput. Harmon. Anal. **29**, 97–103 (2010).
6. Garg, R., Khandekar, R.: Gradient descent with sparsification: An iterative algorithm for sparse recovery with restricted isometry property. In: Bottou, L., Littman, M. (eds.) Proceedings of the 26 th International Confer- ence on Machine Learning, pp. 337-344.
7. Needell, D., Tropp, J.A.: CoSaMP: Iterative signal recovery from incomplete and inaccurate samples. Appl. Comput. Harmon. Anal. **26** 301–321 (2009).

Lagrange Interpolation and New Asymptotic Formulae for the Riemann Zeta Function

Michael I. Ganzburg

Abstract An asymptotic representation for the Riemann zeta function $\zeta(s)$ in terms of the Lagrange interpolation error of some function $f_{s,2N}$ at the Chebyshev nodes is found. The representation is based on new error formulae for the Lagrange polynomial interpolation to a function of the form $f(y) = \int_{\mathbb{R}} \frac{\varphi(t)}{t-iy} \mathrm{d}t$. As the major application of this result, new criteria for $\zeta(s) = 0$ and $\zeta(s) \neq 0$ in the critical strip $0 < \mathrm{Re}\, s < 1$ are given.

1 Introduction

The Riemann zeta function can be defined by the following integral representation [3, Sect. 1.12]:

$$\zeta(s) = \frac{2^{s-1}}{(1-2^{1-s})\Gamma(s)} \int_0^\infty \frac{t^{s-1}\mathrm{e}^{-t}}{\cosh t} \mathrm{d}t, \qquad 0 < \mathrm{Re}\, s < 1. \tag{1}$$

There are several other integral representations for $\zeta(s)$ in the critical strip $0 < \mathrm{Re}\, s < 1$ [3, Sect. 1.12].

In this paper, we find an asymptotic representation for $\zeta(s)$ in terms of the Lagrange interpolation error of some function $f_{s,2N}$ at the Chebyshev nodes. This formula is given in Sect. 4 (Corollary 9). The representation is based on new error formulae for the Lagrange polynomial interpolation to a function of the form $f(y) = \int_{\mathbb{R}} \frac{\varphi(t)}{t-iy} \mathrm{d}t$. These results are discussed in Sect. 2 (Theorem 1, Corollary 3, and Example 4). Some technical asymptotics that are needed for the proof of Corollary 9 are presented in Sect. 3.

Michael I. Ganzburg
Hampton University, Hampton, VA 23668, USA,
e-mail: michael.ganzburg@hamptonu.edu

M. Neamtu and L. Schumaker (eds.), *Approximation Theory XIII: San Antonio 2010*,
Springer Proceedings in Mathematics 13, DOI 10.1007/978-1-4614-0772-0_6,

As the major application of Corollary 9, we obtain in Sect. 5 (Corollary 14) new criteria for $\zeta(s)=0$ and $\zeta(s)\neq 0$ in the critical strip. In addition, several other asymptotic representations for $\zeta(s)$ are given in Theorem 13 and Corollaries 10 and 11. Historic remarks are given in Sect. 6.

Notation. Let $\mathbb{R}$ be the real axis and let $\mathbb{C}=\mathbb{R}+i\mathbb{R}$ be the complex plane. In addition, we use the standard notation

$$||f||_{L_p(-a,a)} := \left(\int_{-a}^{a} |f(y)|^p \mathrm{d}y\right)^{1/p}, \qquad 0<p<\infty, \quad 0<a\leq\infty.$$

Throughout the paper $C, C_1, C_2, \dots$ are positive constants independent of essential parameters, and $C(d_1,\dots,d_k)$, $C_1(d_1,\dots,d_k)$, $C_2(d_1,\dots,d_k),\dots$ denote positive constants that depend only on the parameters $d_1,\dots,d_k$. The same symbol does not necessarily denote the same constant in different occurrences.

2 Lagrange Interpolation

Let $\{y_k\}_{k=1}^{n+1}\subset\mathbb{R}$ be a set of distinct Lagrange interpolation nodes. Let us set $H_{n+1}(y):=\prod_{k=1}^{n+1}(y-y_k)$. For a continuous function $f:\mathbb{R}\to\mathbb{C}$, let

$$L_n(y) = L_n(y,f,H_{n+1})$$

be the unique Lagrange interpolation polynomial of degree n to f at the zeros of H_{n+1}.

We consider a class of functions f of the form

$$f(y) = \int_{\mathbb{R}} \frac{\varphi(t)}{t-iy}\mathrm{d}t, \tag{2}$$

where $\varphi:\mathbb{R}\to\mathbb{C}$ is a measurable function satisfying the condition

$$\int_{\mathbb{R}} \left|\frac{\varphi(t)}{t}\right| \mathrm{d}t < \infty. \tag{3}$$

Note that by the Lebesgue Domination Theorem, the function f defined by (2) is continuous on $\mathbb{R}$. Then the following result holds:

Theorem 1. *If a function $f:\mathbb{R}\to\mathbb{C}$ given by (2) satisfies (3), then for any $y\in\mathbb{R}$,*

$$f(y)-L_n(y,f,H_{n+1}) = H_{n+1}(y)\int_{\mathbb{R}} \frac{\varphi(t)}{(t-iy)H_{n+1}(-it)}\mathrm{d}t. \tag{4}$$

Remark 2 *If $0\in\{y_k\}_{k=1}^{n+1}$, then the right-hand side of (4) can be undefined at $y=0$. In this case, identity (4) at $y=0$ should be replaced with*

$$f(0)-L_n(0) = \lim_{y\to 0} H_{n+1}(y)\int_{\mathbb{R}} \frac{\varphi(t)}{(t-iy)H_{n+1}(-it)}\mathrm{d}t.$$

Proof of Theorem 1 We first note that the Lagrange interpolation polynomial to the function $F_t(y) := (t - iy)^{-1}$, where t is a fixed number from $\mathbb{R} \setminus \{0\}$, is

$$L_n(y, F_t, H_{n+1}) = \frac{H_{n+1}(-it) - H_{n+1}(y)}{(t - iy)H_{n+1}(-it)} \tag{5}$$

(cf. [12, Sect. 3.1]). Therefore, for all $y \in \mathbb{R}$ and for all $t \in \mathbb{R} \setminus \{0\}$,

$$\frac{1}{t - iy} - L_n(y, (t - i\cdot)^{-1}, H_{n+1}) = \frac{H_{n+1}(y)}{(t - iy)H_{n+1}(-it)}. \tag{6}$$

Next, we note that the polynomial L_n defined by (5) has coefficients depending on t. Namely, if $H_{n+1}(z) = \sum_{m=0}^{n+1} c_m z^m$, then it follows from (5) that

$$L_n(y, F_t, H_{n+1}) = \sum_{p=0}^{n} \frac{(\sum_{k=0}^{n-p} c_{k+p+1}(-it)^k)y^p}{H_{n+1}(-it)} = \sum_{p=0}^{n} \frac{(\sum_{k=0}^{n-p} c_{k+p+1}y^k)(-it)^p}{H_{n+1}(-it)}. \tag{7}$$

Further, for all $t \in \mathbb{R}$ and $p = 0, 1, \ldots, n$,

$$\frac{|t|^p}{|H_{n+1}(-it)|} \le \begin{cases} C(p,n), & \text{if} \quad 0 \notin \{y_k\}_{k=1}^{n+1} \\ C(p,n)/t, & \text{if} \quad 0 \in \{y_k\}_{k=1}^{n+1}. \end{cases} \tag{8}$$

Then multiplying both sides of (6) by $\varphi(t)$ and integrating over $\mathbb{R}$, we see that the integral $I_n(y) := \int_0^\infty \varphi(t) L_n(y, F_t, H_{n+1}) \mathrm{d}t$ exists for each $y \in \mathbb{R}$, if $0 \notin \{y_k\}_{k=1}^{n+1}$, and it exists for each $y \in \mathbb{R} \setminus \{0\}$, if $0 \in \{y_k\}_{k=1}^{n+1}$, by (3), (7), and (8). Moreover, $I_n(y) = L_n(y, f, H_{n+1})$, by (2). Therefore, taking into account Remark 1 in case of $0 \in \{y_k\}_{k=1}^{n}$, we conclude that (4) holds for all $y \in \mathbb{R}$. □

For even or odd functions f and polynomials H_{n+1}, it is possible to establish more precise representations.

Corollary 3 *(a) If an even function f of the form*

$$f(y) = \int_0^\infty \frac{\varphi_e(t)}{t^2 + y^2} \mathrm{d}t \tag{9}$$

satisfies the condition

$$\int_0^\infty \frac{|\varphi_e(t)|}{t^2} \mathrm{d}t < \infty, \tag{10}$$

then for any $y \in \mathbb{R}$ and every polynomial $H_{n+1}(y) = \prod_{k=1}^{(n+1)/2} (y^2 - y_k^2)$ of even degree $n+1$,

$$f(y) - L_{n-1}(y, f, H_{n+1}) = H_{n+1}(y) \int_{\mathbb{R}} \frac{\varphi_e(t)}{(t^2 + y^2)H_{n+1}(it)} \mathrm{d}t. \tag{11}$$

(b) If an even function f defined by (9) satisfies (10), then for any $y \in \mathbb{R}$ and every polynomial $H_{n+1}(y) = y\prod_{k=1}^{n/2}(y^2 - y_k^2)$ of odd degree $n+1$,

$$f(y) - L_n(y, f, H_{n+1}) = iyH_{n+1}(y)\int_{\mathbb{R}} \frac{\varphi_e(t)}{t(t^2+y^2)H_{n+1}(-it)}\mathrm{d}t. \tag{12}$$

(c) If an odd function f of the form

$$f(y) = y\int_0^\infty \frac{\varphi_o(t)}{t^2+y^2}\mathrm{d}t \tag{13}$$

satisfies the condition

$$\int_0^\infty \frac{|\varphi_o(t)|}{t^2}\mathrm{d}t < \infty, \tag{14}$$

then for any $y \in \mathbb{R}$ and every polynomial $H_{n+1}(y) = \prod_{k=1}^{(n+1)/2}(y^2 - y_k^2)$ of even degree $n+1$,

$$f(y) - L_n(y, f, H_{n+1}) = yH_{n+1}(y)\int_{\mathbb{R}} \frac{\varphi_o(t)}{(t^2+y^2)H_{n+1}(it)}\mathrm{d}t. \tag{15}$$

(d) If an odd function f defined by (13) satisfies (14), then for any $y \in \mathbb{R}$ and every polynomial $H_{n+1}(y) = y\prod_{k=1}^{n/2}(y^2 - y_k^2)$ of odd degree $n+1$,

$$f(y) - L_{n-1}(y, f, H_{n+1}) = -iH_{n+1}(y)\int_{\mathbb{R}} \frac{t\varphi_o(t)}{(t^2+y^2)H_{n+1}(-it)}\mathrm{d}t. \tag{16}$$

Proof. If f given by (9) satisfies (10), then

$$f(y) = \frac{1}{2}\int_{\mathbb{R}} \frac{\varphi_e(|t|)}{t(t-iy)}\mathrm{d}t,$$

and using Theorem 1 for $\varphi(t) = \varphi_e(|t|)/t$, we have

$$\begin{aligned}
f(y) - L_n(y, f, H_{n+1}) &= \frac{1}{2}H_{n+1}(y)\int_{\mathbb{R}} \frac{\varphi_e(|t|)}{t(t-iy)H_{n+1}(it)}\mathrm{d}t \\
&= \frac{1}{2}H_{n+1}(y)\left(\int_{\mathbb{R}} \frac{\varphi_e(|t|)}{(t^2+y^2)H_{n+1}(it)}\mathrm{d}t + iy\int_{\mathbb{R}} \frac{\varphi_e(|t|)}{t(t^2+y^2)H_{n+1}(it)}\mathrm{d}t\right) \\
&= \frac{1}{2}H_{n+1}(y)\int_{\mathbb{R}} \frac{\varphi_e(|t|)}{(t^2+y^2)H_{n+1}(it)}\mathrm{d}t.
\end{aligned}$$

Therefore, (11) holds. Identities (12), (15), and (16) can be proved similarly. □

The following example plays an important role in finding new asymptotic relations for $\zeta(s)$.

Example 4 *For $s \in \mathbb{C}$ with $Re\, s > 0$ and $\mu > 0$, we set*

$$f_{s,\mu}(y) := |y|^s V_{s+2}(2\mu|y|, 0), \qquad y \in \mathbb{R},$$

where $V_\nu(w,z)$ is a Lommel's function of two variables. Using formulae in [13, Sect. 16.5], we have

$$f_{s,\mu}(y) = |y|^s \cos(\mu|y| + s\pi/2) - \mu^{-s} \sum_{m=0}^{\infty} \frac{(-1)^m (\mu y)^{2m}}{\Gamma(1-s+2m)}. \tag{17}$$

The following integral representation for $f_{s,\mu}$ can be found in [9, (3.389.7)]:

$$f_{s,\mu}(y) = -\frac{\sin(s\pi)}{\pi} \int_0^\infty \frac{t^{s+1} e^{-\mu t}}{t^2 + y^2} dt, \qquad 0 < \mathrm{Re}\, s < 1, \quad \mu > 0. \tag{18}$$

Here, the function $\varphi(t) := -\sin(s\pi) t^{s+1} e^{-\mu t}/\pi$ is continuous on $\mathbb{R}$, and it satisfies condition (10). Therefore by (12) ($y \in \mathbb{R}$),

$$f_{s,\mu}(y) - L_n(y, f_{s,\mu}, H_{n+1}) = \frac{i \sin(s\pi) y H_{n+1}(y)}{\pi} \int_{\mathbb{R}} \frac{t^s e^{-\mu t}}{(t^2 + y^2) H_{n+1}(-it)} dt, \tag{19}$$

where n is even and $H_{n+1}(y) = y \prod_{k=1}^{n/2} (y^2 - y_k^2)$ is an odd polynomial with distinct real zeros.

3 Asymptotic Behavior of the Interpolation Error

In this section, we find an asymptotic representation for the interpolation error in (19) in case of $\mu = 2N$ and $H_{n+1}(y) = y T_{2N}(y)$. Here,

$$T_M(y) := (1/2)((y + \sqrt{y^2 - 1})^M + (y - \sqrt{y^2 - 1})^M)$$

is the Chebyshev polynomial of the first kind of degree M. We begin with some technical estimates.

Lemma 5 *For $0 \le t < M < \infty$,*

$$\exp(t - t^3/(6M^2)) \le (t/M + \sqrt{(t/M)^2 + 1})^M \le \exp(t). \tag{20}$$

Proof. The function $\sinh^{-1} v = \log(v + \sqrt{v^2 + 1})$ is odd, and it has the following power series expansion for $|v| < 1$ (see [9, (1.641.2)]):

$$\log(v + \sqrt{v^2 + 1}) = v - v^3/6 + 3v^5/40 - \ldots = \sum_{k=0}^{\infty} (-1)^k a_k v^{2k+1}, \tag{21}$$

where

$$a_k := \frac{(2k)!}{2^{2k}(k!)^2(2k+1)}, \qquad k = 0, 1, \cdots$$

is a decreasing sequence since $a_{k+1}/a_k = (4k^2+4k+1)/(4k^2+10k+6) < 1, k = 0, 1, \ldots$. Therefore, (20) follows from (21) for $v = t/M \in [0, 1)$. □

Lemma 6 *For $M = 2N, N \in \mathbb{N}$, and $0 \leq t < M$,*

$$F_M(t) := \left| \frac{(-1)^{M/2}}{T_M(it/M)} - \frac{1}{\cosh t} \right| \leq \frac{t^3 \exp(t^3/(3M^2))}{3M^2 \cosh t}. \tag{22}$$

Proof. Using Lemma 5, we have

$$\begin{aligned} F_M(t) &= \frac{2|e^t + e^{-t} - (t/M + \sqrt{(t/M)^2+1})^M - (t/M + \sqrt{(t/M)^2+1})^{-M}|}{((t/M + \sqrt{(t/M)^2+1})^M + (t/M + \sqrt{(t/M)^2+1})^{-M}) \cosh t} \\ &\leq \frac{2(\exp(t)(1 - \exp(-t^3/(6M^2))) + \exp(-t)(\exp(t^3/(6M^2)) - 1))}{(\exp(t - t^3/(6M^2)) + \exp(-t)) \cosh t} \\ &\leq \frac{\exp(t^3/(6M^2)(\exp(t)(1 - \exp(-t^3/(6M^2))) + \exp(-t)(\exp(t^3/(6M^2)) - 1))}{\cosh^2 t} \end{aligned} \tag{23}$$

Finally, applying the elementary inequalities

$$1 - e^{-v} < ve^v, \qquad e^v - 1 < ve^v, \qquad v > 0,$$

to the right-hand side of (23), we arrive at (22). □

Next, we find the asymptotic behavior of the integral

$$I_s(M, y) := \int_0^\infty \frac{t^{s-1} e^{-Mt}}{(1 + (t/y)^2) T_M(it)} dt = \frac{1}{M^s} \int_0^\infty \frac{t^{s-1} e^{-t}}{(1 + t^2/(My)^2) T_M(it/M)} dt, \tag{24}$$

where $y \in \mathbb{R}, M = 2N, N \in \mathbb{N}, \operatorname{Re} s > 0$.

We shall approximate $I(M, y)$ by the integral

$$R_s(M, y) := \frac{(-1)^{M/2}}{M^s} \int_0^\infty \frac{t^{s-1} e^{-t}}{(1 + t^2/(My)^2) \cosh t} dt. \tag{25}$$

Lemma 7 *The following asymptotic holds:*

$$I_s(M, y) = R_s(M, y) + \Delta_s(N, y), \qquad y \in \mathbb{R}, M = 2N, N \in \mathbb{N}, \operatorname{Re} s > 0, \tag{26}$$

where

$$\sup_{y\in\mathbb{R}}|\Delta_s(N,y)| \le C(s)N^{-\mathrm{Re}s-2}. \tag{27}$$

Proof. It follows from (24) and (25) that

$$\Delta_s(N,y) = \frac{(-1)^{M/2}}{M^s}\int_0^\infty \frac{t^{s-1}\mathrm{e}^{-t}}{1+t^2/(My)^2}\left(\frac{(-1)^{M/2}}{T_M(it/M)} - \frac{1}{\cosh t}\right)\mathrm{d}t$$

$$= \frac{(-1)^{M/2}}{M^s}\left(\int_0^{M_1} + \int_{M_1}^\infty\right) = \frac{(-1)^{M/2}}{M^s}(I_1(M,M_1,y) + I_2(M,M_1,y), \tag{28}$$

where M_1 is any number from $(0,M)$. Next by Lemma 6,

$$|I_1(M,M_1,y)| \le \int_0^{M_1} \frac{t^{s-1}\mathrm{e}^{-t}}{1+t^2/(My)^2}F_M(t)\mathrm{d}t$$

$$\le \frac{\exp(M_1^3/(3M^2)}{3M^2}\int_0^{M_1}\frac{t^{\mathrm{Re}s+2}\mathrm{e}^{-t}}{\cosh t}\mathrm{d}t \le \frac{C_1(s)\exp(M_1^3/(3M^2))}{M^2}. \tag{29}$$

Further,

$$|I_2(M,M_1,y)| \le \int_{M_1}^\infty \frac{t^{\mathrm{Re}s-1}\mathrm{e}^{-t}}{1+t^2/(My)^2}\left(\frac{1}{|T_M(it/M)|} + \frac{1}{\cosh t}\right)\mathrm{d}t$$

$$\le 2\int_{M_1}^\infty t^{\mathrm{Re}s-1}\mathrm{e}^{-t}\mathrm{d}t = 2\Gamma(\mathrm{Re}s,M_1) \le C_2(s)M_1^{\mathrm{Re}s-1}\mathrm{e}^{-M_1}, \tag{30}$$

where $\Gamma(z,\beta)$ is the upper incomplete gamma function. Therefore, combining relation (28) with (29) and (30), we have

$$\sup_{y\in\mathbb{R}}|\Delta_s(N,y)| \le C_3(s)M^{-\mathrm{Re}s}(M^{-2}\exp(M_1^3/(3M^2)) + M_1^{\mathrm{Re}s-1}\mathrm{e}^{-M_1}). \tag{31}$$

Choosing $M_1 = M^{2/3}$, $M = 2N$, in (31), we obtain (27). This completes the proof of the lemma. □

An asymptotic representation for the interpolation error is given in the following theorem:

Theorem 8 *Let* $N \in \mathbb{N}$, $0 < \mathrm{Re}s < 1$, *and let*

$$f_{s,2N}(y) = |y|^s\cos(2N|y| + s\pi/2) - (2N)^{-s}\sum_{m=0}^\infty \frac{(-1)^m(2Ny)^{2m}}{\Gamma(1-s+2m)} \tag{32}$$

be the function from Example 4 with $\mu = 2N$. Then for $H_{2N+1}(y) = yT_{2N}(y)$ and all $y \in \mathbb{R}$,

$$\begin{aligned} & f_{s,2N}(y) - L_{2N}(y, f_{s,2N}, H_{2N+1}) \\ = \ & (-1)^N \frac{\sin s\pi}{\pi(2N)^s} T_{2N}(y) \int_0^\infty \frac{t^{s-1}\mathrm{e}^{-t}}{(1+t^2/(2Ny)^2)\cosh t}\mathrm{d}t + T_{2N}(y)\Delta_s(N,y), \quad (33) \end{aligned}$$

where

$$\sup_{y\in\mathbb{R}} |\Delta_s(N,y)| \leq C(s)N^{-\mathrm{Re}\, s-2}. \quad (34)$$

Proof. Setting $H_{2N+1}(y) = yT_{2N}(y)$, we have from (19)

$$f_{s,2N}(y) - L_{2N}(y, f_{s,2N}, H_{2N+1}) = \frac{\sin s\pi}{\pi} T_{2N}(y) \int_0^\infty \frac{t^{s-1}\mathrm{e}^{-2Nt}}{(1+(t/y)^2)T_{2N}(it)}\mathrm{d}t \quad (35)$$

Then (33) and (34) follow from (35) and Lemma 7. □

4 Asymptotic Formulae for $\zeta(s)$

It follows from representation (1) that the integral in (33) can be expressed in terms of $\zeta(s)$. Namely,

$$\begin{aligned} & \int_0^\infty \frac{t^{s-1}\mathrm{e}^{-t}}{(1+t^2/(2Ny)^2)\cosh t}\mathrm{d}t \\ = & \int_0^\infty \frac{t^{s-1}\mathrm{e}^{-t}}{\cosh t}\mathrm{d}t - \frac{1}{(2Ny)^2}\int_0^\infty \frac{t^{s+1}\mathrm{e}^{-t}}{(1+t^2/(2Ny)^2)\cosh t}\mathrm{d}t \\ = & \frac{\Gamma(s)(1-2^{1-s})}{2^{s-1}}\zeta(s) - \frac{1}{(2Ny)^2}\int_0^\infty \frac{t^{s+1}\mathrm{e}^{-t}}{(1+t^2/(2Ny)^2)\cosh t}\mathrm{d}t. \quad (36) \end{aligned}$$

Therefore, the following corollary is a direct consequence of (33) and (36).

Corollary 9 *For $0 < \mathrm{Re}\, s < 1$, $H_{2N+1}(y) = yT_{2N}(y)$, and $y \neq 0$,*

$$\begin{aligned} \zeta(s) = & \frac{\pi 2^{s-1}}{\sin s\pi\Gamma(s)(1-2^{1-s})}(-1)^N(2N)^s \frac{(f_{s,2N}(y) - L_{2N}(y, f_{s,2N}, H_{2N+1}))}{T_{2N}(y)} \\ & + \frac{2^{s-1}}{\Gamma(s)(1-2^{1-s})(2Ny)^2}\int_0^\infty \frac{t^{s+1}\mathrm{e}^{-t}}{(1+t^2/(2Ny)^2)\cosh t}\mathrm{d}t + \Delta_s^*(N,y), \quad (37) \end{aligned}$$

where

$$\sup_{y\in\mathbb{R}} |\Delta_s^*(N,y)| \leq C(s)N^{-2}. \quad (38)$$

Following is a simplified version of Corollary 9 for some subsequence $N = N_k(y)$, $k = 1, 2, \ldots$

Corollary 10 *Let* $0 < \mathrm{Re}\, s < 1$ *and* $H_{2N+1}(y) = yT_{2N}(y)$. *Then for any* $y \in [-1,0) \cup (0,1]$, *there exists a subsequence* $\{N_k\}_{k=1}^{\infty}$ *of even positive numbers such that*

$$\begin{aligned}\zeta(s) &= \frac{\pi 2^{s-1}}{\sin s\pi \Gamma(s)(1-2^{1-s})} \\ &\quad \times (2N_k)^s (f_{s,2N_k}(y) - L_{2N_k}(y, f_{s,2N_k}, H_{2N+1})) + O(N_k^{-2}), \quad k \to \infty. \quad (39)\end{aligned}$$

Proof. We first note that for any $y \in [-1,0) \cup (0,1]$ there is a sequence $\{N_k\}_{k=1}^{\infty}$ of even positive numbers such that

$$\lim_{k\to\infty} T_{2N_k}(y) = 1 \qquad (40)$$

Indeed, if $\mu := (\arccos y)/\pi$ is a rational number m/p, then setting $N_k = 2pk$, we have $T_{2N_k}(y) = 1$, $k = 1,2,\ldots$ If μ is irrational, then the sequence $\{2n\mu \,(\mathrm{mod}\, 1)\}_{n=1}^{\infty}$ is dense in $[0,1]$. Therefore, for some sequence $N_k(y) := 2n_k$, $k = 1,2,\ldots$, (40) holds. Then (39) follows from (37) and (38). □

The following corollary contains a more explicit representation for $\zeta(s)$.

Corollary 11 *For* $0 < \mathrm{Re}\, s < 1$,

$$\begin{aligned}\zeta(s) &= \frac{\pi 2^{s}}{\sin s\pi \Gamma(s)(1-2^{1-s})} \\ &\quad \times \lim_{N\to\infty} (-1)^{N+1}(2N)^{s-1} \sum_{k=1}^{N} (-1)^{k+1} f_{s,2N}\left(\cos\frac{2k-1}{4N}\pi\right) \tan\frac{2k-1}{4N}\pi. \quad (41)\end{aligned}$$

Proof. Let $\{y_N\}_{N=1}^{\infty}$ be a sequence of positive numbers with $\lim_{N\to\infty} y_N = \infty$. Then (37) and (38) imply the relation

$$\begin{aligned}\zeta(s) &= \frac{\pi 2^{s-1}}{\sin s\pi \Gamma(s)(1-2^{1-s})} \\ &\quad \times \lim_{N\to\infty} (-1)^{N}(2N)^{s} \frac{f_{s,2N}(y_N) - L_{2N}(y_N, f_{s,2N}, H_{2N+1})}{T_{2N}(y_N)}, \qquad (42)\end{aligned}$$

where $H_{2N+1}(y) = yT_{2N}(y)$. Next, it follows from (18) that

$$\sup_{y\in\mathbb{R}} |f_{s,2N}(y)| \le (|\sin s\pi|\Gamma(\mathrm{Re}\, s)/\pi)(2N)^{-\mathrm{Re}\, s}. \qquad (43)$$

Since $T_{2N}(y_N) \ge (1/2)y_N^{2N}$ for $y_N > 1$, we get from (42) and (43)

$$\zeta(s) = \frac{\pi 2^{s-1}}{\sin s\pi \Gamma(s)(1-2^{1-s})} \lim_{N\to\infty} (-1)^{N+1}(2N)^{s} \frac{L_{2N}(y_N, f_{s,2N}, H_{2N+1})}{T_{2N}(y_N)}. \qquad (44)$$

Further, we consider a sequence of rational functions

$$R_{2N}(y) = \sum_{k=0}^{2N} A_{k,N} y^k / \sum_{k=0}^{2N} B_{k,N} y^k, \qquad N = 1,2,\ldots,$$

such that $\inf_{N\in\mathbb{N}} B_{2N,N} > 0$ and $\lim_{N\to\infty} R(y_{2N})$ exists for any sequence $\{y_N\}_{N=1}^{\infty}$ of positive numbers with $\lim_{N\to\infty} y_N = \infty$. Then $\lim_{N\to\infty} A_{2N,N}/B_{2N,N}$ exists and there exists a sequence $y_N^* \to \infty$ as $N \to \infty$, such that

$$\lim_{N\to\infty} R_{2N}(y_N^*) = \lim_{N\to\infty} \frac{A_{2N,N}}{B_{2N,N}}. \tag{45}$$

Indeed,

$$\lim_{N\to\infty} R_{2N}(y_{2N}) = \lim_{N\to\infty} \frac{A_{2N,N}/B_{2N,N} + \sum_{k=0}^{2N-1} A_{k,N}/(y_N^{2N-k} B_{2N,N})}{1 + \sum_{k=0}^{2N-1} B_{k,N}/(y_N^{2N-k} B_{2N,N})}. \tag{46}$$

Then choosing a sequence $\{y_N^*\}_{N=1}^{\infty}$ with $\lim_{N\to\infty} y_N^* = \infty$, which satisfies the relations

$$\lim_{N\to\infty} \sum_{k=0}^{2N-1} \frac{A_{k,N}}{(y_N^{*2N-k} B_{2N,N})} = \lim_{N\to\infty} \sum_{k=0}^{2N-1} \frac{B_{k,N}}{(y_N^{*2N-k} B_{2N,N})} = 0,$$

we arrive at (45) from (46). Hence, using (45) for the sequence of rational functions

$$R_{2N} = \frac{(-1)^{N+1}(2N)^s L_{2N}(y_N, f_{s,2N}, H_{2N+1})}{T_{2N}(y_N)}$$

defined in the right-hand side of (44), we have

$$\zeta(s) = \frac{\pi 2^s}{\sin s\pi \Gamma(s)(1-2^{1-s})} \lim_{N\to\infty} \frac{(-1)^{N+1}(2N)^s A_{2N}}{2^{2N-1}}, \tag{47}$$

where $A_{2N}(s)$ is the leading coefficient of the interpolation polynomial L_{2N}. Finally, we compute and simplify $A_{2N}(s)/2^{2N-1}$.

$$\begin{aligned}
\frac{A_{2N}(s)}{2^{2N-1}} &= \sum_{k=1}^{2N} \frac{f_{s,2N}(\cos \frac{2k-1}{4N}\pi)}{T'_{2N}(\cos \frac{2k-1}{4N}\pi) \cos \frac{2k-1}{4N}\pi} \\
&= (1/2N) \sum_{k=1}^{2N} (-1)^{k+1} f_{s,2N}\left(\cos \frac{2k-1}{4N}\pi\right) \tan \frac{2k-1}{4N}\pi \\
&= (1/N) \sum_{k=1}^{N} (-1)^{k+1} f_{s,2N}\left(\cos \frac{2k-1}{4N}\pi\right) \tan \frac{2k-1}{4N}\pi.
\end{aligned} \tag{48}$$

Therefore, (41) follows from (47) and (48). □

5 $\mathbf{L_p(-1,1)}$-Asymptotics and Criteria for $\zeta(s) = 0$ and $\zeta(s) \neq 0$

$L_p(-1,1)$-asymptotic representations for $|\zeta(s)|$ are based on asymptotic (37) and estimates for the remainder terms.

It is easy to see that for $0 < p < \infty$,

$$||T_{2N}\Delta_s^*(N,\cdot)||_{L_p(-1,1)} \leq C(s,p)N^{-2}, \tag{49}$$

where $\Delta_s^*(N,y)$ is the second remainder term in (37), satisfying (38). Next, we find the asymptotic behavior of the first remainder term in (37). Let us set

$$\Phi_s(z) := \int_0^\infty \frac{t^{s+1}\mathrm{e}^{-t}}{(z^2+t^2)\cosh t}\mathrm{d}t = \frac{1}{z^2}\int_0^\infty \frac{t^{s+1}\mathrm{e}^{-t}}{(1+(t/z)^2)\cosh t}\mathrm{d}t, \quad \mathrm{Re}\, s > 0, z \in \mathbb{R}.$$

Lemma 12 *For* $\mathrm{Re}\, s > 0$ *and* $p \in (1/2,\infty)$,

$$\lim_{N\to\infty}(2N)^{1/p}||T_{2N}\Phi_s(2N\cdot)||_{L_p(-1,1)} = ||\cos(\cdot)\Phi_s||_{L_p(-\infty,\infty)} < \infty. \tag{50}$$

In particular,

$$||T_{2N}\Phi_s(2N\cdot)||_{L_p(-1,1)} \leq C(s,p)N^{-1/p}, \quad N = 1,2,\ldots \tag{51}$$

Proof. We first note that

$$||\cos(\cdot)\Phi_s||_{L_p(-\infty,\infty)} < ||\Phi_s||_{L_p(-\infty,\infty)} < \infty.$$

Indeed, for $\mathrm{Re}\, s > 0$ and $p \in (1/2,\infty)$,

$$||\Phi_s||^p_{L_p(-\infty,\infty)} = 2\left(\int_0^1 |\Phi_s(z)|^p\mathrm{d}z + \int_1^\infty |\Phi_s(z)|^p\mathrm{d}z\right)$$
$$\leq 2\left(\left(\int_0^1 t^{\mathrm{Re}\, s-1}\mathrm{e}^{-t}\mathrm{d}t\right)^p + \left(\int_0^\infty t^{\mathrm{Re}\, s+1}\mathrm{e}^{-t}\mathrm{d}t\right)^p\int_1^\infty z^{-2p}\mathrm{d}z\right) < \infty.$$

Next, we need the relation

$$\lim_{N\to\infty}(-1)^N T_{2N}(z/(2N)) = \cos z \tag{52}$$

uniformly on any interval $[-B,B]$. This relation follows from the Mehler–Heine formula [11, 8.21] for $\alpha = -1/2$, since $(-1)^N T_{2N}(z/(2N)) = T_N(1-z^2/(2N)^2)$ converges to $\cos z$ uniformly on any interval $[-B,B]$.

Let B be any number from $(0,\infty)$. Then denoting $I_{N,p} := ||T_{2N}\Phi_s(2N\cdot)||^p_{L_p(-1,1)}$ and using (52), we have

$$\begin{aligned}\limsup_{N\to\infty} 2NI_{N,p} &= \limsup_{N\to\infty}\int_{-2N}^{2N}|T_{2N}(z/(2N))\Phi_s(z)|^p\mathrm{d}z\\ &\le \limsup_{N\to\infty}\int_{-B}^{B}|T_{2N}(z/(2N))\Phi_s(z)|^p\mathrm{d}z+\int_{\mathbb{R}\setminus[-B,B]}|\Phi_s(z)|^p\mathrm{d}z\\ &= \int_{-B}^{B}|\cos z\,\Phi_s(z)|^p\mathrm{d}z+\int_{\mathbb{R}\setminus[-B,B]}|\Phi_s(z)|^p\mathrm{d}z.\end{aligned}$$

Next, using (52) we get

$$\liminf_{N\to\infty} 2NI_{N,p} = \liminf_{N\to\infty}\int_{-2N}^{2N}|T_{2N}(z/(2N))\Phi_s(z)|^p\mathrm{d}z \ge \int_{-B}^{B}|\cos z\,\Phi_s(z)|^p\mathrm{d}z.$$

Therefore,

$$\begin{aligned}\int_{-B}^{B}|\cos z\,\Phi_s(z)|^p\mathrm{d}z &\le \liminf_{N\to\infty} 2NI_{N,p} \le \limsup_{N\to\infty} 2NI_{N,p}\\ &\le \int_{-B}^{B}|\cos z\,\Phi_s(z)|^p\mathrm{d}z+\int_{\mathbb{R}\setminus[-B,B]}|\Phi_s(z)|^p\mathrm{d}z. \qquad (53)\end{aligned}$$

Letting $B\to\infty$ in (53), we arrive at (50). Inequality (51) follows directly from (50). □

We are now in a position to prove the following $L_p(-1,1)$-asymptotics for $|\zeta(s)|$.

Theorem 13 *Let* $\operatorname{Re} s>0$, $p\in(1/2,\infty)$, *and* $H_{2N+1}(y)=yT_{2N}(y)$.

(a) The following asymptotic relation holds

$$\begin{aligned}|\zeta(s)| &= \pi\left(\frac{\pi\Gamma(p+1)}{2^{p+1}\Gamma^2((p+1)/2)}\right)^{1/p}\left|\frac{2^{s-1}}{\sin s\pi\,\Gamma(s)(1-2^{1-s})}\right|\\ &\times \lim_{N\to\infty}(2N)^{\operatorname{Re} s}\left(\int_{-1}^{1}|f_{s,2N}(y)-L_{2N}(y,f_{s,2N},H_{2N+1})|^p\mathrm{d}y\right)^{1/p}. \qquad (54)\end{aligned}$$

(b) If $\zeta(s)=0$, *then*

$$\begin{aligned}&\lim_{N\to\infty}(2N)^{\operatorname{Re} s+1/p}\left(\int_{-1}^{1}|f_{s,2N}(y)-L_{2N}(y,f_{s,2N},H_{2N+1})|^p\mathrm{d}y\right)^{1/p}\\ &=(|\sin s\pi|/\pi)\,||\cos(\cdot)\Phi_s||_{L_p(-\infty,\infty)}. \qquad (55)\end{aligned}$$

Proof. We first assume that $p\in[1,\infty)$. Then multiplying both sides of (37) by $T_{2N}(y)$ and using (49) and (51), we have

$$
\begin{aligned}
||T_{2N}||_{L_p(-1,1)}|\zeta(s)| &= \pi\left|\frac{2^{s-1}}{\sin s\pi\Gamma(s)(1-2^{1-s})}\right| \\
&\quad\times(2N)^{\mathrm{Re}s}\left(\int_{-1}^{1}|f_{s,2N}(y)-L_{2N}(y,f_{s,2N},H_{2N+1})|^p\mathrm{d}y\right)^{1/p} \\
&\quad+O(N^{-1/p})+O(N^{-2}), \qquad N\to\infty. \qquad (56)
\end{aligned}
$$

A similar relation holds in case $1/2 < p < 1$ as well. In this case instead of the triangle inequality, we use the inequalities

$$
\begin{aligned}
&\int_{-1}^{1}|f(x)|^p\mathrm{d}x-\int_{-1}^{1}|g(x)|^p\mathrm{d}x \le \int_{-1}^{1}|f(x)+g(x)|^p\mathrm{d}x \\
&\le \int_{-1}^{1}|f(x)|^p\mathrm{d}x+\int_{-1}^{1}|g(x)|^p\mathrm{d}x. \qquad (57)
\end{aligned}
$$

Using (49), (50), and (57),we have for $p\in(1/2,1)$

$$
\begin{aligned}
||T_{2N}||^p_{L_p(-1,1)}|\zeta(s)|^p &= \pi^p\left|\frac{2^{s-1}}{\sin s\pi\Gamma(s)(1-2^{1-s})}\right|^p \\
&\quad\times(2N)^{p\mathrm{Re}s}\int_{-1}^{1}|f_{s,2N}(y)-L_{2N}(y,f_{s,2N},H_{2N+1})|^p\mathrm{d}y \\
&\quad+O(N^{-1})+O(N^{-2p}), \qquad N\to\infty. \qquad (58)
\end{aligned}
$$

Then it follows from (56) and (58) that (54) holds if

$$
\lim_{N\to\infty}||T_{2N}||_{L_p(-1,1)}=\left(\frac{\pi\Gamma(p+1)}{2^{p+1}\Gamma^2((p+1)/2)}\right)^{1/p}. \qquad (59)
$$

Indeed, by the Fejer Lemma [4],

$$
\lim_{N\to\infty}||T_{2N}||^p_{L_p(-1,1)}=\lim_{N\to\infty}\int_0^{\pi}|\cos 2Nz|^p\sin z\mathrm{d}z=(4/\pi)\int_0^{\pi/2}(\cos z)^p\mathrm{d}z.
$$

This implies (59) and completes the proof of statement (a) of the theorem. Statement (b) follows directly from (37) and (50). □

As a corollary of Theorem 13, we obtain new criteria for $\zeta(s)=0$ and $\zeta(s)\neq 0$ in terms of the L_p-interpolation error of $f_{s,2N}$.

Corollary 14 *Let $H_{2N+1}(y)=yT_{2N}(y)$. Then the following statements hold:*

(a) $\zeta(s)=0$ for $0<\mathrm{Re}s<1$ if and only if for all $p\in(1/2,\infty)$ and $N=1,2,\ldots$,

$$
\begin{aligned}
C_1(s,p)N^{-\mathrm{Re}s-1/p} &\le \left(\int_{-1}^{1}|f_{s,2N}(y)-L_{2N}(y,f_{s,2N},H_{2N+1})|^p\mathrm{d}y\right)^{1/p} \\
&\le C_2(s,p)N^{-\mathrm{Re}s-1/p}.
\end{aligned}
$$

(b) $\zeta(s) \neq 0$ *for* $0 < \mathrm{Re}\, s < 1$ *if and only if for all* $p \in (1/2, \infty)$ *and* $N = 1, 2, \ldots$,

$$C_3(s,p)N^{-\mathrm{Re}\, s} \leq \left(\int_{-1}^{1} |f_{s,2N}(y) - L_{2N}(y, f_{s,2N}, H_{2N+1})|^p \mathrm{d}y \right)^{1/p} \leq C_4(s,p)N^{-\mathrm{Re}\, s}.$$

6 Remarks

Remark 1. Representation (4) is an non-analytic analog of the Hermite interpolation error formula for analytic functions [12]. Bernstein [1] was the first author who extended this formula to the non-analytic function $f(y) = (1-y)^s$, $s > 0$ on $[-1,1]$. The author [5–7] discussed various versions of Bernstein's result; in particular, an extension of Bernstein's formula to $\mathrm{Re}\, s > 0$ was given in [7]. Formula (11) for a more general class of even functions $f(y) = \int_0^\infty \frac{\mathrm{d}\mu(t)}{t^2+y^2}$ and for $H_{n+1} = T_{n+1}$ was established by Lubinsky [10].

Note that formula (4) can be extended to Hermite interpolation of functions f of the more general form $f(y) = \int_{\mathbb{R}} \frac{\mathrm{d}\mu(t)}{t-iy}$ (see [8]).

Remark 2. Formula (37) is similar to the asymptotic representation for the Dirichlet beta function $\beta(s) = \sum_{n=0}^{\infty}(-1)^n(2n+1)^{-s}$, $\mathrm{Re}\, s > 0$, in terms of the Lagrange interpolation error of $|y|^s$ at the Chebyshev nodes, which was found in [7] in connection with study of pointwise rapid convergence of polynomial approximation. The case $s > 0$ was discussed earlier in [5].

Remark 3. We believe that criterion (a) for $\zeta(s) = 0$ and criterion (b) for $\zeta(s) \neq 0$ in Corollary 14 are the first ones in terms of the Lagrange interpolation error. There are numerous other criteria that have been developed for the last 150 years in connection with the celebrated Riemann hypothesis (see survey [2]).

Remark 4. A more general approach to finding new asymptotic formulae for $\zeta(s)$ and $\beta(s)$ was developed in [8]. In particular, we introduce in [8] a general class of nodes for Lagrange and Hermite interpolation that allow various asymptotic representations like (37), (39), (41), (54), and (55).

References

1. Bernstein, S.N.: Extremal Properties of Polynomials and the Best Approximation of Continuous Functions of a Single Real Variable. State United Scientific and Technical Publishing House, Moscow (1937) (in Russian).
2. Conrey, B.: The Riemann Hypothesis. Notices of AMS, **50**, No. 3, 341-353 (2003)
3. Erdelyi, A., Magnus, W., Oberhettinger, F., Tricomi, F.G.: Higher Transcendental Functions, Vol. I. McGraw-Hill, New York (1953)

4. Fejér, L.: Lebesguesche Konstanten und devergente Fourierreihen. JRAM, **138**, 22-53 (1910)
5. Ganzburg, M.I.: The Bernstein constant and polynomial interpolation at the Chebyshev nodes. J. Approx. Theory, **119**, 193-213 (2002)
6. Ganzburg, M.I.: Strong asymptotics in Lagrange interpolation with equidistant nodes. J. Approx. Theory, **122**, 224-240 (2003)
7. Ganzburg, M.I.: Polynomial interpolation, an L-function, and pointwise approximation of continuous functions with equidistant nodes. J. Approx. Theory, **153**, 1-18 (2008)
8. Ganzburg, M.I.: Polynomial interpolation formulae and asymptotic representations of zeta functions. Manuscript (2010)
9. Gradshtein, I.S., Ryzhik, I.M.: Tables of Integrals, Series, and Products. 5th Edition, Academic Press, San Diego (1994)
10. Lubinsky, D.S.: Best approximation and interpolation of $(1+(ax)^2)^{-1}$ and its transforms, J. Approx. Theory **125**, 106-115 (2003).
11. Szegö, G.: Orthogonal Polynomials. Colloquium Publications, **23**, Amer Math. Soc., Providence, RI (1975)
12. Walsh, J.L.: Interpolation and Approximation by Rational Functions in the Complex Domain. 5th Edition, Amer. Math. Soc. Colloq. Publ. **20**, Providence, RI (1969)
13. Watson, G.N.: A Treatise on the Theory of Bessel Functions. Cambridge University Press, Cambridge, UK (1966)

Active Geometric Wavelets

Itai Gershtansky and Shai Dekel

Abstract We present an algorithm for highly geometric sparse representation. The algorithm combines the adaptive Geometric Wavelets method with the Active Contour segmentation to overcome limitations of both algorithms. It generalizes the Geometric Wavelets by allowing to adaptively construct wavelets supported on curved domains. It also improves upon the Active Contour method that can only be used to segment a limited number of objects. We show applications of this new method in medical image segmentation.

1 Introduction

The *Active Contour (Level-Set)* method is a well known approach for image segmentation [2, 10, 11]. It is general enough to allow definition of different cost functions, in order to identify different types of objects in the image, but at the same time, it is also relatively simple to implement. It is also popular because it provides actual segmentation represented by continuous curves, whereas other "edge detection" methods only compute the probability that a pixel is an "edge" pixel or that a pixel belongs to an "object."

The main problem of the existing Active Contour methods is that it is limited to segmenting out a small number of objects (see an attempt to fix this in [5]). Our approach tries to overcome this issue, by applying local segmentations locally and recursively. The result is a multiresolution tree structure of disjoint sub-regions over which one may construct a highly geometric wavelet representation of the image.

Itai Gershtansky
School of Mathematical Sciences, Tel Aviv University, Tel Aviv 69978, Israel,
e-mail: itai.gershtansky@gmail.com

Shai Dekel
GE Healthcare and School of Mathematical Sciences, Tel Aviv University, Tel Aviv 69978, Israel,
e-mail: Shai.Dekel@ge.com

M. Neamtu and L. Schumaker (eds.), *Approximation Theory XIII: San Antonio 2010*, Springer Proceedings in Mathematics 13, DOI 10.1007/978-1-4614-0772-0_7,

This approach generalizes the previous construction of *Geometric Wavelets (GW)* [5], where the recursive subdivision was applied using only straight lines, producing convex polygonal regions. We now recall the GW algorithm:

Given a function $f:[0,1]^2 \to [0,1]$ over the unit cube, it is subdivided using a line segment to two sub-regions Ω_1, Ω_2 such that

$$\|f - Q_{\Omega'}\|^2_{L_2(\Omega')} + \|f - Q_{\Omega''}\|^2_{L_2(\Omega'')}, \tag{1}$$

is minimized, where $Q_{\Omega'}, Q_{\Omega''}$ are polynomials of some fixed low order. Note that for each candidate bisection, the optimal polynomials are given by the least squares method. This process continues recursively, until a stopping criterion is met, typically when (1) is below a given threshold. Observe that the sub-regions are always convex polyhedral domains, which is a crucial property when approximating with piecewise polynomials (see discussion in Sect. 2).

The result of this algorithm is a *Binary Space Partition (BSP)* tree $\mathscr{P}$, composed of pairs $\{(\Omega, Q_\Omega)\}$: the sub-regions and the approximating polynomials constructed over them. The root of the tree is $\left([0,1]^2, Q_{[0,1]^2}\right)$, where $Q_{[0,1]^2}$ is the approximation of the function over the unit cube. This tree can be used to define an adaptive Geometric Wavelet decomposition of the function in the following way. If (Ω, Q_Ω) is the father of $(\Omega', Q_{\Omega'})$, define

$$\psi_{\Omega'} := \psi_{\Omega'}(f) := 1_{\Omega'}(Q_{\Omega'} - Q_\Omega), \tag{2}$$

as the geometric wavelet associated with the sub-region Ω' and the function f. The low resolution component, associated with the root of the BSP tree is

$$\psi_{[0,1]^2} := Q_{[0,1]^2}. \tag{3}$$

The wavelets (2) are in fact a "local difference" components that belong to the detail space between two levels in the BSP tree, a "low resolution" level associated with Q_Ω and a "higher resolution" level associated with $Q_{\Omega'}$. The GW method follows the classical procedure of n-term wavelet approximation [4, 6]: The importance of the wavelet is measured by its L_2 norm, and so we reorder:

$$\left\|\psi_{\Omega_{k_1}}\right\|_2 \geq \left\|\psi_{\Omega_{k_2}}\right\|_2 \geq \left\|\psi_{\Omega_{k_3}}\right\|_2 \geq \cdots \tag{4}$$

Given an integer $n \in N$, we have the n-term approximation

$$\psi_{[0,1]^2} + \sum_{i=1}^{n} \psi_{\Omega_{k_i}}. \tag{5}$$

It can be shown that under mild condition on the BSP tree and the function f,

$$f = \sum_{\Omega \in \mathscr{P}} \psi_\Omega(f).$$

Since edge singularities in images are in general not line segments, the above method will require bisections at several levels of the BSP tree to approximate them. To this end, we enhance the method of [5], by using more advanced segmentation algorithms at each recursive subdivision step. Instead of minimizing (1), we minimize a Mumford–Shah type functionals such as [2]

$$\|f - Q_{\mathrm{in}(\gamma)}\|^2_{L_2(\mathrm{in}(\gamma))} + \|f - Q_{\mathrm{out}(\gamma)}\|^2_{L_2(\mathrm{out}(\gamma))} + \mu \cdot \mathrm{length}(\gamma), \tag{6}$$

where γ is a closed curve and $\mathrm{in}(\gamma)$ and $\mathrm{out}(\gamma)$ are its inside and outside domains, respectively. The first two terms are the penalties for approximation over the two sub-regions and the third term is the penalty for curve length. Again, for each fixed curve, the approximation polynomials are uniquely determined by the least squares method. There are numerous variants to (6) and numerical algorithms to compute them. These algorithms are all iterative and most of them are highly sensitive to the input initial curve. In our algorithm we also use a more localized level-set variation [10] that works well if the initialization curve is "close" to the solution curve. For some given $\varepsilon > 0$ and any smooth function ϕ, let $\delta(\phi)$ be an approximation to the "Dirac" of ϕ (controlled by the zero level set of ϕ)

$$\delta(\phi) := \begin{cases} \frac{1}{2\varepsilon}\left(1 + \cos\left(\frac{\pi\phi(x)}{\varepsilon}\right)\right), & 0 \le \phi(x) \le \varepsilon, \\ 0, & \text{otherwise}, \end{cases}$$

and denote for some pre-determined radius r

$$\beta_r(x,y) := \begin{cases} 1, & |x-y| \le r, \\ 0, & \text{otherwise}. \end{cases}$$

Then, a "local" energy functional is given by

$$E(\phi) := \int_\Omega \delta\phi(x) \int_\Omega \beta_r(x,y) F(I(y), \phi(y))\,\mathrm{d}y\mathrm{d}x + \mu \int_\Omega \delta\phi(x) |\nabla\phi(x)|\,\mathrm{d}x, \tag{7}$$

where F is an "internal" energy term (see the details of [10]).

Our *Active Geometric Wavelet (AGW)* algorithm for sparse representation is thus composed of 3 steps:

1. *Initialization* – In the first step, we try to find connected groups of pixels with similar values. The outer boundaries of these connected groups are used as initial guesses for the segmentation algorithm in the second step.
2. *Construction of the geometric BSP tree* – Since the contours computed in step 1 are expected to be close to objects in the image, the segmentation is computed with the localized functional (7). In case this segmentation gives an error larger than a given threshold, we switch to the functional (6) and continue the subdivision process. The recursive application of these active contour segmentations creates a geometric BSP tree structure over the image.
3. *Creation of the n-term approximation* – An approximating wavelet sum is created according to (4) and (5).

The paper is organized as follows. In Sect. 2, we provide the theoretical foundation for the AGW method. In Sect. 3, we describe the algorithm in detail and in Sect. 4 we provide numerical examples for Computed Tomography (CT) images.

2 Theoretical Background

2.1 A Jackson Estimate for Piecewise Polynomial Approximation Using Non-convex Domains

Let $\Pi_{r-1}\left(\mathbb{R}^d\right)$ denote the multivariate polynomials of total degree $r-1$ (order r) in d variables. Our objective is to approximate a given function by low order polynomials over a possibly non-convex sub-domains. For polynomial approximation over a single convex domain there is a complete characterization of the degree of approximation by smoothness measures such as the modulus of smoothness and K-functional, where the constants are universal over all convex domains (see [5] and references therein). However, the situation is essentially different when approximating over non-convex domains (see examples in [9]).

In the following we define the notion of an α-class that quantifies how "close" a given domain is to being convex. We then give a Jackson estimate for an n-term approximation using piecewise polynomials over sub-domains all in the same α-class.

Definition 1. Let $\alpha \geq 1$. We say that a bounded domain $\Omega \subset \mathbb{R}^d$ belongs to the α-class if there exist an ellipsoid θ, with center v_θ, such that $\theta \subseteq \Omega \subseteq \theta_\alpha$, where θ_α is the α-blowup of θ

$$\theta_\alpha := \{v_\theta + \alpha\,(x - v_\theta)\ : x \in \theta\}.$$

John's Lemma [7] proves that all bounded convex domains in $\mathbb{R}^d$ are in the $d-$class. In some sense, the notion of the α-class improves upon the "Chunkiness Parameter" [6] which is frequently used in the Finite Element Method literature to evaluate the shape of a given domain for the purpose of local polynomial approximation. The "Chunkiness Parameter" relates to the ratio between a minimal enclosing ball and maximal contained ball, so in this sense using ellipsoids is better for long and thin, but possibly non-convex domains. The following lemma is a generalization of Lemma 2.4b from [5], where it was proved for convex domains.

Lemma 1. *For any Ω that belongs to α-class, $P \in \Pi_{r-1}\left(\mathbb{R}^d\right)$ and $0 < p,q \leq \infty$ we have*

$$\|P\|_{L_q(\Omega)} \sim |\Omega|^{1/q-1/p}\, \|P\|_{L_p(\Omega)},$$

with constants of equivalency depending on d,r,p,q and α.

Proof. By the equivalency of finite dimensional Banach spaces, we have that $\|P\|_{L_p(B(0,1))} \sim \|P\|_{L_q(B(0,\alpha))}$, for any polynomial $P \in \Pi_{r-1}\left(\mathbb{R}^d\right)$, where $B(0,l)$

$= \{x \in \mathbb{R}^n : |x| \le l\}$, with constants of equivalency depending only on p,q,d,r and α. Since Ω is in the α-class, there exists an ellipsoid $\theta \subseteq \Omega$ and an affine transformation A_θ, $A_\theta x = M_\theta x + v_\theta$, satisfying $A_\theta(B(0,1)) = \theta$, for which

$$B(0,1) \subseteq A_\theta^{-1}(\Omega) \subseteq B(0,\alpha).$$

Therefore,

$$\|P\|_{L_q(\Omega)} = |\det M_\theta|^{1/q} \|P(A_\theta \cdot)\|_{L_q(A_\theta^{-1}(\Omega))} \le |\det M_\theta|^{1/q} \|P(A_\theta \cdot)\|_{L_q(B(0,\alpha))}$$

$$\le c\, |\det M_\theta|^{1/q} \|P(A_\theta \cdot)\|_{L_p(B(0,1))} \le c\, |\det M_\theta|^{1/q} \|P(A_\theta \cdot)\|_{L_p(A_\theta^{-1}(\Omega))}$$
$$\le c\, |\det M_\theta|^{1/q-1/p} \|P\|_{L_p(\Omega)}. \qquad \square$$

We can now apply the machinery introduced in [8] to obtain a Jackson estimate. The following theorems are in fact Theorem 3.3 and Theorem 3.4 from [8], formulated in a general enough manner that allows us to apply them for the case of piecewise polynomial approximation over general subdomains.

Theorem 1. *Suppose $\{\Phi_m\}$ is a sequence of functions in $L_p(\mathbb{R}^d)$, $0 < p < \infty$, which satisfies the following additional properties when $1 < p < \infty$*

1. *$\Phi_m \in L_\infty(\mathbb{R}^d)$, supp$(\Phi_m) \subset E_m$ with $0 < |E_m| < \infty$, and* $\|\Phi_m\|_\infty \le c_1 |E_m|^{-1/p} \|\Phi_m\|_p$.
2. *If $x \in E_m$, then*

$$\sum_{x \in E_j, |E_j| \ge |E_m|} \left(\frac{|E_m|}{|E_j|}\right)^{1/p} \le c_1,$$

where the summation is over all indices j for which E_j satisfies the indicated conditions.

Denote (formally) $f := \sum_m \Phi_m$ and assume that for some $0 < \tau < p$

$$N(f) := \left(\sum_m \|\Phi_m\|_p^\tau\right)^{1/\tau} < \infty. \tag{8}$$

Then $\sum_m |\Phi_m(\cdot)| < \infty$ a.e. on $\mathbb{R}^d$, and hence, f is well defined. Furthermore, if $1 \le p < \infty$, condition (8) can be replaced by the weaker condition

$$N(f) := \left\|\left\{\|\Phi_m\|_p\right\}\right\|_{wl_\tau} < \infty, \tag{9}$$

where $\|\{x_m\}\|_{wl_\tau}$ denotes the weak l_τ-norm of the sequence $\{x_m\}$:

$$\|\{x_m\}\|_{wl_\tau} := \inf\left\{M : \#\left\{m : |x_m| > Mn^{-1/\tau}\right\} < n \text{ for } n = 1,2,\dots\right\}.$$

Theorem 2. *Under the hypothesis of Theorem 2.3, suppose* $\{\Phi_m^*\}_{j=1}^{\infty}$ *is a rearrangement of the sequence* $\{\Phi_m\}$ *such that* $\|\Phi_1^*\|_p \geq \|\Phi_2^*\|_p \geq \ldots$. *Denote* $S_n := \sum_{j=1}^{n} \Phi_1^*$. *Then*

$$\|f - S_n\|_p \leq cn^{-\beta} N(f) \text{ with } \beta = 1/\tau - 1/p, \tag{10}$$

where $c = 1$, *if* $0 < p \leq 1$ *and* $c = c(\beta, p, c_1)$, *if* $1 < p < \infty$. *Furthermore, the estimate remains valid if condition (8) can be replaced by (9) when* $1 \leq p < \infty$.

We first observe that if the AGW method uses piecewise constants, then we actually have equality in condition (1) of Theorem 1 for any type of domain. For higher order polynomials, we need to assume that the domains are in the α-class for some fixed α and then we obtain condition (1) by application of Lemma 1. Also, if each step of the recursive subdivision bisects a domain into sub-domains of relatively "substantial" area, then also condition (2) of Theorem 1 is satisfied. The quantity $N(f)$ should be considered as a "geometric sparsity gauge" for the function f. It will be typically very small for cartoon-type images, if the domains of the Active Geometric Wavelets are aligned with the curve singularities. In these settings, Theorem 2 says that a "greedy" n-term approximation based on AGW performs well.

2.2 Adaptive Local Selection of the Weight μ

One of the key elements of the AGW algorithm is a correct choice of the parameter μ in (6). A possible strategy is the following: As an initial guess, choose a large value of the parameter, one that gives an empty segmentation, that is, a segmentation where all the pixels are considered to be "outside." Then, gradually, diminish the value of μ, until some segmentation is achieved. As a motivation for this strategy we consider minimizing the Chan–Vese functional (6) over the simple indicator function of a circle. More formally, suppose we have an function I defined on the cube $[0,1]^2$. We would like to minimize

$$M(\mu) = M(\gamma, c_1, c_2, \mu) = \int_{\text{in}(\gamma)} (I - c_1)^2 + \int_{\text{out}(\gamma)} (I - c_2)^2 + \mu \cdot \text{length}(\gamma), \tag{11}$$

where $\gamma : [0,1] \to [0,1]^2$ is any closed curve. in (γ) is the region (or union of regions) that is (are) inside γ (including the boundary of γ) and out (γ) is the complement of $in(\gamma)$ in $[0,1]^2$. More specifically, we wish to investigate the dependence of the solution on μ.

Theorem 3. *Let* $I : [0,1]^2 \to [0,1]$ *the characteristic function of a circle* $C = \left\{x \in [0,1]^2 : |x - x_0| \leq a\right\} \subset [0,1]^2$, *where* $0 < a < \sqrt{0.5/\pi}$. *Then with* $\mu_0 = 0.5a\left(1 - \pi a^2\right)$ *we have*

$$\min M(\mu) = \begin{cases} \pi a^2 \left(1 - \pi a^2\right), & \mu_0 \leq \mu, \\ 2\mu\pi a, & 0 \leq \mu \leq \mu_0, \end{cases}$$

$$\arg\min M(\mu) = \begin{cases} \gamma_\emptyset, & \mu_0 \leq \mu, \\ \gamma_C, & 0 \leq \mu \leq \mu_0, \end{cases} \tag{12}$$

where $\gamma_\emptyset$ is the empty curve, for which $out(\gamma_\emptyset) = I$, $in(\gamma_\emptyset) = \emptyset$ and $\gamma_C = \partial C$.

Remark 1. We restrict ourselves to the case where the radius of the circle is sufficiently small, i.e. $0 < a < \sqrt{0.5/\pi}$, so as to keep the image boundaries far from the object in question in order not to deal with some geometric issues that arise from such proximity. This is not a significant restriction and allows a simpler proof.

Proof. Proof of Theorem 3. We use the following notation:

$$\Gamma_{\text{in}}(\Omega) := \{\gamma : in(\gamma) \cap \Omega \subseteq \Omega\}, \ \Gamma_{\text{out}}(\Omega) := \{\gamma : in(\gamma) \cap \Omega = \emptyset\},$$

for the set of all curves that are completely inside Ω, and the set of all curves that are completely outside it, respectively. Next, we calculate the penalty for approximation in a region where the function takes two values: an area of k_1 with value a_1 and an area of k_2 with value a_2. The average is $(k_1 a_1 + k_2 a_2)/(k_1 + k_2)$ and the penalty is

$$k_1 \left(\frac{k_1 a_1 + k_2 a_2}{k_1 + k_2} - a_1\right)^2 + k_2 \left(\frac{k_1 a_1 + k_2 a_2}{k_1 + k_2} - a_2\right)^2 = \frac{k_1 k_2 (a_1 - a_2)^2}{(k_1 + k_2)}.$$

Let us first find $\arg\min_{\gamma \in \Gamma_{\text{in}}(C)} M(\gamma, \mu)$. If $\gamma \in \Gamma_{\text{in}}$, then $length(\gamma) \leq 2\pi a$, otherwise $M(\gamma_C) < M(\gamma)$, because the penalty for length for γ_C is smaller than that of γ and the sum of penalties for inner and outer approximation for γ cannot be smaller than that of γ_C, which is 0. Moreover, if the considered curve, γ, is not a circle, then according to the isoperimetric inequality, a circular curve with the same area will give a smaller value for M, because it will have a smaller length penalty. Thus, we may consider only the circular curves in Γ_{in}.

Denote the radius of γ as r and M becomes a function of r with $0 \leq r \leq a$. In this case, according to the above calculation,

$$M(r) = \frac{\left(1 - \pi a^2\right)\left(\pi a^2 - \pi r^2\right)}{1 - \pi r^2} + 2\pi\mu r.$$

For a given μ we need to find $\inf_r M(r)$. We compute

$$M'(r) = \frac{-2\pi r\left(1-\pi a^2\right)\left(1-\pi r^2\right) + 2\pi r\left(1-\pi a^2\right)\left(\pi a^2 - \pi r^2\right)}{\left(1-\pi r^2\right)^2} + 2\pi\mu$$

$$= \frac{2\pi r\left(1-\pi a^2\right)\left(\pi a^2 - 1\right)}{\left(1-\pi r^2\right)^2} + 2\pi\mu = 2\pi\left(\mu - r\left(\frac{1-\pi a^2}{1-\pi r^2}\right)^2\right).$$

and

$$M''(r) = -2\pi \left(\frac{1-\pi a^2}{1-\pi r^2}\right)^2 \left(\frac{1+3\pi r^2}{1-\pi r^2}\right) < 0.$$

The second derivative is always negative, which implies that the first derivative is strictly decreasing from $M'(0) = 2\pi\mu$ to $M'(a) = 2\pi(\mu - a)$. If $\mu \geq a$, M' is always non-negative for $0 \leq r \leq a$, and so M is non decreasing. Therefore, the minimum is obtained at $r = 0$. Since $\mu_0 < a \leq \mu$, this agrees with the first case in (12). If $\mu < a$, there's a parameter r for which $M'(r) = 0$ and M has a local maximum, since $M'' < 0$. Therefore, a global minimum is achieved at one of the end points $r = 0$, with $M(0) = \pi a^2 (1 - \pi a^2)$ or $r = a$, with $M(a) = 2\mu\pi a$. If $M(0) < M(a)$, then $\pi a^2 (1 - \pi a^2) < 2\mu\pi a$ which is exactly the condition $\mu_0 < \mu$, and this again agrees with the first case in (12). If $M(a) \leq M(0)$, then $\mu \leq \mu_0$ and this agrees with the second case in (12).

Next, we deal with the case where γ is not necessarily contained in the circle. We now describe M as a function of three variables: p, q and l, where

$$p := |\text{in}(\gamma) \cap C|, q := \left|\text{in}(\gamma) \cap \left([0,1]^2 \backslash C\right)\right|, l := \text{length}(\gamma).$$

The functional (11) now take the form

$$M(p,q,l) = \frac{(1 - \pi a^2 - q)(\pi a^2 - p)}{1 - \pi a^2} + \frac{pq}{p+q} + \mu \cdot l,$$

where we observe that not all non-negative triplets (p,q,l) are geometrically feasible. Next, we define a functional of two variables

$$Y(p,q) := \frac{(1 - \pi a^2 - q)(\pi a^2 - p)}{1 - \pi a^2} + \frac{pq}{p+q} + \mu \cdot \sqrt{4\pi(p+q)}.$$

Observe that by the isoperimetric inequality, for any curve γ, one has $length(\gamma) \geq \sqrt{4\pi(p+q)}$, with equality for circles, and so $Y(p,q) \leq M(\gamma)$. Observe that if it is possible for some fixed pair p, q, to choose γ to be a circle, then this would imply $Y(p,q) = \min_{p,q,l} M(p,q,l)$ over feasible triplets. However, there are cases where this is not possible, such as where γ contains C and almost all of $I \backslash C$, with $p = \pi a^2$ and $q = 1 - \pi a^2 - \varepsilon$, for some small ε. In this case, γ cannot have the shape of a circle.

We now minimize Y over $A = [0, \pi a^2] \times [0, 1 - \pi a^2]$. We compute

$$\frac{\partial Y}{\partial p} = \left(\frac{q}{p+q}\right)^2 - \left(\frac{1-\pi a^2 - q}{1-p-q}\right)^2 + 2\pi\mu(4\pi(p+q))^{-\frac{1}{2}},$$

$$\frac{\partial Y}{\partial q} = \left(\frac{p}{p+q}\right)^2 - \left(\frac{\pi a^2 - p}{1-p-q}\right)^2 + 2\pi\mu(4\pi(p+q))^{-\frac{1}{2}},$$

and then

$$\frac{\partial^2 Y}{\partial p^2} = -2q^2 (p+q)^{-3} - 2\left(1-\pi a^2 - q\right)(1-p-q)^{-3} - 4\pi^2 \mu \left(4\pi (p+q)\right)^{-3/2},$$

$$\frac{\partial^2 Y}{\partial q^2} = -2p^2 (p+q)^{-3} - 2\left(\pi a^2 - p\right)(1-p-q)^{-3} - 4\pi^2 \mu \left(4\pi (p+q)\right)^{-3/2}.$$

Observe that for $p > 0$ and $q > 0$, we have that $\partial^2 Y / \partial^2 p, \partial^2 Y / \partial^2 p < 0$. For a point to be a minimum it is necessary that $\partial^2 Y / \partial^2 p, \partial^2 Y / \partial^2 p > 0$, so there are no internal minimum points. On the lines $\left[0, \pi a^2\right] \times \left(1 - \pi a^2\right)$ and $\pi a^2 \times \left[0, 1 - \pi a^2\right]$, we have

$$\frac{\partial Y}{\partial p}\left(p, 1 - \pi a^2\right) > 0, \ \frac{\partial Y}{\partial q}\left(\pi a^2, q\right) > 0,$$

so the minimum of Y is on the union of the two lines $A_{\text{in}} = \left[0, \pi a^2\right] \times 0$ and $A_{\text{out}} = 0 \times \left[0, 1 - \pi a^2\right]$. Furthermore, for every point $z = (0, q) \in A_{\text{out}}$, the point $z' = (q, 0) \in A_{\text{in}}$ satisfies $Y\left(z'\right) < Y(z)$, because

$$Y\left(z'\right) = \pi a^2 - q < \pi a^2 - \frac{\pi a^2}{1 - \pi a^2} q = Y(z),$$

where we use the condition $\pi a^2 < 0.5$. The conclusion is that the minimum of Y is attained on A_{in}. On this line, M and Y have the same value, so the minimizer of Y is either $(p,q) = (0,0)$ or $(p,q) = \left(\pi a^2, 0\right)$, as shown before.

Finally, the minimum of Mover A is also attained at $(p,q) = (0,0)$ or $(p,q) = \left(\pi a^2, 0\right)$; otherwise, the minimum for M is attained at some point $z \in A \backslash A_{\text{in}}$. Denote by z_0 the point for which Y attains its minimum on A_{in} (either $(p,q) = (0,0)$ or $(p,q) = \left(\pi a^2, 0\right)$), then, $M(z_0) = Y(z_0) < Y(z) \leq M(z)$, which is a contradiction. □

We conclude from the last result that for the simple case of a characteristic image of a circle, the approach of taking a large value of μ, i.e. $\mu > \mu_0(a)$, then reducing it until a non-empty segmentation is achieved ($\mu \leq \mu_0$), is indeed an approach that gives the required segmentation.

3 Overview of the AGW Algorithm

The algorithm is composed of three stages: Initialization, construction of the BSP tree and building an approximation. In the first step, we try to find contours that will serve as initial guesses for the segmentations. Since the level-set method is sensitive to the initialization, starting from a good initial guess is critical to the success of the algorithm. Therefore, we begin by searching for groups of pixels with relatively similar grey-level and sorts these groups according to their set size. Sets with size smaller than a threshold are discarded. The result of applying this step on standard test images can be seen in Figs. 1 and 2 below.

Fig. 1: Initial pixel groups for the "Peppers" image

Fig. 2: Initial pixel groups for the "Cameraman" image

In Fig. 3, we see a Computed Tomography (CT) image and the pixel groups computed for this image. The goal in medical imaging is to segment correctly the various internal organs and perform certain measurements and analysis. We see that some key organs such as the kidneys and spine were not identified, since the imaging characteristics of these organs have higher variability. Therefore, to correctly identify initial pixel groups associated with these organs, we applied an anisotropic diffusion algorithm [12] to sharpen the edges and smooth the areas between them and then computed the pixel groups on this pre-processed image (Fig. 4).

In the second step, the algorithm builds the BSP tree. It minimizes the "local" functional (7) with the outer contours of the pixel groups found in the first step, starting with the largest group and continuing in a diminishing order. Each of these iterations gives a bisection of a sub-region of the picture to an object and background and adds two new sibling nodes to the BSP tree at an arbitrary level. If a pixel group is segmented, but the approximation error is above some required threshold,

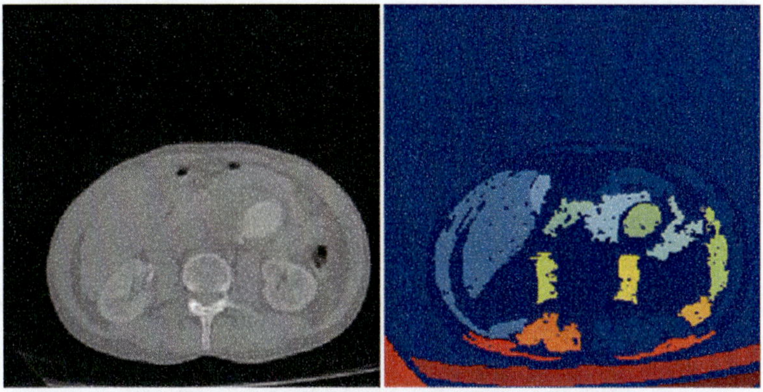

Fig. 3: Initial pixel groups for a CT image

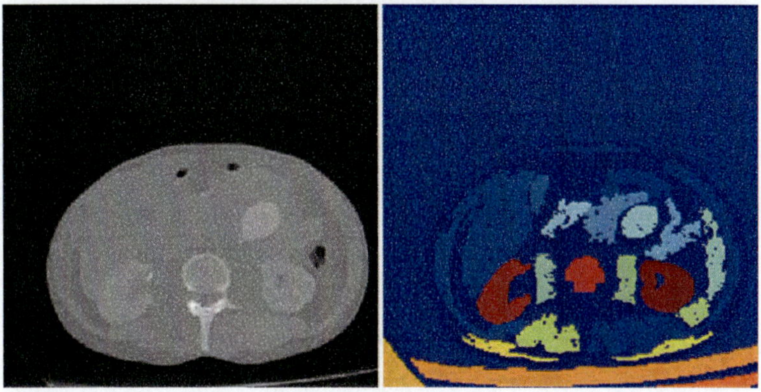

Fig. 4: CT image after anisotropic diffusion and the pixel groups computed from it

the algorithm continues to bisect it using a grid of circles as the initial guess and minimizing the "global" functional (6).

In Fig. 5 we see on the left the initial guess for the first segmentation, obtained from the largest pixel group and on the right, the segmentation computed from this initial guess. We see that the segmentation did not correctly segment the liver. This is exactly the weak point of a regular Active Contour algorithm that our method solves, because since the approximation error in this domain is found to be large, this domain will be further subdivided. In Fig. 6, we see on the left the grid of circles that serves as an initialization for the active contour segmentation of the first domain and on the right we see the correct segmentation appearing at the second level of the BSP tree.

The minimization process is stopped when the value of the functional is fluctuating about a certain value for a number of iterations. The output of this step is a BSP tree and a corresponding set of geometric wavelets. In the final step, the Active

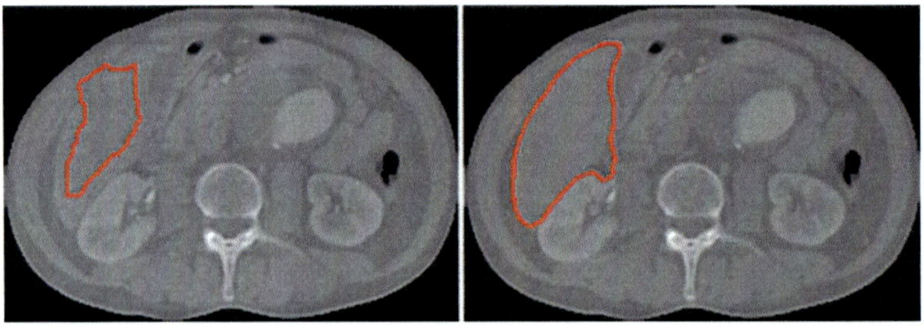

Fig. 5: Initial guess for segmentation and the first level segmentation obtained from it

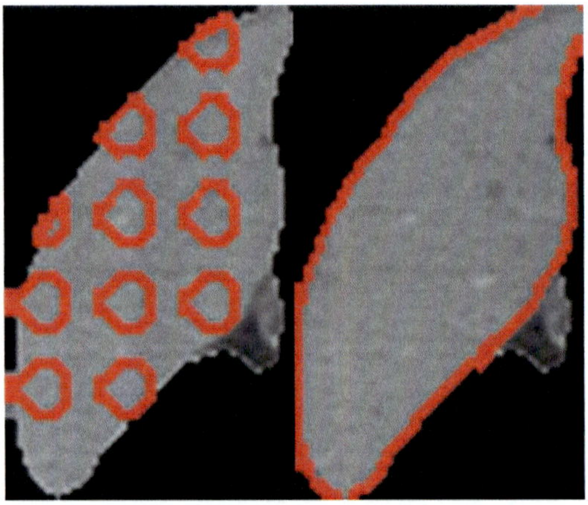

Fig. 6: Initialization of second level segmentation and segmentation at second level obtained from it

Geometric Wavelets are ordered by their norm as in (4) and an approximation is created from the low resolution component and the largest n terms.

To summarize, this is the AGW algorithm:

1. *Creation of initial pixel groups:*
   (a) *(Optional) Create a pre-processed input image for this step by applying anisotropic diffusion to the original image.*
   (b) *For each pixel p in the picture*
      – *If not part of a pixel group, create a new candidate pixel group and add p to the group,*
      – *For every unprocessed pixel q in group: if one of its 4-connected neighbors rdoes not belong to another group and* $|I(r) - I(q)| < \varepsilon$ *(for some threshold* ε*), add rto the group.*
   (c) *Sort groups according to size and discard groups whose size is smaller than a threshold.*

2. *Initialize a BSP tree with the root* $\left\{\left(I, Q_{[0,1]^2}\right), \psi_{[0,1]^2}\right\}$.
3. *For every leaf* $\{\Omega, \psi_\Omega\}$ *in the tree, if the approximation error is larger than a threshold*

   (a) *Create an initialization curve. If the domain* Ω *contains pixel groups from step 1, then the initialization curve is determined from the largest such contained group. Else the initialization is a grid of small circles intersected with* Ω.

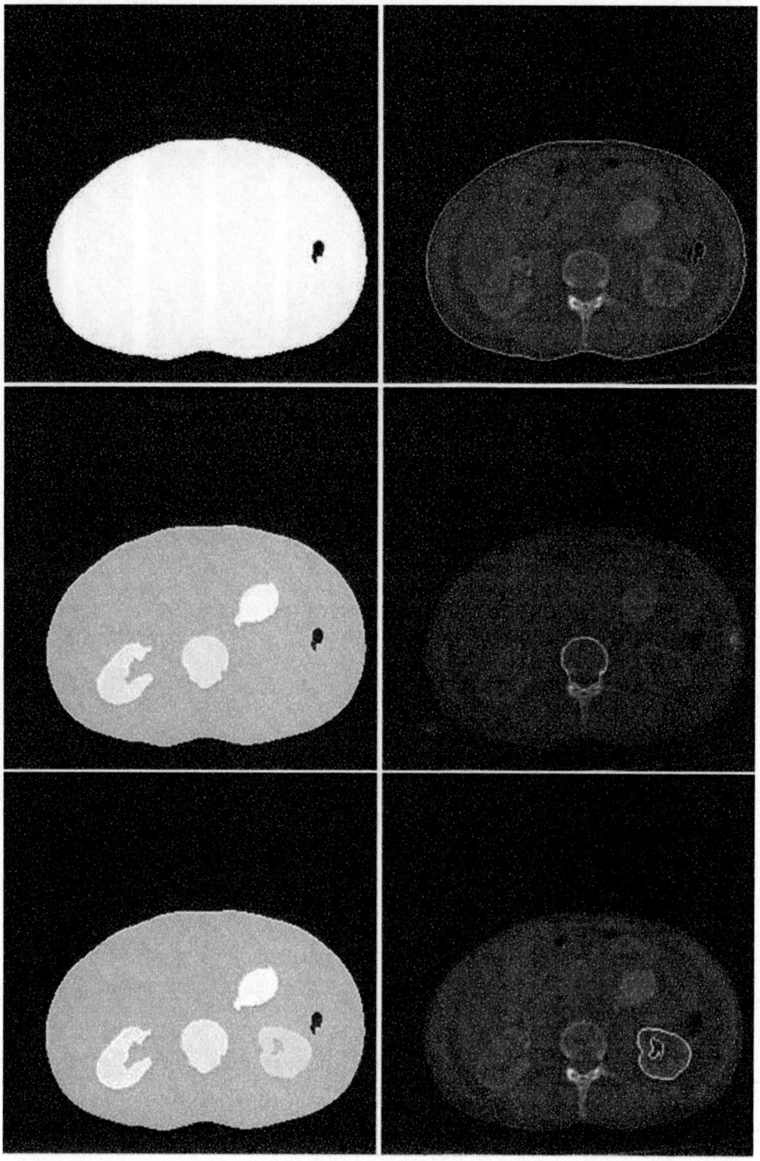

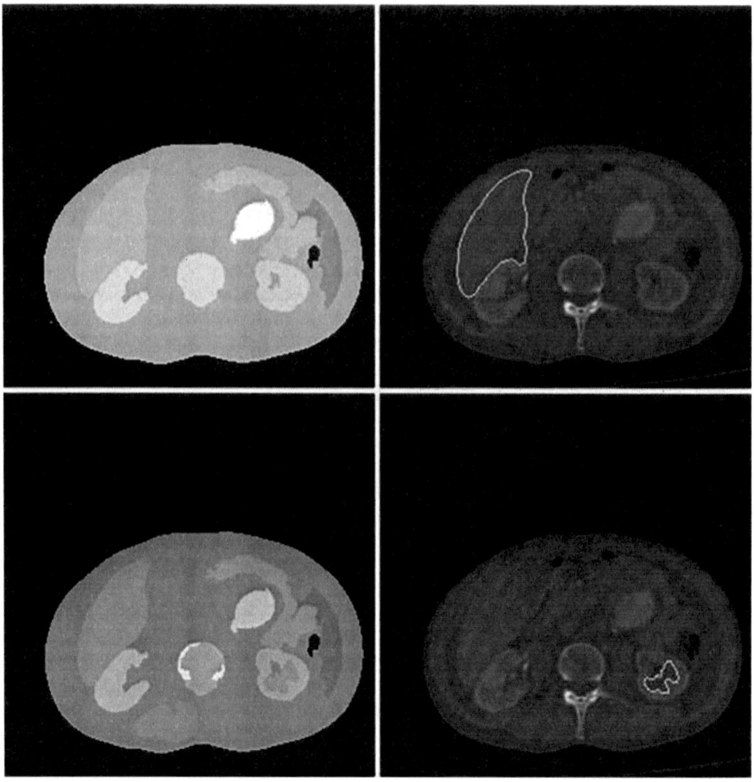

Fig. 7: Adding active geometric wavelets to the approximation of a CT image

(b) *Minimize the local Active Contour functional with large* μ.
(c) *Repeat previous step, diminishing* μ *until a valid non-empty segmentation is found.*
(d) *Create and add to BSP two leaves corresponding to the two sub-regions found in step 3.c.*

4. *Sort the Active Geometric Wavelets according to* (4): $\psi_{\Omega_1}, \psi_{\Omega_2}, \ldots$
5. *For some given n, the output of the algorithm is the n-term approximation* $\psi_{[0,1]^2} + \sum_{i=1}^{n} \psi_{\Omega_{k_i}}$.

4 Experimental Results

In Fig. 7, we show an example of medical image segmentation which is one of the potential applications of our AGW method. On the left we see the segmentation that is derived from the n-term Active Geometric Wavelet approximation with n increasing (for several values of n). We see that with more terms added, the algorithm correctly adds the various organs of the body. On the right, we show the compact support of the wavelet that is added at that particular step, i.e. the n-th term.

References

1. S. Brenner, L. Scott, The mathematical theory of finite elements methods, Springer-Verlag (1994).
2. T. Chan, L. Vese, Active contours without edges, IEEE Trans. Image Processing **10**, 266–277 (2001).
3. G. Chung, L. Vese, Image segmentation using a multilayer level-set approach, Computing and Visualization in Science **12**, 267–285 (2009).
4. A. Cohen, Numerical analysis of wavelet methods, Elsevier Science (2003).
5. S. Dekel, D. Leviatan, Adaptive multivariate approximation using binary space partitions and geometric wavelets, SIAM J. Num. Anal. **43**, 707–732 (2005).
6. R. DeVore, Nonlinear approximation, Acta Numerica **7**, 51–150 (1998).
7. F. John, Extremum problems with inequalities as subsidiary conditions, Studies and Essays Presented to R. Courant on his 60th Birthday, pp. 187-204, Interscience Publishers (1948).
8. B. Karaivanov, P. Petrushev, Nonlinear piecewise polynomials approximation beyond Besov spaces, App. Comp. Harmonic Analysis **15**, 177–223 (2003).
9. R. Kazinnik, S. Dekel, N. Dyn, Low bit-rate image coding using adaptive geometric piecewise polynomial approximation, IEEE Trans. Image Processing **16**, 2225-2233 (2007).
10. S. Lankton, A. Tannenbaum, Localizing Region-Based Active Contours, IEEE Trans. Image processing **17**, 2029–2039 (2008).
11. S. Osher, R. Fedkiw, Level Set Methods and Dynamic Implicit Surfaces. New York: Cambridge Univ. Press (2003).
12. P. Perona, J. Malik, Scale-space and edge detection using anisotropic diffusion, IEEE Trans. Pattern Analysis and Machine Intelligence **12**, 629-639 (1990).
13. J.A. Sethian, Level Set Methods and Fast Marching Methods, 2nd edition, Cambridge University Press (1999).

Interpolating Composite Systems

Philipp Grohs

Abstract Composite systems are multiscale decompositions similar to wavelet MRAs but with several different dilation operations. A notable example of such a system is given by the shearlet transform. In this paper we construct interpolating wavelet-type decompositions for such systems and study their approximation properties. As a main application we give an example of an interpolating shearlet transform.

1 Introduction

Interpolating wavelets are a well-known construction similar to the usual L_2-theory but with L_2-projectors onto the scaling spaces replaced by L_∞-projectors defined by an interpolation procedure. A remarkable result is that this much simpler transform satisfies the same norm-equivalences as L_2-wavelets between Besov-space (or Triebel-Lizorskin-space) norms and discrete norms on the coefficients – provided the space embeds into L_∞, see [4]. Since the interpolating wavelet transform is not stable in an L_2-sense, and since there exist many nice and general L_2 wavelet constructions [2], usually the more complicated L_2-theory is preferred (although L_2-stability is by no means necessary for many applications like for instance PDE-solver [10]).

Nevertheless, there exist situations to which the interpolating wavelet construction can be generalized, whereas the generalization of the L_2 constructions is much more difficult or even impossible.

One such case is that of manifold-valued data where interpolating wavelets can be defined and it can also be shown that they satisfy the same desirable properties

Philipp Grohs
Institute of Geometry, TU Graz. Kopernikusgasse 24, A 8010 Graz, Austria
e-mail: philippgrohs@gmail.com

M. Neamtu and L. Schumaker (eds.), *Approximation Theory XIII: San Antonio 2010*, Springer Proceedings in Mathematics 13, DOI 10.1007/978-1-4614-0772-0_8,

as their linear counterparts [7, 14]. On the other hand, the scope of L_2-wavelet constructions which are amemable to generalization to manifold-valued data is very limited, see [6].

Another such case is if the domain of the data is a manifold. Then one can first build a hierachical triangulation of the manifold and define interpolating wavelets in a straightforward way. By using the general method of *lifting*, see [15], it is then possible to even generate L_2-stable MRAs from this initial simple multiscale decomposition.

In this paper, we study yet another such case, namely the case of composite dilations [9] where there is more than one dilation involved. For this situation, there still does not exist a satisfying construction to produce (tight) frames with nice properties such as compact support and continuity.

A particularly relevant example of a composite dilation system is the recently introduced shearlet system [8, 13].

Our main result here is that for a composite dilation system a natural analogue of the interpolating wavelet construction can be defined with continuous and compactly supported 'wavelets'.

A somewhat related result can be found in [12] where a similar, albeit adaptive construction is presented. The main difference between [12] and our work is that we construct one system encompassing all combinations of dilation matrices simultaneously, while [12] constructs several different systems, one for each path in a 'tree of dilation matrices'.

2 Composite Dilation Systems

In this section, we collect the necessary definitions and preliminaries and assumptions. Our main object of interest are so-called dilation families as defined in [5]. Let us fix a spacial dimension d and a finite probability space (E,μ). A dilation family $\mathscr{W}$ is per definition a mapping $E \to \mathbb{Z}^{d\times d} : e \to W_e$ satisfying the following *compatibility condition*:

There exists an expanding matrix W such that for every $\mathbf{e} = (e_1,\ldots,e_j) \in E^j$ there exists a unimodular matrix $U_\mathbf{e} \in GL(\mathbb{Z},d)$ such that the matrix $W_\mathbf{e}$ defined as

$$W_{(e_j,e_{j-1},\ldots,e_1)} := W_{e_j} W_{(e_{j-1},\ldots,e_1)},$$

can be written as

$$W_\mathbf{e} = U_\mathbf{e} W^j.$$

In [5], we call this the lattice compatibility condition (LCC).

The LCC ensures that for $j \in \mathbb{N}$ and $\mathbf{e} \in E^j$, all the lattices $W_\mathbf{e}^{-1}\mathbb{Z}^d$ agree with the lattice $\Gamma^j := W^{-j}\mathbb{Z}^d$.

Notation. We will write elements of the product space E^j in boldface. For $\mathbf{e} \in E^j$ we sometimes write $|\mathbf{e}| := j$. We use the symbol $|\cdot|$ as either the absolute value or as the sup-norm on $\mathbb{R}^d$. If it is irrelevant which norm we use, we write simply $\|\cdot\|$. The meaning will be clear from the context. We use the symbol $A \lesssim B$ to indicate that A is bounded by a constant times B, where the constant is independent of certain parameters which will also be clear from the context.

We make another assumption ensuring that the lattices $W_{\mathbf{e}}\mathbb{Z}^d$ do not get too warped. This is called *moderation* and defined by

$$\rho(\mathscr{W}) := \limsup_{j \to \infty} \sup_{\mathbf{e} \in E^j} \left\| W_{\mathbf{e}}^{-1} \right\|^{1/j} < 1.$$

The goal of the present paper is to construct functions φ, ψ^l, $l = 1, \ldots, L$ such that the system $\Phi \cup \Psi$ with

$$\Phi := \left\{ \varphi(\cdot - \alpha) : \alpha \in \mathbb{Z}^d \right\} \tag{1}$$

and

$$\Psi := \left\{ \psi^l \left(W_{\mathbf{e}} \cdot - \alpha \right) : \mathbf{e} \in E^j,\ j \in \mathbb{N}, \alpha \in \mathbb{Z}^d, l = 1, \ldots, L \right\} \tag{2}$$

constitutes a set of representatives for $L_\infty(\mathbb{R}^d)$. Let us give some examples:

Example 1. The first example is when E consists of one point and we have an expanding matrix W as our dilation family. Then the system Ψ is just a usual wavelet system, see [2].

Example 2. Here, we let $E = \{0, \ldots, 7\}$ with uniform measure and define our dilation family via the quincunx matrix $W = \begin{pmatrix} 1 & -1 \\ 1 & 1 \end{pmatrix}$ and the matrices

$$U_0 = \begin{pmatrix} 1 & 0 \\ 1 & 0 \end{pmatrix}, U_1 = \begin{pmatrix} 0 & 1 \\ 1 & 0 \end{pmatrix}, U_2 = \begin{pmatrix} -1 & 0 \\ 0 & 1 \end{pmatrix}, U_3 = \begin{pmatrix} 0 & -1 \\ 1 & 0 \end{pmatrix}$$

and $U_{3+i} = -U_i$, $i = 1, \ldots, 4$, representing the symmetries of the unit square in the plane. It is easy to see that the system Ψ now looks as follows:

$$\Psi = \left\{ \psi^l \left(U_e W^j \cdot - \alpha \right) : e \in E,\ j \in \mathbb{N}, \alpha \in \mathbb{Z}^d, l = 1, \ldots, L \right\}.$$

This is the composite dilation system as introduced in [11].

Example 3. Our third example consists of the recently introduced shearlet system [13]. We have $E = \{0, 1\}$ with uniform measure, $W = \operatorname{diag}(4, 2)$, $U_0 = I$ and $U_1 = U = \begin{pmatrix} 1 & 1 \\ 0 & 1 \end{pmatrix}$. A system Ψ now looks as follows:

$$\Psi = \left\{ \psi^l \left(U^l W^j \cdot - \alpha \right) :\ j \in \mathbb{N},\ 0 \le l < 2^j,\ \alpha \in \mathbb{Z}^d, l = 1, \ldots, L \right\}.$$

In [5], it is shown that all these examples (and many more) satisfy the moderation condition and the LCC.

3 Interpolating Systems

In order to construct such representation systems, we follow the principle of *Multiresolution Analysis (MRA)*: We consider the spaces V^j defined as

$$V^j := \left\{ f(W_e \cdot) : f \in V_{j-1},\ e \in E \right\}, \quad V^0 = \mathrm{cls}_{L_\infty} \mathrm{span}\, \Phi.$$

We assume that the spaces V^j form a filtration, i.e. we have

$$V^j \subset V^{j+1}, \quad j \in \mathbb{N}. \tag{3}$$

In order to get this property we require that the function φ is $\mathscr{W}$ – *refinable*, meaning that

$$\varphi(\cdot) = \sum_{e\in E} \mu(\{e\}) \sum_{\alpha\in\mathbb{Z}^d} a_e(\alpha)\varphi(W_e \cdot - \alpha) \tag{4}$$

for some filter family $\mathscr{A} = (a_e)_{e\in E}$, $a_e \in l_0(\mathbb{Z}^d)$ satisfying the *sum rule*

$$\sum_{\beta\in\mathbb{Z}^d} a_e(\alpha - W_e\beta) = 1 \quad \text{for all } \alpha \in \mathbb{Z}^d,\ e \in E. \tag{5}$$

We will now construct a projection operators which map a continuous function onto the scaling spaces V^j. Let us consider a bounded, continuous function f. We can project this function onto V^j via the operator $P^j : C(\mathbb{R}^d) \cap L_\infty \to V^j$ defined by

$$P^j f(\cdot) = \sum_{\mathbf{e}\in E^j} \mu^j(\{\mathbf{e}\}) \sum_{\alpha\in\mathbb{Z}^d} f\left(W_{\mathbf{e}}^{-1}\alpha\right) \varphi\left(W_{\mathbf{e}} \cdot - \alpha\right), \quad j \geq 1,$$
$$P^0 f = \sum_{\alpha\in\mathbb{Z}^d} f(\alpha)\varphi(\cdot - \alpha).$$

We further require that φ is *cardinal*, meaning that

$$\varphi\big|_{\mathbb{Z}^d} = \delta_0,$$

where $\varphi\big|_{\mathbb{Z}^d}$ denotes the restriction of φ to the integer grid and δ_0 denotes the Dirac sequence. This surely is the case if the filter family $\mathscr{A}$ is *cardinal*, which means that

$$a_e(W_e \cdot) = \delta_0(\cdot) \quad \text{for all } e \in E.$$

It follows immediately for this case that P^j is an interpolation operator:

Lemma 1. *For a continuous function f and P^j, Γ^j defined as above and φ cardinal, we have*

$$P^j f|_{\Gamma^j} = f_{\Gamma^j}. \tag{6}$$

Proof. The statement is trivial for $j = 0$. For $j \geq 0$ we can decompose the function $P^j f$ into the sum

$$P^j f = \sum_{\mathbf{e}\in E^j} \mu^j(\{\mathbf{e}\}) g_{\mathbf{e}},$$

where

$$g_{\mathbf{e}} := \sum_{\alpha\in\mathbb{Z}^d} f\left(W_{\mathbf{e}}^{-1}\alpha\right)\varphi\left(W_{\mathbf{e}}\cdot-\alpha\right).$$

We show that $g_{\mathbf{e}}\big|_{\Gamma^j} = f\big|_{\Gamma^j}$. Since $\sum_{\mathbf{e}\in E^j}\mu^j(\{\mathbf{e}\}) = 1$, this implies the statement. Because of the LCC we can write for $x = W^{-j}\beta$, $\beta\in\mathbb{Z}^d$,

$$\begin{aligned} g_{\mathbf{e}}(x) &= \sum_{\alpha\in\mathbb{Z}^d} f\left(W_{\mathbf{e}}^{-1}\alpha\right)\varphi\left(W_{\mathbf{e}}x-\alpha\right) \\ &= \sum_{\alpha\in\mathbb{Z}^d} f\left(W^{-j}U_{\mathbf{e}}^{-1}\alpha\right)\varphi\left(U_{\mathbf{e}}W^j x-\alpha\right) \\ &= \sum_{\alpha\in\mathbb{Z}^d} f\left(W^{-j}\alpha\right)\varphi\left(U_{\mathbf{e}}\beta-U_{\mathbf{e}}\alpha\right) \\ &= f(W^{-j}\beta) = f(x). \end{aligned}$$

□

Now that we have defined a sequence of projection operators onto our scaling spaces V^j, we need to find a representation for the details $Q^j f := P^j f - P^{j-1} f$. In order to do this we use the cardinality of φ as well as (4). Let us define the *subdivision scheme* [1] $S_e : l_\infty(\mathbb{Z}^d)\to l_\infty(\mathbb{Z}^d)$, $e\in E$ as

$$S_e p(\cdot) := \sum_{\alpha\in\mathbb{Z}^d} a_e(\cdot - W_e\alpha)p(\alpha), p\in l_\infty(\mathbb{Z}^d).$$

Lemma 2. *With φ cardinal satisfying (4), we have the following representation:*

$$Q^j f(\cdot) = \sum_{\mathbf{e}\in E^j}\mu^j(\{\mathbf{e}\})\sum_{\alpha\in\mathbb{Z}^d} q_{\mathbf{e}}(f)(\alpha)\varphi\left(W_{\mathbf{e}}\cdot-\alpha\right), \tag{7}$$

with

$$q_{(e_1,\dots,e_j)}(f)(\alpha) := f\left(W_{(e_1,\dots,e_j)}^{-1}\alpha\right) - S_{e_j}\left(f(W_{(e_1,\dots,e_{j-1})}^{-1}\beta)\right)_{\beta\in\mathbb{Z}^d}(\alpha). \tag{8}$$

If the filter family $\mathscr{A}$ is cardinal, then we have that

$$q_{(e_1,\dots,e_j)}(f)(W_{e_j}\alpha) = 0 \quad \textit{for all } \alpha\in\mathbb{Z}^d. \tag{9}$$

Proof. In general, for any $p\in l_\infty(\mathbb{Z}^d)$, the refinability property (4) implies that

$$\sum_{\alpha\in\mathbb{Z}^d} p(\alpha)\varphi(\cdot-\alpha) = \sum_{e\in E}\mu(\{e\})\sum_{\alpha\in\mathbb{Z}^d} S_e p(\alpha)\varphi(W_e\cdot-\alpha). \tag{10}$$

Hence, we can write $Q^j f(\cdot) = P^j f(\cdot) - P^{j-1} f(\cdot)$ as

$$
\begin{aligned}
Q^j f(\cdot) &= \sum_{\mathbf{e}\in E^j} \mu^j(\{\mathbf{e}\}) \sum_{\alpha\in\mathbb{Z}^d} f\left(W_{\mathbf{e}}^{-1}\alpha\right)\varphi\left(W_{\mathbf{e}}\cdot-\alpha\right) \\
&\quad - \sum_{\tilde{\mathbf{e}}\in E^{j-1}} \mu^{j-1}(\{\tilde{\mathbf{e}}\}) \sum_{e\in E} \mu(\{e\}) \sum_{\alpha\in\mathbb{Z}^d} S_e f\left(W_{\tilde{\mathbf{e}}}^{-1}\cdot\right)(\alpha)\varphi\left(W_e W_{\tilde{\mathbf{e}}}\cdot-\alpha\right) \\
&= \sum_{\mathbf{e}\in E^j} \mu^j(\{\mathbf{e}\}) \sum_{\alpha\in\mathbb{Z}^d} q_{\mathbf{e}}(f)(\alpha)\varphi\left(W_{\mathbf{e}}\cdot-\alpha\right).
\end{aligned}
$$

The rest of the statement is clear. □

We now show that, provided the function f has some smoothness properties, the coefficients $|q_{\mathbf{e}}(f)(\alpha)|$ decay exponentially in $|\mathbf{e}|$. This result is crucial if we want to use such a scheme for compression.

Theorem 1. *Consider a cardinal, continuous, and compactly supported φ satisfying (4) with a cardinal filter family $\mathscr{A}$. Assume that f is in the Lipschitz space Lip $(\gamma, \mathbb{R}^d)$, $0<\gamma<1$. Then*

$$
|q_{\mathbf{e}}(f)(\alpha)| \lesssim \rho(\mathscr{W})^{\gamma|\mathbf{e}|},
$$

where the implicit constant is independent of $\mathbf{e}$ *uniformly.*

Proof. Suppose that

$$
\|f(\cdot-y)-f(\cdot)\|_\infty \le D|y|^\gamma.
$$

Consider a coefficient

$$
q_{\mathbf{e}}(f)(\alpha) := f\left(W_{(e_1,\dots,e_j)}^{-1}\alpha\right) - S_{e_j}\left(f(W_{(e_1,\dots,e_{j-1})}^{-1}\beta)\right)_{\beta\in\mathbb{Z}^d}(\alpha).
$$

Assume that all filters a_e are supported in a symmetric set $[-N,N]^d$, $e\in E$. Then we have for $\tilde{\mathbf{e}} = (e_1,\dots,e_{j-1})$ that

$$
S_e\left(f(W_{\tilde{\mathbf{e}}}^{-1}\beta)\right)_{\beta\in\mathbb{Z}^d}(\alpha) = \sum_{\beta\in\mathbb{Z}^d} a_e\left(\alpha - W_e\beta\right) f\left(W_{\tilde{\mathbf{e}}}^{-1}\beta\right).
$$

Using the sum rules, it follows that for *any* point $c\in\mathbb{R}^d$

$$
\begin{aligned}
q_{\mathbf{e}}(f)(\alpha) &= f\left(W_{(e_1,\dots,e_j)}^{-1}\alpha\right) - \sum_{\beta\in\mathbb{Z}^d} a_{e_j}\left(\alpha - W_{e_j}\beta\right) f\left(W_{\tilde{\mathbf{e}}}^{-1}\beta\right) \\
&= \left(f\left(W_{(e_1,\dots,e_j)}^{-1}\alpha\right) - c\right) + \left(\sum_{\beta\in\mathbb{Z}^d} a_{e_j}\left(\alpha - W_{e_j}\beta\right)\left(f\left(W_{\tilde{\mathbf{e}}}^{-1}\beta\right) - c\right)\right).
\end{aligned}
$$

It follows that

$$\begin{aligned}|q_{\mathbf{e}}(f)(\alpha)| &\leq \left|f\left(W_{(e_1,\ldots,e_j)}^{-1}\alpha\right)-c\right| + \left|\sum_{\beta\in\mathbb{Z}^d} a_{e_j}\left(\alpha - W_{e_j}\beta\right)\left(f\left(W_{\tilde{\mathbf{e}}}^{-1}\beta\right)-c\right)\right| \\ &\leq \left|f\left(W_{(e_1,\ldots,e_j)}^{-1}\alpha\right)-c\right| + \|a_e\|_1 \sup_{\beta\in W_{e_j}^{-1}(\alpha+N)} \left|f\left(W_{\tilde{\mathbf{e}}}^{-1}\beta\right)-c\right|.\end{aligned}$$

Picking e.g. $c = f\left(W_{\mathbf{e}}^{-1}\alpha\right)$ we get that

$$\begin{aligned}|q_{\mathbf{e}}(f)(\alpha)| &\leq \|a_e\|_1 \sup_{\alpha+N}\left|f\left(W_{\mathbf{e}}^{-1}x\right)-f\left(W_{\mathbf{e}}^{-1}\alpha\right)\right| \\ &\leq \|a_e\|_1 D \sup_{x\in[-N,N]^d}\|W_{\mathbf{e}}^{-1}x\|^{\gamma} \lesssim \rho(\mathscr{W})^{\gamma j}. \end{aligned} \tag{11}$$

□

Remark 1. The previous result also holds for $\gamma > 0$ arbitrary with only minor modifications in the proof. In this case, we would have to impose higher order sum-rules on the filter family $\mathscr{A}$. We do not know if an inverse to this result holds; it would be somewhat disappointing since it would mean that the spaces we can characterize by the interpolating transform are just spaces that can equally well be described with wavelets.

Theorem 1 immediately implies that the infinite series $P^0 f + \sum_j Q^j f$ converges uniformly to f if f lies in some Lipschitz space. We can also show this under weaker assumptions:

Theorem 2. *Let φ be as in Theorem 1 and $f : \mathbb{R}^d \to \mathbb{R}$ uniformly continuous. Then*

$$\left\|f - \left(P^0 f + \sum_{j=1}^{J} Q^j f\right)\right\|_\infty \to 0, \quad \textit{for } J \to \infty. \tag{12}$$

Proof. The statement is equivalent to the fact that

$$\|f - P^J f\|_\infty \to 0.$$

Consider the modulus of continuity

$$\omega(f,h) := \|f(\cdot + h) - f(\cdot)\|_\infty.$$

We know that $\omega(f,h) \to 0$ for $h \to 0$. Also, using standard arguments and the moderation property it is not difficult to see that

$$\omega(P^j f, h) \lesssim \omega(f,h).$$

Using the fact that P^J interpolates data on Γ^J, it follows that for $x = W^{-J}(\alpha + h)$, $\alpha \in \mathbb{Z}^d$, $h \in [0,1]^d$ we have

$$\begin{aligned} |f(x) - P^J f(x)| &\leq |f(x) - f(W^{-J}\alpha)| + |P^J f(x) - P^J(W^{-J}\alpha)| \\ &\lesssim \omega(f, \lambda^{-J}) \to 0, \end{aligned}$$

where $\lambda > 1$ is the smallest eigenvalue of W. This proves the statement. □

The previous results allow us to construct meaningful representation systems Ψ for any moderate dilation family $\mathscr{W}$ satisfying the LCC. Denote by $\Omega_e^* := \mathbb{Z}^d / W_e \mathbb{Z}^d \setminus \{0\}$. Then we can define $\psi_{e,\omega}$, $e \in E$, $\omega \in \Omega_e^*$ via

$$\psi_{e,\omega}(\cdot) := \varphi(W_e \cdot - \omega). \tag{13}$$

Writing $\mathscr{L} = \{(e,\omega) : \ e \in E, \ \omega \in \Omega_e^*\}$ and $L = |\mathscr{L}|$, we now show that the system

$$\Psi := \left\{ \psi_l (W_{\mathbf{e}} \cdot - \alpha) : \ j \in \mathbb{N}, \ \mathbf{e} \in E^j, \ l \in \mathscr{L} \right\}$$

'spans' the space of bounded, uniformly continuous functions. We use the notation

$$\langle f, \psi_{e,\omega} (W_{\mathbf{e}} \cdot - \alpha) \rangle := \mu^{|\mathbf{e}|+1}(\{(e,\mathbf{e})\}) q_{(e,\mathbf{e})}(f) (W_e \alpha + \omega). \tag{14}$$

Theorem 3. *For φ satisfying the conditions of Theorem 1, ψ^l defined as in (13), and a bounded and uniformly continuous function f, we have*

$$P_0 f(\cdot) + \sum_{j \geq 0} \sum_{\mathbf{e} \in E^j} \sum_{l \in \mathscr{L}} \sum_{\alpha \in \mathbb{Z}^d} \langle f, \psi_l (W_{\mathbf{e}} \cdot - \alpha) \rangle \psi_l (W_{\mathbf{e}} \cdot - \alpha) = f(\cdot)$$

in the sense that the sum on the left-hand side converges uniformly.

Proof. This is just a reformulation of Theorem 2. □

It is now time to ask if there actually exist functions φ which satisfy the assumptions of Theorem 1. One of the main results of [5] is that the answer is yes:

Theorem 4 ([5]). *For any moderate dilation family $\mathscr{W}$ satisfying the LCC there exists a cardinal, continuous and compactly supported function which is $\mathscr{W}$ refinable.*

This implies that for any moderate dilation family $\mathscr{W}$ we can construct systems Φ, Ψ as in (1) and (2) such that these systems form a stable basis for $L_\infty(\mathbb{R}^d)$. As special cases we obtain an interpolating shearlet transform and an interpolating composite wavelet transform associated to Example 2.

4 Shearlets

Let us work out in more detail the construction for the example of the shearlet system. Recall that we have $E = \{0,1\}$ with uniform counting measure. Furthermore, we have

$$W_0 = W = \begin{pmatrix} 4 & 0 \\ 0 & 2 \end{pmatrix}, \quad \text{and } W_1 = \begin{pmatrix} 4 & 2 \\ 0 & 2 \end{pmatrix}.$$

Clearly, $W_1 = U_1 W_0$ with

$$U_1 = U = \begin{pmatrix} 1 & 1 \\ 0 & 1 \end{pmatrix}$$

being unimodular. For this case, we have a very general construction of suitable functions φ. Indeed, consider the univariate filter $b : \mathbb{Z} \to \mathbb{R}$ which is defined as the filter of the Deslauriers–Dubuc scheme [3] of some arbitrary order and with dilation factor 2. We can also define the filter $\tilde{b}$ via

$$\tilde{b}(\cdot) := \sum_{\beta \in \mathbb{Z}} b(\cdot - 2\beta).$$

Then the filter $\tilde{b}$ has dilation factor 4 (see [1] for more information regarding dilation factors). Now define

$$a(\alpha) := \tilde{b}(\alpha_1) b(\alpha_2), \quad \alpha = (\alpha_1, \alpha_2) \in \mathbb{Z}^2.$$

With $a_0(\cdot) = a(\cdot)$ and $a_1(\cdot) := a_0(U^{-1}\cdot)$, we can show [5] that there exists a continuous, compactly supported and cardinal function φ that satisfies the refinement equation (4). Now we define

$$\psi^l(\cdot) := \varphi\left(W \cdot - \beta_l\right), \quad l = 0, \ldots, 6,$$

where

$$\beta_0 = (0,1),\ \beta_1 = (0,2),\ \beta_2 = (0,3),\ \beta_3 = (1,0),\ \beta_4 = (1,1),$$
$$\beta_5 = (1,2),\ \beta_6 = (1,3).$$

Further define

$$\psi^{l+7}(\cdot) := \varphi\left(UW \cdot - U\beta^l\right), \quad l = 0, \ldots, 6.$$

Now, in complete analogy with the conventional case of interpolating wavelet transformations we can represent a continuous function f as a superposition of sheared and dilated translates of a finite set of functions:

$$f(\cdot) = P_0 f(\cdot) + \sum_{j \geq 0} \sum_{k=0}^{2^j - 1} \sum_{l=0}^{13} \sum_{\alpha \in \mathbb{Z}^d} \left\langle f, \psi_l\left(U^k W^j \cdot - \alpha\right) \right\rangle \psi_l\left(U^k W^j \cdot - \alpha\right).$$

This formula can be interpreted as a tight frame representation formula with an L_∞ bilinear product instead of the usual L_2 inner product.

5 Conclusion

In this paper, we gave a flexible construction of interpolating multiscale transforms for composite systems. To illustrate the usefulness of our theoretical results we gave a construction of compactly supported interpolating shearlets. While this construction can be seen as an interpolating tight frame representation, a construction of compactly supported shearlet-type L_2 tight frames is still out of reach at the present time.

References

1. Albert Cavaretta, Wolfgang Dahmen, and Charles Micchelli. *Stationary Subdivision*. American Mathematical Society, 1991.
2. Ingrid Daubechies. *Ten Lectures on Wavelets*. SIAM, 1992.
3. Gilles Deslauriers and Serge Dubuc. Symmetric iterative interpolation processes. *Constructive Approximation*, 5:49–68, 1989.
4. David Donoho. Interpolating wavelet transforms. Technical report, Department of Statistics, Stanford University, 1992.
5. Philipp Grohs. Refinable functions for dilation families. Technical report, TU Graz, 2010.
6. Philipp Grohs and Johannes Wallner. Definability and stability of multiscale decompositions for manifold-valued data. Technical report, TU Graz, 2009.
7. Philipp Grohs and Johannes Wallner. Interpolatory wavelets for manifold-valued data. *Applied and Computational Harmonic Analysis*, 27(3):325–333, 2009.
8. Kanghui Guo, Gitta Kutyniok, and Demetrio Labate. Sparse multidimensional representations using anisotropic dilation and shear operators. In *Wavelets and Splines (Athens, GA, 2005)*, 2006.
9. Kanghui Guo, Wang-Q Lim, Demetrio Labate, Guido Weiss, and Edward Wilson. Wavelets with composite dilations. *Electronic Research Announcments of the American Mathematical Socitey*, 10:78–87, 2004.
10. Mats Holström. Solving hyperbolic pdes using interpolating wavelets. *SIAM Journal on Scientific Computing*, 21:405–420, 2000.
11. Ilya Krishtal, B. Robinson, Guido Weiss, and Edward Wilson. Some simple haar-type wavelets in higher dimensions. *Journal of Geometric Analysis*, 17(1):87–96, 2007.
12. Gitta Kutyniok and Tomas Sauer. Adaptive directional subdivision schemes and shearlet multiresolution analysis. *SIAM Journal on Mathematical Analysis*, 41:1436–1471, 2009.
13. Demetrio Labate, Wang-Q Lim, Gitta Kutyniok, and Guido Weiss. Sparse multidimensional representation using shearlets. In *Proceedings of the SPIE*, pages 254–262, 2005.
14. Iman Ur Rahman, Iddo Droriand, Victoria C. Stodden, David Donoho, and Peter Schröder. Multiscale representations for manifold-valued data. *Multiscale modeling and Simulation*, 4(4):1201–1232, 2006.
15. Wim Sweldens. The lifting scheme: A construction of second generation wavelets. *SIAM Journal on Mathematical Analysis*, 29(2):511, 1998.

Wavelets and Framelets Within the Framework of Nonhomogeneous Wavelet Systems

Bin Han

Abstract In this paper, we shall discuss recent developments in the basic theory of wavelets and framelets within the framework of nonhomogeneous wavelet systems in a natural and simple way. We shall see that nonhomogeneous wavelet systems naturally link many aspects of wavelet analysis together. There are two fundamental issues of the basic theory of wavelets and framelets: frequency-based nonhomogeneous dual framelets in the distribution space and stability of nonhomogeneous wavelet systems in a general function space. For example, without any a priori condition, we show that every dual framelet filter bank derived via the oblique extension principle (OEP) always has an underlying frequency-based nonhomogeneous dual framelet in the distribution space. We show that directional representations to capture edge singularities in high dimensions can be easily achieved by constructing nonstationary nonhomogeneous tight framelets in $L_2(\mathbb{R}^d)$ with the dilation matrix $2I_d$. Moreover, such directional tight framelets are derived from tight framelet filter banks derived via OEP. We also address the algorithmic aspects of wavelets and framelets such as discrete wavelet/framelet transform and its basic properties in the discrete sequence setting. We provide the reader in this paper a more or less complete picture so far on wavelets and framelets within the framework of nonhomogeneous wavelet systems.

1 Introduction

As a multidisciplinary subject, wavelet analysis is a broad area including approximation schemes using shift-invariant spaces in approximation theory, characterization of function spaces by various wavelets in harmonic analysis, curve and surface

Bin Han
Department of Mathematical and Statistical Sciences, University of Alberta, Edmonton, Alberta T6G 2G1, Canada
e-mail: bhan@ualberta.ca

M. Neamtu and L. Schumaker (eds.), *Approximation Theory XIII: San Antonio 2010*, Springer Proceedings in Mathematics 13, DOI 10.1007/978-1-4614-0772-0_9,

generation via subdivision schemes in computer graphics, wavelet-based algorithms for numerical solutions to differential equations, discrete wavelet/framelet transform and filter banks in engineering for signal and image processing, etc. In this paper, we shall look at wavelets (nonredundant wavelet systems) and framelets (redundant wavelet systems) within the framework of nonhomogeneous wavelet systems.

Let us first recall some notation. For a function $f : \mathbb{R}^d \to \mathbb{C}$ and a $d \times d$ real-valued invertible matrix U, throughout the paper we shall adopt the following notation:

$$f_{U;\mathsf{k},\mathsf{n}}(x) := |\det U|^{1/2} e^{-i\mathsf{n}\cdot Ux} f(Ux - \mathsf{k}) \quad \text{and} \quad f_{U;\mathsf{k}} := f_{U;\mathsf{k},\mathbf{0}}, \quad x, \mathsf{k}, \mathsf{n} \in \mathbb{R}^d, \tag{1}$$

where i denotes the imaginary unit satisfying $i^2 = -1$. Let M be a $d \times d$ real-valued invertible matrix, and let Ψ be a subset of square integrable functions in $L_2(\mathbb{R}^d)$. The commonly accepted definition of a wavelet or a framelet is tightly linked to the following *homogeneous* M-*wavelet system*

$$\mathrm{WS}(\Psi) := \{\psi_{\mathsf{M}^j;\mathsf{k}} \mid j \in \mathbb{Z}, \mathsf{k} \in \mathbb{Z}^d, \psi \in \Psi\}, \tag{2}$$

which has been extensively studied in the function space $L_2(\mathbb{R}^d)$ in the literature of wavelet analysis, often with M being an integer expansive matrix. For example, see [1–3,5–43,47–75] and many references therein. Here, we say that M is *an expansive matrix* if all its eigenvalues have modulus greater see than one. If a homogeneous wavelet system $\mathrm{WS}(\Psi)$ in (2) is an orthonormal basis, a Riesz basis, a tight frame, or a frame of $L_2(\mathbb{R}^d)$, in the literature of wavelet analysis, the generating elements in Ψ of (2) are often called orthonormal M-wavelets, Riesz M-wavelets, tight M-framelets, or M-framelets, respectively. In other words, a wavelet or a framelet in almost all the literature of wavelet analysis is simply a synonym of the generators of a homogeneous wavelet system with certain stability property in $L_2(\mathbb{R}^d)$. However, in this paper, we call the generating set Ψ a homogeneous orthonormal M-wavelet, a homogeneous Riesz M-wavelet, a homogeneous tight M-framelet, or a homogeneous M-framelet in $L_2(\mathbb{R}^d)$, if $\mathrm{WS}(\Psi)$ in (2) is an orthonormal basis, a Riesz basis, a tight frame, or a frame of $L_2(\mathbb{R}^d)$, respectively.

It is important to point out that in this paper the elements in a set S of generators are not necessarily distinct, and S may be an infinite set. The notation $h \in S$ in a summation means that h visits every element (with multiplicity) in S once and only once. For a set S, we shall use $\#S$ to denote its cardinality, counting multiplicity. For example, for $\Psi = \{\psi^1, \ldots, \psi^s\}$, its cardinality $\#\Psi$ is s, all the functions $\psi^1, \ldots, \psi^s$ are not necessarily distinct, and $\psi \in \Psi$ in (2) simply means $\psi = \psi^1, \ldots, \psi^s$.

Though a homogeneous wavelet system has been the center of extensive study for many years in the current literature of wavelet analysis [2]–[75], in this paper we focus on another system – a nonhomogeneous wavelet system, which links many aspects of wavelets and framelets together in a natural and simple way. For subsets Φ and Ψ of $L_2(\mathbb{R}^d)$, *a nonhomogeneous* M-*wavelet system* is defined to be

$$\mathrm{WS}_J(\Phi; \Psi) := \{\phi_{\mathsf{M}^J;\mathsf{k}} \mid \mathsf{k} \in \mathbb{Z}^d, \phi \in \Phi\} \cup \{\psi_{\mathsf{M}^j;\mathsf{k}} \mid j \geqslant J, \mathsf{k} \in \mathbb{Z}^d, \psi \in \Psi\}, \tag{3}$$

where J is an integer representing the coarsest scale level.

In this paper we used the words *homogeneous* and *nonhomogeneous* to distinguish the two types of wavelet systems in (2) and (3), due to the following considerations. Note that the scale levels j in (2) are all integers and the scale levels j in (3) are only integers no less than J. First of all, it is easy to see that the system $\mathsf{WS}(\Psi)$ in (2) is invariant (that is, homogeneous of order 0) and the system $\mathsf{WS}_J(\Phi;\Psi)$ in (3) is not invariant (nonhomogeneous) under the dilation operation. More precisely,

$$\{f_{\mathsf{M};\mathbf{0}} \mid f \in \mathsf{WS}(\Psi)\} = \mathsf{WS}(\Psi)$$

and

$$\{f_{\mathsf{M};\mathbf{0}} \mid f \in \mathsf{WS}_J(\Phi;\Psi)\} = \mathsf{WS}_{J+1}(\Phi;\Psi).$$

Secondly, the function spaces that can be characterized by the system $\mathsf{WS}(\Psi)$ in (2) are homogeneous function spaces (see [69]), while the function spaces that can be characterized by the system $\mathsf{WS}_J(\Phi;\Psi)$ in (3) are nonhomogeneous function spaces (also called inhomogeneous function spaces in analysis), see [51]. Thirdly, the dilation and shift operations are uniformly (or homogeneously) applied to all generators of $\mathsf{WS}(\Psi)$ in (2), but are differently (nonhomogeneously) applied to the two sets Φ and Ψ of generators of $\mathsf{WS}_J(\Phi;\Psi)$. Consequently, it seems quite natural for us to call the system $\mathsf{WS}(\Psi)$ in (2) a homogeneous wavelet system, and the system $\mathsf{WS}_J(\Phi;\Psi)$ in (3) a nonhomogeneous wavelet system.

For the particular function space $L_2(\mathbb{R}^d)$, in Sect. 2 we shall show that a nonhomogeneous wavelet system $\mathsf{WS}_J(\Phi;\Psi)$ at a given scale level J will lead to a sequence of nonhomogeneous wavelet systems $\mathsf{WS}_J(\Phi;\Psi)$ at all scale levels J with almost all properties preserved. Moreover, as the scale level J goes to $-\infty$, the limit of the sequence of nonhomogeneous wavelet systems $\mathsf{WS}_J(\Phi;\Psi)$ is a homogeneous wavelet system $\mathsf{WS}(\Psi)$ which shares almost all the properties of the original nonhomogeneous wavelet system. Moreover, we show in Sect. 2 that for nonredundant nonhomogeneous wavelet systems, such as nonhomogeneous orthonormal wavelets and nonhomogeneous biorthogonal wavelets, there are intrinsic connections of a nonredundant nonhomogeneous wavelet system with refinable function vectors and multiresolution analysis. Moreover, the frame approximation property of $\mathsf{WS}_J(\Phi;\Psi)$ is uniquely determined by the set Φ, and is independent of the set Ψ.

To characterize various nonhomogeneous wavelet systems, we show in Sect. 3 that it is very natural to study a frequency-based nonhomogeneous wavelet system in the distribution space. In Sect. 3, we introduce the notion of a pair of frequency-based nonhomogeneous dual framelets in the distribution space, for which we provide a complete characterization in Sect. 3. A similar complete characterization is also established in [44] for a pair of frequency-based fully nonstationary dual framelets with real-valued dilation matrices. As demonstrated in Sect. 3, nonhomogeneous wavelet systems have intrinsic refinable structure and are naturally connected to framelet transform in the function setting. In a certain sense, the notion of a frequency-based nonhomogeneous dual framelet in the distribution space corresponds to the perfect reconstruction property of a dual framelet filter bank in the discrete sequence setting. Without any a priori condition (such as membership in $L_2(\mathbb{R}^d)$, smoothness and vanishing moments of the generators, approximation properties etc.), we prove that any dual framelet filter bank derived via the oblique extension principle always has an underlying frequency-based nonhomogeneous dual framelet in the distribution space. We provide a complete characterization of nonhomogeneous tight M-framelets in $L_2(\mathbb{R}^d)$ for any real-valued expansive matrix M.

One of the fundamental properties of wavelets is the characterization of various function spaces by wavelet coefficients. It turns out that this is closely related to wavelets and framelets in a general function space. In Sect. 4, we shall introduce the notion of wavelets and framelets in a general function space. Then we show that there are two fundamental issues for wavelets and framelets in function spaces: one is a frequency-based dual framelet in the distribution space, which is closely linked to the perfect reconstruction property; the other is the stability of a nonhomogeneous wavelet system in a given function space, which is the frame property of nonhomogeneous wavelet systems. The ability of wavelets and framelets for characterizing function spaces is largely due to the fact that for a frequency-based nonhomogeneous dual framelet in the distribution space, the nonhomogeneous wavelet system often has stability in many function spaces. The notion of vanishing moments will naturally come from the renormalization of the wavelet or framelet.

In Sect. 5, we shall study wavelets and framelets that are derived from filter banks. We provide conditions or characterizations in terms of filters for the two fundamental issues: frequency-based nonhomogeneous dual framelets in the distribution space and its stability in various function spaces such as Sobolev spaces.

In Sect. 6, we shall address the algorithmic aspects of wavelets and framelets by discussing discrete wavelet/framelet transform with a dual framelet filter bank and its basic properties. The properties of discrete wavelet/framelet transform with a filter bank are traditionally derived from wavelets and framelets in the function setting via a multiresolution analysis (MRA). However, in this paper, we shall study discrete wavelet/framelet transform and its properties purely in the discrete sequence setting without any wavelets or framelets in the function setting involved. Though this approach is very natural for algorithms, there has not been much activity in this direction, and there are a few challenging questions to be resolved.

Finally, using the results in previous sections, in Sect. 7 we provide two examples of tight framelets in $L_2(\mathbb{R}^d)$. The first example shows that for any real-valued expansive matrix M, one can always construct a nonhomogeneous tight M-framelet $\mathsf{WS}_J(\Phi;\Psi)$ in $L_2(\mathbb{R}^d)$ such that each of Φ and Ψ contains only one element, which is a C^∞ function in the Schwarz class. The second example shows that the first example can be modified by a simple splitting technique into a directional tight $2I_2$-framelet in $L_2(\mathbb{R}^2)$ so that edge singularities can be captured. Both examples have underlying OEP-based tight framelet filter bank and therefore, it is possible to implement such directional tight framelets by fast algorithms using filter banks. We shall also discuss the projection method for constructing wavelets and framelets.

The main goal of this paper is to outline the most recent developments in the basic theory of wavelets and framelets within the unifying framework of nonhomogeneous wavelet systems in a natural and simple way. Without being either too general or too technical, we strive in this paper to provide the reader a more or less complete picture of this topic. Therefore, most technical proofs are removed in this paper; instead we provide precise references in the literature for the proofs of all the results presented in this paper. But for the convenience of the reader, we provide detailed information for important concepts, and we devote our effort to writing this survey article in a logically coherent way and discussing a daunting wide spectrum of many aspects of wavelets and framelets.

Over more than 20 years, extensive research has been made by many researchers from a wide spectrum of research areas, and many important results on wavelet analysis have been established. However, unavoidably there exist many different notational systems and approaches in the literature. The existence of different notation, which are often inconsistent and even inconsistent within a single notational system, makes it much more difficult for researchers to communicate and appreciate each other's work, since a large amount of time has to be spent on becoming familiar with the different notational systems. Another purpose of this paper is to promote a recently developed, relatively consistent, notational system for a wide spectrum of wavelet analysis. Such a proposed notation system has been constantly tested and improved over more than a year now, and appears adequate and consistent to treat almost all aspects of wavelet analysis in the space/time domain, Fourier/frequency domain, and Laurent/trigonometric polynomial domain (for filter bank design). The proposed notational system is achieved by establishing a macro file (in LaTex) so that one can simply modify the macro file to use one's own favorite symbols easily. This paper is an example written using such a macro file and notational system. Both the macro file and detailed instructions for using it can be freely downloaded at `http://www.ualberta.ca/~bhan/publ.htm`.

2 Nonhomogeneous Wavelet Systems in $L_2(\mathbb{R}^d)$

In this section, we study nonhomogeneous wavelet systems in the square integrable function space $L_2(\mathbb{R}^d)$. We shall see that a nonhomogeneous wavelet system in $L_2(\mathbb{R}^d)$ will naturally lead to a homogeneous wavelet system as its limiting system, with almost all properties preserved. For a nonredundant nonhomogeneous wavelet system in $L_2(\mathbb{R}^d)$, such as a nonhomogeneous orthonormal or biorthogonal wavelet basis of $L_2(\mathbb{R}^d)$, we shall see that it naturally leads to a multiresolution analysis (MRA), refinable structure, and refinable function vectors.

We first study redundant nonhomogeneous wavelet systems in $L_2(\mathbb{R}^d)$ such as nonhomogeneous M-framelets and nonhomogeneous dual M-framelets in $L_2(\mathbb{R}^d)$.

For (not necessarily finite) subsets Φ and Ψ of $L_2(\mathbb{R}^d)$ and an integer J, we say that $\mathrm{WS}_J(\Phi;\Psi)$ in (3) is *a nonhomogeneous* M*-wavelet frame of* $L_2(\mathbb{R}^d)$ if there exist two positive constants C_1 and C_2 such that for all $f \in L_2(\mathbb{R}^d)$,

$$C_1\|f\|^2_{L_2(\mathbb{R}^d)} \leqslant \sum_{\phi\in\Phi}\sum_{\mathsf{k}\in\mathbb{Z}^d} |\langle f, \phi_{\mathsf{M}^J;\mathsf{k}}\rangle|^2 + \sum_{j=J}^{\infty}\sum_{\psi\in\Psi}\sum_{\mathsf{k}\in\mathbb{Z}^d} |\langle f, \psi_{\mathsf{M}^j;\mathsf{k}}\rangle|^2 \leqslant C_2\|f\|^2_{L_2(\mathbb{R}^d)}. \quad (4)$$

For the best possible constants C_1 and C_2 in (4), we call $\sqrt{C_1}$ the lower frame bound of $\mathrm{WS}_J(\Phi;\Psi)$, and $\sqrt{C_2}$ the upper frame bound of $\mathrm{WS}_J(\Phi;\Psi)$ in $L_2(\mathbb{R}^d)$.

For nonhomogeneous M-wavelet frames in $L_2(\mathbb{R}^d)$, we have

Proposition 1 *([44, Proposition 4]) Let M be a $d\times d$ real-valued invertible matrix. Let Φ and Ψ be subsets of $L_2(\mathbb{R}^d)$. Suppose that $\mathrm{WS}_J(\Phi;\Psi)$ is a nonhomogeneous M-wavelet frame of $L_2(\mathbb{R}^d)$ for some integer J satisfying (4) for some positive constants C_1 and C_2. Then (4) holds for all integers J; that is, $\mathrm{WS}_J(\Phi;\Psi)$ is a nonhomogeneous M-wavelet frame of $L_2(\mathbb{R}^d)$ with the same lower and upper frame bounds for all integers J. If in addition M is an expansive matrix and $\sum_{\phi\in\Phi}\|\phi\|^2_{L_2(\mathbb{R}^d)}<\infty$, then the homogeneous M-wavelet system $\mathrm{WS}(\Psi)$ is a (homogeneous M-wavelet) frame of $L_2(\mathbb{R}^d)$ with the same lower and upper frame bounds satisfying*

$$C_1\|f\|^2_{L_2(\mathbb{R}^d)}\leqslant\sum_{j\in\mathbb{Z}}\sum_{\psi\in\Psi}\sum_{\mathsf{k}\in\mathbb{Z}^d}|\langle f,\psi_{\mathsf{M}^j;\mathsf{k}}\rangle|^2\leqslant C_2\|f\|^2_{L_2(\mathbb{R}^d)},\qquad\forall f\in L_2(\mathbb{R}^d).\quad(5)$$

Due to Proposition 1, if $\mathrm{WS}_0(\Phi;\Psi)$ is a nonhomogeneous M-wavelet frame of $L_2(\mathbb{R}^d)$ satisfying (4) with $J=0$, then we say that $\{\Phi;\Psi\}$ is a (nonhomogeneous) M-framelet of $L_2(\mathbb{R}^d)$. If $\mathrm{WS}(\Psi)$ satisfies (5) for some positive constants C_1 and C_2, then we say that Ψ is a homogeneous M-framelet of $L_2(\mathbb{R}^d)$.

Next, we recall the definition of a pair of nonhomogeneous dual M-wavelet frames in $L_2(\mathbb{R}^d)$. Let M be a $d\times d$ real-valued invertible matrix. Let

$$\Phi=\{\phi^1,\dots,\phi^r\},\ \Psi=\{\psi^1,\dots,\psi^s\},\ \tilde\Phi=\{\tilde\phi^1,\dots,\tilde\phi^r\},\ \tilde\Psi=\{\tilde\psi^1,\dots,\tilde\psi^s\}\quad(6)$$

be subsets of $L_2(\mathbb{R}^d)$, where $r,s\in\mathbb{N}\cup\{0,+\infty\}$. Let $\mathrm{WS}_J(\Phi;\Psi)$ be defined in (3) and $\mathrm{WS}_J(\tilde\Phi;\tilde\Psi)$ be defined similarly. We say that the pair $(\mathrm{WS}_J(\Phi;\Psi),\mathrm{WS}_J(\tilde\Phi;\tilde\Psi))$ is *a pair of nonhomogeneous dual M-wavelet frames of $L_2(\mathbb{R}^d)$* if each of $\mathrm{WS}_J(\Phi;\Psi)$ and $\mathrm{WS}_J(\tilde\Phi;\tilde\Psi)$ is a nonhomogeneous M-wavelet frame of $L_2(\mathbb{R}^d)$ and the following identity holds

$$\sum_{\ell=1}^{r}\sum_{\mathsf{k}\in\mathbb{Z}^d}\langle f,\phi^\ell_{\mathsf{M}^J;\mathsf{k}}\rangle\langle\tilde\phi^\ell_{\mathsf{M}^J;\mathsf{k}},g\rangle+\sum_{j=J}^{\infty}\sum_{\ell=1}^{s}\sum_{\mathsf{k}\in\mathbb{Z}^d}\langle f,\psi^\ell_{\mathsf{M}^j;\mathsf{k}}\rangle\langle\tilde\psi^\ell_{\mathsf{M}^j;\mathsf{k}},g\rangle=\langle f,g\rangle,\quad(7)$$

for all $f,g\in L_2(\mathbb{R}^d)$. Note that all the series in (7) converge absolutely.

If the notation ϕ^ℓ (a function indexed by ℓ) in (6) appears confusing with the ℓ-th power of a function ϕ (which is very rare in wavelet analysis), we suggest the alternative $\phi^{[\ell]}$. Note that the subsets in (6) can have repeated elements, and the above definition of nonhomogeneous dual framelets can be further generalized to uncountable subsets in (6). For this purpose, we suggest the notation $\Phi=\{\phi^\ell\mid\ell\in\Lambda_\Phi\},\Psi=\{\psi^\ell\mid\ell\in\Lambda_\Psi\}$ and $\tilde\Phi=\{\tilde\phi^\ell\mid\ell\in\Lambda_\Phi\},\tilde\Psi=\{\tilde\psi^\ell\mid\ell\in\Lambda_\Psi\}$ for general index sets Λ_Φ and Λ_Ψ.

For pairs of nonhomogeneous dual M-wavelet frames of $L_2(\mathbb{R}^d)$, we have

Proposition 2 *([44, Proposition 5]) Let M be a $d\times d$ real-valued invertible matrix. Let $\Phi,\Psi,\tilde\Phi,\tilde\Psi$ in (6) be subsets of $L_2(\mathbb{R}^d)$. Suppose that for some integer J, $(\mathrm{WS}_J(\Phi;\Psi),\mathrm{WS}_J(\tilde\Phi;\tilde\Psi))$ is a pair of nonhomogeneous dual M-wavelet frames of*

$L_2(\mathbb{R}^d)$. Then it is a pair of nonhomogeneous dual M*-wavelet frames of $L_2(\mathbb{R}^d)$ for all integers J. If in addition* M *is an expansive matrix and $\sum_{\phi\in\Phi}\|\phi\|^2_{L_2(\mathbb{R}^d)} + \sum_{\tilde\phi\in\tilde\Phi}\|\tilde\phi\|^2_{L_2(\mathbb{R}^d)} < \infty$, then the pair $(\mathrm{WS}(\Psi),\mathrm{WS}(\tilde\Psi))$ is a pair of homogeneous dual* M*-wavelet frames of $L_2(\mathbb{R}^d)$, that is, each of $\mathrm{WS}(\Psi)$ and $\mathrm{WS}(\tilde\Psi)$ is a (homogeneous* M*-wavelet) frame of $L_2(\mathbb{R}^d)$ and the following identity holds:*

$$\sum_{j\in\mathbb{Z}}\sum_{\ell=1}^{s}\sum_{\mathrm{k}\in\mathbb{Z}^d}\langle f,\psi^\ell_{\mathsf{M}^j;\mathrm{k}}\rangle\langle\tilde\psi^\ell_{\mathsf{M}^j;\mathrm{k}},g\rangle=\langle f,g\rangle,\qquad \forall\, f,g\in L_2(\mathbb{R}^d). \tag{8}$$

Due to Proposition 2, we say that $(\{\Phi;\Psi\},\{\tilde\Phi;\tilde\Psi\})$ is a (nonhomogeneous) dual M-framelet in $L_2(\mathbb{R}^d)$ if $(\mathrm{WS}_0(\Phi;\Psi),\mathrm{WS}_0(\tilde\Phi;\tilde\Psi))$ is a pair of nonhomogeneous dual M-wavelet frames of $L_2(\mathbb{R}^d)$; furthermore, $\{\Phi;\Psi\}$ is called a nonhomogeneous tight M-framelet in $L_2(\mathbb{R}^d)$ if $\tilde\Phi=\Phi$ and $\tilde\Psi=\Psi$. Similarly, we say that $(\Psi,\tilde\Psi)$ is a homogeneous dual M-framelet of $L_2(\mathbb{R}^d)$ if each of $\mathrm{WS}(\Psi)$ and $\mathrm{WS}(\tilde\Psi)$ is a frame in $L_2(\mathbb{R}^d)$ and (8) holds; furthermore, Ψ is called a homogeneous tight M-framelet in $L_2(\mathbb{R}^d)$ if $\tilde\Psi=\Psi$.

Propositions 1 and 2 show that for a general redundant nonhomogeneous wavelet system at a given scale level J, there is a sequence of similar nonhomogeneous wavelet systems at all scale levels with almost all the properties preserved. Moreover, when the scale level goes to $-\infty$, its limiting system is a redundant homogeneous wavelet system with the corresponding properties preserved.

Let $(\mathrm{WS}_J(\Phi;\Psi),\mathrm{WS}_J(\tilde\Phi;\tilde\Psi))$ be a pair of nonhomogeneous dual M-wavelet frames of $L_2(\mathbb{R}^d)$. Then it is easy to deduce (see [43, 44] and Sect. 3) that for all $j\geqslant J$ and $f,g\in L_2(\mathbb{R}^d)$,

$$\begin{aligned}\sum_{\ell=1}^{r}\sum_{\mathrm{k}\in\mathbb{Z}^d}&\langle f,\phi^\ell_{\mathsf{M}^{j+1};\mathrm{k}}\rangle\langle\tilde\phi^\ell_{\mathsf{M}^{j+1};\mathrm{k}},g\rangle\\&=\sum_{\ell=1}^{r}\sum_{\mathrm{k}\in\mathbb{Z}^d}\langle f,\phi^\ell_{\mathsf{M}^j;\mathrm{k}}\rangle\langle\tilde\phi^\ell_{\mathsf{M}^j;\mathrm{k}},g\rangle+\sum_{\ell=1}^{s}\sum_{\mathrm{k}\in\mathbb{Z}^d}\langle f,\psi^\ell_{\mathsf{M}^j;\mathrm{k}}\rangle\langle\tilde\psi^\ell_{\mathsf{M}^j;\mathrm{k}},g\rangle,\end{aligned} \tag{9}$$

which is simply the framelet transform in the function setting.

In the following, let us discuss its frame approximation order in Sobolev spaces. For $\tau\in\mathbb{R}$, we denote by $H^\tau(\mathbb{R}^d)$ the Sobolev space consisting of all tempered distributions f such that $\|f\|^2_{H^\tau(\mathbb{R}^d)}:=\langle f,f\rangle_{H^\tau(\mathbb{R}^d)}<\infty$, where the inner product on $H^\tau(\mathbb{R}^d)$ is defined to be

$$\langle f,g\rangle_{H^\tau(\mathbb{R}^d)}:=\int_{\mathbb{R}^d}\hat f(\xi)\overline{\hat g(\xi)}(1+\|\xi\|^2)^\tau \mathrm{d}\xi,\qquad f,g\in H^\tau(\mathbb{R}^d). \tag{10}$$

For $\tau\geqslant 0$, we define the semi-norm $|f|^2_{H^\tau(\mathbb{R}^d)}:=\int_{\mathbb{R}^d}|\hat f(\xi)|^2\|\xi\|^{2\tau}\mathrm{d}\xi$ for $f\in H^\tau(\mathbb{R}^d)$. Associated with $(\mathrm{WS}_J(\Phi;\Psi),\mathrm{WS}_J(\tilde\Phi;\tilde\Psi))$ at the scale level $n\geqslant J$, the truncated frame approximation/projection operator $\mathrm{P}_n: L_2(\mathbb{R}^d)\to L_2(\mathbb{R}^d)$ is defined to be

$$\mathrm{P}_n(f):=\sum_{\ell=1}^{r}\sum_{\mathrm{k}\in\mathbb{Z}^d}\langle f,\phi^\ell_{\mathsf{M}^J;\mathrm{k}}\rangle\tilde\phi^\ell_{\mathsf{M}^J;\mathrm{k}}+\sum_{j=J}^{n-1}\sum_{\ell=1}^{s}\sum_{\mathrm{k}\in\mathbb{Z}^d}\langle f,\psi^\ell_{\mathsf{M}^j;\mathrm{k}}\rangle\tilde\psi^\ell_{\mathsf{M}^j;\mathrm{k}}.$$

For $\tau \geqslant 0$, we say that $(\mathrm{WS}_J(\Phi;\Psi), \mathrm{WS}_J(\tilde{\Phi};\tilde{\Psi}))$ has τ *frame approximation order* ([23]) if there exist a positive constant C and a positive integer N such that

$$\|\mathrm{P}_n(f) - f\|_{L_2(\mathbb{R}^d)} \leqslant C|\det \mathsf{M}|^{-\tau n/d}|f|_{H^\tau(\mathbb{R}^d)}, \qquad \forall f \in H^\tau(\mathbb{R}^d) \quad \text{and} \quad n \geqslant N.$$

By (9), we see that

$$\mathrm{P}_n(f) = [\mathscr{Q}(f_{\mathsf{M}^{-n};\mathbf{0}})]_{\mathsf{M}^n;\mathbf{0}} = [\mathscr{Q}(f(\mathsf{M}^{-n}\cdot))](\mathsf{M}^n\cdot), \qquad n \geqslant J,\ f \in L_2(\mathbb{R}^d),$$

where the quasi-interpolation operator $\mathscr{Q} : L_2(\mathbb{R}^d) \to L_2(\mathbb{R}^d)$ is defined to be

$$\mathscr{Q}(f) := \sum_{\ell=1}^{r} \sum_{\mathsf{k}\in\mathbb{Z}^d} \langle f, \phi^\ell(\cdot - \mathsf{k})\rangle \tilde{\phi}^\ell(\cdot - \mathsf{k}).$$

Therefore, for a pair $(\mathrm{WS}_J(\Phi;\Psi), \mathrm{WS}_J(\tilde{\Phi};\tilde{\Psi}))$ of nonhomogeneous dual M-wavelet frames of $L_2(\mathbb{R}^d)$, its frame approximation order is completely determined by the sets $\Phi, \tilde{\Phi}$, and has nothing to do with the generating sets $\Psi, \tilde{\Psi}$. The approximation property of the quasi-interpolation operator $\mathscr{Q}$ is well studied in approximation theory, for example, see [55, 66] and references therein.

Next, we discuss nonredundant nonhomogeneous wavelet systems such as nonhomogeneous Riesz M-wavelet bases and nonhomogeneous biorthogonal M-wavelet bases. For subsets Φ and Ψ of $L_2(\mathbb{R}^d)$ and an integer J, we say that $\mathrm{WS}_J(\Phi;\Psi)$ is *a nonhomogeneous Riesz* M*-wavelet basis* of $L_2(\mathbb{R}^d)$ if

1. The linear span of $\mathrm{WS}_J(\Phi;\Psi)$ is dense in $L_2(\mathbb{R}^d)$
2. There exist two positive constants C_3 and C_4 such that

$$C_3 \sum_{h\in \mathrm{WS}_J(\Phi;\Psi)} |w_h|^2 \leqslant \Big\| \sum_{h\in \mathrm{WS}_J(\Phi;\Psi)} w_h h \Big\|^2_{L_2(\mathbb{R}^d)} \leqslant C_4 \sum_{h\in \mathrm{WS}_J(\Phi;\Psi)} |w_h|^2 \qquad (11)$$

for all finitely supported sequences $\{w_h\}_{h\in \mathrm{WS}_J(\Phi;\Psi)}$. Note that the sequences here are indexed by the elements of the set $\mathrm{WS}_J(\Phi;\Psi)$.

For the best possible constants C_3 and C_4 in (11), we call $\sqrt{C_3}$ the lower Riesz bound of $\mathrm{WS}_J(\Phi;\Psi)$ and $\sqrt{C_4}$ the upper Riesz bound of $\mathrm{WS}_J(\Phi;\Psi)$ in $L_2(\mathbb{R}^d)$.

The following result on the relation between a frame and a Riesz basis of $L_2(\mathbb{R}^d)$ is well known in functional analysis and harmonic analysis.

Lemma 1. *Let Φ and Ψ be subsets of $L_2(\mathbb{R}^d)$ and let J be an integer. Then $\mathrm{WS}_J(\Phi;\Psi)$ is a Riesz basis of $L_2(\mathbb{R}^d)$, if and only if, it is a frame of $L_2(\mathbb{R}^d)$ and it is $l_2(\mathrm{WS}_J(\Phi;\Psi))$-linearly independent, that is, if $\sum_{h\in \mathrm{WS}_J(\Phi;\Psi)} w_h h = 0$ in $L_2(\mathbb{R}^d)$ for a square summable sequence $\{w_h\}_{h\in \mathrm{WS}_J(\Phi;\Psi)} \in l_2(\mathrm{WS}_J(\Phi;\Psi))$, then $w_h = 0$ for all $h \in \mathrm{WS}_J(\Phi;\Psi)$.*

Proof. Though the proof can be found in standard textbooks, for the convenience of the reader, we sketch a proof here. Define a framelet reconstruction/synthesis operator $\mathscr{V} : l_2(\mathrm{WS}_J(\Phi;\Psi)) \to L_2(\mathbb{R}^d)$, $\mathscr{V}(\{w_h\}_{h\in \mathrm{WS}_J(\Phi;\Psi)}) := \sum_{h\in \mathrm{WS}_J(\Phi;\Psi)} w_h h$. By the definition of a Riesz basis, it is easy to see that $\mathrm{WS}_J(\Phi;\Psi)$ is a Riesz basis of

$L_2(\mathbb{R}^d)$ if and only if $\mathscr{V}$ is a well-defined bounded and invertible operator. Similarly, by the definition of a frame in (5), $\mathrm{WS}_J(\Phi;\Psi)$ is a frame of $L_2(\mathbb{R}^d)$ if and only if $\mathscr{V}$ is a well-defined bounded onto operator with the range of its adjoint operator $\mathscr{W}$ being closed, where $\mathscr{W}: L_2(\mathbb{R}^d) \to l_2(\mathrm{WS}_J(\Phi;\Psi)), \mathscr{W}(f) := \{\langle f,h\rangle\}_{h\in \mathrm{WS}_J(\Phi;\Psi)}$ is the framelet decomposition/analysis operator. Now it is straightforward to see that $\mathrm{WS}_J(\Phi;\Psi)$ is a Riesz basis of $L_2(\mathbb{R}^d)$ if and only if it is a frame of $L_2(\mathbb{R}^d)$ and it is $l_2(\mathrm{WS}_J(\Phi;\Psi))$-linearly independent, that is, $\mathscr{V}$ is one-to-one. □

According to Lemma 1, a nonhomogeneous Riesz M-wavelet basis of $L_2(\mathbb{R}^d)$ is simply a nonredundant M-wavelet frame of $L_2(\mathbb{R}^d)$. A nonhomogeneous orthonormal M-wavelet basis is a special case of a nonhomogeneous Riesz M-wavelet basis.

For nonhomogeneous Riesz wavelet bases of $L_2(\mathbb{R}^d)$, we have

Proposition 3 *([44, Theorem 6]) Let* M *be a* $d\times d$ *real-valued invertible matrix. Let* Φ *and* Ψ *be subsets of* $L_2(\mathbb{R}^d)$. *Suppose that* $\mathrm{WS}_J(\Phi;\Psi)$ *is a nonhomogeneous Riesz* M*-wavelet basis of* $L_2(\mathbb{R}^d)$ *satisfying* (11) *for some integer* J. *Then* (11) *holds for all integers* J *and* $\mathrm{WS}_J(\Phi;\Psi)$ *is a nonhomogeneous Riesz* M*-wavelet basis of* $L_2(\mathbb{R}^d)$ *with the same lower and upper Riesz bounds for all integers* J. *If in addition* M *is expansive and* $\sum_{\phi\in\Phi}\|\phi\|^2_{L_2(\mathbb{R}^d)}<\infty$, *then* $\mathrm{WS}(\Psi)$ *is a homogeneous Riesz* M*-wavelet basis of* $L_2(\mathbb{R}^d)$ *with the same lower and upper Riesz bounds, that is, the linear span of* $\mathrm{WS}(\Psi)$ *is dense in* $L_2(\mathbb{R}^d)$ *and*

$$C_3 \sum_{h\in \mathrm{WS}(\Psi)} |w_h|^2 \leqslant \Big\| \sum_{h\in \mathrm{WS}(\Psi)} w_h h\Big\|^2_{L_2(\mathbb{R}^d)} \leqslant C_4 \sum_{h\in \mathrm{WS}(\Psi)} |w_h|^2 \tag{12}$$

for all finitely supported sequences $\{w_h\}_{h\in \mathrm{WS}(\Psi)}$.

Due to Proposition 3, we say that $\{\Phi;\Psi\}$ is a (nonhomogeneous Riesz) M-wavelet in $L_2(\mathbb{R}^d)$ if $\mathrm{WS}_0(\Phi;\Psi)$ is a nonhomogeneous Riesz M-wavelet basis of $L_2(\mathbb{R}^d)$. Similarly, we say that Ψ is a homogeneous (Riesz) M-wavelet in $L_2(\mathbb{R}^d)$ if $\mathrm{WS}(\Psi)$ is a homogeneous Riesz M-wavelet basis of $L_2(\mathbb{R}^d)$.

Let δ denote the *Dirac sequence* such that $\delta(0)=1$ and $\delta(\mathsf{k})=0$ for all $\mathsf{k}\neq 0$. Let $\Phi,\Psi,\tilde{\Phi},\tilde{\Psi}$ in (6) be subsets of $L_2(\mathbb{R}^d)$. Let J be an integer. We say that $(\mathrm{WS}_J(\Phi;\Psi),\mathrm{WS}_J(\tilde{\Phi};\tilde{\Psi}))$ is *a pair of nonhomogeneous biorthogonal* M*-wavelet bases* of $L_2(\mathbb{R}^d)$ if

1. Each of $\mathrm{WS}_J(\Phi;\Psi)$ and $\mathrm{WS}_J(\tilde{\Phi};\tilde{\Psi})$ is a Riesz M-wavelet basis of $L_2(\mathbb{R}^d)$;
2. The following biorthogonality relations hold:

$$\langle \phi^\ell_{\mathsf{M}^J;\mathsf{k}}, \tilde{\phi}^{\ell'}_{\mathsf{M}^J;\mathsf{k}'}\rangle = \delta(\mathsf{k}-\mathsf{k}')\delta(\ell-\ell'), \quad \langle \psi^n_{\mathsf{M}^j;\mathsf{k}}, \tilde{\phi}^{\ell'}_{\mathsf{M}^J;\mathsf{k}'}\rangle = 0,$$
$$\langle \phi^\ell_{\mathsf{M}^J;\mathsf{k}}, \tilde{\psi}^{n'}_{\mathsf{M}^{j'};\mathsf{k}'}\rangle = 0, \quad \langle \psi^n_{\mathsf{M}^j;\mathsf{k}}, \tilde{\psi}^{n'}_{\mathsf{M}^{j'};\mathsf{k}'}\rangle = \delta(\mathsf{k}-\mathsf{k}')\delta(n-n')\delta(j-j'), \tag{13}$$

for all $\mathsf{k},\mathsf{k}'\in\mathbb{Z}^d$, $j,j'\in\mathbb{Z}\cap[J,\infty)$, $\ell,\ell'=1,\dots,r$, and $n,n'=1,\dots,s$.

The following is a standard result which can be easily proved using Lemma 1.

Lemma 2. *Let Φ and Ψ be subsets of $L_2(\mathbb{R}^d)$. $(\mathsf{WS}_J(\Phi;\Psi),\mathsf{WS}_J(\tilde{\Phi};\tilde{\Psi}))$ is a pair of nonhomogeneous biorthogonal M-wavelet bases of $L_2(\mathbb{R}^d)$ if and only if it is a pair of nonhomogeneous dual M-wavelet frames of $L_2(\mathbb{R}^d)$ and the biorthogonality conditions in* (13) *are satisfied.*

For pairs of nonhomogeneous biorthogonal M-wavelet bases of $L_2(\mathbb{R}^d)$, we have

Theorem 1. *([44, Theorem 7]) Let M be a $d\times d$ real-valued invertible matrix. Let $\Phi,\Psi,\ \tilde{\Phi},\tilde{\Psi}$ in* (6) *be finite subsets of $L_2(\mathbb{R}^d)$. Suppose $(\mathsf{WS}_J(\Phi;\Psi),\mathsf{WS}_J(\tilde{\Phi};\tilde{\Psi}))$ is a pair of nonhomogeneous biorthogonal M-wavelet bases of $L_2(\mathbb{R}^d)$ for some integer J. Then it is a pair of nonhomogeneous biorthogonal M-wavelet bases of $L_2(\mathbb{R}^d)$ for all integers J. Moreover, there exist $r\times r$ matrices $\mathbf{a}_{\mathsf{k}},\tilde{\mathbf{a}}_{\mathsf{k}}$ and $s\times r$ matrices $\mathbf{b}_{\mathsf{k}},\tilde{\mathbf{b}}_{\mathsf{k}}$, $\mathsf{k}\in\mathbb{Z}^d$ of $2\pi\mathbb{Z}^d$-periodic functions in $L_2(\mathbb{T}^d)$ such that for all $\mathsf{k}\in\mathbb{Z}^d$,*

$$e^{-i\mathsf{k}\cdot\mathsf{M}^{\mathsf{T}}\xi}\hat{\phi}(\mathsf{M}^{\mathsf{T}}\xi)=\mathbf{a}_{\mathsf{k}}(\xi)\hat{\phi}(\xi) \quad\text{and}\quad e^{-i\mathsf{k}\cdot\mathsf{M}^{\mathsf{T}}\xi}\hat{\psi}(\mathsf{M}^{\mathsf{T}}\xi)=\mathbf{b}_{\mathsf{k}}(\xi)\hat{\phi}(\xi), \tag{14}$$

$$e^{-i\mathsf{k}\cdot\mathsf{M}^{\mathsf{T}}\xi}\hat{\tilde{\phi}}(\mathsf{M}^{\mathsf{T}}\xi)=\tilde{\mathbf{a}}_{\mathsf{k}}(\xi)\hat{\tilde{\phi}}(\xi) \quad\text{and}\quad e^{-i\mathsf{k}\cdot\mathsf{M}^{\mathsf{T}}\xi}\hat{\tilde{\psi}}(\mathsf{M}^{\mathsf{T}}\xi)=\tilde{\mathbf{b}}_{\mathsf{k}}(\xi)\hat{\tilde{\phi}}(\xi), \tag{15}$$

for almost every $\xi\in\mathbb{R}^d$, where

$$\phi=[\phi^1,\ldots,\phi^r]^{\mathsf{T}},\ \psi=[\psi^1,\ldots,\psi^s]^{\mathsf{T}},\ \tilde{\phi}=[\tilde{\phi}^1,\ldots,\tilde{\phi}^r]^{\mathsf{T}},\ \tilde{\psi}=[\tilde{\psi}^1,\ldots,\tilde{\psi}^s]^{\mathsf{T}}. \tag{16}$$

If M is a $d\times d$ integer invertible matrix with $\mathsf{d}_{\mathsf{M}}:=|\det\mathsf{M}|$, then $s=r(\mathsf{d}_{\mathsf{M}}-1)$ and

$$\overline{\mathbf{P}_{[\tilde{\mathbf{a}}_0,\tilde{\mathbf{b}}_0]}(\xi)}^{\mathsf{T}}\mathbf{P}_{[\mathbf{a}_0,\mathbf{b}_0]}(\xi)=I_{r\mathsf{d}_{\mathsf{M}}},\qquad a.e.\,\xi\in\mathbb{R}^d,$$

where $I_{r\mathsf{d}_{\mathsf{M}}}$ denotes the $(r\mathsf{d}_{\mathsf{M}})\times(r\mathsf{d}_{\mathsf{M}})$ identity matrix and

$$\mathbf{P}_{[\mathbf{a}_0,\mathbf{b}_0]}(\xi):=\begin{bmatrix}\mathbf{a}_0(\xi+2\pi\omega_0) & \mathbf{a}_0(\xi+2\pi\omega_1) & \cdots & \mathbf{a}_0(\xi+2\pi\omega_{\mathsf{d}_{\mathsf{M}}-1})\\ \mathbf{b}_0(\xi+2\pi\omega_0) & \mathbf{b}_0(\xi+2\pi\omega_1) & \cdots & \mathbf{b}_0(\xi+2\pi\omega_{\mathsf{d}_{\mathsf{M}}-1})\end{bmatrix}$$

and $\{\omega_0,\ldots,\omega_{\mathsf{d}_{\mathsf{M}}-1}\}:=[(\mathsf{M}^{\mathsf{T}})^{-1}\mathbb{Z}^d]\cap[0,1)^d$. If M is a $d\times d$ real-valued expansive matrix, then $(\mathsf{WS}(\Psi),\mathsf{WS}(\tilde{\Psi}))$ is a pair of homogeneous biorthogonal M-wavelet bases of $L_2(\mathbb{R}^d)$, that is, each of $\mathsf{WS}(\Psi)$ and $\mathsf{WS}(\tilde{\Psi})$ is a Riesz basis of $L_2(\mathbb{R}^d)$ and the last identity of (13) *holds for all $\mathsf{k},\mathsf{k}'\in\mathbb{Z}^d$, $j,j'\in\mathbb{Z}$, and $n,n'=1,\ldots,s$.*

In view of Theorem 1, we say that $(\{\Phi;\Psi\},\{\tilde{\Phi};\tilde{\Psi}\})$ is a (nonhomogeneous) biorthogonal M-wavelet in $L_2(\mathbb{R}^d)$ if $(\mathsf{WS}_0(\Phi;\Psi),\mathsf{WS}_0(\tilde{\Phi};\tilde{\Psi}))$ is a pair of nonhomogeneous Riesz M-wavelet bases of $L_2(\mathbb{R}^d)$; furthermore, $\{\Phi;\Psi\}$ is called a (nonhomogeneous) orthonormal M-wavelet in $L_2(\mathbb{R}^d)$ if $\tilde{\Phi}=\Phi$ and $\tilde{\Psi}=\Psi$. We refer to the relations in (14) as the *refinable structure* of the nonhomogeneous wavelet system $\mathsf{WS}_J(\Phi;\Psi)$. When M is an integer matrix, (14) is equivalent to the following well-known relations:

$$\hat{\phi}(\mathsf{M}^{\mathsf{T}}\xi)=\mathbf{a}(\xi)\hat{\phi}(\xi) \quad\text{and}\quad \hat{\psi}(\mathsf{M}^{\mathsf{T}}\xi)=\mathbf{b}(\xi)\hat{\phi}(\xi), \tag{17}$$

where $\mathbf{a}:=\mathbf{a}_0$ and $\mathbf{b}:=\mathbf{b}_0$ in (14). Theorem 1 shows that for nonhomogeneous biorthogonal M-wavelets, the refinable structure in (14) and (15) is intrinsically built into nonhomogeneous wavelet systems. Consequently, all nonhomogeneous orthonormal M-wavelets and nonhomogeneous biorthogonal M-wavelets must come from the refinable structures in (14) and (15), and they are intrinsically connected to multiresolution analysis and refinable function vectors.

A similar result is also true for nonhomogeneous M-wavelets.

Proposition 4 *([44, Theorem 8]) Let* M *be a* $d \times d$ *integer invertible matrix. Let* Φ *and* Ψ *in* (6) *be finite subsets of* $L_2(\mathbb{R}^d)$. *Suppose that* $\mathrm{WS}_J(\Phi;\Psi)$ *is a nonhomogeneous Riesz* M*-wavelet basis of* $L_2(\mathbb{R}^d)$ *for some integer* J. *Then the following statements are equivalent to each other:*

(i) *there exist subsets* $\tilde{\Phi}$ *and* $\tilde{\Psi}$ *in* (6) *of* $L_2(\mathbb{R}^d)$ *such that* $(\mathrm{WS}_J(\Phi;\Psi), \mathrm{WS}_J(\tilde{\Phi};\tilde{\Psi}))$ *is a pair of nonhomogeneous biorthogonal* M*-wavelet bases of* $L_2(\mathbb{R}^d)$*;*
(ii) *there exist an* $r \times r$ *matrix* **a** *and an* $s \times r$ *matrix* **b** *of* $2\pi\mathbb{Z}^d$*-periodic measurable functions in* $L_2(\mathbb{T}^d)$ *such that* (6) *holds with* ϕ *and* ψ *being defined in* (16).

In the following, we present a simple example to show that not every homogeneous M-wavelet has the property in item (ii) of Proposition 4. In fact, assume that $\mathrm{WS}_0(\Phi;\Psi)$ is a nonhomogeneous orthonormal M-wavelet basis of $L_2(\mathbb{R}^d)$ with Φ and Ψ in (6). For a real number ε, define

$$\mathring{\phi}^\ell := \phi^\ell + \varepsilon\psi^\ell, \quad \ell = 1,\dots,r \quad \text{and} \quad \mathring{\psi}^\ell := \psi^\ell + \varepsilon\psi^\ell(\mathsf{M}\cdot), \qquad \ell = 1,\dots,s.$$

Let $\mathring{\Phi} := \{\mathring{\phi}^1,\dots,\mathring{\phi}^r\}$ and $\mathring{\Psi} := \{\mathring{\psi}^1,\dots,\mathring{\psi}^s\}$. By a simple argument, for $0 < \varepsilon < 1$, $\mathrm{WS}_0(\mathring{\Phi};\mathring{\Psi})$ is a nonhomogeneous Riesz M-wavelet basis of $L_2(\mathbb{R}^d)$, but item (ii) of Proposition 4 cannot be true.

The results in this section have been initiated in [43] for dimension one, and can be generalized to a general function space instead of $L_2(\mathbb{R}^d)$.

For a nonhomogeneous M-wavelet system $\mathrm{WS}_J(\Phi;\Psi)$, the set Φ generally cannot be the empty set and M is just a real-valued invertible matrix. However, to consider the limit of $\mathrm{WS}_J(\Phi;\Psi)$ as $J \to -\infty$, we have to assume that M is expansive and this is largely due to the following well-known result in the literature.

Lemma 3. *(e.g. [44, Lemma 3]) Let* M *be a* $d \times d$ *real-valued expansive matrix and* Φ *be a (not necessarily finite) subset of* $L_2(\mathbb{R}^d)$ *such that* $\sum_{\phi\in\Phi} \|\phi\|^2_{L_2(\mathbb{R}^d)} < \infty$. *Suppose that there exists a positive constant* C *such that* $\sum_{\phi\in\Phi}\sum_{\mathsf{k}\in\mathbb{Z}^d} |\langle f, \phi(\cdot - \mathsf{k})\rangle|^2 \leqslant C\|f\|^2_{L_2(\mathbb{R}^d)}$ *for all* $f \in L_2(\mathbb{R}^d)$. *Then*

$$\lim_{j\to-\infty} \sum_{\phi\in\Phi}\sum_{\mathsf{k}\in\mathbb{Z}^d} |\langle f, \phi_{\mathsf{M}^j;\mathsf{k}}\rangle|^2 = 0 \qquad \forall f \in L_2(\mathbb{R}^d). \tag{18}$$

It remains unclear whether the property in (18) will force M to be expansive in some sense. It is also unclear to us at this moment whether there exists a nontrivial non-expansive (integer) dilation matrix M such that there is a nonhomogeneous M-wavelet system which is a tight frame or a Riesz basis of $L_2(\mathbb{R}^d)$. As in [21], for a nonhomogeneous framelet, it is also of interest to study the structure of the canonical dual and its other dual systems with the wavelet structure.

3 Frequency-Based Nonhomogeneous Dual Framelets in the Distribution Space

To characterize nonhomogeneous dual or tight framelets in $L_2(\mathbb{R}^d)$, we shall take a frequency-based approach by studying frequency-based (nonstationary) nonhomogeneous wavelet systems in the distribution space. More precisely, we shall introduce and characterize a pair of frequency-based nonstationary nonhomogeneous dual wavelet frames in the distribution space. Such a notion is closely related to the perfect reconstruction property in the function setting of nonhomogeneous wavelet systems. We shall see the importance of such a notion in this and later sections.

Following the standard notation, we denote by $\mathscr{D}(\mathbb{R}^d)$ the linear space of all compactly supported C^∞ (test) functions with the usual topology, and $\mathscr{D}'(\mathbb{R}^d)$ the linear space of all distributions, that is, $\mathscr{D}'(\mathbb{R}^d)$ is the dual space of $\mathscr{D}(\mathbb{R}^d)$. By duality, it is easy to see that translation, dilation and modulation in (1) can be naturally extended to distributions in $\mathscr{D}'(\mathbb{R}^d)$. For a tempered distribution f, by the definition of the notation $f_{U;\mathsf{k},\mathsf{n}}$ in (1), we have

$$\widehat{f_{U;\mathsf{k},\mathsf{n}}} = \mathrm{e}^{-\mathrm{i}\mathsf{k}\cdot\mathsf{n}}\hat{f}_{(U^\mathsf{T})^{-1};-\mathsf{n},\mathsf{k}} \quad \text{and} \quad \widehat{f_{U;\mathsf{k}}} = \hat{f}_{(U^\mathsf{T})^{-1};\mathbf{0},\mathsf{k}}, \tag{19}$$

where the Fourier transform is defined to be $\hat{f}(\xi) := \int_{\mathbb{R}^d} f(x)\mathrm{e}^{-\mathrm{i}x\cdot\xi}\mathrm{d}x, \xi \in \mathbb{R}^d$ for $f \in L_1(\mathbb{R}^d)$ and can be naturally extended to tempered distributions.

By $L_p^{\mathrm{loc}}(\mathbb{R}^d)$ we denote the linear space of all measurable functions f such that $\int_K |f(x)|^p\mathrm{d}x < \infty$ for every compact subset K of $\mathbb{R}^d$ with the usual modification for $p = \infty$. Note that $L_1^{\mathrm{loc}}(\mathbb{R}^d)$ is just the set of all locally integrable functions that can be globally identified as distributions, that is, $L_p^{\mathrm{loc}}(\mathbb{R}^d) \subseteq L_1^{\mathrm{loc}}(\mathbb{R}^d) \subseteq \mathscr{D}'(\mathbb{R}^d)$ for all $1 \leqslant p \leqslant \infty$. For $\mathbf{f} \in \mathscr{D}(\mathbb{R}^d)$ and $\psi \in L_1^{\mathrm{loc}}(\mathbb{R}^d)$, we shall use the following pairing

$$\langle \mathbf{f}, \psi \rangle := \int_{\mathbb{R}^d} \mathbf{f}(\xi)\overline{\psi(\xi)}\mathrm{d}\xi \quad \text{and} \quad \langle \psi, \mathbf{f} \rangle := \overline{\langle \mathbf{f}, \psi \rangle} = \int_{\mathbb{R}^d} \psi(\xi)\overline{\mathbf{f}(\xi)}\mathrm{d}\xi.$$

When $\mathbf{f} \in \mathscr{D}(\mathbb{R}^d)$ and $\psi \in \mathscr{D}'(\mathbb{R}^d)$, the duality pairings $\langle \mathbf{f}, \psi \rangle$ and $\langle \psi, \mathbf{f} \rangle$ are understood as $\langle \mathbf{f}, \psi \rangle := \overline{\langle \psi, \mathbf{f} \rangle} := \overline{\psi(\overline{\mathbf{f}})}$, where here bar refers to the complex conjugate.

Let J be an integer and N be a $d \times d$ real-valued invertible matrix. Let Φ and Ψ_j, $j \geqslant J$ be subsets of distributions. *A frequency-based nonstationary nonhomogeneous* N*-wavelet system* is defined to be

$$\begin{aligned} \mathrm{FWS}_J(\Phi; \{\Psi_j\}_{j=J}^\infty) = {} & \{\varphi_{\mathsf{N}^J;\mathbf{0},\mathsf{k}} \mid \mathsf{k} \in \mathbb{Z}^d, \varphi \in \Phi\} \cup \\ & \bigcup_{j=J}^\infty \{\psi_{\mathsf{N}^j;\mathbf{0},\mathsf{k}} \mid \mathsf{k} \in \mathbb{Z}^d, \psi \in \Psi_j\}. \end{aligned} \tag{20}$$

For the particular case $\Psi_j = \Psi$ for all $j \geqslant J$, a frequency-based nonstationary nonhomogeneous N-wavelet system in (20) becomes *a frequency-based (stationary) nonhomogeneous* N*-wavelet system*:

$$\mathrm{FWS}_J(\Phi; \Psi) = \{\varphi_{\mathsf{N}^J;\mathbf{0},\mathsf{k}} \mid \mathsf{k} \in \mathbb{Z}^d, \varphi \in \Phi\} \cup \{\psi_{\mathsf{N}^j;\mathbf{0},\mathsf{k}} \mid j \geqslant J, \mathsf{k} \in \mathbb{Z}^d, \psi \in \Psi\}.$$

For a nonhomogeneous M-wavelet system $\mathrm{WS}_J(\Phi;\Psi)$ such that all the generators in Φ and Ψ are tempered distributions, by (19), the image of the nonhomogeneous M-wavelet system $\mathrm{WS}_J(\Phi;\Psi)$ under the Fourier transform simply becomes the frequency-based nonhomogeneous $(\mathsf{M}^{\mathsf{T}})^{-1}$-wavelet system $\mathrm{FWS}_J(\widehat{\Phi};\widehat{\Psi})$, where $\widehat{H} := \{\hat{h} \mid h \in H\}$ for a subset H of tempered distributions.

For analysis of wavelets and framelets, as argued in [43, 44], it is often easier to work with frequency-based nonhomogeneous wavelet systems $\mathrm{FWS}_J(\widehat{\Phi};\widehat{\Psi})$ instead of space/time-based nonhomogeneous wavelet systems $\mathrm{WS}_J(\Phi;\Psi)$, though both are equivalent to each other within the framework of tempered distributions. Since we consider frequency-based nonhomogeneous wavelets and framelets in the distribution space $\mathscr{D}'(\mathbb{R}^d)$, it is natural for us to consider $\mathrm{FWS}_J(\Phi;\Psi) \subseteq \mathscr{D}'(\mathbb{R}^d)$.

Let N be a $d \times d$ real-valued invertible matrix. Let

$$\Phi = \{\varphi^1,\ldots,\varphi^r\} \quad \text{and} \quad \tilde{\Phi} = \{\tilde{\varphi}^1,\ldots,\tilde{\varphi}^r\} \tag{21}$$

and

$$\Psi_j = \{\psi^{j,1},\ldots,\psi^{j,s_j}\} \quad \text{and} \quad \tilde{\Psi}_j = \{\tilde{\psi}^{j,1},\ldots,\tilde{\psi}^{j,s_j}\} \tag{22}$$

be subsets of $\mathscr{D}'(\mathbb{R}^d)$ for $j \geqslant J$, where $r, s_j \in \mathbb{N} \cup \{0,+\infty\}$. Let $\mathrm{FWS}_J(\Phi;\{\Psi_j\}_{j=J}^{\infty})$ be defined in (20) and $\mathrm{FWS}_J(\tilde{\Phi};\{\tilde{\Psi}_j\}_{j=J}^{\infty})$ be defined similarly. As in [43, 44], we say that the pair

$$(\mathrm{FWS}_J(\Phi;\{\Psi_j\}_{j=J}^{\infty}),\mathrm{FWS}_J(\tilde{\Phi};\{\tilde{\Psi}_j\}_{j=J}^{\infty})) \tag{23}$$

is *a pair of frequency-based nonstationary nonhomogeneous dual* N-*wavelet frames in the distribution space* $\mathscr{D}'(\mathbb{R}^d)$ if the following identity holds

$$\sum_{\ell=1}^{r}\sum_{\mathsf{k}\in\mathbb{Z}^d}\langle \mathbf{f},\varphi^{\ell}_{\mathsf{N}^J;\mathbf{0},\mathsf{k}}\rangle\langle\tilde{\varphi}^{\ell}_{\mathsf{N}^J;\mathbf{0},\mathsf{k}},\mathbf{g}\rangle+\sum_{j=J}^{\infty}\sum_{\ell=1}^{s_j}\sum_{\mathsf{k}\in\mathbb{Z}^d}\langle \mathbf{f},\psi^{j,\ell}_{\mathsf{N}^j;\mathbf{0},\mathsf{k}}\rangle\langle\tilde{\psi}^{j,\ell}_{\mathsf{N}^j;\mathbf{0},\mathsf{k}},\mathbf{g}\rangle=(2\pi)^d\langle\mathbf{f},\mathbf{g}\rangle \tag{24}$$

for all $\mathbf{f},\mathbf{g} \in \mathscr{D}(\mathbb{R}^d)$, where the infinite series in (24) converge in the following sense

(i) for every $\mathbf{f},\mathbf{g} \in \mathscr{D}(\mathbb{R}^d)$, all the series

$$\sum_{\ell=1}^{r}\sum_{\mathsf{k}\in\mathbb{Z}^d}\langle \mathbf{f},\varphi^{\ell}_{\mathsf{N}^J;\mathbf{0},\mathsf{k}}\rangle\langle\tilde{\varphi}^{\ell}_{\mathsf{N}^J;\mathbf{0},\mathsf{k}},\mathbf{g}\rangle \quad \text{and} \quad \sum_{\ell=1}^{s_j}\sum_{\mathsf{k}\in\mathbb{Z}^d}\langle \mathbf{f},\psi^{j,\ell}_{\mathsf{N}^j;\mathbf{0},\mathsf{k}}\rangle\langle\tilde{\psi}^{j,\ell}_{\mathsf{N}^j;\mathbf{0},\mathsf{k}},\mathbf{g}\rangle \tag{25}$$

converge absolutely for all integers $j \geqslant J$;

(ii) for every $\mathbf{f},\mathbf{g} \in \mathscr{D}(\mathbb{R}^d)$, the following limit exists and

$$\lim_{J_+\to+\infty}\Big(\sum_{\ell=1}^{r}\sum_{\mathsf{k}\in\mathbb{Z}^d}\langle \mathbf{f},\varphi^{\ell}_{\mathsf{N}^J;\mathbf{0},\mathsf{k}}\rangle\langle\tilde{\varphi}^{\ell}_{\mathsf{N}^J;\mathbf{0},\mathsf{k}},\mathbf{g}\rangle + \sum_{j=J}^{J_+-1}\sum_{\ell=1}^{s_j}\sum_{\mathsf{k}\in\mathbb{Z}^d}\langle \mathbf{f},\psi^{j,\ell}_{\mathsf{N}^j;\mathbf{0},\mathsf{k}}\rangle\langle\tilde{\psi}^{j,\ell}_{\mathsf{N}^j;\mathbf{0},\mathsf{k}},\mathbf{g}\rangle\Big) = (2\pi)^d\langle\mathbf{f},\mathbf{g}\rangle. \tag{26}$$

As shown in [43,44], the condition in the above item (i) is automatically satisfied if $\Phi, \Psi_j, \tilde{\Phi}, \tilde{\Psi}_j$ are subsets of $L_2^{\mathrm{loc}}(\mathbb{R}^d)$ and r, s_j are finite. The condition in item (ii) is simply the perfect reconstruction property for the test function space $\mathscr{D}(\mathbb{R}^d)$ of a pair of frequency-based nonhomogeneous wavelet systems in (23).

The following result shows an intrinsic refinable structure for a pair of frequency-based nonhomogeneous dual N-wavelet frames in the distribution space $\mathscr{D}'(\mathbb{R}^d)$.

Proposition 5 *([44, Corollary 15]) Let* N *be a* $d \times d$ *real-valued invertible matrix. Let* $\Phi, \tilde{\Phi}$ *be as in* (21), *and*

$$\Psi = \{\psi^1, \dots, \psi^s\} \quad and \quad \tilde{\Psi} = \{\tilde{\psi}^1, \dots, \tilde{\psi}^s\} \tag{27}$$

be subsets of distributions in $\mathscr{D}'(\mathbb{R}^d)$. *Then* $(\mathrm{FWS}_J(\Phi;\Psi), \mathrm{FWS}_J(\tilde{\Phi};\tilde{\Psi}))$ *is a pair of frequency-based nonhomogeneous dual* N*-wavelet frames in the distribution space* $\mathscr{D}'(\mathbb{R}^d)$ *for some integer* J, *if and only if, for all* $\mathbf{f}, \mathbf{g} \in \mathscr{D}(\mathbb{R}^d)$,

$$\lim_{j \to +\infty} \sum_{\ell=1}^{r} \sum_{\mathsf{k} \in \mathbb{Z}^d} \langle \mathbf{f}, \varphi^{\ell}_{\mathsf{N}^j;\mathbf{0},\mathsf{k}} \rangle \langle \tilde{\varphi}^{\ell}_{\mathsf{N}^j;\mathbf{0},\mathsf{k}}, \mathbf{g} \rangle = (2\pi)^d \langle \mathbf{f}, \mathbf{g} \rangle \tag{28}$$

and

$$\sum_{\ell=1}^{r} \sum_{\mathsf{k} \in \mathbb{Z}^d} \langle \mathbf{f}, \varphi^{\ell}_{I_d;\mathbf{0},\mathsf{k}} \rangle \langle \tilde{\varphi}^{\ell}_{I_d;\mathbf{0},\mathsf{k}}, \mathbf{g} \rangle + \sum_{\ell=1}^{s} \sum_{\mathsf{k} \in \mathbb{Z}^d} \langle \mathbf{f}, \psi^{\ell}_{I_d;\mathbf{0},\mathsf{k}}, \rangle \langle \tilde{\psi}^{\ell}_{I_d;\mathbf{0},\mathsf{k}}, \mathbf{g} \rangle \tag{29}$$
$$= \sum_{\ell=1}^{r} \sum_{\mathsf{k} \in \mathbb{Z}^d} \langle \mathbf{f}, \varphi^{\ell}_{\mathsf{N};\mathbf{0},\mathsf{k}} \rangle \langle \tilde{\varphi}^{\ell}_{\mathsf{N};\mathbf{0},\mathsf{k}}, \mathbf{g} \rangle.$$

By Proposition 5, if $(\mathrm{FWS}_J(\Phi;\Psi), \mathrm{FWS}_J(\tilde{\Phi};\tilde{\Psi}))$ is a pair of frequency-based nonhomogeneous dual N-wavelet frames in the distribution space $\mathscr{D}'(\mathbb{R}^d)$ for some integer J, then it is true for all integers J. Consequently, we call $(\{\Phi;\Psi\}, \{\tilde{\Phi};\tilde{\Psi}\})$ a frequency-based (nonhomogeneous) dual N-framelet in the distribution space. The condition in (28) is just a normalization condition. The condition in (29) shares similarity to the refinable structure in a multiresolution analysis, and implies that for all $J_- \leqslant J_+$ and for all $\mathbf{f}, \mathbf{g} \in \mathscr{D}(\mathbb{R}^d)$,

$$\sum_{\ell=1}^{r} \sum_{\mathsf{k} \in \mathbb{Z}^d} \langle \mathbf{f}, \varphi^{\ell}_{\mathsf{N}^{J_-};\mathbf{0},\mathsf{k}} \rangle \langle \tilde{\varphi}^{\ell}_{\mathsf{N}^{J_-};\mathbf{0},\mathsf{k}}, \mathbf{g} \rangle + \sum_{j=J_-}^{J_+-1} \sum_{\ell=1}^{s} \sum_{\mathsf{k} \in \mathbb{Z}^d} \langle \mathbf{f}, \psi^{\ell}_{\mathsf{N}^j;\mathbf{0},\mathsf{k}} \rangle \langle \tilde{\psi}^{\ell}_{\mathsf{N}^j;\mathbf{0},\mathsf{k}}, \mathbf{g} \rangle$$
$$= \sum_{\ell=1}^{r} \sum_{\mathsf{k} \in \mathbb{Z}^d} \langle \mathbf{f}, \varphi^{\ell}_{\mathsf{N}^{J_+};\mathbf{0},\mathsf{k}} \rangle \langle \tilde{\varphi}^{\ell}_{\mathsf{N}^{J_+};\mathbf{0},\mathsf{k}}, \mathbf{g} \rangle.$$

The following result completely characterizes a pair of frequency-based nonstationary nonhomogeneous dual N-wavelet frames in the distribution space.

Theorem 2. *([44, Theorem 11]) Let* J *be an integer. Let* N *be a* $d \times d$ *real-valued invertible matrix such that* N^{-1} *is expansive. Let* $\Phi, \tilde{\Phi}$ *in* (21) *and* $\Psi_j, \tilde{\Psi}_j$ *in* (22) *be finite subsets of* $L_2^{\mathrm{loc}}(\mathbb{R})$ *for all integers* $j \geqslant J$. *Then the pair in* (23) *is a pair* *of frequency-based nonstationary nonhomogeneous dual* N*-wavelet frames in the distribution space* $\mathscr{D}'(\mathbb{R}^d)$ *if and only if*

$$\lim_{J_+\to+\infty}\left\langle \mathscr{I}_{\Phi}^{\mathbf{0}}(\mathsf{N}^J\cdot)+\sum_{j=J}^{J_+-1}\mathscr{I}_{\Psi_j}^{\mathbf{0}}(\mathsf{N}^j\cdot),\mathbf{h}\right\rangle=\langle 1,\mathbf{h}\rangle \qquad \forall\,\mathbf{h}\in\mathscr{D}(\mathbb{R}^d) \tag{30}$$

and

$$\mathscr{I}_{\Phi}^{\mathsf{N}^J\mathsf{k}}(\mathsf{N}^J\xi)+\sum_{j=J}^{\infty}\mathscr{I}_{\Psi_j}^{\mathsf{N}^j\mathsf{k}}(\mathsf{N}^j\xi)=0, \qquad a.e.\,\xi\in\mathbb{R}^d, 0\neq\mathsf{k}\in\cup_{j=J}^{\infty}[\mathsf{N}^{-j}\mathbb{Z}^d], \tag{31}$$

(the infinite sums in (31) *are in fact finite for* ξ *on any bounded set.) where*

$$\mathscr{I}_{\Phi}^{\mathsf{k}}(\xi):=\sum_{\ell=1}^{r}\overline{\varphi^\ell(\xi)}\tilde{\varphi}^\ell(\xi+2\pi\mathsf{k}),\ \mathsf{k}\in\mathbb{Z}^d \ \textit{and}\ \mathscr{I}_{\Phi}^{\mathsf{k}}:=0,\ \ \mathsf{k}\in\mathbb{R}^d\backslash\mathbb{Z}^d, \tag{32}$$

$$\mathscr{I}_{\Psi_j}^{\mathsf{k}}(\xi):=\sum_{\ell=1}^{s_j}\overline{\psi^{j,\ell}(\xi)}\tilde{\psi}^{j,\ell}(\xi+2\pi\mathsf{k}),\ \mathsf{k}\in\mathbb{Z}^d \ \textit{and}\ \mathscr{I}_{\Psi_j}^{\mathsf{k}}:=0,\ \ \mathsf{k}\in\mathbb{R}^d\backslash\mathbb{Z}^d. \tag{33}$$

As a direct consequence of Theorem 2, we have the following result on frequency-based nonhomogeneous dual N-wavelet frames in the distribution space.

Corollary 1 *([44, Corollary 15]) Let* N *be a* $d\times d$ *real-valued invertible matrix such that* N^{-1} *is expansive. Let* $\Phi,\tilde{\Phi}$ *in* (21) *and* $\Psi,\tilde{\Psi}$ *in* (27) *be finite subsets of* $L_2^{\mathrm{loc}}(\mathbb{R}^d)$*. Then the following statements are equivalent:*

(i) $(\{\Phi;\Psi\},\{\tilde{\Phi};\tilde{\Psi}\})$ *is a frequency-based nonhomogeneous dual* N*-framelet in the distribution space* $\mathscr{D}'(\mathbb{R}^d)$*;*
(ii) (28) *and* (29) *are satisfied for all* $\mathbf{f},\mathbf{g}\in\mathscr{D}(\mathbb{R}^d)$*;*
(iii) *the following relations are satisfied:*

$$\lim_{j\to+\infty}\left\langle \sum_{\ell=1}^{r}\overline{\varphi^\ell(\mathsf{N}^j\cdot)}\tilde{\varphi}^\ell(\mathsf{N}^j\cdot),\mathbf{h}\right\rangle=\langle 1,\mathbf{h}\rangle, \qquad \forall\,\mathbf{h}\in\mathscr{D}(\mathbb{R}^d); \tag{34}$$

and

$$\mathscr{I}_{\Phi}^{\mathsf{k}}(\xi)+\mathscr{I}_{\Psi}^{\mathsf{k}}(\xi)=\mathscr{I}_{\Phi}^{\mathsf{N}\mathsf{k}}(\mathsf{N}\xi), \qquad a.e.\,\xi\in\mathbb{R}^d, \mathsf{k}\in\mathbb{Z}^d\cup[\mathsf{N}^{-1}\mathbb{Z}^d], \tag{35}$$

where $\mathscr{I}_{\Phi}^{\mathsf{k}}$ *is defined in* (32) *and*

$$\mathscr{I}_{\Psi}^{\mathsf{k}}(\xi):=\sum_{\ell=1}^{s}\overline{\psi^\ell(\xi)}\tilde{\psi}^\ell(\xi+2\pi\mathsf{k}),\ \mathsf{k}\in\mathbb{Z}^d \quad \textit{and}\quad \mathscr{I}_{\Psi}^{\mathsf{k}}:=0,\ \mathsf{k}\in\mathbb{R}^d\backslash\mathbb{Z}^d. \tag{36}$$

There is a more general result than Theorem 2. In fact, we have a complete characterization (see [44, Theorem 11]) for a pair of frequency-based fully nonstationary nonhomogeneous dual wavelet frames in the distribution space; that is, not only the generating set Ψ_j can depend on the scale level j, but also the dilation N^j (which is the j-th power of N) at the scale level j could be a general arbitrary matrix N_j. The assumption that N^{-1} is expansive in Corollary 1 is only used for proving the equivalence between (28) and (34). To appreciate the results in Theorem 2 and Corollary 1 for nonhomogeneous framelets, we shall compare them with the characterization of homogeneous framelets in $L_2(\mathbb{R}^d)$ at the end of this section.

In the following result and later sections, we shall see the importance of the notion of a frequency-based nonhomogeneous dual framelet in the distribution space.

Theorem 3. *Let M be a $d \times d$ integer expansive matrix. Define $\mathsf{N} := (\mathsf{M}^\mathsf{T})^{-1}$. Let* $\mathbf{a}$ *and* $\tilde{\mathbf{a}}$ *be $2\pi\mathbb{Z}^d$-periodic measurable functions such that*

$$|1-\mathbf{a}(\xi)| \leqslant C\|\xi\|^{\varepsilon}, \qquad |1-\tilde{\mathbf{a}}(\xi)| \leqslant \tilde{C}\|\xi\|^{\tilde{\varepsilon}}, \qquad a.e.\ \xi \in \mathbb{R}^d \tag{37}$$

for some positive numbers $\varepsilon, \tilde{\varepsilon}, C, \tilde{C}$. Define

$$\varphi(\xi) := \prod_{j=1}^{\infty} \mathbf{a}(\mathsf{N}^j\xi) \quad and \quad \tilde{\varphi}(\xi) := \prod_{j=1}^{\infty} \tilde{\mathbf{a}}(\mathsf{N}^j\xi), \qquad \xi \in \mathbb{R}^d. \tag{38}$$

Then φ and $\tilde{\varphi}$ are well-defined functions in $L_\infty^{\mathrm{loc}}(\mathbb{R}^d)$ satisfying

$$\varphi(\mathsf{M}^\mathsf{T}\xi) = \mathbf{a}(\xi)\varphi(\xi) \quad and \quad \tilde{\varphi}(\mathsf{M}^\mathsf{T}\xi) = \tilde{\mathbf{a}}(\xi)\tilde{\varphi}(\xi), \qquad a.e. \xi \in \mathbb{R}^d. \tag{39}$$

Let $\theta_1, \ldots, \theta_r, \mathbf{b}_{j,1}, \ldots, \mathbf{b}_{j,s_{j-1}}, \tilde{\theta}_1, \ldots, \tilde{\theta}_r, \tilde{\mathbf{b}}_{j,1}, \ldots, \tilde{\mathbf{b}}_{j,s_{j-1}}, r, s_{j-1} \in \mathbb{N} \cup \{0\}$ and $j \in \mathbb{N}$, be $2\pi\mathbb{Z}^d$-periodic measurable functions in $L_2^{\mathrm{loc}}(\mathbb{R}^d)$ (that is, in $L_2(\mathbb{T}^d)$). Define

$$\begin{aligned} \varphi^\ell(\xi) &:= \theta_\ell(\xi)\varphi(\xi) \quad and \quad \tilde{\varphi}^\ell(\xi) := \tilde{\theta}_\ell(\xi)\tilde{\varphi}(\xi), \quad \ell = 1, \ldots, r, \\ \psi^{j-1,\ell}(\mathsf{M}^\mathsf{T}\xi) &:= \mathbf{b}_{j,\ell}(\xi)\varphi(\xi) \quad and \quad \tilde{\psi}^{j-1,\ell}(\mathsf{M}^\mathsf{T}\xi) := \tilde{\mathbf{b}}_{j,\ell}(\xi)\tilde{\varphi}(\xi), \end{aligned}$$

for $\ell = 1, \ldots, s_{j-1}$ and $j \in \mathbb{N}$. Then $\Phi, \tilde{\Phi}$ in (21) and $\Psi_j, \tilde{\Psi}_j$ in (22) are subsets of $L_2^{\mathrm{loc}}(\mathbb{R})$ for all $j \in \mathbb{N} \cup \{0\}$. The pair in (23) is a pair of frequency-based nonstationary nonhomogeneous dual N-wavelet frames in the distribution space $\mathscr{D}'(\mathbb{R})$ for every integer $J \geqslant 0$, if and only if, for all $j \in \mathbb{N}$ and for all $\omega \in \Omega_\mathsf{N} := [\mathsf{N}\mathbb{Z}^d] \cap [0,1)^d$,

$$\Theta(\mathsf{M}^\mathsf{T}\xi)\overline{\mathbf{a}(\xi+2\pi\omega)}\tilde{\mathbf{a}}(\xi) + \sum_{\ell=1}^{s_{j-1}} \overline{\mathbf{b}_{j,\ell}(\xi+2\pi\omega)}\tilde{\mathbf{b}}_{j,\ell}(\xi) = \Theta(\xi)\delta(\omega), \tag{40}$$

for almost every $\xi \in \sigma_\varphi \cap (\sigma_{\tilde{\varphi}} - 2\pi\omega)$, and

$$\lim_{j\to+\infty} \langle \Theta(\mathsf{N}^j\cdot), \mathbf{h}\rangle = \langle 1, \mathbf{h}\rangle \quad \forall\, \mathbf{h} \in \mathscr{D}(\mathbb{R}^d) \quad with \quad \Theta(\xi) := \sum_{\ell=1}^{r} \overline{\theta_\ell(\xi)}\tilde{\theta}_\ell(\xi), \tag{41}$$

where $\sigma_\varphi := \{\xi \mid \sum_{\mathsf{k}\in\mathbb{Z}^d} |\varphi(\xi+2\pi\mathsf{k})| \neq 0\}$, $\sigma_{\tilde{\varphi}} := \{\xi \mid \sum_{\mathsf{k}\in\mathbb{Z}^d} |\tilde{\varphi}(\xi+2\pi\mathsf{k})| \neq 0\}$.

If $\mathbf{a}$ and $\tilde{\mathbf{a}}$ are $2\pi\mathbb{Z}^d$-periodic trigonometric polynomials satisfying $\mathbf{a}(0) = \tilde{\mathbf{a}}(0) = 1$, then it is evident that (37) holds with $\varepsilon = \tilde{\varepsilon} = 1$ and $\sigma_\varphi = \sigma_{\tilde{\varphi}} = \mathbb{R}^d$. Therefore, without a priori condition, Theorem 3 shows that every dual framelet filter bank has an underlying frequency-based dual framelet in the distribution space.

In the rest of this section, we characterize nonhomogeneous framelets in the function space $L_2(\mathbb{R}^d)$. We say that the pair in (23) is *a pair of frequency-based nonstationary nonhomogeneous dual* N*-wavelet frames of* $L_2(\mathbb{R}^d)$ if (i) all elements in the two systems of the pair belong to $L_2(\mathbb{R}^d)$, (ii) each system in the pair is a frame in $L_2(\mathbb{R}^d)$, and (iii) (24) holds for all $\mathbf{f},\mathbf{g} \in L_2(\mathbb{R}^d)$ with the series converging absolutely. By the Plancherel Theorem $(2\pi)^d\langle f,g\rangle = \langle \hat{f},\hat{g}\rangle$ and (19), it is straightforward to see that a space/time-based pair $(\mathrm{WS}_J(\Phi;\Psi),\mathrm{WS}_J(\tilde{\Phi};\tilde{\Psi}))$ is a pair of nonhomogeneous dual M-wavelet frames in $L_2(\mathbb{R}^d)$ if and only if the frequency-based pair $(\mathrm{FWS}_J(\widehat{\Phi};\widehat{\Psi}),\mathrm{FWS}_J(\widehat{\tilde{\Phi}};\widehat{\tilde{\Psi}}))$ is a pair of frequency-based nonhomogeneous dual $(\mathsf{M}^\mathsf{T})^{-1}$-wavelet frames in $L_2(\mathbb{R}^d)$, where $\widehat{H} := \{\hat{h} \mid h \in H\}$ for a subset H of tempered distributions.

The following result, which is a special case of Theorem 5, characterizes frequency-based dual framelets in $L_2(\mathbb{R}^d)$.

Corollary 2 *([44, Theorem 9]) Let* N *be a* $d\times d$ *real-valued invertible matrix and* J *be an integer. Let* $\Phi,\tilde{\Phi}$ *and* $\Psi_j,\tilde{\Psi}_j$ *be at most countable subsets of distributions on* $\mathbb{R}^d$ *for all integers* $j \geqslant J$*. Then the pair in* (23) *is a pair of frequency-based nonstationary nonhomogeneous dual* N*-wavelet frames in* $L_2(\mathbb{R}^d)$ *if and only if*

(i) *there exists a positive constant* C *such that for all* $\mathbf{f},\mathbf{g} \in \mathscr{D}(\mathbb{R}^d)$,

$$\sum_{\varphi\in\Phi}\sum_{\mathsf{k}\in\mathbb{Z}^d}|\langle \mathbf{f},\varphi_{\mathsf{N}^J;\mathbf{0},\mathsf{k}}\rangle|^2+\sum_{j=J}^{\infty}\sum_{\psi\in\Psi_j}\sum_{\mathsf{k}\in\mathbb{Z}^d}|\langle \mathbf{f},\psi_{\mathsf{N}^j;\mathbf{0},\mathsf{k}}\rangle|^2 \leqslant C\|\mathbf{f}\|^2_{L_2(\mathbb{R}^d)}, \tag{42}$$

$$\sum_{\tilde{\varphi}\in\tilde{\Phi}}\sum_{\mathsf{k}\in\mathbb{Z}^d}|\langle \mathbf{g},\tilde{\varphi}_{\mathsf{N}^J;\mathbf{0},\mathsf{k}}\rangle|^2+\sum_{j=J}^{\infty}\sum_{\tilde{\psi}\in\tilde{\Psi}_j}\sum_{\mathsf{k}\in\mathbb{Z}^d}|\langle \mathbf{g},\tilde{\psi}_{\mathsf{N}^j;\mathbf{0},\mathsf{k}}\rangle|^2 \leqslant C\|\mathbf{g}\|^2_{L_2(\mathbb{R}^d)}; \tag{43}$$

(ii) *the pair in* (23) *is a pair of frequency-based nonstationary nonhomogeneous dual* N*-wavelet frames in the distribution space* $\mathscr{D}'(\mathbb{R}^d)$.

By Lemma 2, we say that $(\mathrm{FWS}_J(\Phi;\{\Psi_j\}_{j=J}^{\infty}),\mathrm{FWS}_J(\tilde{\Phi};\{\tilde{\Psi}_j\}_{j=J}^{\infty}))$ is a pair of frequency-based nonstationary nonhomogeneous biorthogonal N-wavelet bases of $L_2(\mathbb{R}^d)$ if it is a pair of frequency-based nonstationary nonhomogeneous dual N-wavelet frames of $L_2(\mathbb{R}^d)$ and the frequency-based biorthogonality relations hold:

$$\begin{aligned}&\langle \varphi^{\ell}_{\mathsf{N}^J;\mathbf{0},\mathsf{k}},\tilde{\varphi}^{\ell'}_{\mathsf{N}^J;\mathbf{0},\mathsf{k}'}\rangle = \delta(\mathsf{k}-\mathsf{k}')\delta(\ell-\ell'), \quad \langle \psi^{j,n}_{\mathsf{N}^j;\mathbf{0},\mathsf{k}},\tilde{\varphi}^{\ell'}_{\mathsf{N}^J;\mathbf{0},\mathsf{k}'}\rangle = 0,\\ &\langle \varphi^{\ell}_{\mathsf{N}^J;\mathbf{0},\mathsf{k}},\tilde{\psi}^{j',n'}_{\mathsf{N}^{j'};\mathbf{0},\mathsf{k}'}\rangle = 0,\ \langle \psi^{j,n}_{\mathsf{N}^j;\mathbf{0},\mathsf{k}},\tilde{\psi}^{j',n'}_{\mathsf{N}^{j'};\mathbf{0},\mathsf{k}'}\rangle = (2\pi)^d\delta(\mathsf{k}-\mathsf{k}')\delta(n-n')\delta(j-j')\end{aligned} \tag{44}$$

for all $\mathsf{k},\mathsf{k}' \in \mathbb{Z}^d$, $j,j' \in \mathbb{Z}\cap[J,+\infty)$, $\ell,\ell' = 1,\ldots,r$, and $n,n' = 1,\ldots,s$. With the frequency-based biorthogonality conditions in (44), Corollary 2 also characterizes frequency-based nonstationary nonhomogeneous biorthogonal N-wavelets in $L_2(\mathbb{R}^d)$.

As a direct consequence of results in Theorem 2 and Corollary 2, we have

Proposition 6 *([44, Corollary 17]) Let* M *be a* $d\times d$ *real-valued expansive matrix and define* $\mathsf{N} := (\mathsf{M}^{\mathsf{T}})^{-1}$*. Let* J_0 *be an integer. Let* Φ *in* (21) *and* Ψ_j *in* (22) *be finite subsets of* $L_2^{\mathrm{loc}}(\mathbb{R}^d)$ *for all* $j \geqslant J_0$*. Then the following are equivalent:*

1. $\mathrm{FWS}_J(\Phi;\{\Psi_j\}_{j=J}^{\infty})$ *is a frequency-based nonstationary nonhomogeneous tight* N*-wavelet frame of* $L_2(\mathbb{R}^d)$ *for every integer* $J \geqslant J_0$*, that is,* $\Phi,\Psi_j \subseteq L_2(\mathbb{R}^d)$ *for all* $j \geqslant J_0$*, and for all* $J \geqslant J_0$ *and* $\mathbf{f} \in L_2(\mathbb{R}^d)$*,*

$$\sum_{\ell=1}^{r}\sum_{\mathsf{k}\in\mathbb{Z}^d}|\langle \mathbf{f},\varphi^{\ell}_{\mathsf{N}^J;\mathbf{0},\mathsf{k}}\rangle|^2+\sum_{j=J}^{\infty}\sum_{\ell=1}^{s_j}\sum_{\mathsf{k}\in\mathbb{Z}^d}|\langle \mathbf{f},\psi^{j,\ell}_{\mathsf{N}^j;\mathbf{0},\mathsf{k}}\rangle|^2=(2\pi)^d\|\mathbf{f}\|^2_{L_2(\mathbb{R}^d)}; \tag{45}$$

2. $(\mathrm{FWS}_J(\Phi;\{\Psi_j\}_{j=J}^{\infty}),\mathrm{FWS}_J(\Phi;\{\Psi_j\}_{j=J}^{\infty}))$ *is a pair of frequency-based nonstationary nonhomogeneous dual* N*-wavelet frames in the distribution space for every integer* $J \geqslant J_0$*;*
3. $\lim_{j\to+\infty}\sum_{\ell=1}^{r}\langle|\varphi^{\ell}(\mathsf{N}^j\cdot)|^2,\mathbf{h}\rangle=\langle 1,\mathbf{h}\rangle$ *for all* $\mathbf{h}\in\mathscr{D}(\mathbb{R}^d)$ *and for all* $j \geqslant J_0$ *and almost every* $\xi\in\mathbb{R}^d$*,*

$$\begin{aligned}\sum_{\ell=1}^{r}\overline{\varphi^{\ell}(\xi)}\varphi^{\ell}(\xi+2\pi\mathsf{k})&+\sum_{\ell=1}^{s_j}\overline{\psi^{j,\ell}(\xi)}\psi^{j,\ell}(\xi+2\pi\mathsf{k})\\&=\sum_{\ell=1}^{r}\overline{\varphi^{\ell}(\mathsf{N}\xi)}\varphi^{\ell}(\mathsf{N}(\xi+2\pi\mathsf{k})),\quad \mathsf{k}\in\mathbb{Z}^d\cap[\mathsf{N}^{-1}\mathbb{Z}^d],\end{aligned} \tag{46}$$

$$\sum_{\ell=1}^{r}\overline{\varphi^{\ell}(\xi)}\varphi^{\ell}(\xi+2\pi\mathsf{k})+\sum_{\ell=1}^{s_j}\overline{\psi^{j,\ell}(\xi)}\psi^{j,\ell}(\xi+2\pi\mathsf{k})=0,\ \mathsf{k}\in\mathbb{Z}^d\backslash[\mathsf{N}^{-1}\mathbb{Z}^d], \tag{47}$$

$$\sum_{\ell=1}^{r}\overline{\varphi^{\ell}(\mathsf{N}\xi)}\varphi^{\ell}(\mathsf{N}(\xi+2\pi\mathsf{k}))=0,\quad \mathsf{k}\in[\mathsf{N}^{-1}\mathbb{Z}^d]\backslash\mathbb{Z}^d. \tag{48}$$

Moreover, if the following additional property holds:

$$\mathbf{h}(\xi)\mathbf{h}(\xi+2\pi\mathsf{k})=0\qquad a.e.\ \xi\in\mathbb{R}^d,\mathsf{k}\in\mathbb{Z}^d\backslash\{\mathbf{0}\},\mathbf{h}\in\Phi\cup(\cup_{j=J_0}^{\infty}\Psi_j),$$

then all the conditions in (46)–(48) *are reduced to the following simple condition*

$$\sum_{\ell=1}^{r}|\varphi^{\ell}(\xi)|^2+\sum_{\ell=1}^{s_j}|\psi^{j,\ell}(\xi)|^2=\sum_{\ell=1}^{r}|\varphi^{\ell}(\mathsf{N}\xi)|^2,\quad a.e.\,\xi\in\mathbb{R}^d, j\geqslant J_0. \tag{49}$$

To appreciate the results in Theorem 2 and Proposition 6 for nonhomogeneous framelets, let us recall the characterization of homogeneous dual or tight framelets in $L_2(\mathbb{R}^d)$ in the literature. Many researchers have contributed to this problem, to mention a few references here, see [3, 9, 20, 29–31, 72, 73] and references therein. For the convenience of presentation, we only state the frequency-based versions here.

Theorem 4. *([30, Theorem 2.7], [31, Theorem 2.5], [72, Corollary 2]) Let* M *be a* $d \times d$ *integer expansive matrix. Define* $\mathsf{N} := (\mathsf{M}^{\mathsf{T}})^{-1}$. *Let* Ψ *and* $\tilde{\Psi}$ *in* (27) *be finite subsets of* $L_2(\mathbb{R}^d)$. *Then* $(\mathrm{FWS}(\Psi), \mathrm{FWS}(\tilde{\Psi}))$, *with* $\mathrm{FWS}(\Psi) := \{\psi_{\mathsf{N}^j;\mathbf{0},\mathsf{k}} \mid j \in \mathbb{Z}, \mathsf{k} \in \mathbb{Z}^d, \psi \in \Psi\}$, *is a pair of frequency-based homogeneous dual* N*-wavelet frames of* $L_2(\mathbb{R}^d)$ *if and only if*

(i) *there exists a positive constant* C *such that*

$$\sum_{j\in\mathbb{Z}}\sum_{\ell=1}^{s}\sum_{\mathsf{k}\in\mathbb{Z}^d}\left(|\langle \mathbf{f}, \psi^{\ell}_{\mathsf{N}^j;\mathbf{0},\mathsf{k}}\rangle|^2 + |\langle \mathbf{f}, \tilde{\psi}^{\ell}_{\mathsf{N}^j;\mathbf{0},\mathsf{k}}\rangle|^2\right) \leqslant C\|\mathbf{f}\|^2_{L_2(\mathbb{R}^d)}, \quad \forall\, \mathbf{f} \in L_2(\mathbb{R}^d); \tag{50}$$

(ii) *the following identities hold: for almost every* $\xi \in \mathbb{R}^d$,

$$\sum_{j\in\mathbb{Z}}\sum_{\ell=1}^{s}\overline{\psi^{\ell}(\mathsf{N}^j\xi)}\tilde{\psi}^{\ell}(\mathsf{N}^j\xi) = 1, \tag{51}$$

$$\sum_{\ell=1}^{s}\sum_{j=0}^{+\infty}\overline{\psi^{\ell}(\mathsf{N}^{-j}\xi)}\tilde{\psi}^{\ell}(\mathsf{N}^{-j}(\xi+2\pi\gamma)) = 0, \qquad \gamma \in \mathbb{Z}^d\backslash[\mathsf{N}^{-1}\mathbb{Z}^d]. \tag{52}$$

The absolute convergence of all the series in (51) and (52) is implicitly guaranteed by item (i) of Theorem 4. For homogeneous tight framelets, we have

Corollary 3 *([30, Theorem 2.8], [31, Theorem 2.7], [73, Corollary 1.3]) Let* M *be a* $d \times d$ *integer expansive matrix. Define* $\mathsf{N} := (\mathsf{M}^{\mathsf{T}})^{-1}$. *Let* Ψ *in* (27) *be a finite subset of* $L_2(\mathbb{R}^d)$. *Then* $\mathrm{FWS}(\Psi)$ *is a frequency-based homogeneous tight* N*-wavelet frame of* $L_2(\mathbb{R}^d)$, *that is,* $\sum_{j\in\mathbb{Z}}\sum_{\ell=1}^{s}\sum_{\mathsf{k}\in\mathbb{Z}^d}|\langle \mathbf{f}, \psi^{\ell}_{\mathsf{N}^j;\mathbf{0},\mathsf{k}}\rangle|^2 = 2\pi\|\mathbf{f}\|^2_{L_2(\mathbb{R}^d)}$ *for all* $\mathbf{f} \in L_2(\mathbb{R}^d)$, *if and only if,* (51) *and* (52) *hold with* $\tilde{\psi}^{\ell} := \psi^{\ell}, \ell = 1, \ldots, s$.

In fact, Theorem 4 and Corollary 3 have been established in [30, 31] for homogeneous dual or tight framelets in general subspaces of $L_2(\mathbb{R}^d)$ (see [31, page 381] and [30, page 4]). Theorem 4 and Corollary 3 are consequences of general Grammian analysis in [72, 73]. Nevertheless, comparing with the above results on homogeneous framelets, our results on nonhomogeneous framelets have several important features. Firstly, only finite sums are involved in our characterizations and therefore, there is no issue on the convergence of the series. Infinite series are unavoidable in the characterization of homogeneous framelets in Theorem 4 and Corollary 3 and their convergence has to be guaranteed by extra conditions such as the Bessel condition in (50). Secondly, our characterization handles any arbitrary dilation matrix and fully nonstationary nonhomogeneous framelets. However, for homogeneous framelets with a non-integer expansive matrix, it is known (for example, [9]) that the characterization of homogeneous tight framelets even for rational dilation factor in dimension one is extremely complicated, not to mention a general dilation matrix in high dimensions. In fact, there are barely results in the literature on characterization of homogeneous tight framelets for a general real-valued dilation matrix and for nonstationary homogeneous framelets. Thirdly, our characterizations do not impose any extra conditions such as membership in $L_2(\mathbb{R}^d)$, Bessel stability condition, vanishing moments, etc.

4 Wavelets and Framelets in Function Spaces

In this section, we shall introduce (nonhomogeneous) wavelets and framelets in a general function space. We shall see that frequency-based nonhomogeneous dual framelets in the distribution space play a critical role in the study of wavelets and framelets in various function spaces.

In Sect. 2, we have discussed nonhomogeneous wavelets and framelets in the commonly-used function space $L_2(\mathbb{R}^d)$. We have established in Corollary 2 a characterization of nonhomogeneous wavelets and framelets in $L_2(\mathbb{R}^d)$. In this section, we generalize the definition in the space/time domain of wavelets and framelets from $L_2(\mathbb{R}^d)$ to a general function space. Then we shall look at them for some particular function spaces such as Sobolev spaces.

Let $(\mathscr{B}, \|\cdot\|_{\mathscr{B}})$ denote a (Banach) function space with $(\mathscr{B}', \|\cdot\|_{\mathscr{B}'})$ being its dual function space. We always assume that $\mathscr{D}(\mathbb{R}^d) \subseteq \mathscr{B} \cap \mathscr{B}'$ and $(\mathscr{B}')' = \mathscr{B}$. The pairing between $\mathscr{B}$ and $\mathscr{B}'$ is understood in the usual sense as $\langle f, g\rangle = \int_{\mathbb{R}^d} f(x)\overline{g(x)}\mathrm{d}x$ for $f, g \in \mathscr{D}(\mathbb{R}^d)$. For example, if $\mathscr{B}$ is the Besov space $B^{\tau}_{p,q}(\mathbb{R}^d)$ for some $\tau \in \mathbb{R}$ and $1 < p, q < \infty$, then $\mathscr{B}' = B^{-\tau}_{p',q'}(\mathbb{R}^d)$ with $1/p + 1/p' = 1/q + 1/q' = 1$. Similarly, if $\mathscr{B}$ is a Triebel–Lizorkin space $F^{\tau}_{p,q}(\mathbb{R}^d)$, then $\mathscr{B}' = F^{-\tau}_{p',q'}(\mathbb{R}^d)$.

Let Φ and $\Psi_j, j \geqslant J$ be subsets of $\mathscr{B}$. For a $d \times d$ real-valued invertible matrix M, we define a nonstationary nonhomogeneous M-wavelet system in $\mathscr{B}$ as follows:

$$\begin{aligned} \mathrm{WS}_J(\Phi; \{\Psi_j\}_{j=J}^{\infty}) :=& \{\phi_{\mathsf{M}^J;\mathsf{k}} \mid \mathsf{k} \in \mathbb{Z}^d, \phi \in \Phi\} \\ &\cup \{\psi_{\mathsf{M}^j;\mathsf{k}} \mid j \geqslant J, \mathsf{k} \in \mathbb{Z}^d, \psi \in \Psi_j\}. \end{aligned} \tag{53}$$

We assume that $\mathrm{WS}_J(\Phi; \{\Psi_j\}_{j=J}^{\infty}) \subseteq \mathscr{B}$. For a function space $\mathscr{B}$, we assume that there exists a sequence space $(\mathfrak{b}_{\mathscr{B}}, \|\cdot\|_{\mathfrak{b}_{\mathscr{B}}})$, whose sequences are indexed by the elements of $\mathrm{WS}_J(\Phi; \{\Psi_j\}_{j=J}^{\infty})$. Furthermore, we also assume the following conditions on the sequence space $(\mathfrak{b}_{\mathscr{B}}, \|\cdot\|_{\mathfrak{b}_{\mathscr{B}}})$:

1. for every $h_0 \in \mathrm{WS}_J(\Phi; \{\Psi_j\}_{j=J}^{\infty})$, there exists a positive constant C_{h_0} such that

$$|w_{h_0}| \leqslant C_{h_0} \|\{w_h\}_{h \in \mathrm{WS}_J(\Phi;\{\Psi_j\}_{j=J}^{\infty})}\|_{\mathfrak{b}_{\mathscr{B}}} \qquad \forall\, \{w_h\}_{h \in \mathrm{WS}_J(\Phi;\{\Psi_j\}_{j=J}^{\infty})} \in \mathfrak{b}_{\mathscr{B}}; \tag{54}$$

2. for every finite subset $K \subset \mathrm{WS}_J(\Phi; \{\Psi_j\}_{j=J}^{\infty})$ and for every sequence $\{f_n\}_{n=1}^{\infty}$ in $\mathscr{D}(\mathbb{R}^d)$ such that $\lim_{n\to\infty} f_n = f$ in $(\mathscr{B}', \|\cdot\|_{\mathscr{B}'})$,

$$\|\{\langle f, h\rangle\}_{h \in K}\|_{\mathfrak{b}_{\mathscr{B}}} \leqslant \limsup_{n\to\infty} \|\{\langle f_n, h\rangle\}_{h \in K}\|_{\mathfrak{b}_{\mathscr{B}}}; \tag{55}$$

   where $\{\langle f, h\rangle\}_{h \in K}$ is a sequence by setting zero for indices outside K;
3. for every increasing sequence $\{K_n\}_{n=1}^{\infty}$ of finite subsets of $\mathrm{WS}_J(\Phi; \{\Psi_j\}_{j=J}^{\infty})$ such that $\cup_{n=1}^{\infty} K_n = \mathrm{WS}_J(\Phi; \{\Psi_j\}_{j=J}^{\infty})$,

$$\lim_{n\to\infty} \|\{w_h\}_{h \in \mathrm{WS}_J(\Phi;\{\Psi_j\}_{j=J}^{\infty}) \backslash K_n}\|_{\mathfrak{b}_{\mathscr{B}}} = 0, \qquad \forall\, \{w_h\}_{h \in \mathrm{WS}_J(\Phi;\{\Psi_j\}_{j=J}^{\infty})} \in \mathfrak{b}_{\mathscr{B}}. \tag{56}$$

Between the two sequence spaces $(\mathfrak{b}_{\mathscr{B}}, \|\cdot\|_{\mathfrak{b}_{\mathscr{B}}})$ and $(\mathfrak{b}_{\mathscr{B}'}, \|\cdot\|_{\mathfrak{b}_{\mathscr{B}'}})$, we also assume a natural relation which mimics the Cauchy–Schwarz inequality: there exists a positive constant $C_{\mathfrak{b}}$ such that the series $\sum_{h\in \mathrm{WS}_J} \overline{w_h}\tilde{w}_h$ converges absolutely and

$$\Big|\sum_{h\in \mathrm{WS}_J} \overline{w_h}\tilde{w}_h\Big| \leqslant C_{\mathfrak{b}} \|\{w_h\}_{h\in \mathrm{WS}_J}\|_{\mathfrak{b}_{\mathscr{B}}} \|\{\tilde{w}_h\}_{h\in \mathrm{WS}_J}\|_{\mathfrak{b}_{\mathscr{B}'}} \tag{57}$$

for all $\{w_h\}_{h\in \mathrm{WS}_J} \in \mathfrak{b}_{\mathscr{B}}$ and $\{\tilde{w}_h\}_{h\in \mathrm{WS}_J} \in \mathfrak{b}_{\mathscr{B}'}$, where $\mathrm{WS}_J := \mathrm{WS}_J(\Phi; \{\Psi_j\}_{j=J}^{\infty})$.

We say that $\mathrm{WS}_J(\Phi; \{\Psi_j\}_{j=J}^{\infty}) \subset \mathscr{B}$ is *a preframe of* $\mathscr{B}$ (or has stability in $\mathscr{B}$) with respect to $\mathfrak{b}_{\mathscr{B}'}$ if there exist positive constants C_5 and C_6 such that

$$C_5\|f\|_{\mathscr{B}'} \leqslant \|\{\langle f,h\rangle\}_{h\in \mathrm{WS}_J(\Phi;\{\Psi_j\}_{j=J}^{\infty})}\|_{\mathfrak{b}_{\mathscr{B}'}} \leqslant C_6\|f\|_{\mathscr{B}'}, \qquad \forall f \in \mathscr{B}'. \tag{58}$$

(58) indicates that up to equivalence the norm $\|\cdot\|_{\mathfrak{b}_{\mathscr{B}'}}$ is often uniquely determined by $\mathscr{B}'$ and is independent of the actual elements of the index set $\mathrm{WS}_J(\Phi; \{\Psi_j\}_{j=J}^{\infty})$. We say that $\mathrm{WS}_J(\Phi; \{\Psi_j\}_{j=J}^{\infty})$ is *a prebasis of* $\mathscr{B}$ with respect to $\mathfrak{b}_{\mathscr{B}'}$ if it is a preframe of $\mathscr{B}$ with respect to $\mathfrak{b}_{\mathscr{B}'}$ and it is $\mathfrak{b}_{\mathscr{B}}$-linearly independent, that is, if for a sequence $\{w_h\}_{h\in \mathrm{WS}_J(\Phi;\{\Psi_j\}_{j=J}^{\infty})} \in \mathfrak{b}_{\mathscr{B}}$, $\sum_{h\in \mathrm{WS}_J(\Phi;\{\Psi_j\}_{j=J}^{\infty})} w_h\langle h,f\rangle = 0$ for all $f \in \mathscr{B}'$, then we must have $w_h = 0$ for all $h \in \mathrm{WS}_J(\Phi; \{\Psi_j\}_{j=J}^{\infty})$.

For subsets $\Phi = \{\phi^1,\dots,\phi^r\}, \Psi_j = \{\psi^{j,1},\dots,\psi^{j,s_j}\}, j \geqslant J$ of $\mathscr{B}$ and subsets $\tilde{\Phi} = \{\tilde{\phi}^1,\dots,\tilde{\phi}^r\}, \tilde{\Psi}_j = \{\tilde{\psi}^{j,1},\dots,\tilde{\psi}^{j,s_j}\}, j \geqslant J$ of $\mathscr{B}'$, we say that

$$(\mathrm{WS}_J(\Phi; \{\Psi_j\}_{j=J}^{\infty}), \mathrm{WS}_J(\tilde{\Phi}; \{\tilde{\Psi}_j\}_{j=J}^{\infty}))$$

is *a pair of nonstationary nonhomogeneous dual* M*-wavelet frames in* $(\mathscr{B}, \mathscr{B}')$ if

1. $\mathrm{WS}_J(\Phi; \{\Psi_j\}_{j=J}^{\infty})$ is a preframe of $\mathscr{B}$ with respect to $\mathfrak{b}_{\mathscr{B}'}$;
2. $\mathrm{WS}_J(\tilde{\Phi}; \{\tilde{\Psi}_j\}_{j=J}^{\infty})$ is a preframe of $\mathscr{B}'$ with respect to $\mathfrak{b}_{\mathscr{B}}$;
3. for all $f \in \mathscr{B}'$ and $g \in \mathscr{B}$, the following identity holds:

$$\sum_{\ell=1}^{r}\sum_{\mathsf{k}\in\mathbb{Z}^d} \langle f, \phi^{\ell}_{\mathsf{M}^J;\mathsf{k}}\rangle\langle \tilde{\phi}^{\ell}_{\mathsf{M}^J;\mathsf{k}}, g\rangle + \sum_{j=J}^{\infty}\sum_{\ell=1}^{s_j}\sum_{\mathsf{k}\in\mathbb{Z}^d} \langle f, \psi^{j,\ell}_{\mathsf{M}^j;\mathsf{k}}\rangle\langle \tilde{\psi}^{j,\ell}_{\mathsf{M}^j;\mathsf{k}}, g\rangle = \langle f,g\rangle. \tag{59}$$

By the assumption in (57), all the series on the left-hand side of (59) converge absolutely. From the above definition, we have

$$f = \sum_{\ell=1}^{r}\sum_{\mathsf{k}\in\mathbb{Z}^d} \langle f, \phi^{\ell}_{\mathsf{M}^J;\mathsf{k}}\rangle\tilde{\phi}^{\ell}_{\mathsf{M}^J;\mathsf{k}} + \sum_{j=J}^{\infty}\sum_{\ell=1}^{s_j}\sum_{\mathsf{k}\in\mathbb{Z}^d} \langle f, \psi^{j,\ell}_{\mathsf{M}^j;\mathsf{k}}\rangle\tilde{\psi}^{j,\ell}_{\mathsf{M}^j;\mathsf{k}}, \qquad f \in \mathscr{B}' \tag{60}$$

and (58) holds (that is, $\|f\|_{\mathscr{B}'}$ is characterized by its framelet coefficients). Similarly,

$$g = \sum_{\ell=1}^{r}\sum_{\mathsf{k}\in\mathbb{Z}^d} \langle g, \tilde{\phi}^{\ell}_{\mathsf{M}^J;\mathsf{k}}\rangle\phi^{\ell}_{\mathsf{M}^J;\mathsf{k}} + \sum_{j=J}^{\infty}\sum_{\ell=1}^{s_j}\sum_{\mathsf{k}\in\mathbb{Z}^d} \langle g, \tilde{\psi}^{j,\ell}_{\mathsf{M}^j;\mathsf{k}}\rangle\psi^{j,\ell}_{\mathsf{M}^j;\mathsf{k}}, \qquad g \in \mathscr{B}, \tag{61}$$

$$C_6^{-1}\|g\|_{\mathscr{B}} \leqslant \|\{\langle g,\tilde{h}\rangle\}_{\tilde{h}\in \mathrm{WS}_J(\tilde{\Phi};\{\tilde{\Psi}_j\}_{j=J}^{\infty})}\|_{\mathfrak{b}_{\mathscr{B}}} \leqslant C_5^{-1}\|g\|_{\mathscr{B}}, \qquad \forall\, g \in \mathscr{B}. \tag{62}$$

We say that $(\mathrm{WS}_J(\Phi;\{\Psi_j\}_{j=J}^\infty),\mathrm{WS}_J(\tilde\Phi;\{\tilde\Psi_j\}_{j=J}^\infty))$ is *a pair of nonstationary nonhomogeneous biorthogonal* M*-wavelet bases in* $(\mathscr{B},\mathscr{B}')$ if it is a pair of nonstationary nonhomogeneous dual M-wavelet frames in $(\mathscr{B},\mathscr{B}')$ and the biorthogonality relations similar to (13) hold, that is, the two systems are biorthogonal to each other.

In the following, we discuss the particular case of Sobolev spaces which are Hilbert spaces. For a Sobolev space $\mathscr{B}=H^\tau(\mathbb{R}^d)$, we have $\mathscr{B}'=H^{-\tau}(\mathbb{R}^d)$ and we define a normed sequence space $(\mathfrak{b}_{H^\tau(\mathbb{R}^d)},\|\cdot\|_{\mathfrak{b}_{H^\tau(\mathbb{R}^d)}})$, indexed by the elements of $\mathrm{WS}_J(\Phi;\{\Psi_j\}_{j=J}^\infty)$, with the weighted norm $\|\cdot\|_{\mathfrak{b}_{H^\tau(\mathbb{R}^d)}}$ as follows:

$$\Big\|\{w_h\}_{h\in\mathrm{WS}_J(\Phi;\{\Psi_j\}_{j=J}^\infty)}\Big\|^2_{\mathfrak{b}_{H^\tau(\mathbb{R}^d)}}:=\sum_{\phi\in\Phi}\sum_{\mathsf{k}\in\mathbb{Z}^d}|\det\mathsf{M}|^{2\tau J/d}|w_{\phi_{\mathsf{M}^J;\mathsf{k}}}|^2 \\ +\sum_{j=J}^\infty\sum_{\psi\in\Psi_j}\sum_{\mathsf{k}\in\mathbb{Z}^d}|\det\mathsf{M}|^{2\tau j/d}|w_{\psi_{\mathsf{M}^j;\mathsf{k}}}|^2 \quad (63)$$

and $\mathfrak{b}_{H^\tau(\mathbb{R}^d)}:=\{\{w_h\}_{h\in\mathrm{WS}_J(\Phi;\{\Psi_j\}_{j=J}^\infty)}\ |\ \|\{w_h\}_{h\in\mathrm{WS}_J(\Phi;\{\Psi_j\}_{j=J}^\infty)}\|_{\mathfrak{b}_{H^\tau(\mathbb{R}^d)}}<\infty\}$.

$\mathrm{WS}_J(\Phi;\{\Psi_j\}_{j=J}^\infty)\subset H^\tau(\mathbb{R}^d)$ is a preframe of $H^\tau(\mathbb{R}^d)$ with respect to $\mathfrak{b}_{H^{-\tau}(\mathbb{R}^d)}$ satisfying (58) with $\mathscr{B}=H^\tau(\mathbb{R}^d)$ if and only if for all $g\in H^\tau(\mathbb{R}^d)$,

$$C_5\|g\|^2_{H^\tau(\mathbb{R}^d)}\leqslant\sum_{\phi\in\Phi}\sum_{\mathsf{k}\in\mathbb{Z}^d}|\langle g,|\det\mathsf{M}|^{-\tau J/d}\phi_{\mathsf{M}^J;\mathsf{k}}\rangle_{H^\tau(\mathbb{R}^d)}|^2 \\ +\sum_{j=J}^\infty\sum_{\psi\in\Psi_j}\sum_{\mathsf{k}\in\mathbb{Z}^d}|\langle g,|\det\mathsf{M}|^{-\tau j/d}\psi_{\mathsf{M}^j;\mathsf{k}}\rangle_{H^\tau(\mathbb{R}^d)}|^2\leqslant C_6\|g\|^2_{H^\tau(\mathbb{R}^d)}, \quad (64)$$

with $\langle\cdot,\cdot\rangle_{H^\tau(\mathbb{R}^d)}$ in (10). Define $\hat g(\xi):=\hat f(\xi)(1+\|\xi\|^2)^\tau$ for $f\in H^{-\tau}(\mathbb{R}^d)$. Then $\|f\|_{H^{-\tau}(\mathbb{R}^d)}=\|g\|_{H^\tau(\mathbb{R}^d)}$ and $\langle f,\psi_{\mathsf{M}^j;\mathsf{k}}\rangle=\langle g,\psi_{\mathsf{M}^j;\mathsf{k}}\rangle_{H^\tau(\mathbb{R}^d)}$. We see that (58) with $\mathscr{B}=H^\tau(\mathbb{R}^d)$ is equivalent to (64) (see [51, Proposition 2.1] with $\mathsf{M}=2I_d$).

Hence, $\mathrm{WS}_J(\Phi;\{\Psi_j\}_{j=J}^\infty)\subset H^\tau(\mathbb{R}^d)$ is a preframe of $H^\tau(\mathbb{R}^d)$ with respect to $\mathfrak{b}_{H^{-\tau}(\mathbb{R}^d)}$ is equivalent to saying that after a renormalization of $\mathrm{WS}_J(\Phi;\{\Psi_j\}_{j=J}^\infty)$,

$$\mathrm{WS}_J^\tau(\Phi;\{\Psi_j\}_{j=J}^\infty):=\{|\det\mathsf{M}|^{-\tau J/d}\phi_{\mathsf{M}^J;\mathsf{k}}\ :\ \mathsf{k}\in\mathbb{Z}^d,\phi\in\Phi\} \\ \cup\{|\det\mathsf{M}|^{-\tau j/d}\psi_{\mathsf{M}^j;\mathsf{k}}\ :\ j\geqslant J,\mathsf{k}\in\mathbb{Z}^d,\psi\in\Psi_j\} \quad (65)$$

is a frame of the Hilbert space $H^\tau(\mathbb{R}^d)$ in the classical sense. If $\mathrm{WS}_J(\Phi;\{\Psi_j\}_{j=J}^\infty)$ is a prebasis of $H^\tau(\mathbb{R}^d)$ with respect to $\mathfrak{b}_{H^{-\tau}(\mathbb{R}^d)}$, then $\mathrm{WS}_J^\tau(\Phi;\{\Psi_j\}_{j=J}^\infty)$ is a Riesz basis of the Hilbert space $H^\tau(\mathbb{R}^d)$ in the classical sense.

For a Besov space $B^\tau_{p,q}(\mathbb{R}^d)$ with $\tau\in\mathbb{R}$ and $1<p,q<\infty$, define

$$\|\{w_h\}_{h\in\mathrm{WS}_J(\Phi;\{\Psi_j\}_{j=J}^\infty)}\|^q_{\mathfrak{b}_{B^\tau_{p,q}(\mathbb{R}^d)}}:=\Big(\sum_{\phi\in\Phi}\sum_{\mathsf{k}\in\mathbb{Z}^d}|w_{\phi_{\mathsf{M}^J;\mathsf{k}}}|^p\Big)^{q/p} \\ +\sum_{j=J}^\infty|\det\mathsf{M}|^{(1/2-1/p+\tau/d)jq}\Big(\sum_{\psi\in\Psi_j}\sum_{\mathsf{k}\in\mathbb{Z}^d}|w_{\psi_{\mathsf{M}^j;\mathsf{k}}}|^p\Big)^{q/p}. \quad (66)$$

For $\mathscr{B} = \mathfrak{b}_{H^\tau(\mathbb{R}^d)}$ or more generally $\mathscr{B} = \mathfrak{b}_{B^\tau_{p,q}(\mathbb{R}^d)}$, it is easy to check that (57) and all the conditions for $\mathfrak{b}_{\mathscr{B}}$ are satisfied.

In the rest of this section, we shall use the frequency-based definition instead. For a function space $(\mathscr{B}, \|\cdot\|_{\mathscr{B}})$ such that all the elements in $\mathscr{B}$ are tempered distributions, we define its corresponding frequency-based function space $(\widehat{\mathscr{B}}, \|\cdot\|_{\widehat{\mathscr{B}}})$ as follows:

$$\widehat{\mathscr{B}} := \{\hat{f} : f \in \mathscr{B}\} \quad \text{and} \quad \|\hat{f}\|_{\widehat{\mathscr{B}}} := \|f\|_{\mathscr{B}}, \qquad f \in \mathscr{B}. \tag{67}$$

That is, the Fourier transform is an isometry from $(\mathscr{B}, \|\cdot\|_{\mathscr{B}})$ to $(\widehat{\mathscr{B}}, \|\cdot\|_{\widehat{\mathscr{B}}})$. Also, by Plancherel theorem $(2\pi)^d\langle f,g\rangle = \langle \hat{f},\hat{g}\rangle$, we simply take $\mathfrak{b}_{\widehat{\mathscr{B}}} = \mathfrak{b}_{\mathscr{B}}$.

Let $\mathfrak{B}$ be a (frequency-based) function space and $\mathfrak{B}'$ be its dual space. We assume that $\mathscr{D}(\mathbb{R}^d) \subset \mathfrak{B} \cap \mathfrak{B}'$, $(\mathfrak{B}')' = \mathfrak{B}$, and $\mathscr{D}(\mathbb{R}^d)$ is dense in both $\mathfrak{B}$ and $\mathfrak{B}'$. For subsets $\Phi, \Psi_j, j \geqslant J$ of $\mathfrak{B}$, we say that $\mathrm{FWS}_J(\Phi; \{\Psi_j\}_{j=J}^\infty)$ is *a frequency-based preframe of* $\mathfrak{B}$ with respect to $\mathfrak{b}_{\mathfrak{B}'}$ if there exist positive constants C_7 and C_8 such that

$$C_7\|\mathbf{f}\|_{\mathfrak{B}'} \leqslant \|\{\langle \mathbf{f},\mathbf{h}\rangle_{\mathbf{h}\in \mathrm{FWS}_J(\Phi;\{\Psi_j\}_{j=J}^\infty)}\|_{\mathfrak{b}_{\mathfrak{B}'}} \leqslant C_8\|\mathbf{f}\|_{\mathfrak{B}'}, \qquad \forall\, \mathbf{f} \in \mathfrak{B}'. \tag{68}$$

Similarly, $\mathrm{FWS}_J(\Phi; \{\Psi_j\}_{j=J}^\infty)$ is *a frequency-based prebasis of* $\mathfrak{B}$ with respect to $\mathfrak{b}_{\mathfrak{B}'}$ if it is a frequency-based preframe of $\mathfrak{B}$ with respect to $\mathfrak{b}_{\mathfrak{B}'}$ and it is $\mathfrak{b}_{\mathfrak{B}}$-linearly independent, that is, if $\sum_{\mathbf{h}\in \mathrm{FWS}_J(\Phi;\{\Psi_j\}_{j=J}^\infty)} w_{\mathbf{h}}\langle f,\mathbf{h}\rangle = 0\ \forall\, f \in \mathfrak{B}'$ for $\{w_{\mathbf{h}}\}_{\mathbf{h}\in \mathrm{FWS}_J(\Phi;\{\Psi_j\}_{j=J}^\infty)} \in \mathfrak{b}_{\mathfrak{B}}$, then $w_{\mathbf{h}} = 0$ for all $\mathbf{h} \in \mathrm{FWS}_J(\Phi; \{\Psi_j\}_{j=J}^\infty)$.

For subsets $\Phi, \Psi_j, j \geqslant J$ of $\mathfrak{B}$ and subsets $\tilde{\Phi}, \tilde{\Psi}_j, j \geqslant J$ of $\mathfrak{B}'$ in (21) and (22), we say that $(\mathrm{FWS}_J(\Phi; \{\Psi_j\}_{j=J}^\infty), \mathrm{FWS}_J(\tilde{\Phi}; \{\tilde{\Psi}_j\}_{j=J}^\infty))$ is *a pair of frequency-based nonstationary nonhomogeneous dual* N*-wavelet frames in* $(\mathfrak{B}, \mathfrak{B}')$ if:

1. $\mathrm{FWS}_J(\Phi; \{\Psi_j\}_{j=J}^\infty)$ is a frequency-based preframe of $\mathfrak{B}$ with respect to $\mathfrak{b}_{\mathfrak{B}'}$.
2. $\mathrm{FWS}_J(\tilde{\Phi}; \{\tilde{\Psi}_j\}_{j=J}^\infty)$ is a frequency-based preframe of $\mathfrak{B}'$ with respect to $\mathfrak{b}_{\mathfrak{B}}$.
3. for all $f \in \mathfrak{B}'$ and $g \in \mathfrak{B}$, the following identity holds:

$$\sum_{\ell=1}^{r}\sum_{\mathsf{k}\in\mathbb{Z}^d} \langle \mathbf{f}, \varphi^\ell_{\mathsf{N}^J;\mathbf{0},\mathsf{k}}\rangle\langle \tilde{\varphi}^\ell_{\mathsf{N}^J;\mathbf{0},\mathsf{k}}, \mathbf{g}\rangle + \sum_{j=J}^{\infty}\sum_{\ell=1}^{s_j}\sum_{\mathsf{k}\in\mathbb{Z}^d} \langle \mathbf{f}, \psi^{j,\ell}_{\mathsf{N}^j;\mathbf{0},\mathsf{k}}\rangle\langle \tilde{\psi}^{j,\ell}_{\mathsf{N}^j;\mathbf{0},\mathsf{k}}, \mathbf{g}\rangle = (2\pi)^d\langle \mathbf{f},\mathbf{g}\rangle. \tag{69}$$

It is evident that $(\mathrm{WS}_J(\Phi; \{\Psi_j\}_{j=J}^\infty), \mathrm{WS}_J(\tilde{\Phi}; \{\tilde{\Psi}_j\}_{j=J}^\infty))$ is a pair of nonstationary nonhomogeneous dual M-wavelet frames in $(\mathscr{B}, \mathscr{B}')$ if and only if

$$(\mathrm{FWS}_J(\widehat{\Phi}; \{\widehat{\Psi_j}\}_{j=J}^\infty), \mathrm{FWS}_J(\widehat{\tilde{\Phi}}; \{\widehat{\tilde{\Psi}_j}\}_{j=J}^\infty))$$

is a pair of frequency-based nonstationary nonhomogeneous dual $(\mathsf{M}^\mathsf{T})^{-1}$-wavelet frames in $(\widehat{\mathscr{B}}, \widehat{\mathscr{B}'})$. Similarly, we say that the pair in (23) is a pair of frequency-based nonstationary nonhomogeneous biorthogonal N-wavelet bases in $(\mathfrak{B}, \mathfrak{B}')$ if it is a pair of frequency-based nonstationary nonhomogeneous dual N-wavelet frames in $(\mathfrak{B}, \mathfrak{B}')$ and the frequency-based biorthogonality relations in (44) hold.

By the following result, we see that the notion of a frequency-based nonhomogeneous dual wavelet frames in the distribution space plays a basic role in the study of pairs of nonhomogeneous dual wavelet frames in a pair of dual function spaces.

Theorem 5. *Let* N *be a* $d \times d$ *real-valued invertible matrix. Let* $\Phi, \tilde{\Phi}$ *in* (21) *and* $\Psi_j, \tilde{\Psi}_j, j \geqslant J$ *in* (22) *be subsets of distributions. Then the pair in* (23) *is a pair of frequency-based nonstationary nonhomogeneous dual* N*-wavelet frames in a pair of dual function spaces* $(\mathfrak{B}, \mathfrak{B}')$*, if and only if,*

(i) there exists a positive constant C such that

$$\left\|\{\langle \mathbf{f}, \mathbf{h}\rangle\}_{\mathbf{h}\in \mathrm{FWS}_J(\Phi;\{\Psi_j\}_{j=J}^{\infty})}\right\|_{\mathfrak{b}_{\mathfrak{B}'}} \leqslant C\|\mathbf{f}\|_{\mathfrak{B}'}, \qquad \mathbf{f}\in \mathscr{D}(\mathbb{R}^d) \tag{70}$$

and

$$\left\|\{\langle \mathbf{g}, \tilde{\mathbf{h}}\rangle\}_{\tilde{\mathbf{h}}\in \mathrm{FWS}_J(\tilde{\Phi};\{\tilde{\Psi}_j\}_{j=J}^{\infty})}\right\|_{\mathfrak{b}_{\mathfrak{B}}} \leqslant C\|\mathbf{g}\|_{\mathfrak{B}}, \qquad \mathbf{g}\in \mathscr{D}(\mathbb{R}^d); \tag{71}$$

(ii) the pair $(\mathrm{FWS}_J(\Phi;\{\Psi_j\}_{j=J}^{\infty}), \mathrm{FWS}_J(\tilde{\Phi};\{\tilde{\Psi}_j\}_{j=J}^{\infty}))$ *is a pair of frequency-based nonstationary nonhomogeneous dual* N*-wavelet frames in the distribution space.*

Proof. We sketch a proof which is parallel to the proof of [43, Theorem 7] for the Sobolev space $\mathfrak{B} = \widehat{H^\tau(\mathbb{R})}$. For more details on a complete proof, see [43, Theorem 7]. Since the necessity part is trivial, we only prove the sufficiency part.

For $\mathbf{h}_0 \in \mathrm{FWS}_J(\Phi;\{\Psi_j\}_{j=J}^{\infty})$, by our assumption in (54) and (70), we have $|\langle \mathbf{f}, \mathbf{h}_0\rangle| \leqslant CC_{\mathbf{h}_0}\|\mathbf{f}\|_{\mathfrak{B}'}$ for all $\mathbf{f}\in\mathscr{D}(\mathbb{R}^d)$. Hence, $\langle \cdot, \mathbf{h}_0\rangle$ can be extended into a continuous linear functional on $\mathfrak{B}'$, and therefore, $\mathbf{h}_0$ can be identified with an element in $\mathfrak{B}$. By our assumption in (55) and (56), we can deduce that (70) holds for all $\mathbf{f}\in\mathfrak{B}'$. Similarly, we have $\tilde{\mathbf{h}}_0 \in \mathfrak{B}'$ for every $\tilde{\mathbf{h}}_0 \in \mathrm{FWS}_J(\tilde{\Phi};\{\tilde{\Psi}_j\}_{j=J}^{\infty})$ and (71) holds for all $\mathbf{g}\in\mathfrak{B}$. By the relation in (57), we deduce that (68) holds and

$$\tilde{C}_7\|\mathbf{g}\|_{\mathfrak{B}} \leqslant \|\{\langle \mathbf{g}, \tilde{\mathbf{h}}\rangle_{\tilde{\mathbf{h}}\in \mathrm{FWS}_J(\tilde{\Phi};\{\tilde{\Psi}_j\}_{j=J}^{\infty})}\|_{\mathfrak{b}_{\mathfrak{B}}} \leqslant \tilde{C}_8\|\mathbf{g}\|_{\mathfrak{B}}, \qquad \forall\, \mathbf{g}\in\mathfrak{B}$$

with $\tilde{C}_7 = C_7 := (C_0C)^{-1}$ and $\tilde{C}_8 = C_8 := C$.

By item (ii), we see that (24) holds for all $\mathbf{f}, \mathbf{g} \in \mathscr{D}(\mathbb{R}^d)$. Since $\mathscr{D}(\mathbb{R}^d)$ is dense in $\mathfrak{B}$ and $\mathfrak{B}'$, by what has been proved and by a standard density argument, we see that identity (69) holds for all $\mathbf{f}\in\mathfrak{B}'$ and $\mathbf{g}\in\mathfrak{B}$. Therefore, the pair in (23) is a pair of frequency-based nonstationary nonhomogeneous dual N-wavelet frames in a pair of dual function spaces $(\mathfrak{B}, \mathfrak{B}')$. □

As discussed in Sect. 3, we have a complete picture about item (ii) of Theorem 5 on frequency-based nonstationary nonhomogeneous dual framelets in the distribution space. Though we have some sufficient conditions in [51] for item (i) of Theorem 5 for the particular case $\mathfrak{B} = \widehat{H^\tau(\mathbb{R}^d)}$, more effort is needed to get a complete understanding and characterization of item (i) of Theorem 5 on the stability of a frequency-based nonstationary nonhomogeneous wavelet systems in various function spaces. In Sect. 5, we shall address the stability issue of a nonhomogeneous wavelet system which is derived from a filter bank.

5 Wavelets and Framelets Derived from Filter Banks

Many nonhomogeneous wavelets and framelets are derived from filter banks. In this section, we shall study the two fundamental issues on frequency-based nonhomogeneous wavelets and framelets that come from filter banks: frequency-based nonhomogeneous dual framelets in the distribution space and stability of a frequency-based nonhomogeneous wavelet system in a general function space. For simplicity, we only discuss stationary nonhomogeneous wavelets and framelets in Sobolev spaces.

As a particular case of Theorem 3, we have

Corollary 4 *Let* M *be a* $d \times d$ *integer expansive matrix. Define* $\mathsf{N} := (\mathsf{M}^{\mathsf{T}})^{-1}$. *Let* $\mathbf{a}$ *and* $\tilde{\mathbf{a}}$ *be* $2\pi\mathbb{Z}^d$*-periodic trigonometric polynomials satisfying* $\mathbf{a}(0) = \tilde{\mathbf{a}}(0) = 1$. *Define* φ *and* $\tilde{\varphi}$ *as in* (38). *For* $2\pi\mathbb{Z}^d$*-periodic trigonometric polynomials* $\Theta, \mathbf{b}_1, \ldots, \mathbf{b}_s, \tilde{\mathbf{b}}_1, \ldots, \tilde{\mathbf{b}}_s$, *define* $\eta(\xi) := \Theta(\xi)\tilde{\varphi}(\xi)$ *and*

$$\psi^{\ell}(\mathsf{M}^{\mathsf{T}}\xi) := \mathbf{b}_{\ell}(\xi)\varphi(\xi) \quad \text{and} \quad \tilde{\psi}^{\ell}(\mathsf{M}^{\mathsf{T}}\xi) := \tilde{\mathbf{b}}_{\ell}(\xi)\tilde{\varphi}(\xi), \qquad \ell = 1, \ldots, s. \tag{72}$$

Define

$$\Phi := \{\varphi\}, \quad \Psi := \{\psi^1, \ldots, \psi^s\}, \quad \tilde{\Phi} := \{\eta\}, \quad \tilde{\Psi} := \{\tilde{\psi}^1, \ldots, \tilde{\psi}^s\}. \tag{73}$$

Then, $(\{\Phi;\Psi\}, \{\tilde{\Phi};\tilde{\Psi}\})$ *is a frequency-based nonhomogeneous dual* N*-framelet in the distribution space* $\mathscr{D}'(\mathbb{R}^d)$, *if and only if,* $\Theta(0) = 1$ *and*

$$\Theta(\mathsf{M}^{\mathsf{T}}\xi)\overline{\mathbf{a}(\xi + 2\pi\omega)}\tilde{\mathbf{a}}(\xi) + \sum_{\ell=1}^{s} \overline{\mathbf{b}_{\ell}(\xi + 2\pi\omega)}\tilde{\mathbf{b}}_{\ell}(\xi) = \Theta(\xi)\delta(\omega) \tag{74}$$

for all $\omega \in \Omega_{\mathsf{N}} := [\mathsf{N}\mathbb{Z}^d] \cap [0,1)^d$.

We shall see in Sect. 6 that (74) is equivalent to the perfect reconstruction property of the dual framelet filter bank $(\{\mathbf{a};\mathbf{b}_1,\ldots,\mathbf{b}_s\}, \{\tilde{\mathbf{a}};\tilde{\mathbf{b}}_1,\ldots,\tilde{\mathbf{b}}_s\})_{\Theta}$.

To study the stability of $\mathrm{FWS}_J(\Phi;\Psi)$ in a function space $\mathfrak{B}$ and the stability of $\mathrm{FWS}_J(\tilde{\Phi};\tilde{\Psi})$ in $\mathfrak{B}'$, we have to introduce some definitions.

By ∂_j we denote the partial derivative with respect to the j-th coordinate. Define $\mathbb{N}_0 := \mathbb{N} \cup \{0\}$. For $\beta = (\beta_1, \ldots, \beta_d)^{\mathsf{T}} \in \mathbb{N}_0^d$, we define $|\beta| := \beta_1 + \cdots + \beta_d$, $\partial := (\partial_1, \ldots, \partial_d)^{\mathsf{T}}$, and $\partial^{\beta} := \partial_1^{\beta_1} \cdots \partial_d^{\beta_d}$. For a nonnegative integer m and two smooth functions $\mathbf{f}$ and $\mathbf{g}$, the notation $\mathbf{f}(\xi) = \mathbf{g}(\xi) + O(\|\xi - \xi_0\|^m)$ as $\xi \to \xi_0$ means that $\partial^{\beta}\mathbf{f}(\xi_0) = \partial^{\beta}\mathbf{g}(\xi_0)$ for all $\beta \in \mathbb{N}_0^d$ such that $|\beta| < m$.

For $1 \leqslant p \leqslant \infty$, the L_p smoothness of a distribution f is measured by its L_p critical exponent $\nu_p(f)$ defined by $\nu_p(f) = \sup\{\tau \in \mathbb{R} \mid f \in B^{\tau}_{p,p}(\mathbb{R}^d)\}$. For the particular case $p = 2$, we have $\nu_2(f) = \sup\{\tau \in \mathbb{R} \mid f \in H^{\tau}(\mathbb{R}^d)\}$. More generally, for a family of function spaces $\mathfrak{B}^{\tau}$ with smooth exponent $\tau \in \mathbb{R}$ and for any distribution f, we define $\nu(f \mid \mathfrak{B}^{\cdot}) := \sup\{\tau \in \mathbb{R} \mid f \in \mathfrak{B}^{\tau}\}$. In case that $f \notin \mathfrak{B}^{\tau}$ for all τ, we simply use the convention $\nu(f \mid \mathfrak{B}^{\cdot}) := -\infty$.

It has been shown in [51, Theorem 2.3 and Proposition 2.6] that

Proposition 7 *Let $\tau \in \mathbb{R}$ be a real number. Let $\mathbf{b}_1, \dots, \mathbf{b}_s$ be $2\pi\mathbb{Z}^d$-periodic trigonometric polynomials such that there is a nonnegative integer $m > -\tau$ satisfying*

$$\mathbf{b}_\ell(\xi) = O(\|\xi\|^m), \qquad \xi \to 0, \quad \ell = 1, \dots, s,$$

Let ϕ be a compactly supported tempered distribution such that $\nu_2(\phi) > \tau$. Define $\varphi := \hat{\phi}$ and $\psi^1, \dots, \psi^s$ as in (72) with $\mathsf{M} = 2I_d$. Then there exists a positive constant C (depending on J and $\varphi, \psi^1, \dots, \psi^s$) such that

$$\|\{\langle \mathbf{f}, \mathbf{h}\rangle\}_{\mathbf{h}\in \mathrm{FWS}_J(\{\varphi\};\{\psi^1,\dots,\psi^s\})}\|_{\mathfrak{b}_{\widehat{H^{-\tau}(\mathbb{R}^d)}}} \leqslant C\|\mathbf{f}\|_{\widehat{H^{-\tau}(\mathbb{R}^d)}}, \qquad \mathbf{f} \in \widehat{H^{-\tau}(\mathbb{R}^d)}.$$

With a little bit more work, one can show that Proposition 7 holds for a general integer isotropic expansive matrix M. Here we say that M is isotropic if M is similar to a diagonal matrix with all the diagonal entries having the same modulus. For the particular case $\tau = 0$, the following result says that if a compactly supported function $\phi \in L_2(\mathbb{R}^d)$ is refinable, then it automatically implies $\nu_2(\phi) > 0$.

Theorem 6. *([37, Theorem 2.2]) Let M be a $d \times d$ integer expansive matrix and $\mathbf{a}$ be an $r \times r$ matrix of $2\pi\mathbb{Z}^d$-periodic trigonometric polynomials. If $\phi = [\phi^1, \dots, \phi^r]^\mathsf{T}$ is an $r \times 1$ column vector of compactly supported functions in $L_2(\mathbb{R}^d)$ and satisfies $\hat{\phi}(\mathsf{M}^\mathsf{T}\xi) = \mathbf{a}(\xi)\hat{\phi}(\xi)$ for almost every $\xi \in \mathbb{R}^d$, then $\nu_2(\phi^\ell) > 0$ for all $\ell = 1, \dots, r$.*

Consequently, combining Corollary 4 and Proposition 7, we have

Theorem 7. *([51, Theorem 1.1 with $\mathsf{M} = 2$]) Under the same notation as in Corollary 4. Suppose that M is a $d \times d$ integer isotropic expansive matrix. Let $\phi, \tilde{\phi}$ be compactly supported tempered distributions. Define $\varphi := \hat{\phi}$ and $\tilde{\varphi} := \hat{\tilde{\phi}}$. Let $\Phi, \Psi, \tilde{\Phi}, \tilde{\Psi}$ be defined in (73). For any given $\tau \in \mathbb{R}$, if $\Theta(0) = 1$ and the following three conditions are satisfied:*

1. *(74) holds for the filter bank $(\{\mathbf{a}; \mathbf{b}_1, \dots, \mathbf{b}_s\}, \{\tilde{\mathbf{a}}; \tilde{\mathbf{b}}_1, \dots, \tilde{\mathbf{b}}_s\})_\Theta$;*
2. *$\nu_2(\phi) > \tau$ and $\nu_2(\tilde{\phi}) > -\tau$;*
3. *there exist nonnegative integers $m > -\tau$ and $\tilde{m} > \tau$ satisfying*

$$\mathbf{b}_\ell(\xi) = O(\|\xi\|^m) \quad \text{and} \quad \tilde{\mathbf{b}}_\ell(\xi) = O(\|\xi\|^{\tilde{m}}), \qquad \xi \to 0, \quad \ell = 1, \dots, s,$$

then $(\mathrm{FWS}_J(\Phi;\Psi), \mathrm{FWS}_J(\tilde{\Phi};\tilde{\Psi}))$ is a pair of frequency-based nonhomogeneous dual N-wavelet frames in the pair of dual Sobolev spaces $(H^\tau(\mathbb{R}^d), H^{-\tau}(\mathbb{R}^d))$.

The quantity $\nu_2(\phi)$ can be estimated from its refinement mask $\mathbf{a}$. To do so, let us introduce some definitions. We say that $\mathbf{a}$ has *m sum rules* with respect to M if

$$\partial^\beta \mathbf{a}(2\pi\omega) = 0, \qquad \beta \in \mathbb{N}_0^d \quad \text{with} \quad |\beta| < m, \ \omega \in \Omega_\mathsf{N} \backslash \{\mathbf{0}\}, \tag{75}$$

where $\Omega_\mathsf{N} := [\mathsf{N}\mathbb{Z}^d] \cap [0,1)^d$ and $\mathsf{N} := (\mathsf{M}^\mathsf{T})^{-1}$. Moreover, if $\mathbf{a}$ has m sum rules but not $m+1$ with respect to M, then we denote by $\mathrm{sr}(\mathbf{a}, \mathsf{M}) := m$.

We provide here an example of nonhomogeneous framelets using box splines. Box splines are important examples of refinable functions in high dimensions. For a given $d \times m$ (direction) matrix Ξ of full rank with integer entries and $m \geqslant d$, the Fourier transform of the box spline M_Ξ and the mask of M_Ξ are given by

$$\widehat{M_\Xi}(\xi) := \prod_{\mathsf{k}\in\Xi} \frac{1-e^{-i\mathsf{k}\cdot\xi}}{i\mathsf{k}\cdot\xi} \quad \text{and} \quad \mathbf{a}_\Xi(\xi) = \prod_{\mathsf{k}\in\Xi} \frac{1+e^{-i\mathsf{k}\cdot\xi}}{2}, \qquad \xi \in \mathbb{R}^d, \tag{76}$$

satisfying $\widehat{M_\Xi}(2\xi) = \mathbf{a}_\Xi(\xi)\widehat{M_\Xi}(\xi)$, where $\mathsf{k} \in \Xi$ means that k is a column vector of Ξ and k goes through all the columns of Ξ once and only once.

The box spline M_Ξ belongs to $C^{m(\Xi)-1}$, where $m(\Xi)+1$ is the minimum number of columns that can be discarded from Ξ to obtain a matrix of rank $< d$. In other words, we have $\nu_2(M_\Xi) = m(\Xi) + 1/2$. When Ξ is a $1 \times m$ row vector with all its components being 1, the box spline M_Ξ is the well-known B-spline of order m and has the mask $2^{-m}(1+e^{-i\xi})^m$. See [24] for more details on box splines. Theorem 7 can be applied to the box splines to obtain framelets in Sobolev spaces. For any $\tau \in \mathbb{R}$ such that $0 < \tau < \min(m(\Xi)+1/2, \mathrm{sr}(\mathbf{a}_\Xi, 2I_d))$, then $\mathrm{WS}_0(\{M_\Xi\};\{M_\Xi\})$ is a nonhomogeneous $2I_d$-framelet in $H^\tau(\mathbb{R}^d)$, see [51] for more detail.

To study wavelets in Sobolev spaces, in the following we recall a key quantity in wavelet analysis. Let M be a $d \times d$ integer invertible matrix and define $\mathsf{N} := (\mathsf{M}^\mathsf{T})^{-1}$. Suppose that $\mathbf{a}(0) = 1$ and $\mathbf{a}$ has m sum rules but not $m+1$ sum rules with respect to M, that is, $m := \mathrm{sr}(\mathbf{a}, \mathsf{M})$. For $\beta = (\beta_1, \ldots, \beta_d)^\mathsf{T} \in \mathbb{N}_0^d$ with $|\beta| = m$ and $n \in \mathbb{N}$, we define a sequence $u^{n,\beta} = \{u^{n,\beta}(\mathsf{k})\}_{\mathsf{k}\in\mathbb{Z}^d}$ which is uniquely determined by

$$\sum_{\mathsf{k}\in\mathbb{Z}^d} u^{n,\beta}(\mathsf{k})e^{-i\mathsf{k}\cdot\xi} = (1-e^{-i\xi_1})^{\beta_1}\cdots(1-e^{-i\xi_d})^{\beta_d}\mathbf{a}((\mathsf{M}^\mathsf{T})^{n-1}\xi)\cdots\mathbf{a}(\mathsf{M}^\mathsf{T}\xi)\mathbf{a}(\xi),$$

where $\xi = (\xi_1, \ldots, \xi_d)^\mathsf{T} \in \mathbb{R}^d$. For $1 \leqslant p \leqslant \infty$, we define

$$\rho(\mathbf{a}, \mathsf{M}, p) := \max\{\limsup_{n\to\infty} \|u^{n,\beta}\|_{l_p(\mathbb{Z}^d)}^{1/n} \mid \beta \in \mathbb{N}_0^d, |\beta| = m\}.$$

Define a fundamental quantity $\nu_p(\mathbf{a}, \mathsf{M})$ in wavelet analysis (see [36, page 61]) by

$$\nu_p(\mathbf{a}, \mathsf{M}) := -\log_{\rho(\mathsf{M})}[|\det \mathsf{M}|^{1-1/p}\rho(\mathbf{a}, \mathsf{M}, p)], \tag{77}$$

where $\rho(\mathsf{M})$ denotes the spectral radius of M. $\nu_p(\mathbf{a}, \mathsf{M})$ is called the L_p *smoothness exponent* of a mask $\mathbf{a}$ with a dilation matrix M.

The above quantity $\nu_p(\mathbf{a}, \mathsf{M})$ plays a very important role in characterizing the convergence of a cascade algorithm in a Sobolev space, and in characterizing the L_p critical exponent of a refinable function vector. It was showed in [36, Theorem 4.3] that the vector cascade algorithm associated with mask $\mathbf{a}$ and an isotropic dilation matrix M converges in the Sobolev space $H_p^\tau(\mathbb{R}^d) := \{f \in L_p(\mathbb{R}^d) \mid \partial^\beta f \in L_p(\mathbb{R}^d)\ \forall\, |\beta| \leqslant \tau\}$ for a nonnegative integer τ, if and only if, $\nu_p(\mathbf{a}, \mathsf{M}) > \tau$. In general, $\nu_p(\mathbf{a}, \mathsf{M})$ provides a lower bound for the L_p critical exponent of a compactly supported refinable function ϕ with a mask $\mathbf{a}$ and a dilation matrix M, that is, $\nu_p(\mathbf{a},\mathsf{M}) \leqslant \nu_p(\phi)$ always holds if $\hat{\phi}(\mathsf{M}^\mathsf{T}\xi) = \mathbf{a}(\xi)\hat{\phi}(\xi)$ and ϕ is compactly supported. Moreover, if the shifts of the refinable function ϕ associated with a mask $\mathbf{a}$ and an integer isotropic expansive matrix M are stable (that is, for every $\xi \in \mathbb{R}^d$, there exists $\mathsf{k}_\xi \in \mathbb{Z}^d$ such that $\hat{\phi}(\xi + 2\pi\mathsf{k}_\xi) \neq 0$), then $\nu_p(\phi) = \nu_p(\mathbf{a},\mathsf{M})$. That is, $\nu_p(\mathbf{a},\mathsf{M})$ indeed characterizes the L_p smoothness exponent of a compactly supported nontrivial refinable function ϕ with a mask $\mathbf{a}$ and an integer isotropic expansive matrix M. Furthermore, we also have $\nu_p(\mathbf{a},\mathsf{M}) \geqslant \nu_q(\mathbf{a},\mathsf{M}) \geqslant \nu_p(\mathbf{a},\mathsf{M}) + (1/q - 1/p)\log_{\rho(\mathsf{M})}|\det \mathsf{M}|$ for $1 \leqslant p \leqslant q \leqslant \infty$. For a trigonometric polynomial mask $\mathbf{a}$ and an integer dilation matrix M, the quantity $\nu_2(\mathbf{a},\mathsf{M})$ can be numerically computed using symmetry of the mask $\mathbf{a}$ by finding the spectral radius of certain finite matrix ([35, Algorithm 2.1]). Interested readers should consult [13,14,32,35,36,46,50,57–59,61,64,70,75] and references therein on the convergence of cascade algorithms and smoothness of refinable functions.

For frequency-based nonhomogeneous biorthogonal wavelet bases, we have

Theorem 8. *Let* M *be a* $d \times d$ *integer isotropic expansive matrix. Define* $\mathsf{N} := (\mathsf{M}^\mathsf{T})^{-1}$. *Let* $\mathbf{a}$ *and* $\tilde{\mathbf{a}}$ *be* $2\pi\mathbb{Z}^d$*-periodic trigonometric polynomials satisfying* $\mathbf{a}(0) = \tilde{\mathbf{a}}(0) = 1$. *Define* φ *and* $\tilde{\varphi}$ *as in* (38). *For* $2\pi\mathbb{Z}^d$*-periodic trigonometric polynomials* $\mathbf{b}_1,\dots,\mathbf{b}_s, \tilde{\mathbf{b}}_1,\dots,\tilde{\mathbf{b}}_s$, *define* $\psi^1,\dots,\psi^s,\tilde{\psi}^1,\dots,\tilde{\psi}^s$ *as in* (72). *Define* $\Phi,\Psi,\tilde{\Phi},\tilde{\Psi}$ *as in* (73) *with* $\Theta = 1$. *Then* $(\mathrm{FWS}_J(\Phi;\Psi), \mathrm{FWS}_J(\tilde{\Phi};\tilde{\Psi}))$ *is a pair of frequency-based nonhomogeneous Riesz* N*-wavelet bases in the pair of dual Sobolev spaces* $(H^\tau(\mathbb{R}^d), H^{-\tau}(\mathbb{R}^d))$, *if*

$$\nu_2(\mathbf{a},\mathsf{M}) > \tau, \qquad \nu_2(\tilde{\mathbf{a}},\mathsf{M}) > -\tau \tag{78}$$

and

$$\overline{\mathbf{P}_{[\tilde{\mathbf{a}},\tilde{\mathbf{b}}_1,\dots,\tilde{\mathbf{b}}_{\mathsf{d}_\mathsf{M}-1}]}(\xi)}^\mathsf{T} \mathbf{P}_{[\mathbf{a},\mathbf{b}_1,\dots,\mathbf{b}_{\mathsf{d}_\mathsf{M}-1}]}(\xi) = I_{\mathsf{d}_\mathsf{M}}, \qquad \xi \in \mathbb{R}^d, \tag{79}$$

where $\{\omega_0,\dots,\omega_{\mathsf{d}_\mathsf{M}-1}\} = \Omega_\mathsf{N} := [\mathsf{N}\mathbb{Z}^d] \cap [0,1)^d$, $\mathsf{d}_\mathsf{M} := |\det \mathsf{M}|$, *and the matrix* $\mathbf{P}_{[\mathbf{a},\mathbf{b}_1,\dots,\mathbf{b}_{\mathsf{d}_\mathsf{M}-1}]}(\xi)$ *is defined to be*

$$\begin{bmatrix} \mathbf{a}(\xi+2\pi\omega_0) & \mathbf{a}(\xi+2\pi\omega_1) & \cdots & \mathbf{a}(\xi+2\pi\omega_{\mathsf{d}_\mathsf{M}-1}) \\ \mathbf{b}_1(\xi+2\pi\omega_0) & \mathbf{b}_1(\xi+2\pi\omega_1) & \cdots & \mathbf{b}_1(\xi+2\pi\omega_{\mathsf{d}_\mathsf{M}-1}) \\ \vdots & \vdots & \ddots & \vdots \\ \mathbf{b}_{\mathsf{d}_\mathsf{M}-1}(\xi+2\pi\omega_0) & \mathbf{b}_{\mathsf{d}_\mathsf{M}-1}(\xi+2\pi\omega_1) & \cdots & \mathbf{b}_{\mathsf{d}_\mathsf{M}-1}(\xi+2\pi\omega_{\mathsf{d}_\mathsf{M}-1}) \end{bmatrix}.$$

Theorem 8 has been proved in [51, Theorem 3.1] with $\mathsf{M} = 2I_d$. A more technical proof can be used to show that Theorem 8 holds. Moreover, when $d = 1$, the conditions in (78) and (79) in Theorem 8 are also necessary ([51]).

In the following, we present an example of wavelets in Sobolev spaces. Let m be a positive integer. Let $d = 1$ and $\mathsf{M} = 2$. Define $\tilde{\mathbf{a}}(\xi) = 1, \mathbf{b}(\xi) = e^{-i\xi}$,

$$\mathbf{a}(\xi) := \cos^{2m}(\xi/2)\sum_{n=0}^{m-1}\frac{(m+n-1)!}{n!(m-1)!}\sin^{2n}(\xi/2) \quad \text{and} \quad \tilde{\mathbf{b}}(\xi) = e^{-i\xi}\overline{\mathbf{a}(\xi+\pi)}.$$

Then $(\mathrm{FWS}_0(\{\varphi\};\{\psi\}),\mathrm{FWS}_0(\{\tilde{\varphi}\};\{\tilde{\psi}\}))$ is a pair of frequency-based nonhomogeneous biorthogonal 2^{-1}-wavelet bases in $(H^\tau(\mathbb{R}),H^{-\tau}(\mathbb{R}))$ for all $\tau \in (1/2, \nu_2(\mathbf{a},2))$. Note that $\nu_2(\mathbf{a},2) > 1/2$ for all $m \in \mathbb{N}$. For more detail, see [51].

6 Discrete Framelet Transform and Its Basic Properties

Algorithmic aspect of wavelets and framelets is a fundamental part of wavelet analysis. Though we have seen the connections of frequency-based wavelets and framelets in the distribution space with filter banks in previous sections, it is very natural and important to study the discrete wavelet/framelet transform and its basic properties purely from the discrete setting. The properties of discrete wavelet/framelet transform with a filter bank are traditionally derived from wavelets and framelets in the function setting via a multiresolution analysis [6, 19, 68, 69]. However, in this section, we shall study discrete wavelet/framelet transform and its properties purely in the discrete sequence setting without any wavelets or framelets in the function setting involved. Though this approach is very natural for algorithms, little attention has been paid to this direction until a few years ago in the literature. We shall review the recent developments on this topic in this section.

By $l(\mathbb{Z}^d)$ we denote the linear space of all sequences $v : \mathbb{Z}^d \to \mathbb{C}$ of complex numbers on $\mathbb{Z}^d$. For $v \in l(\mathbb{Z}^d)$, we often write $v = \{v(\mathsf{k})\}_{\mathsf{k}\in\mathbb{Z}^d}$, and we shall model signals by $l(\mathbb{Z}^d)$. Similarly, by $l_0(\mathbb{Z}^d)$ we denote the linear space of all finitely supported sequences on $\mathbb{Z}^d$. An element in $l_0(\mathbb{Z}^d)$ is often regarded as a finite-impulse-response (FIR) filter (also called a finitely supported mask in the literature of wavelet analysis). We often use u for a general filter, and v for a general signal or dataset.

A discrete framelet transform can be described using two linear operators – the subdivision operator and the transition operator. More precisely, for a filter $u \in l_0(\mathbb{Z}^d)$ and a $d \times d$ integer invertible matrix M, the *subdivision operator* $\mathscr{S}_{u,\mathsf{M}} : l(\mathbb{Z}^d) \to l(\mathbb{Z}^d)$ is defined to be

$$[\mathscr{S}_{u,\mathsf{M}}v](\mathsf{n}) := |\det \mathsf{M}| \sum_{\mathsf{k}\in\mathbb{Z}^d} v(\mathsf{k})u(\mathsf{n}-\mathsf{M}\mathsf{k}), \qquad \mathsf{n} \in \mathbb{Z}^d \tag{80}$$

for $v \in l(\mathbb{Z}^d)$, and the *transition operator* $\mathscr{T}_{u,\mathsf{M}} : l(\mathbb{Z}^d) \to l(\mathbb{Z}^d)$ is defined to be

$$[\mathscr{T}_{u,\mathsf{M}}v](\mathsf{n}) := |\det \mathsf{M}| \sum_{\mathsf{k}\in\mathbb{Z}^d} v(\mathsf{k})\overline{u(\mathsf{k}-\mathsf{M}\mathsf{n})}, \qquad \mathsf{n} \in \mathbb{Z}^d \tag{81}$$

for $v \in l(\mathbb{Z}^d)$. For $u \in l_0(\mathbb{Z}^d)$ and $v \in l(\mathbb{Z}^d)$, the convolution $u * v$ of u and v is defined to be $[u * v](\mathsf{n}) := \sum_{\mathsf{k}\in\mathbb{Z}^d} u(\mathsf{k})v(\mathsf{n}-\mathsf{k})$, $\mathsf{n} \in \mathbb{Z}^d$.

In the following, we introduce a multi-level discrete framelet transform employing a filter bank derived from the oblique extension principle. Let

$$(\{a;b_1,\ldots,b_s\},\{\tilde{a};\tilde{b}_1,\ldots,b_s\})_\Theta \tag{82}$$

be a filter bank with all filters in $l_0(\mathbb{Z}^d)$. For a positive integer J, a J-level discrete framelet transform consists of two parts: a J-level framelet decomposition and a J-level framelet reconstruction.

Let M be a $d \times d$ integer invertible matrix. Using the primal low-pass filter a and primal high-pass filters $b_1, \ldots, b_s$ for decomposition, *a J-level framelet decomposition* is given by: for $j = J, \ldots, 1$,

$$v_{j-1} := |\det \mathsf{M}|^{-1/2} \mathscr{T}_{a,\mathsf{M}} v_j, \quad w_{j-1;\ell} := |\det \mathsf{M}|^{-1/2} \mathscr{T}_{b_\ell,\mathsf{M}} v_j, \qquad \ell = 1, \ldots, s, \tag{83}$$

where $v_J : \mathbb{Z}^d \to \mathbb{C}$ is an input signal. After a J-level framelet decomposition, the original input signal v_J is decomposed into one sequence v_0 of low-pass framelet coefficients and sJ sequences $w_{j;\ell}$ of high-pass framelet coefficients for $\ell = 1, \ldots, s$ and $j = 0, \ldots, J-1$. Such framelet coefficients are often processed for various purposes. One of the most commonly employed operations is thresholding so that the low-pass framelet coefficients v_0 and high-pass framelet coefficients $w_{j;\ell}$ become $\mathring{v}_0$ and $\mathring{w}_{j;\ell}$, respectively. More precisely, $\mathring{w}_{j;\ell}(\mathsf{k}) = \eta(w_{j;\ell}(\mathsf{k})), \mathsf{k} \in \mathbb{Z}^d$, where $\eta : \mathbb{C} \to \mathbb{C}$ is a thresholding function. For example, for a given threshold value $\varepsilon > 0$, the hard thresholding function η^{hard} and soft-threshold function η^{soft} are defined to be

$$\eta^{\text{hard}}(z) = \begin{cases} z, & \text{if } |z| \geqslant \varepsilon; \\ 0, & \text{otherwise} \end{cases} \quad \text{and} \quad \eta^{\text{soft}}(z) = \begin{cases} z - \varepsilon \frac{z}{|z|}, & \text{if } |z| \geqslant \varepsilon; \\ 0, & \text{otherwise.} \end{cases}$$

Using the dual low-pass filter $\tilde{a}$ and dual high-pass filters $\tilde{b}_1, \ldots, \tilde{b}_s$ for reconstruction, *a J-level framelet reconstruction* is

$$\check{v}_0 := \Theta * \mathring{v}_0, \tag{84}$$

$$\check{v}_j := |\det \mathsf{M}|^{-1/2} \mathscr{S}_{\tilde{a},\mathsf{M}} \check{v}_{j-1} + |\det \mathsf{M}|^{-1/2} \sum_{\ell=1}^{s} \mathscr{S}_{\tilde{b}_\ell,\mathsf{M}} \mathring{w}_{j-1;\ell}, \qquad j = 1, \ldots, J, \tag{85}$$

$$\text{recover } \mathring{v}_J \text{ from } \check{v}_J \text{ via the relation } \mathring{v}_J = \Theta * \check{v}_J. \tag{86}$$

For a multi-level discrete framelet transform, there are three fundamental properties: perfect reconstruction, stability, and sparsity. In the following, we address them one by one. When nothing is performed on the framelet coefficients, that is, $\mathring{v}_0 = v_0$ and $\mathring{w}_{j;\ell} = w_{j;\ell}$ for all $\ell = 1, \ldots, s, j = 0, \ldots, J-1$, we say that the above J-level discrete framelet transform has the perfect reconstruction property if $\mathring{v}_J = v_J$. Note that a J-level discrete framelet transform recursively employs one-level discrete framelet transforms J times. Therefore, to study the perfect reconstruction property of a J-level discrete framelet transform for all positive integers J, it suffices to study it for $J = 1$. For simplicity of presentation, we often use the formal Fourier series $\hat{v}$ of a sequence $v = \{v(\mathsf{k})\}_{\mathsf{k} \in \mathbb{Z}^d}$, which is defined by

$$\hat{v}(\xi) := \sum_{\mathsf{k} \in \mathbb{Z}^d} v(\mathsf{k}) e^{-i\mathsf{k} \cdot \xi}, \qquad \xi \in \mathbb{R}^d.$$

Now we have the following result on perfect reconstruction of a discrete framelet transform with a filter bank in (82) (see [45, Theorem 1.4.2] and [7, 22, 23, 40, 42]).

Theorem 9. *(Oblique Extension Principle) Let $a \in l_0(\mathbb{Z}^d)$ be a primal low-pass filter and $b_1,\dots,b_s \in l_0(\mathbb{Z}^d)$ be primal high-pass filters for decomposition. Let $\tilde{a} \in l_0(\mathbb{Z}^d)$ be a dual low-pass filter and $\tilde{b}_1,\dots,\tilde{b}_s \in l_0(\mathbb{Z}^d)$ be dual high-pass filters for reconstruction. Let M be a $d \times d$ integer invertible matrix and $\Theta \in l_0(\mathbb{Z}^d)$. Then the filter bank in* (82) *has the following perfect reconstruction property:*

$$\Theta * v = |\det \mathsf{M}|^{-1} \mathscr{S}_{\tilde{a},\mathsf{M}}(\Theta * \mathscr{T}_{a,\mathsf{M}} v) + |\det \mathsf{M}|^{-1} \sum_{\ell=1}^{s} \mathscr{S}_{\tilde{b}_\ell,\mathsf{M}} \mathscr{T}_{b_\ell,\mathsf{M}} v, \ \forall\, v \in l(\mathbb{Z}^d), \quad (87)$$

if and only if, for all $\xi \in \mathbb{R}^d$,

$$\hat{\Theta}(\mathsf{M}^\mathsf{T}\xi)\overline{\hat{a}(\xi+2\pi\omega)}\hat{\tilde{a}}(\xi) + \sum_{\ell=1}^{s} \overline{\widehat{b_\ell}(\xi+2\pi\omega)}\widehat{\tilde{b}_\ell}(\xi) = \hat{\Theta}(\xi)\delta(\omega), \quad (88)$$

for all $\omega \in \Omega_{(\mathsf{M}^\mathsf{T})^{-1}} := [(\mathsf{M}^\mathsf{T})^{-1}\mathbb{Z}^d] \cap [0,1)^d$.

If (88) holds, then we say that the filter bank in (82) is *a dual M-framelet filter bank* (derived from OEP). The role played by the factor $|\det \mathsf{M}|^{-1/2}$ in (83) and (85) is explained by the following result [45, Proposition 1.4.5]:

Proposition 8 *Let $\theta, a, b_1,\dots,b_s \in l_0(\mathbb{Z}^d)$. Then for all $v \in l_2(\mathbb{Z}^d)$,*

$$\|\theta * \mathscr{T}_{a,\mathsf{M}} v\|^2_{l_2(\mathbb{Z}^d)} + \sum_{\ell=1}^{s} \|\mathscr{T}_{b_\ell,\mathsf{M}} v\|^2_{l_2(\mathbb{Z}^d)} = |\det \mathsf{M}| \|\theta * v\|^2_{l_2(\mathbb{Z}^d)}, \quad (89)$$

if and only if, $\{a; b_1,\dots,b_s\}_\Theta$ is a tight M-framelet filter bank, where $\hat{\Theta}(\xi) := |\hat{\theta}(\xi)|^2$; that is, $(\{a; b_1,\dots,b_s\}, \{a; b_1,\dots,b_s\})_\Theta$ is a dual M-framelet filter bank.

For analysis of a multi-level discrete framelet transform, it is convenient to rewrite the J-level framelet decomposition using *a J-level decomposition/analysis operator* $\mathscr{W}_J : l(\mathbb{Z}^d) \to (l(\mathbb{Z}^d))^{1\times(sJ+1)}$ by

$$\mathscr{W}_J v := (v_0, w_{0;1},\dots,w_{0;s},\dots,w_{J-1;1},\dots,w_{J-1;s}), \quad (90)$$

where $w_{j-1;\ell}$ and v_0 are defined in (83). Similarly, *a J-level reconstruction/synthesis operator* $\mathscr{V}_J : (l(\mathbb{Z}^d))^{1\times(sJ+1)} \to l(\mathbb{Z}^d)$ is defined by

$$\mathscr{V}_J(\mathring{v}_0, \mathring{w}_{0;1},\dots,\mathring{w}_{0;s},\dots,\mathring{w}_{J-1;1},\dots,\mathring{w}_{J-1;s}) = \mathring{v}_J, \quad (91)$$

where $\mathring{v}_J$ is computed via the recursive formulas in (84)–(86). Due to the recursive cascade structure of the operators $\mathscr{W}_J$ and $\mathscr{V}_J$ in (83) and (84)–(86), a multi-level discrete framelet transform is often called *a fast framelet transform.*

Note that a J-level discrete framelet transform has the perfect reconstruction property if and only if $\mathscr{V}_J\mathscr{W}_J v = v$ for all $v \in l(\mathbb{Z}^d)$. By Theorem 9, we have a complete characterization of the perfect reconstruction property.

Beyond the perfect reconstruction property, another fundamental property of a multi-level discrete framelet transform is its stability. We say that a multi-level discrete framelet transform with a dual M-framelet filter bank in (82) has *stability* in the space $l_2(\mathbb{Z}^d)$ if there exists a positive constant C such that

$$\|\mathscr{W}_J v\|_{(l_2(\mathbb{Z}^d))^{1\times(sJ+1)}} \leqslant C\|v\|_{l_2(\mathbb{Z}^d)}, \qquad \forall\, v \in l_2(\mathbb{Z}^d),\ J \in \mathbb{N}, \tag{92}$$

$$\|\mathscr{V}_J \mathbf{w}\|_{l_2(\mathbb{Z}^d)} \leqslant C\|\mathbf{w}\|_{(l_2(\mathbb{Z}^d))^{1\times(sJ+1)}}, \qquad \forall\, \mathbf{w} \in (l_2(\mathbb{Z}^d))^{1\times(sJ+1)},\ J \in \mathbb{N}. \tag{93}$$

Obviously, (92) implies that a small change in an input data v induces a small change of all framelet coefficients. Similarly, (93) means that a small perturbation of all framelet coefficients results in a small perturbation of a reconstructed signal. The notion of stability of a multi-level discrete framelet transform can be extended into other sequence spaces. Under the assumption that $\hat{a}(0) = \hat{\tilde{a}}(0) = 1$, if its underlying frequency-based nonhomogeneous dual framelet in Theorem 3 has stability in $L_2(\mathbb{R}^d)$, then it is not difficult to deduce that the multi-level discrete framelet transform with the dual framelet filter bank in (82) has *stability* in the space $l_2(\mathbb{Z})$. However, so far, there is no necessary and sufficient condition available for the stability of a multi-level discrete framelet transform in the discrete sequence setting. This problem is currently under investigation, and we shall report the results elsewhere. The characterization of the stability of a multilevel discrete framelet transform in the sequence space $l_2(\mathbb{Z}^d)$ is tightly linked to the existence of an associated compactly supported refinable function in $L_2(\mathbb{R}^d)$.

It has been shown in [45, Proposition 1.4.8] for $\mathsf{M} = 2$ (also see [23]) that for a dual M-framelet filter bank in (82) we must have $s \geqslant |\det \mathsf{M}| - 1$. Moreover, if $s = |\det \mathsf{M}| - 1$, then it is essentially a biorthogonal wavelet filter bank. We say that a filter bank in (82) is *a biorthogonal* M*-wavelet filter bank* if $\hat{\Theta} = 1$ and $s = |\det \mathsf{M}| - 1$. The difference between wavelets and framelets is mainly due to the following result (see [45, Proposition 1.1.2]).

Proposition 9 *Let* $(\{a; b_1, \ldots, b_s\}, \{\tilde{a}; \tilde{b}_1, \ldots, \tilde{b}_s\})_\Theta$ *be a dual* M*-framelet filter bank such that* $\hat{\Theta} = 1$. *Let* $\mathscr{W} := \mathscr{W}_1$ *and* $\mathscr{V} := \mathscr{V}_1$, *where* $\mathscr{W}_J$ *and* $\mathscr{V}_J$ *are defined in* (90) *and* (91), *respectively. Then* $\mathscr{W}$ *is onto* $\iff$ $\mathscr{V}$ *is one-one* $\iff$ $\mathscr{V}\mathscr{W} = \mathrm{Id}_{l(\mathbb{Z}^d)}$ *and* $\mathscr{W}\mathscr{V} = \mathrm{Id}_{(l(\mathbb{Z}^d))^{1\times(s+1)}}$ $\iff$ $s = |\det \mathsf{M}| - 1$.

Note that the perfect reconstruction property is simply $\mathscr{V}\mathscr{W} = \mathrm{Id}_{l(\mathbb{Z}^d)}$. Hence, a discrete wavelet transform refers to invertible framelet operators $\mathscr{W}$ and $\mathscr{V}$. Therefore, for a wavelet transform, we can first perform the synthesis operator $\mathscr{V}$ followed by the analysis operator $\mathscr{W}$; the perfect reconstruction property is still preserved.

Next, we study the sparsity of a discrete framelet transform. One key feature of a discrete framelet transform is its sparse representation for smooth or piecewise smooth signals. It is desirable to have as many as possible negligible framelet coefficients for smooth signals. Smooth signals are theoretically modeled by polynomials of various degrees. Denote $\mathbb{N}_0 := \mathbb{N} \cup \{0\}$, the set of all nonnegative integers. Let $\mathsf{p} : \mathbb{R}^d \to \mathbb{C}$ be a polynomial in d-variables, that is, $\mathsf{p} = \sum_{\beta \in \mathbb{N}_0^d} p_\beta x^\beta$. Sampling the polynomial p on the integer lattice $\mathbb{Z}^d$, we have a polynomial sequence $\mathsf{p}|_{\mathbb{Z}^d} : \mathbb{Z}^d \to \mathbb{C}$ which is given by $[\mathsf{p}|_{\mathbb{Z}^d}](\mathsf{n}) = \mathsf{p}(\mathsf{n}), \mathsf{n} \in \mathbb{Z}^d$. If a sequence $v = \{v(\mathsf{k})\}_{\mathsf{k} \in \mathbb{Z}^d}$ is a polynomial sequence, then a polynomial p, satisfying $v(\mathsf{k}) = \mathsf{p}(\mathsf{k})$ for all $\mathsf{k} \in \mathbb{Z}^d$, is uniquely determined. Therefore, for simplicity of presentation, we shall use p to denote both a polynomial function p on $\mathbb{R}^d$ and its induced polynomial sequence $\mathsf{p}|_{\mathbb{Z}^d}$ on $\mathbb{Z}^d$. One can easily tell them apart from the context.

For a nonnegative integer $m \in \mathbb{N}_0$, Π_m denotes the space of all polynomials in d-variables of (total) degree no more than m. In particular, $\Pi := \cup_{m=0}^{\infty} \Pi_m$ denotes the space of all polynomials on $\mathbb{R}^d$. For $\mathsf{p}(x) = \sum_{\beta \in \mathbb{N}_0^d} p_\beta x^\beta$ and a smooth function $\mathbf{f}(\xi)$, we shall use the following polynomial differentiation operator in this section:

$$\mathsf{p}(x - \mathrm{i}\partial)\mathbf{f}(\xi) := \sum_{\beta \in \mathbb{N}_0^d} p_\beta (x - \mathrm{i}\partial)^\beta \mathbf{f}(\xi), \qquad x, \xi \in \mathbb{R}^d,$$

where $\partial := (\partial_1, \ldots, \partial_d)^\mathsf{T}$ acts on the variable ξ and $x^\beta := x_1^{\beta_1} \cdots x_d^{\beta_d}$ for $x = (x_1, \ldots, x_d)^\mathsf{T}$ and $\beta = (\beta_1, \ldots, \beta_d)^\mathsf{T} \in \mathbb{N}_0^d$.

Proposition 10 *([42, Propositions 2.1 and 3.1]) Let M be a $d \times d$ integer invertible matrix. Let $u = \{u(\mathsf{k})\}_{\mathsf{k} \in \mathbb{Z}^d} \in l_0(\mathbb{Z}^d)$. Let $\mathsf{p} \in \Pi_m$. Then*

1. *$\mathsf{p} * u$ is a polynomial sequence in Π_m given by*

$$[\mathsf{p} * u] := \sum_{\mathsf{k} \in \mathbb{Z}^d} \mathsf{p}(\cdot - \mathsf{k})u(\mathsf{k}) = [\mathsf{p}(\cdot - \mathrm{i}\partial)\hat{u}(\xi)]|_{\xi=0} = \sum_{\beta \in \mathbb{N}_0^d} \frac{(-\mathrm{i})^{|\beta|}}{\beta!} \partial^\beta \mathsf{p}(\cdot) \partial^\beta \hat{u}(0).$$

   Also, $\partial^\beta(\mathsf{p} * u) = (\partial^\beta \mathsf{p}) * u$, $\mathsf{p}(\cdot - y) * u = (\mathsf{p} * u)(\cdot - y)$ for all $\beta \in \mathbb{N}_0^d$ and $y \in \mathbb{R}^d$;

2. *$\mathscr{T}_{u,\mathsf{M}}\mathsf{p} = \mathsf{p}(\mathsf{M}\cdot) * \mathring{u} \in \Pi_m$, where $\mathring{u} \in l_0(\mathbb{Z}^d)$ satisfies*

$$\hat{\mathring{u}}(\xi) = |\det \mathsf{M}| \overline{\hat{u}((\mathsf{M}^\mathsf{T})^{-1}\xi)} + O(\|\xi\|^{m+1}), \qquad \xi \to 0.$$

Let $\mathsf{c} \in \mathbb{R}^d$. By Proposition 10, we see that $\mathsf{p} * u = \mathsf{p}(\cdot - \mathsf{c})$ for all $\mathsf{p} \in \Pi_m$ if and only if $\hat{u}$ has $m+1$ *linear-phase moments with phase* c: $\hat{u}(\xi) = \mathrm{e}^{-\mathrm{i}\mathsf{c}\cdot\xi} + O(\|\xi\|^{m+1})$ as $\xi \to 0$. Proposition 10 also implies that $\mathscr{T}_{u,\mathsf{M}}\mathsf{p} = 0$ for all $\mathsf{p} \in \Pi_m$ if and only if $\hat{u}(\xi) = O(\|\xi\|^{m+1})$ as $\xi \to 0$, that is, u has $m+1$ *vanishing moments*.

Now we proceed to investigate the subdivision operator acting on polynomial spaces. In contrast to the case of the convolution operator and the transition operator, $\mathscr{S}_{u,\mathsf{M}}\mathsf{p}$ is not always a polynomial sequence for an input polynomial sequence p.

Theorem 10. *([42, Lemma 3.2]) Let $u = \{u(\mathsf{k})\}_{\mathsf{k} \in \mathbb{Z}^d} \in l_0(\mathbb{Z}^d)$ be a finitely supported sequence on $\mathbb{Z}^d$ and $\mathsf{p} \in \Pi$ be a polynomial in d-variables. Then the following statements are equivalent:*

1. *$\mathscr{S}_{u,\mathsf{M}}\mathsf{p}$ is a polynomial sequence, that is, $\mathscr{S}_{u,\mathsf{M}}\mathsf{p} \in \Pi$;*
2. *$\sum_{\mathsf{k} \in \mathbb{Z}^d} (\partial^\beta \mathsf{p})(-k - \mathsf{M}^{-1}\gamma) u(\mathsf{M}\mathsf{k} + \gamma) = \sum_{\mathsf{k} \in \mathbb{Z}^d} (\partial^\beta \mathsf{p})(-\mathsf{k}) u(\mathsf{M}\mathsf{k})$ for all $\beta \in \mathbb{N}_0^d$ and $\gamma \in \Gamma_\mathsf{M} := \mathbb{Z}^d \cap [\mathsf{M}[0,1)^d]$;*
3. *$[(\partial^\beta \mathsf{p})(-\mathrm{i}\partial)(\widehat{u^{[\gamma]}}(\xi) \mathrm{e}^{-\mathrm{i}(\mathsf{M}^{-1}\gamma \cdot \xi)})]|_{\xi=0} = [(\partial^\beta \mathsf{p})(-\mathrm{i}\partial)\widehat{u^{[0]}}(\xi)]|_{\xi=0}$ for all $\beta \in \mathbb{N}_0^d$ and $\gamma \in \Gamma_\mathsf{M}$, where $\widehat{u^{[\gamma]}}(\xi) := \sum_{\mathsf{k} \in \mathbb{Z}^d} u(\mathsf{M}\mathsf{k} + \gamma) \mathrm{e}^{-\mathrm{i}\mathsf{k}\cdot\xi}$ is a coset sequence at γ;*

4. $[\partial^\beta \mathsf{p}(-i\mathsf{M}^{-1}\partial)\hat{u}(\xi)]|_{\xi=2\pi\omega} = 0$ *for all* $\beta \in \mathbb{N}_0^d$ *and* $\omega \in \Omega_{(\mathsf{M}^\mathsf{T})^{-1}} \backslash \{\mathbf{0}\}$, *where* $\Omega_{(\mathsf{M}^\mathsf{T})^{-1}} := [(\mathsf{M}^\mathsf{T})^{-1}\mathbb{Z}^d] \cap [0,1)^d$.

Moreover, if any of the above items (1)–(4) holds, then for every $\beta \in \mathbb{N}_0^d$, $y \in \mathbb{R}^d$,

$$\mathscr{S}_{u,\mathsf{M}}(\partial^\beta \mathsf{p}) = [(\partial^\beta \mathsf{p})(\mathsf{M}^{-1}\cdot)] * u \quad \textit{and} \quad \mathscr{S}_{u,\mathsf{M}}(\mathsf{p}(\cdot - y)) = [\mathscr{S}_{u,\mathsf{M}}\mathsf{p}](\cdot - \mathsf{M}y). \tag{94}$$

By Theorem 10, we see that $\mathscr{S}_{u,\mathsf{M}}\mathsf{p} \in \Pi_m$ for all $\mathsf{p} \in \Pi_m$ if and only if $\partial^\beta \hat{u}(2\pi\omega) = 0$ for all $\beta \in \mathbb{N}_0^d$ with $|\beta| \leqslant m$ and $\omega \in \Omega_{(\mathsf{M}^\mathsf{T})^{-1}} \backslash \{\mathbf{0}\}$. That is, $\mathscr{S}_{u,\mathsf{M}}\mathsf{p} \in \Pi_m$ for all $\mathsf{p} \in \Pi_m$ if and only if u has $m+1$ sum rules with respect to M.

7 Directional Tight Framelets in $L_2(\mathbb{R}^d)$ and Projection Method

As an application of the theory developed in previous sections, we study in this section frequency-based nonstationary nonhomogeneous tight framelets in $L_2(\mathbb{R}^d)$ with directionality. We also discuss the projection method for tight framelets.

As an application of Proposition 6, we have the following constructive result:

Theorem 11. *([44, Theorem 2]) Let* M *be a* $d \times d$ *real-valued expansive matrix. Then there exist two real-valued functions* ϕ, ψ *in the Schwarz class such that*

(i) $\{\{\phi\};\{\psi\}\}$ *is a nonhomogeneous tight* M*-framelet in* $L_2(\mathbb{R}^d)$*: for all* $J \in \mathbb{Z}$,

$$\sum_{\mathsf{k}\in\mathbb{Z}^d} |\langle f, \phi_{\mathsf{M}^J;\mathsf{k}}\rangle|^2 + \sum_{j=J}^{\infty}\sum_{\mathsf{k}\in\mathbb{Z}^d} |\langle f, \psi_{\mathsf{M}^j;\mathsf{k}}\rangle|^2 = \|f\|^2_{L_2(\mathbb{R}^d)} \qquad \forall\, f \in L_2(\mathbb{R}^d); \tag{95}$$

(ii) $\hat{\phi}$ *and* $\hat{\psi}$ *are compactly supported* C^∞ *even functions;*

(iii) ψ *has infinite vanishing moments;* $\hat{\psi}$ *vanishes in a neighborhood of the origin;*

(iv) *there exist* $2\pi\mathbb{Z}^d$*-periodic measurable functions* $\mathbf{a}_\mathsf{k}, \mathbf{b}_\mathsf{k}, \mathsf{k} \in \mathbb{Z}^d$ *in* $C^\infty(\mathbb{T}^d)$ *such that for all* $\xi \in \mathbb{R}^d$ *and* $\mathsf{k} \in \mathbb{Z}^d$,

$$e^{-i\mathsf{k}\cdot\mathsf{M}^\mathsf{T}\xi}\hat{\phi}(\mathsf{M}^\mathsf{T}\xi) = \mathbf{a}_\mathsf{k}(\xi)\hat{\phi}(\xi) \quad \textit{and} \quad e^{-i\mathsf{k}\cdot\mathsf{M}^\mathsf{T}\xi}\hat{\psi}(\mathsf{M}^\mathsf{T}\xi) = \mathbf{b}_\mathsf{k}(\xi)\hat{\phi}(\xi).$$

Moreover, $\{\psi\}$ *is a homogeneous tight* M*-framelet in* $L_2(\mathbb{R}^d)$ *satisfying*

$$\sum_{j\in\mathbb{Z}}\sum_{\mathsf{k}\in\mathbb{Z}^d} |\langle f, \psi_{\mathsf{M}^j;\mathsf{k}}\rangle|^2 = \|f\|^2_{L_2(\mathbb{R}^d)} \qquad \forall\, f \in L_2(\mathbb{R}^d).$$

As in [44], we can modify examples in Theorem 11 by a splitting technique in [31] to achieve directionality. Here we only provide an example for $\mathsf{M} = 2I_2$.

Corollary 5 *([44, Theorem 18]) Let* $\mathsf{M} = 2I_2$*. Let* m *be a positive integer and* $0 \leqslant \rho < 1$*. Then there exist two real-valued functions* ϕ *and* η *in* $L_2(\mathbb{R}^d) \cap C^\infty(\mathbb{R}^2)$ *satisfying all the following properties:*

(i) $\hat{\phi}$ *is a compactly supported radial basis function in* $C^\infty(\mathbb{R}^2)$ *and there exists a* $2\pi\mathbb{Z}^2$*-periodic function* $\mathbf{a}$ *in* $C^\infty(\mathbb{T}^2)$ *such that* $\hat{\phi}(2\xi) = \mathbf{a}(\xi)\hat{\phi}(\xi)$ *for all* $\xi \in \mathbb{R}^2$*;*

(ii) η *has the tensor-product structure in polar coordinates and there exist positive real numbers* $\mathsf{r}_1, \mathsf{r}_2, \theta_0$ *such that* $\theta_0 < \frac{\pi}{m}$ *and*

$$\operatorname{supp} \eta = \{\mathsf{r}e^{[iu\theta} \mid \mathsf{r}_1 \leqslant \mathsf{r} \leqslant \mathsf{r}_2, -\theta_0 \leqslant \theta \leqslant \theta_0\};$$

(iii) *for* $J \in \mathbb{N}_0$*, a nonstationary nonhomogeneous tight* $2I_2$*-framelet in* $L_2(\mathbb{R}^2)$*:*

$$\sum_{\mathsf{k} \in \mathbb{Z}^2} |\langle f, \phi_{2^J I_2;\mathsf{k}} \rangle|^2 + \sum_{j=J}^{\infty} \sum_{\ell=1}^{s_j} \sum_{\mathsf{k} \in \mathbb{Z}^2} |\langle f, \psi^{j,\ell}_{2^j I_2;\mathsf{k}} \rangle|^2 = \|f\|^2_{L_2(\mathbb{R}^2)}, \ \forall f \in L_2(\mathbb{R}^2), \quad (96)$$

where $s_j := m2^{\lfloor \rho j \rfloor}$*,* $\lfloor \cdot \rfloor$ *is the floor function, and all* $\psi^{j,\ell}$ *are real-valued functions in the Schwarz class satisfying the following properties:*

1. $\psi^{j,0}$ *is defined as follows: for* $\mathsf{r} \geqslant 0$ *and* $\theta \in [-\pi, \pi)$*,*

$$\widehat{\psi^{j,0}}(\mathsf{r}e^{\mathrm{i}\theta}) := \begin{cases} \eta(\mathsf{r}e^{\mathrm{i}2^{\lfloor \rho j \rfloor}\theta}) + \eta(-\mathsf{r}e^{\mathrm{i}2^{\lfloor \rho j \rfloor}\theta}), & \text{if } \theta \in [-2^{-\lfloor \rho j \rfloor}\pi, 2^{-\lfloor \rho j \rfloor}\pi), \\ 0, & \text{if } \theta \in [-\pi, \pi) \backslash [-2^{-\lfloor \rho j \rfloor}\pi, 2^{-\lfloor \rho j \rfloor}\pi); \end{cases}$$

2. *other* $\psi^{j,\ell}$ *are obtained via rotations from* $\psi^{j,0}$*: for all* $\ell = 1, \ldots, s_j$*,*

$$\psi^{j,\ell}(\mathsf{r}e^{\mathrm{i}\theta}) := \psi^{j,0}(\mathsf{r}e^{\mathrm{i}\theta} e^{\mathrm{i}2^{-\lfloor \rho j \rfloor}\pi(\ell-1)/m}), \qquad \mathsf{r} \geqslant 0, \theta \in [-\pi, \pi); \quad (97)$$

3. *the support of* $\widehat{\psi^{j,0}_{2^j I_2;\mathsf{k}}}$ *has two parts which are symmetric about the origin and each part obeys width* $\approx$ *length*$^{1-\rho}$*. More precisely, for all* $\mathsf{k} \in \mathbb{Z}^2$ *and* $j \geqslant 0$*,*

$$\operatorname{supp} \widehat{\psi^{j,0}_{2^j I_2;\mathsf{k}}} = \{\mathsf{r}e^{\mathrm{i}\theta}, -\mathsf{r}e^{\mathrm{i}\theta} \mid 2^j \mathsf{r}_1 \leqslant \mathsf{r} \leqslant 2^j \mathsf{r}_2, -2^{j-\lfloor \rho j \rfloor}\theta_0 \leqslant \theta \leqslant 2^{j-\lfloor \rho j \rfloor}\theta_0\};$$

4. $\widehat{\psi^{j,\ell}}$ *are compactly supported functions in* $C^\infty(\mathbb{R}^2)$ *vanishing in a neighborhood of the origin and there exist* $2\pi\mathbb{Z}^2$*-periodic functions* $\mathbf{b}_{j,\ell}$ *in* $C^\infty(\mathbb{T}^2)$ *such that*

$$\widehat{\psi^{j,\ell}}(2\xi) = \mathbf{b}_{j,\ell}(\xi)\hat{\phi}(\xi), \qquad \xi \in \mathbb{R}^2, \ \ell = 1, \ldots, s_j, \ j \in \mathbb{N} \cup \{0\}.$$

See Figs. 1 and 2 for the graphs of the directional framelets in Corollary 5. As discussed in [44], the two families of examples of tight framelets have associated tight framelet filter banks derived from OEP and can be also derived from Theorem 3 through such tight framelet filter banks. the relation width $\approx$ length$^{1-\rho}$ with $\rho = 1/2$ in item (3) is called hyperbolic scaling in [4], which is claimed to be important for directional representations to capture edge singularity. Though the tight framelets in Theorem 11 and Corollary 5 are compactly supported in the frequency domain, using FIR filters and similar ideas as in Corollary 5, it is not very difficult to

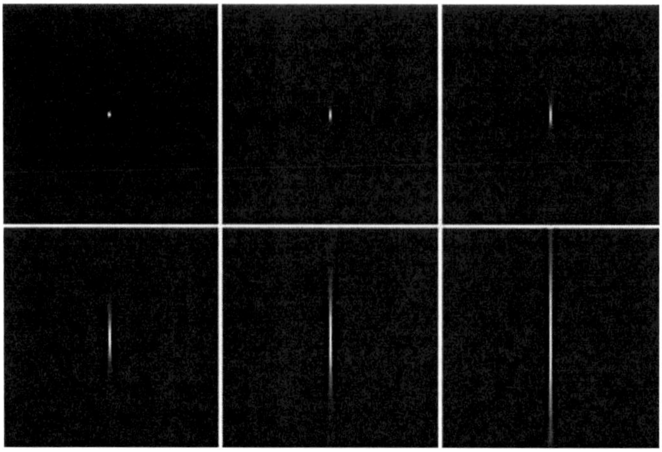

Fig. 1: Tight framelets at different scale levels: $\psi^{2,0}, \psi^{4,0}, \psi^{8,0}, \psi^{16,0}, \psi^{32,0}, \psi^{64,0}$ in Corollary 5

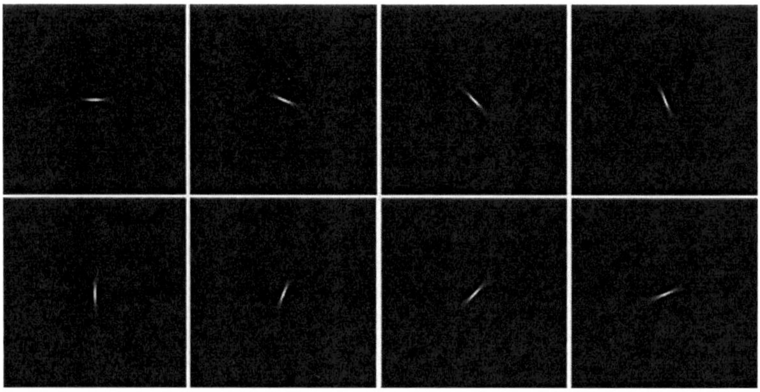

Fig. 2: Tight framelets at the scale level $j = 8$ with different rotation directions: $\psi^{8,\ell}, \ell = 0, \dots, 7$ in Corollary 5. The framelet $\psi^{8,\ell}$ is obtained from $\psi^{8,0}$ by rotation via (97) for all $\ell = 1, \dots, 7$

construct compactly supported nonstationary dual $2I_2$-framelets in $L_2(\mathbb{R}^2)$ with directionality and vanishing moments in the time domain. We shall address this issue elsewhere.

We finish this paper by discussing the projection method on framelets and refinable functions. For a block diagonal $d \times d$ dilation matrix M, a simple way of constructing M-wavelets and M-framelets in d dimensions is the tensor product of lower dimensional ones. The projection method, which is sort of "an inverse" of the tensor product method, is useful for deriving low-dimensional wavelets and framelets from higher dimension ones [32, 34, 38]. In fact, one can first use the tensor product method to construct high dimensional wavelets and framelets from one-dimensional ones, then one can use the projection method to derive new low-dimensional wavelets and framelets. Though projection method works well for homogeneous/nonhomogeneous wavelets or framelets in function spaces, for simplicity, we only discuss here the projection method for stationary nonhomogeneous tight framelets in $L_2(\mathbb{R}^d)$.

Let P be a $d \times m$ real-valued matrix with $d \leqslant m$. Let $f : \mathbb{R}^m \to \mathbb{C}$ be a tempered distribution and $u : \mathbb{Z}^m \to \mathbb{C}$ be a filter. Now we "define" a projected function Pf and a projected filter Pu in the lower dimension d as follows:

$$\widehat{\mathrm{P}f}(\xi) := \hat{f}(\mathrm{P}^{\mathsf{T}}\xi) \quad \text{and} \quad \widehat{\mathrm{P}u}(\xi) := \hat{u}(\mathrm{P}^{\mathsf{T}}\xi), \qquad \xi \in \mathbb{R}^d. \tag{98}$$

The projection operator is well-defined under some mild conditions [38]. Here we assume that $\hat{f}$ and $\hat{u}$ are continuous (e.g., f and u are compactly supported) so that $\widehat{\mathrm{P}f}$ and $\widehat{\mathrm{P}u}$ are continuous and therefore are distributions, even though Pf and Pu may not make sense. Note that the box spline M_{Ξ} defined in (76) is simply a projected function $\Xi\chi_{[0,1)^m}$, where $\chi_{[0,1)^m}$ is the characteristic function of $[0,1)^m$.

The following is a consequence of Proposition 6 and [38, Theorem 2.2].

Proposition 11 *Let $m \geqslant d$ and $\mathsf{M}, \mathring{\mathsf{M}}$ be $d \times d$ and $m \times m$ real-valued expansive matrices, respectively. Suppose that there is a $d \times m$ real-valued matrix* P *such that*

$$\mathrm{P}\mathring{\mathsf{M}} = \mathsf{M}\mathrm{P} \quad \text{and} \quad \mathrm{P}^{\mathsf{T}}(\mathbb{Z}^d \backslash [\mathsf{M}^{\mathsf{T}}\mathbb{Z}^d]) \subseteq \mathbb{Z}^m \backslash [\mathring{\mathsf{M}}^{\mathsf{T}}\mathbb{Z}^m]. \tag{99}$$

If $(\{\Phi;\Psi\},\{\tilde{\Phi};\tilde{\Psi}\})$ is a nonhomogeneous tight $\mathring{\mathsf{M}}$-framelet in $L_2(\mathbb{R}^m)$ such that $\hat{f}$ is continuous for every element f in $\Phi,\Psi,\tilde{\Phi}$, and $\tilde{\Psi}$, then $(\{\mathrm{P}\Phi;\mathrm{P}\Psi\},\{\mathrm{P}\tilde{\Phi};\mathrm{P}\tilde{\Psi}\})$ is a nonhomogeneous tight M*-framelet in $L_2(\mathbb{R}^d)$, where $\mathrm{P}\Phi := \{\mathrm{P}\phi \mid \phi \in \Phi\}$.*

Let $\{\{\phi\};\{\psi^1,\ldots,\psi^{2^m-1}\}\}$ be the tensor product Haar orthonormal $2I_m$-wavelet in $L_2(\mathbb{R}^m)$. Let $\mathrm{P} = [1,\ldots,1]$. Then $\{\{\mathrm{P}\phi\};\{\mathrm{P}\psi^1,\ldots,\mathrm{P}\psi^{2^m-1}\}\}$ is a tight 2-framelet in $L_2(\mathbb{R})$. Note that $\mathrm{P}\phi$ is the B-spline of order m. After a simple procedure for reducing the number of generators [38], this projected tight framelet becomes the spline tight 2-framelet obtained in [73]. The projection method is also applicable to filter banks. Let a filter bank in (82) be a dual $\mathring{\mathsf{M}}$-framelet filter bank. Let P be a $d \times m$ integer projection matrix satisfying (99). By [38, Theorem 4.9], $(\{\mathrm{P}a;\mathrm{P}b_1,\ldots,\mathrm{P}b_s\},\{\mathrm{P}\tilde{a};\mathrm{P}\tilde{b}_1,\ldots,\mathrm{P}\tilde{b}_s\})_{\mathrm{P}\Theta}$ is a dual M-framelet filter bank. The projection method also works well with refinable functions and refinable structure:

Proposition 12 *([38, Theorem 4.1]) Let $m \geqslant d$ and $\mathsf{M}, \mathring{\mathsf{M}}$ be $d \times d$ and $m \times m$ real-valued expansive matrices, respectively. Suppose that* P *is a $d \times m$ integer matrix satisfying $\mathrm{P}\mathring{\mathsf{M}} = \mathsf{M}\mathrm{P}$ and $\mathrm{P}\mathbb{Z}^m = \mathbb{Z}^d$. Let ϕ be a compactly supported function satisfying $\hat{\phi}(\mathring{\mathsf{M}}^{\mathsf{T}}\xi) = \hat{a}(\xi)\hat{\phi}(\xi)$ for all $\xi \in \mathbb{R}^m$ for some finitely supported mask $a \in l_0(\mathbb{Z}^m)$. Then $\widehat{\mathrm{P}\phi}(\mathsf{M}^{\mathsf{T}}\xi) = \widehat{\mathrm{P}a}(\xi)\widehat{\mathrm{P}\phi}(\xi)$ for all $\xi \in \mathbb{R}^d$ and $sr(a,\mathring{\mathsf{M}}) \leqslant sr(\mathrm{P}a,\mathsf{M})$. Moreover, for all $1 \leqslant p \leqslant \infty$, $\nu_p(\phi) \leqslant \nu_p(\mathrm{P}\phi)$ and*

$$|\det \mathsf{M}|^{1-1/p}\rho(\mathrm{P}a,\mathsf{M},p) \leqslant |\det \mathring{\mathsf{M}}|^{1-1/p}\rho(a,\mathring{\mathsf{M}},p).$$

If in addition $\rho(\mathsf{M}) = \rho(\mathring{\mathsf{M}})$, then $\nu_p(a,\mathring{\mathsf{M}}) \leqslant \nu_p(\mathrm{P}a,\mathsf{M})$ for all $1 \leqslant p \leqslant \infty$.

We provide an example to demonstrate the usefulness of the projection method. Let $\mathsf{M} = 2I_d$ and a be an interpolatory mask (that is, $a(0) = 2^{-d}$ and $a(2k) = 0$ for all $k \in \mathbb{Z}^d \backslash \{0\}$) such that the support of the mask a is contained inside $[-3,3]^d$. Then $\nu_\infty(a, 2I_d) \leqslant 2$, and consequently, its associated compactly supported refinable function cannot be in $C^2(\mathbb{R}^d)$. The argument is as follows (see [32]). Suppose that $\nu_\infty(a, 2I_d) > 2$. Then we must have $\mathrm{sr}(a, 2I_d) \geqslant 3$. Let $\mathrm{P} = [1, 0, \ldots, 0]$. Combining the fact that a is interpolatory, we can easily conclude that $\mathrm{P}a$ must be an interpolatory mask with four sum rules and support contained inside $[-3,3]$. Such a mask is unique and is given by $\mathrm{P}a(-3) = \mathrm{P}a(3) = -1/32, \mathrm{P}a(-1) = \mathrm{P}a(1) = 9/32$, $\mathrm{P}a(0) = 1/2$, and $\mathrm{P}a(k) = 0$ for all $k \in \mathbb{Z} \backslash \{-3, -1, 0, 1, 3\}$. However, $\nu_\infty(\mathrm{P}a, 2) = 2$. Consequently, by Proposition 11, we deduce that $\nu_\infty(a, 2I_d) \leqslant \nu_\infty(\mathrm{P}a, 2) = 2$, a contradiction. Hence, $\nu_\infty(a, 2I_d) \leqslant 2$.

Acknowledgements This research was supported by NSERC Canada.

References

1. J. J. Benedetto and S. Li, The theory of multiresolution analysis frames and applications to filter banks, *Appl. Comput. Harmon. Anal.*, **5** (1998), 389–427.
2. L. Borup, R. Gribonval and M. Nielsen, Bi-framelet systems with few vanishing moments characterize Besov spaces. *Appl. Comput. Harmon. Anal* **17** (2004), 3–28.
3. M. Bownik, A characterization of affine dual frames in $L^2(\mathbb{R}^n)$, *Appl. Comput. Harmon. Anal.* **8** (2000), 203-221.
4. E. J. Candès and D. L. Donoho, New tight frames of curvelets and optimal representations of objects with C^2 singularities, *Comm. Pure Appl. Math.* **56** (2004), 219–266.
5. A. S. Cavaretta, W. Dahmen, and C. A. Micchelli, Stationary subdivision. *Mem. Amer. Math. Soc.* **93** (1991), no. 453.
6. C. K. Chui, An introduction to wavelets. Academic Press, Inc., Boston, MA, 1992.
7. C. K. Chui, W. He and J. Stöckler, Compactly supported tight and sibling frames with maximum vanishing moments, *Appl. Comput. Harmon. Anal.* **13** (2002), 224–262.
8. C. K. Chui, W. He, and J. Stöckler, Nonstationary tight wavelet frames. II. Unbounded intervals. *Appl. Comput. Harmon. Anal.* **18** (2005), 25–66.
9. C. K. Chui and X. Shi, Orthonormal wavelets and tight frames with arbitrary real dilations. *Appl. Comput. Harmon. Anal.* **9** (2000), 243–264.
10. A. Cohen and I. Daubechies, A stability criterion for biorthogonal wavelet bases and their related subband coding scheme. *Duke Math. J.* **68** (1992), 313–335.
11. A. Cohen and I. Daubechies, A new technique to estimate the regularity of refinable functions, *Rev. Mat. Iberoamericana* **12** (1996), 527–591.
12. A. Cohen, I. Daubechies, and J.-C. Feauveau, Biorthogonal bases of compactly supported wavelets. *Comm. Pure Appl. Math.* **45** (1992), 485–560.
13. A. Cohen, I. Daubechies, and G. Plonka, Regularity of refinable function vectors. *J. Fourier Anal. Appl.* **3** (1997), 295–324.
14. A. Cohen, K. Gröchenig, and L. F. Villemoes, Regularity of multivariate refinable functions. *Constr. Approx.* **15** (1999), 241–255.
15. W. Dahmen, Stability of multiscale transformations, *J. Fourier Anal. Appl.* **2** (1996), 341–361.
16. W. Dahmen, Multiscale and wavelet methods for operator equations, in *Multiscale problems and methods in numerical simulations*, 31–96, Lecture Notes in Math. 1825, Springer, (2003).
17. X. Dai, D. R. Larson, and D. M. Speegle, Wavelet sets in $\mathbb{R}^n$, *J. Fourier Anal. Appl.* **3** (1997), 451–456.

18. I. Daubechies, Orthonormal bases of compactly supported wavelets. *Comm. Pure Appl. Math.* **41** (1988), 909–996.
19. I. Daubechies, Ten lectures on wavelets, SIAM, CBMS Series, 1992.
20. I. Daubechies, A. Grossmann, and Y. Meyer, Painless nonorthogonal expansions. *J. Math. Phys.* **27** (1986), 1271–1283.
21. I. Daubechies and B. Han, The canonical dual frame of a wavelet frame, *Appl. Comput. Harmon. Anal.*, **12**, (2002), 269–285.
22. I. Daubechies and B. Han, Pairs of dual wavelet frames from any two refinable functions, *Constr. Approx.*, **20** (2004), 325–352.
23. I. Daubechies, B. Han, A. Ron, and Z. Shen, Framelets: MRA-based constructions of wavelet frames, *Appl. Comput. Harmon. Anal.* **14** (2003), 1–46.
24. C. de Boor, K. Höllig, and S. Riemenschneider, *Box Splines*, Springer-Verlag, (1993).
25. R. A. DeVore, B. Jawerth, and P. Popov, Compression of wavelet decompositions. *Amer. J. Math.* **114** (1992), 737–785.
26. N. Dyn and D. Levin, Subdivision schemes in geometric modelling. *Acta Numer.* **11** (2002), 73–144.
27. M. Ehler, On multivariate compactly supported bi-frames, *J. Fourier Anal. Appl.* **13** (2007), 511–532.
28. M. Ehler and B. Han, Wavelet bi-frames with few generators from multivariate refinable functions, *Appl. Computat. Harmon. Anal.*, **25** (2008), 407–414.
29. M. Frazier, G. Garrigós, K. Wang, and G. Weiss, A characterization of functions that generate wavelet and related expansion, *J. Fourier Anal. Appl.* **3** (1997), 883–906.
30. B. Han, Wavelets, M.Sc. thesis at Institute of Mathematics, the Chinese Academy of Sciences, June 1994.
31. B. Han, On dual wavelet tight frames, *Appl. Comput. Harmon. Anal.*, **4** (1997), 380–413.
32. B. Han, Analysis and construction of optimal multivariate biorthogonal wavelets with compact support, *SIAM Math. Anal.* **31** (2000), 274–304.
33. B. Han, Approximation properties and construction of Hermite interpolants and biorthogonal multiwavelets, *J. Approx. Theory*, **110**, (2001), 18–53.
34. B. Han, Projectable multivariate refinable functions and biorthogonal wavelets, *Appl. Comput. Harmon. Anal.*, **13**, (2002), 89–102.
35. B. Han, Computing the smoothness exponent of a symmetric multivariate refinable function, *SIAM J. Matrix Anal. Appl.*, **24** (2003), 693–714.
36. B. Han, Vector cascade algorithms and refinable function vectors in Sobolev spaces, *J. Approx. Theory*, **124** (2003), 44–88.
37. B. Han, Compactly supported tight wavelet frames and orthonormal wavelets of exponential decay with a general dilation matrix, *J. Comput. Appl. Math.*, **155** (2003), 43–67.
38. B. Han, Construction of wavelets and framelets by the projection method, *Intern. J. Appl. Math. Appl.*, **1** (2008), 1–40.
39. B. Han, Refinable functions and cascade algorithms in weighted spaces with Hölder continuous masks. *SIAM J. Math. Anal.* **40** (2008), 70–102.
40. B. Han, Dual multiwavelet frames with high balancing order and compact fast frame transform, *Appl. Comput. Harmon. Anal.* **26** (2009), 14–42.
41. B. Han, Matrix extension with symmetry and applications to symmetric orthonormal complex M-wavelets, *J. Fourier Anal. Appl.* **15** (2009), 684–705.
42. B. Han, The structure of balanced multivariate biorthogonal multiwavelets and dual multiframelets, *Math. Comp.*, **79** (2010), 917–951.
43. B. Han, Pairs of frequency-based nonhomogeneous dual wavelet frames in the distribution space, *Appl. Comput. Harmon. Anal.* **29** (2010), 330–353.
44. B. Han, Nonhomogeneous wavelet systems in high dimensions, *Appl. Comput. Harmon. Anal.* doi:10.1016/j.acha.2011.04.002, published online.
45. B. Han, Famelets and Wavelets: Algorithms and Basic Theory, book manuscript in preparation.
46. B. Han and R. Q. Jia, Multivariate refinement equations and convergence of subdivision schemes. *SIAM J. Math. Anal.* **29** (1998), 1177–1999.

47. B. Han and Q. Mo, Symmetric MRA tight wavelet frames with three generators and high vanishing moments, *Appl. Comput. Harmon. Anal.*, **18** (2005), 67–93.
48. B. Han and Z. Shen, Wavelets from the Loop scheme, *J. Fourier Anal. Appl.*, **11** (2005), 615–637.
49. B. Han and Z. Shen, Characterization of Sobolev spaces of arbitrary smoothness using nonstationary tight wavelet frames, *Israel J. Math.* 172 (2009), 371–398.
50. B. Han and Z. Shen, Compactly supported symmetric C^∞ wavelets with spectral approximation order, *SIAM J. Math. Anal.*, **40** (2008), 905–938.
51. B. Han and Z. Shen, Dual wavelet frames and Riesz bases in Sobolev spaces, *Constr. Approx.*, **29** (2009), 369–406.
52. B. Han and X. S. Zhuang, Analysis and construction of multivariate interpoalting refinable function vectors, *Acta Appl. Math.*, **107** (2009), 143–171.
53. B. Han and X. S. Zhuang, Matrix extension with symmetry and its application to symmetric orthonormal multiwavelets, *SIAM J. Math. Anal.* **42** (2010), 2297–2317.
54. D. P. Hardin, B. Kessler, Bruce, and P. R. Massopust, Multiresolution analyses based on fractal functions. *J. Approx. Theory* **71** (1992), 104–120.
55. K. Jetter, D. X. Zhou, Order of linear approximation from shift-invariant spaces, *Constr. Approx.* **11** (1995), 423–438.
56. R. Q. Jia, Approximation properties of multivariate wavelets. *Math. Comp.* **67** (1998), 647–665.
57. R. Q. Jia, Characterization of smoothness of multivariate refinable functions in Sobolev spaces, *Trans. Amer. Math. Soc.* **351** (1999), 4089–4112.
58. R. Q. Jia and Q. T. Jiang, Spectral analysis of the transition operator and its applications to smoothness analysis of wavelets. *SIAM J. Matrix Anal. Appl.* **24** (2003), 1071–1109.
59. R. Q. Jia, S. D. Riemenschneider, and D. X. Zhou, Smoothness of multiple refinable functions and multiple wavelets. *SIAM J. Matrix Anal. Appl.* **21** (1999), 1–28.
60. R. Q. Jia, J. Z. Wang, and D. X. Zhou, Compactly supported wavelet bases for Sobolev spaces, *Appl. Comput. Harmon. Anal.* **15** (2003), 224–241.
61. Q. T. Jiang, On the regularity of matrix refinable functions. *SIAM J. Math. Anal.* **29** (1998), 1157–1176.
62. M. J. Lai and A. Petukhov, Method of virtual components for constructing redundant filter banks and wavelet frames. *Appl. Comput. Harmon. Anal.* **22** (2007), 304–318.
63. M. J. Lai and J. Stöckler, Construction of multivariate compactly supported tight wavelet frames, *Appl. Comput. Harmon. Anal.* **21** (2006), 324–348.
64. W. Lawton, S. L. Lee, and Z. Shen, Convergence of multidimensional cascade algorithm. *Numer. Math.* **78** (1998), 427–438.
65. W. Lawton, S. L. Lee, and Z. Shen, Stability and orthonormality of multivariate refinable functions, *SIAM J. Math. Anal.* **28** (1997), 999–1014.
66. J. J. Lei, R. Q. Jia and E. W. Cheney, Approximation from shift-invariant spaces by integral operators, *SIAM J. Math. Anal.* **28** (1997), 481-498.
67. S. Li, Characterization of smoothness of multivariate refinable functions and convergence of cascade algorithms of nonhomogeneous refinement equations. *Adv. Comput. Math.* **20** (2004), 311–331.
68. S. Mallat, A wavelet tour of signal processing. Third edition. Elsevier/Academic Press, Amsterdam, 2009.
69. Y. Meyer, Wavelets and operators. Cambridge University Press, Cambridge, 1992.
70. C. A. Micchelli and T. Sauer, Regularity of multiwavelets. *Adv. Comput. Math.* **7** (1997), 455–545.
71. S. D. Riemenschneider and Z. Shen, Wavelets and pre-wavelets in low dimensions, *J. Approx. Theory*, **71** (1992), 18–38.
72. A. Ron and Z. Shen, Affine systems in $L_2(\mathbb{R}^d)$ II. Dual systems. *J. Fourier Anal. Appl.* **3** (1997), 617–637.
73. A. Ron and Z. Shen, Affine systems in $L_2(\mathbb{R}^d)$: the analysis of the analysis operator. *J. Funct. Anal.* **148** (1997), 408–447.

74. Q. Y. Sun, Convergence and boundedness of cascade algorithm in Besov spaces and Triebel-Lizorkin spaces. II. *Adv. Math.* **30** (2001), 22–36.
75. D. X. Zhou, Norms concerning subdivision sequences and their applications in wavelets. *Appl. Comput. Harmon. Anal.* **11** (2001), 329–346.

Compactly Supported Shearlets

Gitta Kutyniok, Jakob Lemvig, and Wang-Q Lim

Abstract Shearlet theory has become a central tool in analyzing and representing 2D data with anisotropic features. Shearlet systems are systems of functions generated by one single generator with parabolic scaling, shearing, and translation operators applied to it, in much the same way wavelet systems are dyadic scalings and translations of a single function, but including a precise control of directionality. Of the many directional representation systems proposed in the last decade, shearlets are among the most versatile and successful systems. The reason for this being an extensive list of desirable properties: shearlet systems can be generated by one function, they provide precise resolution of wavefront sets, they allow compactly supported analyzing elements, they are associated with fast decomposition algorithms, and they provide a unified treatment of the continuum and the digital realm.

The aim of this paper is to introduce some key concepts in directional representation systems and to shed some light on the success of shearlet systems as directional representation systems. In particular, we will give an overview of the different paths taken in shearlet theory with focus on separable and compactly supported shearlets in 2D and 3D. We will present constructions of compactly supported shearlet frames in those dimensions as well as discuss recent results on the ability of compactly supported shearlet frames satisfying weak decay, smoothness, and directional moment conditions to provide optimally sparse approximations of cartoon-like images in 2D as well as in 3D. Finally, we will show that these compactly supported shearlet

Gitta Kutyniok
Department of Mathematics, Technische Universitüt Berlin, 10623 Berlin
e-mail: kutyniok@math.tu-berlin.de

Jakob Lemvig
Department of Mathematics, Technical University of Denmark, Matematiktorvet, Building 303S, 2800 Kgs. Lyngby, Denmark
e-mail: j.lemvig@mat.dtu.dk

Wang-Q Lim
Department of Mathematics, Technische Universitüt Berlin, 10623 Berlin
e-mail: wlim@uni-osnabrueck.de

M. Neamtu and L. Schumaker (eds.), *Approximation Theory XIII: San Antonio 2010*, Springer Proceedings in Mathematics 13, DOI 10.1007/978-1-4614-0772-0_10,

systems provide optimally sparse approximations of an even generalized model of cartoon-like images comprising of C^2 functions that are smooth apart from piecewise C^2 discontinuity edges.

1 Introduction

Recent advances in modern technology have created a brave new world of enormous, multi-dimensional data structures. In medical imaging, seismic imaging, astronomical imaging, computer vision, and video processing, the capabilities of modern computers and high-precision measuring devices have generated 2D, 3D, and even higher dimensional data sets of sizes that were infeasible just a few years ago. The need to efficiently handle such diverse types and huge amounts of data initiated an intense study in developing efficient multivariate encoding methodologies in the applied harmonic analysis research community.

In medical imaging, e.g., CT lung scans, the discontinuity curves of the image are important specific features since one often wants to distinguish between the image "objects" (e.g., the lungs) and the "background"; i.e., it is important to precisely capture the *edges*. This observation holds for various other applications than medical imaging and illustrates that important classes of multivariate problems are governed by *anisotropic features*. Moreover, in high-dimensional data most information is typically contained in lower-dimensional embedded manifolds, thereby also presenting itself as anisotropic features. The anisotropic structures can be distinguished by location and orientation/direction which indicates that our way of analyzing and representing the data should capture not only location, but also directional information.

In applied harmonic analysis, data is typically modeled in a continuum setting as square-integrable functions or, more generally, as distributions. Recently, a novel directional representation system – so-called shearlets – has emerged which provides a unified treatment of such continuum models as well as digital models, allowing, for instance, a precise resolution of wavefront sets, optimally sparse representations of cartoon-like images, and associated fast decomposition algorithms. Shearlet systems are systems generated by one single generator with parabolic scaling, shearing, and translation operators applied to it, in the same way wavelet systems are dyadic scalings and translations of a single function, but including a directionality characteristic owing to the additional shearing operation (and the anisotropic scaling).

The aim of this survey paper is to introduce the key concepts in directional representation systems and, in particular, to shed some light on the success of shearlet systems. Moreover, we will give an overview of the different paths taken in shearlet theory with focus on separable and compactly supported shearlets, since these systems are most well-suited for applications in, e.g., image processing and the theory of partial differential equations.

1.1 Directional Representation Systems

In recent years, numerous approaches for efficiently representing directional features of two-dimensional data have been proposed. A perfunctory list includes: *steerable pyramid* by Simoncelli et al. [40], *directional filter banks* by Bamberger and Smith [2], *2D directional wavelets* by Antoine et al. [1], *curvelets* by Candès and Donoho [4], *contourlets* by Do and Vetterli [10], *bandlets* by LePennec and Mallat [39], and *shearlets* by Labate, Weiss, and two of the authors [37]. Of these, shearlets are among the most versatile and successful systems which owes to the many desirable properties possessed by shearlet systems: they are generated by one function, they provide optimally sparse approximation of so-called cartoon-like images, they allow compactly supported analyzing elements, they are associated with fast decomposition algorithms, and they provide a unified treatment of continuum and digital data.

Cartoon-like images are functions that are C^2 apart from C^2 singularity curves, and the problem of sparsely representing such singularities using 2D representation systems has been extensively studied; only curvelets [1], contourlets [10], and shearlets [21] are known to succeed in this task in an optimal way (see also Sect. 3). We describe contourlets and curvelets in more details in Sect. 1.4 and will here just mention some differences to shearlets. Contourlets are constructed from a discrete filter bank and have therefore, unlike shearlets, no continuum theory. Curvelets, on the other hand, are a continuum-domain system which, unlike shearlets, does not transfer in a uniform way to the digital world. It is fair to say that shearlet theory is a comprehensive theory with a mathematically rich structure as well as a superior connection between the continuum and digital realm.

The missing link between the continuum and digital world for curvelets is caused by the use of rotation as a means to parameterize directions. One of the distinctive features of shearlets is the use of shearing in place of rotation; this is, in fact, decisive for a clear link between the continuum and digital world which stems from the fact that the shear matrix preserves the integer lattice. Traditionally, the shear parameter ranges over a non-bounded interval. This has the effect that the directions are not treated uniformly, which is particularly important in applications. On the other hand, rotations clearly do not suffer from this deficiency. To overcome this shortcoming of shearing, Guo, Labate, and Weiss together with two of the authors [37] (see also [20]) introduced the so-called cone-adapted shearlet systems, where the frequency plane is partitioned into a horizontal and a vertical cone which allows restriction of the shear parameter to bounded intervals (Sect. 2.1), thereby guaranteeing uniform treatment of directions.

Shearlet systems therefore come in two ways: One class being generated by a unitary representation of the shearlet group and equipped with a particularly 'nice' mathematical structure however, causes a bias towards one direction, which makes it unattractive for applications; the other class being generated by a quite similar procedure, but restricted to cones in frequency domain, thereby ensuring an equal treatment of all directions. To be precise this treatment of directions is only "almost equal" since there still is a slight, but controllable, bias towards directions of the coordinate axes, see also Fig. 4 in Sect. 2.2. For both classes, the *continuous* shearlet systems are associated with a 4-dimensional parameter space consisting of a scale parameter measuring the resolution, a shear parameter measuring the orientation, and a translation parameter measuring the position of the shearlet (Sect. 1.3). A sampling of this parameter space leads to *discrete* shearlet systems, and it is obvious that the possibilities for this are numerous. Using dyadic sampling leads to so-called regular shearlet systems which are those discrete systems mainly considered in this paper. It should be mentioned that also irregular shearlet systems have attracted some attention, and we refer to the papers [27–29]. We end this section by remarking that these discrete shearlet systems belong to a larger class of representation systems – the so-called composite wavelets [23–25].

1.2 Anisotropic Features, Discrete Shearlet Systems, and Quest for Sparse Approximations

In many applications in 2D and 3D imaging the important information is often located around *edges* separating "image objects" from "background." These features correspond precisely to the anisotropic structures in the data. Two-dimensional shearlet systems are carefully designed to efficiently encode such anisotropic features. In order to do this effectively, shearlets are scaled according to a parabolic scaling law, thereby exhibiting a spatial footprint of size 2^{-j} times $2^{-j/2}$, where 2^j is the (discrete) scale parameter; this should be compared to the size of wavelet footprints: 2^{-j} times 2^{-j}. These elongated, scaled needle-like shearlets then parametrize directions by slope encoded in a shear matrix. As mentioned in the previous section, such carefully designed shearlets do, in fact, perform optimally when representing and analyzing anisotropic features in 2D data (Sect. 3).

In 3D, the situation changes somewhat. While in 2D we "only" have to handle one type of anisotropic structures, namely curves, in 3D a much more complex situation can occur, since we find two geometrically very different anisotropic structures: Curves as one-dimensional features and surfaces as two-dimensional anisotropic features. Our 3D shearlet elements in spatial domain will be of size 2^{-j} times $2^{-j/2}$ times $2^{-j/2}$ which corresponds to "plate-like" elements as $j \to \infty$. This indicates that these 3D shearlet systems have been designed to efficiently capture two-dimensional anisotropic structures, but neglecting one-dimensional structures. Nonetheless, surprisingly, these 3D shearlet systems still perform optimally when representing and analyzing 3D data that contain both curve and surface singularities (Sect. 4).

Of course, before we can talk of optimally sparse approximations, we need to actually have these 2D and 3D shearlet systems at hand. Several constructions of discrete band-limited 2D shearlet frames are already known, see [6, 20, 28, 29]. But since spatial localization of the analyzing elements of the encoding system is immensely important both for a precise detection of geometric features as well as for a fast decomposition algorithm, we will mainly follow the sufficient conditions for and construction of compactly supported cone-adapted 2D shearlet systems by Kittipoom and two of the authors [27] (Sect. 2.3). These results provide a large class of separable, compactly supported shearlet systems with good frame bounds, optimally sparse approximation properties, and associated numerically stable algorithms.

1.3 Continuous Shearlet Systems

Discrete shearlet systems are, as mentioned, a sampled version of the so-called continuous shearlet systems. These continuous shearlets come, of course, also in two different flavors, and we will briefly describe these in this section.

1.3.1 Cone-Adapted Shearlet Systems

Anisotropic features in multivariate data can be modeled in many different ways. One possibility is the cartoon-like image class discussed above, but one can also model such directional singularities through distributions. One would, e.g., model a one-dimensional anisotropic structure as the delta distribution of a curve. The so-called *cone-adapted continuous shearlet transform* associated with *cone-adapted continuous shearlet systems* was introduced by Labate and the first author in [30] in the study of resolutions of the wavefront set for such distributions. It was shown that the continuous shearlet transform is not only able to identify the singular support of a distribution, but also the *orientation* of distributed singularities along curves. More precisely, for a class of band-limited shearlet generators $\psi \in L^2(\mathbb{R}^2)$, the first author and Labate [30] showed that the wavefront set of a (tempered) distribution f is precisely the closure of the set of points (t,s), where the shearlet transform of f

$$(a,s,t) \mapsto \left\langle f, a^{-3/4}\psi(A_a^{-1}S_s^{-1}(\cdot - t))\right\rangle, \quad \text{where } A_a = \begin{pmatrix} a & 0 \\ 0 & a^{1/2} \end{pmatrix} \text{ and } S_s = \begin{pmatrix} 1 & s \\ 0 & 1 \end{pmatrix},$$

is *not* of fast decay as the scale parameter $a \to 0$. Later Grohs [18] extended this result to Schwartz-class generators with infinitely many directional vanishing moments, in particular, not necessarily band-limited generators. In other words, these results demonstrate that the wavefront set of a distribution can be *precisely captured* by continuous shearlets. For constructions of continuous shearlet frames with compact support, we refer to [19].

1.3.2 Shearlets from Group Representations

Cone-adapted continuous shearlet systems and their associated cone-adapted continuous transforms described in the previous section have only very recently – in 2009 – attracted attention. Historically, the continuous shearlet transform was first introduced in [20] without restriction to cones in frequency domain. Later, it was shown in [7] that the associated continuous shearlet systems are generated by a strongly continuous, irreducible, square-integrable representation of a locally compact group, the so-called *shearlet group*. This implies that these shearlet systems possess a rich mathematical structure, which in [7] was used to derive uncertainty principles to tune the accuracy of the shearlet transform, and which in [6] allowed the usage of coorbit theory to study smoothness spaces associated with the decay of the shearlet coefficients.

Dahlke, Steidl, and Teschke generalized the shearlet group and the associated continuous shearlet transform to higher dimensions $\mathbb{R}^n$ in the paper [8]. Furthermore, in [8] they showed that, for certain band-limited generators, the continuous shearlet transform is able to identify hyperplane and tetrahedron singularities. Since this transform originates from a unitary group representation, it is not able to capture all directions, in particular, it will not capture the delta distribution on the x_1-axis (and more generally, any singularity with "x_1-directions"). We also remark that the extension in [8] uses another scaling matrix as compared to the one used for the three-dimensional shearlets considered in this paper; we refer to Sect. 4 for a more detailed description of this issue.

1.4 Applications

Shearlet theory has applications in various areas. In this section, we will present two examples of such: Denoising of images and geometric separation of data. Before, in order to show the reader the advantages of digital shearlets, we first give a short overview of the numerical aspects of shearlets and two similar implementations of directional representation systems, namely contourlets and curvelets, discussed in Sect. 1.1.

Curvelets [3]. This approach builds on directional frequency partitioning and the use of the Fast Fourier transform. The algorithm can be efficiently implemented using (in frequency domain) multiplication with the frequency response of a filter and frequency wrapping in place of convolution and down-sampling. However, curvelets need to be band-limited and can only have very good spatial localization if one allows high redundancy.

Contourlets [10]. This approach uses a directional filter bank, which produces directional frequency partitioning similar to those of curvelets. As the main advantage of this approach, it allows a tree-structured filter bank implementation, in which aliasing due to subsampling is allowed to exist. Consequently, one can achieve great efficiency in terms of redundancy and good spatial localization. However, the directional selectivity in this approach is artificially imposed by the special sampling rule of a filter bank which introduces various artifacts. We remark that also the recently introduced *Hybrid Wavelets* [17] suffer from this deficiency.

Shearlets [38]. Using a shear matrix instead of rotation, directionality is naturally adapted for the digital setting in the sense that the shear matrix preserves the structure of the integer grid. Furthermore, excellent spatial localization is achieved by using compactly supported shearlets. The only drawback is that these compactly supported shearlets are not tight frames and, accordingly, the synthesis process needs to be performed by iterative methods.

To illustrate how two of these implementations perform, we have included a denoising example of the Goldhill image using both curvelets[1] and shearlets, see

[1] Produced using Curvelab (Version 2.1.2), which is available from `http://curvelet.org`.

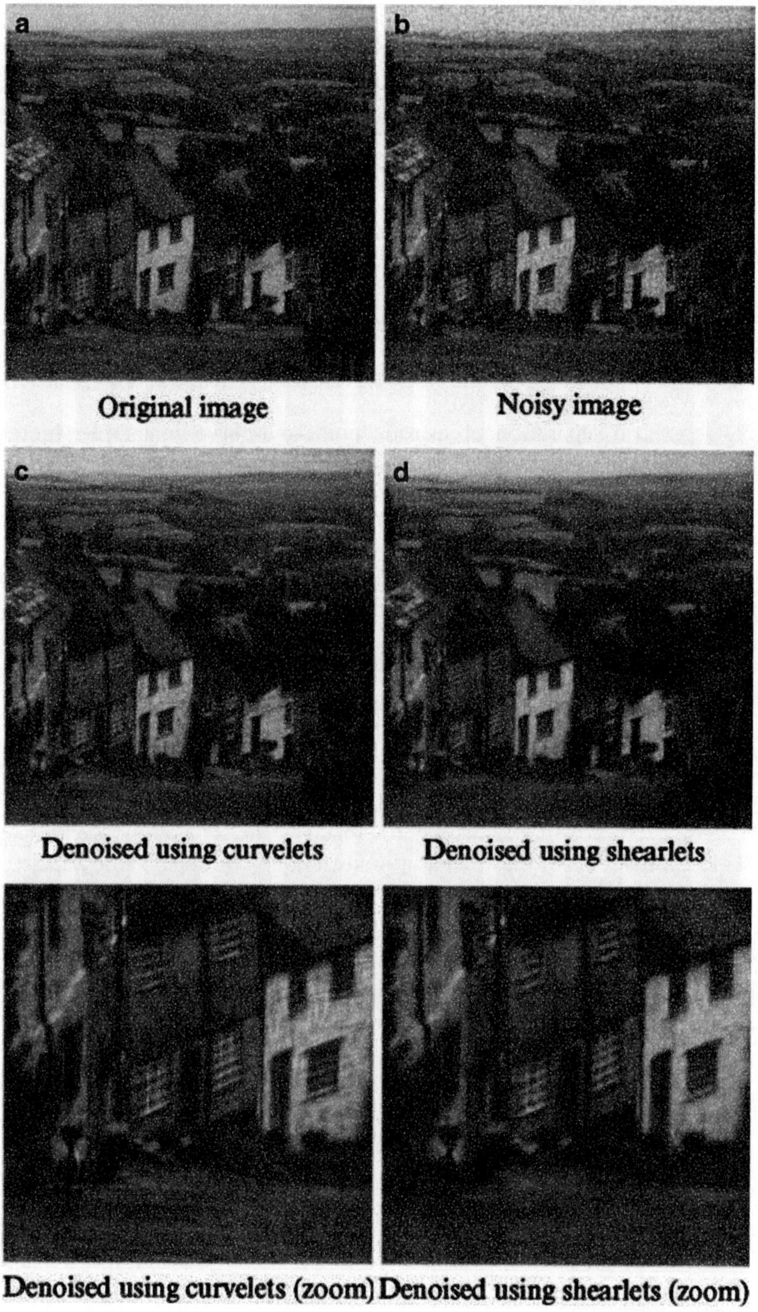

Fig. 1: Denoising of the Goldhill image (512×512) using shearlets and curvelets. The noisy image in (**b**) has a peak signal-to-noise ratio of 20.17 dB. The curvelet-denoised image in (**c**) and (**e**) has a PSNR of 28.70 dB, while the shearlet-denoised image in (**d**) and (**f**) has a PSNR of only 29.20 dB

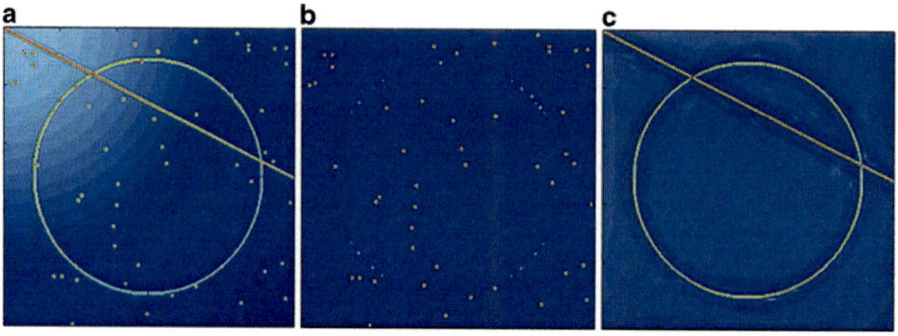

Original point-curve data of size 256 × 256. Separated point-like data (captured by wavelets). Separated curve-like data (captured by shearlets).

Fig. 2: Geometric separation of mixed "point-and-curve" data. (**a**): Input data. (**b**) and (**c**): The output of the separation algorithm

Fig. 1. We omit a detailed analysis of the denoising results and leave the visual comparison to the reader. For a detailed review of the shearlet transform and associated aspects, we refer to [14, 16, 36, 38]. We also refer to [26, 35] for MRA based algorithmic approaches to the shearlet transform.

The shearlet transform, in companion with the wavelet transform, has also been applied to accomplish geometric separation of "point-and-curve"-like data. An artificially made example of such data can be seen in Fig. 2a. For a theoretical account of these separation ideas we refer to the recent papers by Donoho and the first author [12, 13]. Here, we simply display the result of the separation, see Fig. 1d. For real-world applications of these separation techniques we refer to the paper [33] on neurobiological imaging.

In the spirit of reproducible research [15], we wish to mention that Figs. 1d, f and 2, If and 1d have been produced by the discrete shearlet transform implemented in the Matlab toolbox *Shearlab* which has recently been released under a GNU license and is freely available at `http://www.shearlab.org`.

1.5 Outline

In Sect. 2 we present a review of shearlet theory in $L^2(\mathbb{R}^2)$, where we focus on discrete shearlet systems. We describe the classical band-limited construction (Sect. 2.2) and a more recent construction of compactly supported shearlets (Sect. 2.3). In Sect. 3 we present results on the ability of shearlets to optimally sparsely approximate cartoon-like images. Section 4 is dedicated to a discussion on similar properties of 3D shearlet systems.

2 2D Shearlets

In this section, we summarize what is known about constructions of discrete shearlet systems in 2D. Although all results in this section can easily be extended to (irregular) shearlet systems associated with a general irregular set of parameters for scaling, shear, and translation, we will only focus on the discrete shearlet systems associated with a regular set of parameters as described in the next section. For a detailed analysis of irregular shearlet systems, we refer to [27]. We first start with various notations and definitions for later use.

2.1 Preliminaries

For $j \geq 0, k \in \mathbb{Z}$, let

$$A_{2^j} = \begin{pmatrix} 2^j & 0 \\ 0 & 2^{j/2} \end{pmatrix}, \; S_k = \begin{pmatrix} 1 & k \\ 0 & 1 \end{pmatrix}, \quad \text{and} \quad M_c = \begin{pmatrix} c_1 & 0 \\ 0 & c_2 \end{pmatrix},$$

where $c = (c_1, c_2)$ and c_1, c_2 are some positive constants. Similarly, we define

$$\tilde{A}_{2^j} = \begin{pmatrix} 2^{j/2} & 0 \\ 0 & 2^j \end{pmatrix}, \; \tilde{S}_k = \begin{pmatrix} 1 & 0 \\ k & 1 \end{pmatrix}, \quad \text{and} \quad \tilde{M}_c = \begin{pmatrix} c_2 & 0 \\ 0 & c_1 \end{pmatrix}.$$

Next, we define discrete shearlet systems in 2D.

Definition 1. Let $c = (c_1, c_2) \in (\mathbb{R}_+)^2$. For $\phi, \psi, \tilde{\psi} \in L^2(\mathbb{R}^2)$ the *cone-adapted 2D discrete shearlet system* $SH(\phi, \psi, \tilde{\psi}; c)$ is defined by

$$SH(\phi, \psi, \tilde{\psi}; c) = \Phi(\phi; c_1) \cup \Psi(\psi; c) \cup \tilde{\Psi}(\tilde{\psi}; c),$$

where

$$\Phi(\phi; c_1) = \{\phi(\cdot - m) : m \in c_1\mathbb{Z}^2\},$$

$$\Psi(\psi; c) = \{2^{\frac{3}{4}j}\psi(S_k A_{2^j} \cdot -m) : j \geq 0, -\lceil 2^{j/2} \rceil \leq k \leq \lceil 2^{j/2} \rceil, m \in M_c\mathbb{Z}^2\},$$

and

$$\tilde{\Psi}(\tilde{\psi}; c) = \{2^{\frac{3}{4}j}\tilde{\psi}(\tilde{S}_k \tilde{A}_{2^j} \cdot -m) : j \geq 0, -\lceil 2^{j/2} \rceil \leq k \leq \lceil 2^{j/2} \rceil, m \in \tilde{M}_c\mathbb{Z}^2\}.$$

If $SH(\phi, \psi, \tilde{\psi}; c)$ is a frame for $L^2(\mathbb{R}^2)$, we refer to ϕ as a *scaling function* and ψ and $\tilde{\psi}$ as *shearlets*.

Our aim is to construct compactly supported functions ϕ, ψ, and $\tilde{\psi}$ to obtain compactly supported shearlets in 2D. For this, we will describe general sufficient conditions on the shearlet generators ψ and $\tilde{\psi}$, which lead to the construction of compactly supported shearlets. To formulate our sufficient conditions on ψ and $\tilde{\psi}$ (Sect. 2.3), we will first need to introduce the necessary notational concepts.

For functions $\phi, \psi, \tilde{\psi} \in L^2(\mathbb{R}^2)$, we define $\Theta : \mathbb{R}^2 \times \mathbb{R}^2 \to \mathbb{R}$ by

$$\Theta(\xi,\omega) = |\hat{\phi}(\xi)||\hat{\phi}(\xi+\omega)| + \Theta_1(\xi,\omega) + \Theta_2(\xi,\omega), \tag{1}$$

where

$$\Theta_1(\xi,\omega) = \sum_{j\geq 0} \sum_{|k|\leq \lceil 2^{j/2} \rceil} \left|\hat{\psi}(S_k^T A_{2^{-j}}\xi)\right| \left|\hat{\psi}(S_k{}^T A_{2^{-j}}\xi+\omega)\right|$$

and

$$\Theta_2(\xi,\omega) = \sum_{j\geq 0} \sum_{|k|\leq \lceil 2^{j/2} \rceil} \left|\hat{\tilde{\psi}}(S_k \tilde{A}_{2^{-j}}\xi)\right| \left|\hat{\tilde{\psi}}(S_k \tilde{A}_{2^{-j}}\xi+\omega)\right|.$$

Also, for $c = (c_1, c_2) \in (\mathbb{R}_+)^2$, let

$$R(c) = \sum_{m\in\mathbb{Z}^2\setminus\{0\}} \left(\Gamma_0(c_1^{-1}m)\Gamma_0(-c_1^{-1}m)\right)^{\frac{1}{2}} + \left(\Gamma_1(M_c^{-1}m)\Gamma_1(-M_c^{-1}m)\right)^{\frac{1}{2}} + \left(\Gamma_2(\tilde{M}_c^{-1}m)\Gamma_2(-\tilde{M}_c^{-1}m)\right)^{\frac{1}{2}},$$

where

$$\Gamma_0(\omega) = \operatorname*{ess\,sup}_{\xi\in\mathbb{R}^2} |\hat{\phi}(\xi)||\hat{\phi}(\xi+\omega)| \quad \text{and} \quad \Gamma_i(\omega) = \operatorname*{ess\,sup}_{\xi\in\mathbb{R}^2} \Theta_i(\xi,\omega) \quad \text{for } i = 1,2.$$

2.2 *Classical Construction*

We now first describe the construction of band-limited shearlets which provides tight frames for $L^2(\mathbb{R}^2)$. Constructions of this type were first introduced by Labate, Weiss, and two of the authors in [37]. The *classical example* of a generating shearlet is a function $\psi \in L^2(\mathbb{R}^2)$ satisfying

$$\hat{\psi}(\xi) = \hat{\psi}(\xi_1,\xi_2) = \hat{\psi}_1(\xi_1)\,\hat{\psi}_2(\tfrac{\xi_2}{\xi_1}),$$

where $\psi_1 \in L^2(\mathbb{R})$ is a discrete wavelet, i.e., satisfies the discrete Calderón condition given by

$$\sum_{j\in\mathbb{Z}} |\hat{\psi}_1(2^{-j}\xi)|^2 = 1 \quad \text{for a.e. } \xi \in \mathbb{R},$$

with $\hat{\psi}_1 \in C^\infty(\mathbb{R})$ and $\operatorname{supp}\hat{\psi}_1 \subseteq [-\frac{5}{4}, -\frac{1}{4}] \cup [\frac{1}{4}, \frac{5}{4}]$, and $\psi_2 \in L^2(\mathbb{R})$ is a bump function, namely

$$\sum_{k=-1}^{1} |\hat{\psi}_2(\xi+k)|^2 = 1 \quad \text{for a.e. } \xi \in [-1,1],$$

satisfying $\hat{\psi}_2 \in C^\infty(\mathbb{R})$ and $\operatorname{supp}\hat{\psi}_2 \subseteq [-1,1]$. There are several choices of ψ_1 and ψ_2 satisfying those conditions, and we refer to [20] for further details. The tiling of the frequency domain given by these band-limited generators and choosing

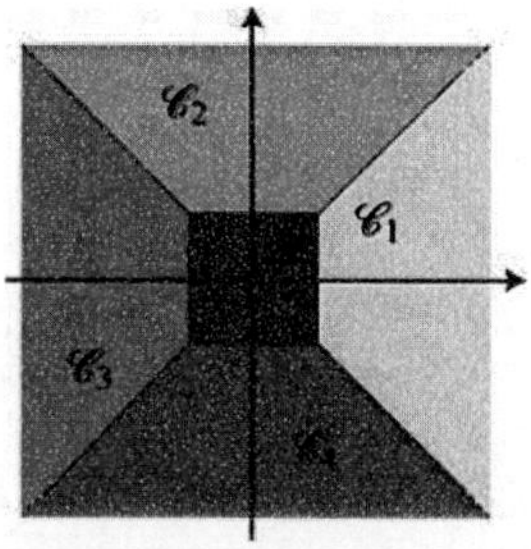

Fig. 3: The cones $\mathscr{C}_1$–$\mathscr{C}_4$ and the centered rectangle $\mathscr{R}$ in the frequency domain

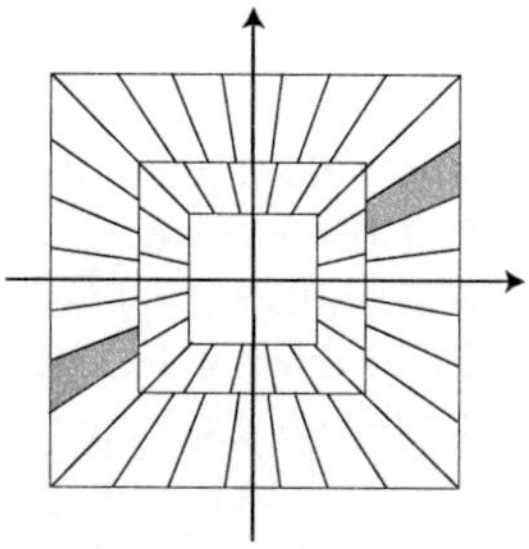

Fig. 4: Tiling of the frequency domain induced by band-limited shearlets

$\tilde{\psi}(x_1,x_2) = \psi(x_2,x_1)$ is illustrated in Fig. 4. As described in Fig. 3, a conic region $\mathscr{C}_1 \cup \mathscr{C}_3$ is covered by the frequency support of shearlets in $\Psi(\psi;c)$ while $\mathscr{C}_2 \cup \mathscr{C}_4$ is covered by $\tilde{\Psi}(\tilde{\psi};c)$. For this particular choice, using an appropriate scaling function ϕ for the centered rectangle $\mathscr{R}$ (see Fig. 3), it was proved in [20, Theorem 3] that the associated cone-adapted discrete shearlet system $SH(\phi,\psi,\tilde{\psi};(1,1))$ forms a Parseval frame for $L^2(\mathbb{R}^2)$.

2.3 Constructing Compactly Supported Shearlets

We are now ready to state general sufficient conditions for the construction of shearlet frames.

Theorem 1 ([27]). *Let $\phi,\psi \in L^2(\mathbb{R}^2)$ be functions such that*

$$\hat{\phi}(\xi_1,\xi_2) \le C_1 \cdot \min\{1,|\xi_1|^{-\gamma}\} \cdot \min\{1,|\xi_2|^{-\gamma}\}$$

and

$$|\hat{\psi}(\xi_1,\xi_2)| \le C_2 \cdot \min\{1,|\xi_1|^{\alpha}\} \cdot \min\{1,|\xi_1|^{-\gamma}\} \cdot \min\{1,|\xi_2|^{-\gamma}\}, \tag{2}$$

for some positive constants $C_1, C_2 < \infty$ and $\alpha > \gamma > 3$. Define $\tilde{\psi}(x_1,x_2) = \psi(x_2,x_1)$, and let $L_{\inf}, L_{\sup}$ be defined by

$$L_{\inf} = \operatorname*{ess\,inf}_{\xi\in\mathbb{R}^2} \Theta(\xi,0) \quad \text{and} \quad L_{\sup} = \operatorname*{ess\,sup}_{\xi\in\mathbb{R}^2} \Theta(\xi,0).$$

Suppose that there is a constant $\tilde{L}_{\inf} > 0$ such that $0 < \tilde{L}_{\inf} \le L_{\inf}$. Then there exist a sampling parameter $c = (c_1,c_2)$ with $c_1 = c_2$ and a constant $\tilde{L}_{\sup} < \infty$ such that

$$R(c) < \tilde{L}_{\inf} \le L_{\inf} \text{ and } L_{\sup} \le \tilde{L}_{\sup},$$

and, further, $SH(\phi,\psi,\tilde{\psi};c)$ *forms a frame for* $L^2(\mathbb{R}^2)$ *with frame bounds* A *and* B *satisfying*

$$\frac{1}{|\det M_c|}[\tilde{L}_{\inf} - R(c)] \leq A \leq B \leq \frac{1}{|\det M_c|}[\tilde{L}_{\sup} + R(c)]. \tag{3}$$

For a detailed proof, we refer to the paper [27] by Kittipoom and two of the authors.

Obviously, band-limited shearlets (from Sect. 2.2) satisfy condition (2). More interestingly, also a large class of spatially compactly supported functions satisfies this condition. In fact, in [27], various constructions of compactly supported shearlets are presented using Theorem 1 and generalized low-pass filters; an example of such a construction procedure is given in Theorem 2 below. In Theorem 1 we assumed $c_1 = c_2$ for the sampling matrix M_c (or $\tilde{M}_c$), the only reason for this being the simplification of the estimates for the frame bounds A,B in (3). In fact, the estimate (3) generalizes easily to non-uniform sampling constants c_1, c_2 with $c_1 \neq c_2$. For explicit estimates of the form (3) in the case of non-uniform sampling, we refer to [27].

The following result provides a specific family of functions satisfying the general sufficiency condition from Theorem 1.

Theorem 2 ([27]). *Let* $K, L > 0$ *be such that* $L \geq 10$ *and* $\frac{3L}{2} \leq K \leq 3L-2$*, and define a shearlet* $\psi \in L^2(\mathbb{R}^2)$ *by*

$$\hat{\psi}(\xi) = m_1(4\xi_1)\hat{\phi}(\xi_1)\hat{\phi}(2\xi_2), \quad \xi = (\xi_1,\xi_2) \in \mathbb{R}^2,$$

where m_0 *is the low pass filter satisfying*

$$|m_0(\xi_1)|^2 = (\cos(\pi\xi_1))^{2K} \sum_{n=0}^{L-1} \binom{K-1+n}{n} (\sin(\pi\xi_1))^{2n}, \quad \xi_1 \in \mathbb{R},$$

m_1 *is the associated bandpass filter defined by*

$$|m_1(\xi_1)|^2 = |m_0(\xi_1 + \tfrac{1}{2})|^2, \quad \xi_1 \in \mathbb{R},$$

and ϕ *is the scaling function given by*

$$\hat{\phi}(\xi_1) = \prod_{j=0}^{\infty} m_0(2^{-j}\xi_1), \quad \xi_1 \in \mathbb{R}.$$

Then there exists a sampling constant $\hat{c}_1 > 0$ *such that the shearlet system* $\Psi(\psi;c)$ *forms a frame for* $\check{L}^2(\mathscr{C}_1 \cup \mathscr{C}_3) := \{f \in L^2(\mathbb{R}^2) : \operatorname{supp} \hat{f} \subset \mathscr{C}_1 \cup \mathscr{C}_3\}$ *for any sampling matrix* M_c *with* $c = (c_1, c_2) \in (\mathbb{R}_+)^2$ *and* $c_2 \leq c_1 \leq \hat{c}_1$.

For these shearlet systems, there is a bias towards the vertical axis, especially at coarse scales, since they are defined for $\check{L}^2(\mathscr{C}_1 \cup \mathscr{C}_3)$, and hence, the frequency support of the shearlet elements overlaps more significantly along the vertical axis. In order to control the upper frame bound, it is therefore desirable to apply a denser sampling along the vertical axis than along the horizontal axis, i.e., $c_1 > c_2$.

Having compactly supported (separable) shearlet frames for $\check{L}^2(\mathscr{C}_1 \cup \mathscr{C}_3)$ at hand by Theorem 2, we can easily construct shearlet frames for the whole space $L^2(\mathbb{R}^2)$. The exact procedure is described in the following theorem from [27].

Theorem 3 ([27]). *Let $\psi \in L^2(\mathbb{R}^2)$ be the shearlet with associated scaling function $\phi_1 \in L^2(\mathbb{R})$ both introduced in Theorem 2, and set $\phi(x_1,x_2) = \phi_1(x_1)\phi_1(x_2)$ and $\tilde{\psi}(x_1,x_2) = \psi(x_2,x_1)$. Then the corresponding shearlet system $SH(\phi,\psi,\tilde{\psi};c)$ forms a frame for $L^2(\mathbb{R}^2)$ for any sampling matrices M_c and $\tilde{M}_c$ with $c = (c_1,c_2) \in (\mathbb{R}_+)^2$ and $c_2 \leq c_1 \leq \hat{c}_1$.*

For the horizontal cone $\mathscr{C}_1 \cup \mathscr{C}_3$ we allow for a denser sampling by M_c along the vertical axis, i.e., $c_2 \leq c_1$, precisely as in Theorem 2. For the vertical cone $\mathscr{C}_2 \cup \mathscr{C}_4$ we analogously allow for a denser sampling along the horizontal axis; since the position of c_1 and c_2 is reversed in $\tilde{M}_c$ compared to M_c, this still corresponds to $c_2 \leq c_1$.

We wish to mention that there is a trade-off between *compact support* of the shearlet generators, *tightness* of the associated frame, and *separability* of the shearlet generators. The known constructions of tight shearlet frames do not use separable generators (Sect. 2.2), and these constructions can be shown to *not* be applicable to compactly supported generators. Tightness is difficult to obtain while allowing for compactly supported generators, but we can gain separability as in Theorem 3, hence fast algorithmic realizations. On the other hand, when allowing non-compactly supported generators, tightness is possible, but separability seems to be out of reach, which makes fast algorithmic realizations very difficult.

We end this section by remarking that the construction results above even generalize to constructions of irregular shearlet systems [28, 29].

3 Sparse Approximations

After having introduced compactly supported shearlet systems in the previous section, we now aim for optimally sparse approximations. To be precise, we will show that these compactly supported shearlet systems provide optimally sparse approximations when representing and analyzing anisotropic features in 2D data.

3.1 Cartoon-like Image Model

Following [11], we introduce $\mathrm{STAR}^2(\nu)$, a class of sets B with C^2 boundaries ∂B and curvature bounded by ν, as well as $\mathscr{E}^2_\nu(\mathbb{R}^2)$, a class of cartoon-like images. For this, in polar coordinates, we let $\rho : [0,2\pi) \to [0,1]$ be a radius function and define the set B by

$$B = \{x \in \mathbb{R}^2 : |x| \leq \rho(\theta), x = (|x|, \theta) \text{ in polar coordinates}\}.$$

In particular, we will require that the boundary ∂B of B is given by the curve

$$\beta(\theta) = \begin{pmatrix} \rho(\theta)\cos(\theta) \\ \rho(\theta)\sin(\theta) \end{pmatrix}, \tag{4}$$

and the class of boundaries of interest to us are defined by

$$\sup|\rho''(\theta)| \le \nu, \quad \rho \le \rho_0 < 1, \tag{5}$$

where $\rho_0 < 1$ needs to be chosen so that $y + B \subset [0,1]^2$ for some $y \in \mathbb{R}^2$.

The following definition now introduces a class of cartoon-like images.

Definition 2. For $\nu > 0$, the set $\mathrm{STAR}^2(\nu)$ is defined to be the set of all $B \subset [0,1]^2$ such that B is a translate of a set obeying (1) and (2). Further, $\mathcal{E}_\nu^2(\mathbb{R}^2)$ denotes the set of functions $f \in L^2(\mathbb{R}^2)$ of the form

$$f = f_0 + f_1 \chi_B,$$

where $B \in \mathrm{STAR}^2(\nu)$ and $f_0, f_1 \in C_0^2(\mathbb{R}^2)$ with $\operatorname{supp} f_i \subset [0,1]^2$ and $\|f_i\|_{C^2} = \sum_{|\alpha| \le 2} \|D^\alpha f_i\|_\infty \le 1$ for $i = 1,2$.

One can also consider a more sophisticated class of cartoon-like images, where the boundary of B is allowed to be *piecewise* C^2, and we refer to the recent paper by two of the authors [34] and to similar considerations for the 3D case in Sect. 4.2.

Donoho [11] proved that the optimal approximation rate for such cartoon-like image models $f \in \mathcal{E}_\nu^2(\mathbb{R}^2)$ which can be achieved for almost any representation system under a so-called polynomial depth search selection procedure of the selected system elements is

$$\|f - f_N\|_2^2 \le C \cdot N^{-2} \quad \text{as } N \to \infty,$$

where f_N is the best N-term approximation of f. As discussed in the next section shearlets in 2D do indeed deliver this optimal approximation rate.

3.2 Optimally Sparse Approximation of Cartoon-like Images

Let $SH(\phi, \psi, \tilde{\psi}; c)$ be a shearlet frame for $L^2(\mathbb{R}^2)$. Since this is a countable set of functions, we can denote it by $SH(\phi, \psi, \tilde{\psi}; c) = (\sigma_i)_{i \in I}$. We let $(\tilde{\sigma}_i)_{i \in I}$ be a dual frame of $(\sigma_i)_{i \in I}$. As our N-term approximation f_N of a cartoon-like image $f \in \mathcal{E}_\nu^2(\mathbb{R}^2)$ by the frame $SH(\phi, \psi, \tilde{\psi}; c)$, we then take

$$f_N = \sum_{i \in I_N} \langle f, \sigma_i \rangle \tilde{\sigma}_i,$$

where $(\langle f, \sigma_i \rangle)_{i \in I_N}$ are the N largest coefficients $\langle f, \sigma_i \rangle$ in magnitude. As in the tight frame case, this procedure does not always yield the *best* N-term approximation, but, surprisingly, even with this rather crude selection procedure, we can prove an (almost) optimally sparse approximation rate. We speak of "almost" optimality due to the (negligible) log-factor in (6). The following result shows that our "new" compactly supported shearlets (see Sect. 2.3) deliver the same approximation rate as *band-limited* curvelets [1], contourlets [10], and shearlets [21].

Theorem 4 ([32]). *Let $c > 0$, and let $\phi, \psi, \tilde{\psi} \in L^2(\mathbb{R}^2)$ be compactly supported. Suppose that, in addition, for all $\xi = (\xi_1, \xi_2) \in \mathbb{R}^2$, the shearlet ψ satisfies*

(i) $|\hat{\psi}(\xi)| \le C_1 \cdot \min\{1, |\xi_1|^{\alpha}\} \cdot \min\{1, |\xi_1|^{-\gamma}\} \cdot \min\{1, |\xi_2|^{-\gamma}\}$ *and*

(ii) $\left|\frac{\partial}{\partial \xi_2} \hat{\psi}(\xi)\right| \le |h(\xi_1)| \cdot \left(1 + \frac{|\xi_2|}{|\xi_1|}\right)^{-\gamma}$,

where $\alpha > 5$, $\gamma \ge 4$, $h \in L^1(\mathbb{R})$, and C_1 is a constant, and suppose that the shearlet $\tilde{\psi}$ satisfies (i) and (ii) with the roles of ξ_1 and ξ_2 reversed. Further, suppose that $SH(\phi, \psi, \tilde{\psi}; c)$ forms a frame for $L^2(\mathbb{R}^2)$.

Then, for any $\nu > 0$, the shearlet frame $SH(\phi, \psi, \tilde{\psi}; c)$ provides (almost) optimally sparse approximations of functions $f \in \mathcal{E}^2_\nu(\mathbb{R}^2)$ in the sense that there exists some $C > 0$ such that

$$\|f - f_N\|_2^2 \le C \cdot N^{-2} \cdot (\log N)^3 \qquad \text{as } N \to \infty, \tag{6}$$

where f_N is the nonlinear N-term approximation obtained by choosing the N largest shearlet coefficients of f.

Condition (i) can be interpreted as both a condition ensuring (almost) separable behavior as well as a moment condition along the horizontal axis, hence enforcing directional selectivity. This condition ensures that the support of shearlets in frequency domain is essentially of the form indicated in Fig. 4. Condition (ii) (together with (i)) is a weak version of a directional vanishing moment condition,[2] which is crucial for having fast decay of the shearlet coefficients when the corresponding shearlet intersects the discontinuity curve. Conditions (i) and (ii) are rather mild conditions on the generators; in particular, shearlets constructed by Theorem 2 and 3, with extra assumptions on the parameters K and L, will indeed satisfy (i) and (ii) in Theorem 1. To compare with the optimality result for band-limited generators we wish to point out that conditions (i) and (ii) are obviously satisfied for band-limited generators.

We remark that this kind of approximation result is not available for shearlet systems coming directly from the shearlet group. One reason for this being that these systems, as mentioned several times, do not treat directions in a uniform way.

4 Shearlets in 3D and Beyond

Shearlet theory has traditionally only dealt with representation systems for two-dimensional data. In the recent paper [8] (and the accompanying paper [9]) this

[2] For the precise definition of directional vanishing moments, we refer to [10].

was changed when Dahlke, Steidl, and Teschke generalized the continuous shearlet transform (see [7, 30]) to higher dimensions. The shearlet transform on $L^2(\mathbb{R}^n)$ by Dahlke, Steidl, and Teschke is associated with the so-called shearlet group in $\mathbb{R}\setminus\{0\}\times\mathbb{R}^{n-1}\times\mathbb{R}^n$, with a dilation matrix of the form

$$A_a = \mathrm{diag}\,(a, \mathrm{sgn}(a)\,|a|^{1/n}, \ldots, \mathrm{sgn}(a)\,|a|^{1/n}), \qquad a \in \mathbb{R}\setminus\{0\},$$

and with a shearing matrix with $n-1$ shear parameters $s = (s_1, \ldots, s_{n-1}) \in \mathbb{R}^{n-1}$ of the form

$$S_s = \begin{bmatrix} 1 & s \\ 0 & I_{n-1} \end{bmatrix},$$

where I_n denotes the $n \times n$ identity matrix. This type of shearing matrix gives rise to shearlets consisting of wedges of size $a^{-1} \times a^{-1/n} \times \cdots \times a^{-1/n}$ *in frequency domain*, where $a^{-1} \gg a^{-1/n}$ for small $a > 0$. Hence, for small $a > 0$, the spatial appearance is a surface-like element of co-dimension one.

In the following section, we will consider shearlet systems in $L^2(\mathbb{R}^3)$ associated with a sightly different shearing matrix. More importantly, we will consider *pyramid-adapted* 3D shearlet systems, since these systems treat directions in a uniform way as opposed to the shearlet systems coming from the shearlet group; this design, of course, parallels the idea behind cone-adapted 2D shearlets. In [22], the continuous version of the pyramid-adapted shearlet system was introduced, and it was shown that the location and the local orientation of the boundary set of certain three-dimensional solid regions can be precisely identified by this continuous shearlet transform. The pyramid-adapted shearlet system can easily be generalized to higher dimensions, but for brevity we only consider the three-dimensional setup and newly introduce it now in the discrete setting.

4.1 Pyramid-Adapted Shearlet Systems

We will scale according to *paraboloidal scaling matrices* A_{2^j}, $\tilde{A}_{2^j}$ or $\breve{A}_{2^j}$, $j \in \mathbb{Z}$, and encode directionality by the *shear matrices* S_k, $\tilde{S}_k$, or $\breve{S}_k$, $k = (k_1, k_2) \in \mathbb{Z}^2$, defined by

$$A_{2^j} = \begin{pmatrix} 2^j & 0 & 0 \\ 0 & 2^{j/2} & 0 \\ 0 & 0 & 2^{j/2} \end{pmatrix}, \quad \tilde{A}_{2^j} = \begin{pmatrix} 2^{j/2} & 0 & 0 \\ 0 & 2^j & 0 \\ 0 & 0 & 2^{j/2} \end{pmatrix}, \quad \text{and} \quad \breve{A}_{2^j} = \begin{pmatrix} 2^{j/2} & 0 & 0 \\ 0 & 2^{j/2} & 0 \\ 0 & 0 & 2^j \end{pmatrix},$$

and

$$S_k = \begin{pmatrix} 1 & k_1 & k_2 \\ 0 & 1 & 0 \\ 0 & 0 & 1 \end{pmatrix}, \qquad \tilde{S}_k = \begin{pmatrix} 1 & 0 & 0 \\ k_1 & 1 & k_2 \\ 0 & 0 & 1 \end{pmatrix}, \qquad \text{and} \quad \breve{S}_k = \begin{pmatrix} 1 & 0 & 0 \\ 0 & 1 & 0 \\ k_1 & k_2 & 1 \end{pmatrix},$$

respectively. The translation lattices will be defined through the following matrices: $M_c = \mathrm{diag}(c_1, c_2, c_2)$, $\tilde{M}_c = \mathrm{diag}(c_2, c_1, c_2)$, and $\breve{M}_c = \mathrm{diag}(c_2, c_2, c_1)$, where $c_1 > 0$ and $c_2 > 0$.

We next partition the frequency domain into the following six pyramids:

$$\mathscr{P}_\iota = \begin{cases} \{(\xi_1,\xi_2,\xi_3) \in \mathbb{R}^3 : \xi_1 \geq 1, |\xi_2/\xi_1| \leq 1, |\xi_3/\xi_1| \leq 1\} : \iota = 1, \\ \{(\xi_1,\xi_2,\xi_3) \in \mathbb{R}^3 : \xi_2 \geq 1, |\xi_1/\xi_2| \leq 1, |\xi_3/\xi_2| \leq 1\} : \iota = 2, \\ \{(\xi_1,\xi_2,\xi_3) \in \mathbb{R}^3 : \xi_3 \geq 1, |\xi_1/\xi_3| \leq 1, |\xi_2/\xi_3| \leq 1\} : \iota = 3, \\ \{(\xi_1,\xi_2,\xi_3) \in \mathbb{R}^3 : \xi_1 \leq -1, |\xi_2/\xi_1| \leq 1, |\xi_3/\xi_1| \leq 1\} : \iota = 4, \\ \{(\xi_1,\xi_2,\xi_3) \in \mathbb{R}^3 : \xi_2 \leq -1, |\xi_1/\xi_2| \leq 1, |\xi_3/\xi_2| \leq 1\} : \iota = 5, \\ \{(\xi_1,\xi_2,\xi_3) \in \mathbb{R}^3 : \xi_3 \leq -1, |\xi_1/\xi_3| \leq 1, |\xi_2/\xi_3| \leq 1\} : \iota = 6, \end{cases}$$

and a centered rectangle

$$\mathscr{R} = \{(\xi_1,\xi_2,\xi_3) \in \mathbb{R}^3 : \|(\xi_1,\xi_2,\xi_3)\|_\infty < 1\}.$$

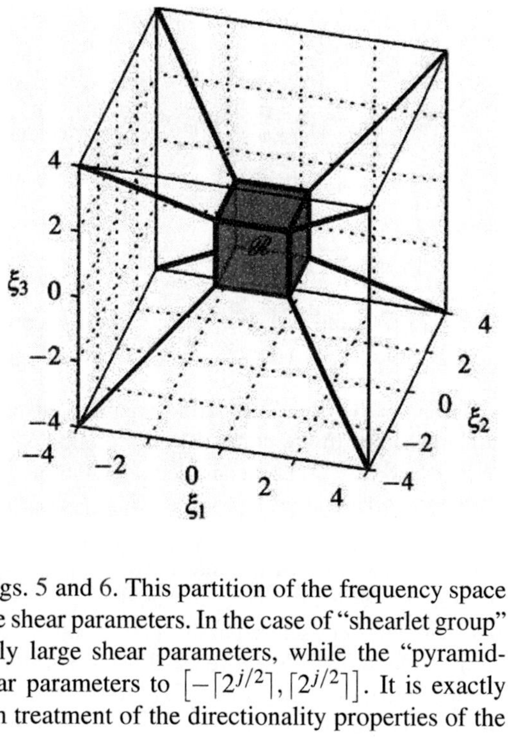

Fig. 5 The partition of the frequency domain: The centered *rectangle* $\mathscr{R}$. The arrangement of the six pyramids is indicated by the "diagonal" lines. See Fig. 6 for a sketch of the pyramids

The partition is illustrated in Figs. 5 and 6. This partition of the frequency space allows us to restrict the range of the shear parameters. In the case of "shearlet group" systems one must allow arbitrarily large shear parameters, while the "pyramid-adapted" systems restrict the shear parameters to $\left[-\lceil 2^{j/2}\rceil, \lceil 2^{j/2}\rceil\right]$. It is exactly this fact that gives a more uniform treatment of the directionality properties of the shearlet system.

These considerations are now made precise in the following definition.

Definition 3. For $c = (c_1, c_2) \in (\mathbb{R}_+)^2$, the *pyramid-adapted 3D shearlet system* $SH(\phi, \psi, \tilde{\psi}, \breve{\psi}; c)$ generated by $\phi, \psi, \tilde{\psi}, \breve{\psi} \in L^2(\mathbb{R}^3)$ is defined by

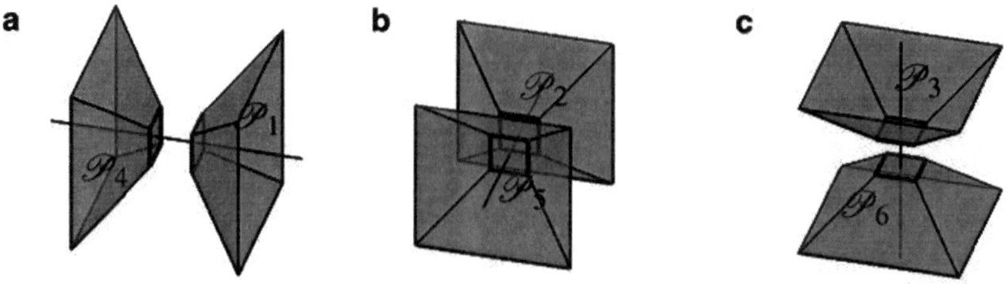

Pyramids $\mathscr{P}_1$ and $\mathscr{P}_4$ and the ξ_1 axis. Pyramids $\mathscr{P}_2$ and $\mathscr{P}_5$ and the ξ_2 axis. Pyramids $\mathscr{P}_3$ and $\mathscr{P}_6$ and the ξ_3 axis.

Fig. 6: The partition of the frequency domain: The "top" of the six pyramids

$$SH(\phi,\psi,\tilde{\psi},\breve{\psi};c) = \Phi(\phi;c_1) \cup \Psi(\psi;c) \cup \tilde{\Psi}(\tilde{\psi};c) \cup \breve{\Psi}(\breve{\psi};c),$$

where

$$\Phi(\phi;c_1) = \{\phi_m = \phi(\cdot - m) : m \in c_1\mathbb{Z}^3\},$$
$$\Psi(\psi;c) = \left\{\psi_{j,k,m} = 2^j \psi(S_k A_{2^j} \cdot - m) : j \geq 0, |k| \leq \lceil 2^{j/2} \rceil, m \in M_c\mathbb{Z}^3\right\},$$
$$\tilde{\Psi}(\tilde{\psi};c) = \{\tilde{\psi}_{j,k,m} = 2^j \tilde{\psi}(\tilde{S}_k \tilde{A}_{2^j} \cdot - m) : j \geq 0, |k| \leq \lceil 2^{j/2} \rceil, m \in \tilde{M}_c\mathbb{Z}^3\},$$

and

$$\breve{\Psi}(\breve{\psi};c) = \{\breve{\psi}_{j,k,m} = 2^j \breve{\psi}(\breve{S}_k \breve{A}_{2^j} \cdot - m) : j \geq 0, |k| \leq \lceil 2^{j/2} \rceil, m \in \breve{M}_c\mathbb{Z}^3\},$$

where $j \in \mathbb{N}_0$ and $k \in \mathbb{Z}^2$. Here we have used the vector notation $|k| \leq K$ for $k = (k_1,k_2)$ and $K > 0$ to denote $|k_1| \leq K$ and $|k_2| \leq K$.

The construction of pyramid-adapted shearlet systems $SH(\phi,\psi,\tilde{\psi},\breve{\psi};c)$ runs along the lines of the construction of cone-adapted shearlet systems in $L^2(\mathbb{R}^2)$ described in Sect. 2.3. For a detailed description, we refer to [31].

We remark that the shearlets in *spatial domain* are of size $2^{-j/2}$ times $2^{-j/2}$ times 2^{-j} which shows that the shearlet elements will become "plate-like" as $j \to \infty$. One could also use the scaling matrix $A_{2^j} = \text{diag}\,(2^j, 2^j, 2^{j/2})$ with similar changes for $\tilde{A}_{2^j}$ and $\breve{A}_{2^j}$. This would lead to "needle-like" shearlet elements instead of the "plate-like" elements considered in this paper, but we will not pursue this further here, and simply refer to [31]. More generally, it is possible to even consider non-paraboloidal scaling matrices of the form $A_j = \text{diag}\,(2^j, 2^{\alpha j}, 2^{\beta j})$ for $0 < \alpha, \beta \leq 1$. One drawback of allowing such general scaling matrices is the lack of fast algorithms for non-dyadic multiscale systems. On the other hand, the parameters α and β allow us to precisely shape the shearlet elements, ranging from very plate-like to very needle-like, according to the application at hand, i.e., choosing the shearlet-shape that is the best "fit" for the geometric characteristics of the considered data.

4.2 Sparse Approximations of 3D Data

We now consider approximations of three-dimensional cartoon-like images using shearlets introduced in the previous section. The three-dimensional cartoon-like images $\mathscr{E}_\nu^2(\mathbb{R}^3)$ will be piecewise C^2 functions with discontinuities on a closed C^2 *surface* whose principal curvatures are bounded by ν. In [31], it was shown that the optimal approximation rate for such 3D cartoon-like image models $f \in \mathscr{E}_\nu^2(\mathbb{R}^3)$ which can be achieved for almost any representation system (under polynomial depth search selection procedure of the approximating coefficients) is

$$\|f - f_N\|_2^2 \leq C \cdot N^{-1} \quad \text{as } N \to \infty,$$

where f_N is the best N-term approximation of f. The following result shows that compactly supported pyramid-adapted shearlets do (almost) deliver this approximation rate.

Theorem 5 ([31]). *Let $c \in (\mathbb{R}_+)^2$, and let $\phi, \psi, \tilde{\psi}, \breve{\psi} \in L^2(\mathbb{R}^3)$ be compactly supported. Suppose that, for all $\xi = (\xi_1, \xi_2, \xi_3) \in \mathbb{R}^3$, the function ψ satisfies:*

(i) $|\hat{\psi}(\xi)| \leq C_1 \cdot \min\{1, |\xi_1|^\alpha\} \cdot \min\{1, |\xi_1|^{-\gamma}\} \cdot \min\{1, |\xi_2|^{-\gamma}\} \cdot \min\{1, |\xi_3|^{-\gamma}\}$,

(ii) $\left|\frac{\partial}{\partial \xi_i} \hat{\psi}(\xi)\right| \leq |h(\xi_1)| \cdot \left(1 + \frac{|\xi_2|}{|\xi_1|}\right)^{-\gamma} \left(1 + \frac{|\xi_3|}{|\xi_1|}\right)^{-\gamma}, \quad i = 2, 3,$

where $\alpha > 8$, $\gamma \geq 4$, $t \mapsto th(t) \in L^1(\mathbb{R})$, and C_1 a constant, and suppose that $\tilde{\psi}$ and $\breve{\psi}$ satisfy analogous conditions with the obvious change of coordinates. Further, suppose that the shearlet system $SH(\phi, \psi, \tilde{\psi}, \breve{\psi}; c)$ forms a frame for $L^2(\mathbb{R}^3)$.

Then, for any $\nu > 0$, the shearlet frame $SH(\phi, \psi, \tilde{\psi}, \breve{\psi}; c)$ provides (almost) optimally sparse approximations of functions $f \in \mathscr{E}_\nu^2(\mathbb{R}^3)$ in the sense that there exists some $C > 0$ such that

$$\|f - f_N\|_2^2 \leq C \cdot N^{-1} \cdot (\log N)^2 \qquad \text{as } N \to \infty. \tag{7}$$

In the following we will give a sketch of the proof of Theorem 5 and, in particular, give a heuristic argument (inspired by a similar one for 2D curvelets in [1]) to explain the exponent N^{-1} in (7).

Proof (Theorem 5, Sketch). Let $f \in \mathscr{E}_\nu^2(\mathbb{R}^3)$ be a 3D cartoon-like image. The main concern is to derive appropriate estimates for the shearlet coefficients $\langle f, \psi_{j,k,m} \rangle$. We first observe that we can assume the scaling index j to be sufficiently large, since f as well as all shearlet elements are compactly supported and since a finite number does not contribute to the asymptotic estimate we are aiming for. In particular, this implies that we do not need to take frame elements from the "scaling" system $\Phi(\phi; c_1)$ into account. Also, we are allowed to restrict our analysis to shearlets $\psi_{j,k,m}$, since the frame elements $\tilde{\psi}_{j,k,m}$ and $\breve{\psi}_{j,k,m}$ can be handled in a similar way.

Letting $|\theta(f)|_n$ denote the nth largest shearlet coefficient $\langle f, \psi_{j,k,m} \rangle$ in absolute value and using the frame property of $SH(\phi, \psi, \tilde{\psi}, \breve{\psi}; c)$, we conclude that

$$\|f - f_N\|_2^2 \leq \frac{1}{A} \sum_{n>N} |\theta(f)|_n^2,$$

for any positive integer N, where A denotes the lower frame bound of the shearlet frame $SH(\phi, \psi, \tilde{\psi}, \breve{\psi}; c)$. Thus, for completing the proof, it therefore suffices to show that

$$\sum_{n>N} |\theta(f)|_n^2 \leq C \cdot N^{-1} \cdot (\log N)^2 \qquad \text{as } N \to \infty. \tag{8}$$

For the following heuristic argument, we need to make some simplifications. We will assume to have a shearlet of the form $\psi(x) = \eta(x_1)\varphi(x_2)\varphi(x_3)$, where η is a wavelet and φ a bump (or a scaling) function. Note that the wavelet "points" in the short direction of the plate-like shearlet. We now consider three cases of coefficients $\langle f, \psi_{j,k,m} \rangle$ (Fig. 7):

(a) Shearlets $\psi_{j,k,m}$ whose support does not overlap with the boundary ∂B.
(b) Shearlets $\psi_{j,k,m}$ whose support overlaps with ∂B and is nearly tangent.
(c) Shearlets $\psi_{j,k,m}$ whose support overlaps with ∂B, but not tangentially.

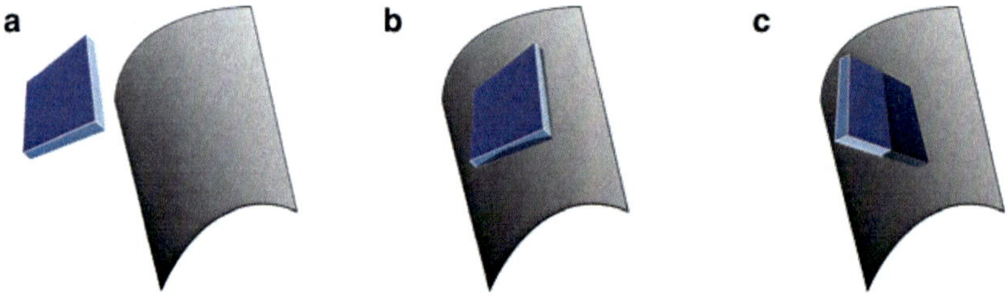

Sketch of shearlets whose support does not overlap with ∂B.

Sketch of shearlets whose support overlaps with ∂B and is nearly tangent.

Sketch of shearlets whose support overlaps with ∂B, but not tangentially.

Fig. 7: The three types of shearlet $\psi_{j,k,m}$ and boundary ∂B interactions considered in the heuristic argument (explaining the approximation rate N^{-1}). Note that only a section of ∂B is shown

As we argue in the following, only coefficients from case (b) will be significant. Case (b) is – loosely speaking – the situation in which the wavelet η breaches, in an almost normal direction, through the discontinuity surface; as is well known from wavelet theory, 1D wavelets efficiently handle such a "jump" discontinuity.

Case (a). Since f is C^2 smooth away from ∂B, the coefficients $|\langle f, \psi_{j,k,m} \rangle|$ will be sufficiently small owing to the wavelet η (and the fast decay of wavelet coefficients of smooth functions).

Case (b). At scale $j > 0$, there are at most $O(2^j)$ coefficients, since the plate-like elements are of size $2^{-j/2}$ times $2^{-j/2}$ (and "thickness" 2^{-j}). By assumptions on f and the support size of $\psi_{j,k,m}$, we obtain the estimate

$$|\langle f, \psi_{j,k,m} \rangle| \leq \|f\|_\infty \left\|\psi_{j,k,m}\right\|_1 \leq C_1 \, (2^{-2j})^{1/2} \left\|\psi_{j,k,m}\right\|_2^{1/2} \leq C_2 \cdot 2^{-j}$$

for some constants $C_1, C_2 > 0$. In other words, we have $O(2^j)$ coefficients bounded by $C_2 \cdot 2^{-j}$. Assuming the case (a) and (c) coefficients are negligible, the nth largest coefficient $|\theta(f)|_n$ is then bounded by

$$|\theta(f)|_n \leq C \cdot n^{-1}.$$

Therefore,

$$\sum_{n>N} |\theta(f)|_n^2 \leq \sum_{n>N} C \cdot n^{-2} \leq C \cdot \int_N^\infty x^{-2} \mathrm{d}x \leq C \cdot N^{-1}$$

and we arrive at (3), but without the log-factor. This in turn shows (7), at least heuristically, and still without the log-factor.

Case (c). Finally, when the shearlets are sheared away from the tangent position in case (b), they will again be small. This is due to the vanishing moment conditions in condition (i) and (ii). □

Clearly, Theorem 5 is an "obvious" three-dimensional version of Theorem 1. However, as opposed to the two-dimensional setting, anisotropic structures in three-dimensional data comprise of *two* morphologically different types of structure, namely surfaces *and* curves. It would, therefore, be desirable to allow our 3D image class to also contain cartoon-like images with *curve* singularities. On the other hand, the pyramid-adapted shearlets introduced in Sect. 4.1 are plate-like and thus, a priori, not optimal for capturing such one-dimensional singularities. Surprisingly, these plate-like shearlet systems still deliver the optimal rate N^{-1} for three-dimensional cartoon-like images $\mathcal{E}^2_{\nu,L}(\mathbb{R}^3)$, where L indicates that we allow our discontinuity surface ∂B to be *piecewise* C^2 smooth; $L \in \mathbb{N}$ is the maximal number of C^2 pieces and $\nu > 0$ is an upper estimate for the principal curvatures on each piece. In other words, for any $\nu > 0$ and $L \in \mathbb{N}$, the shearlet frame $SH(\phi, \psi, \tilde{\psi}, \breve{\psi}; c)$ provides (almost) optimally sparse approximations of functions $f \in \mathcal{E}^2_{\nu,L}(\mathbb{R}^3)$ in the sense that there exists some $C > 0$ such that

$$\|f - f_N\|_2^2 \leq C \cdot N^{-1} \cdot (\log N)^2 \qquad \text{as } N \to \infty. \tag{9}$$

The conditions on the shearlets $\psi, \tilde{\psi}, \breve{\psi}$ are similar to these in Theorem 5, but more technical, and we refer to [31] for the precise statements and definitions as well as the proof of the optimal approximation error rate. Here, we simply remark that there exist numerous examples of shearlets ψ, $\tilde{\psi}$, and $\breve{\psi}$ satisfying these conditions, which lead to (9); one large class of examples are separable generators $\psi, \tilde{\psi}, \breve{\psi} \in L^2(\mathbb{R}^3)$, i.e.,

$$\psi(x) = \eta(x_1)\varphi(x_2)\varphi(x_3), \quad \tilde{\psi}(x) = \varphi(x_1)\eta(x_2)\varphi(x_3), \quad \breve{\psi}(x) = \varphi(x_1)\varphi(x_2)\eta(x_3),$$

where $\eta, \varphi \in L^2(\mathbb{R})$ are compactly supported functions satisfying:

(i) $|\hat{\eta}(\omega)| \leq C_1 \cdot \min\{1, |\omega|^\alpha\} \cdot \min\{1, |\omega|^{-\gamma}\}$,
(ii) $\left| \left(\frac{\partial}{\partial \omega}\right)^\ell \hat{\varphi}(\omega) \right| \leq C_2 \cdot \min\{1, |\omega|^{-\gamma}\} \quad$ for $\ell = 0, 1$,

for $\omega \in \mathbb{R}$, where $\alpha > 8$, $\gamma \geq 4$, and C_1, C_2 are constants.

5 Conclusions

Designing a directional representation system that efficiently handles data with anisotropic features is quite challenging since it needs to satisfy a long list of desired properties: it should have a simple mathematical structure, it should provide optimally sparse approximations of certain image classes, it should allow compactly supported generators, it should be associated with fast decomposition algorithms, and it should provide a unified treatment of the continuum and digital realm.

In this paper, we argue that shearlets meet all these challenges, and are, therefore, one of the most satisfying directional systems. To be more precise, let us briefly review our findings for 2D and 3D data:

- *2D Data.* In Sect. 2, we constructed 2D shearlet systems that efficiently capture anisotropic features and satisfy all the above requirements.
- *3D Data.* In 3D, as opposed to 2D, we face the difficulty that there might exist two geometrically different anisotropic features; 1D and 2D singularities. The main difficulty in extending shearlet systems from the 2D to 3D setting lies, therefore, in introducing a system that is able to represent both these geometrically different structures efficiently. As shown in Sect. 4, a class of plate-like shearlets is able to meet these requirements. In other words, the extension from 2D shearlets to 3D shearlets has been successful in terms of preserving the desirable properties, e.g., optimally sparse approximations. It does, therefore, seem that an extension to 4D or even higher dimensions is, if not straightforward then, at the very least, feasible. In particular, the step to 4D now "only" requires the efficient handling of yet "another" type of anisotropic feature.

Acknowledgements The first and third author acknowledge support from DFG Grant SPP-1324, KU 1446/13. The first author also acknowledges support from DFG Grant KU 1446/14.

References

1. J. P. Antoine, P. Carrette, R. Murenzi, and B. Piette, *Image analysis with two-dimensional continuous wavelet transform*, Signal Process. **31** (1993), 241–272.
2. R. H. Bamberger and M. J. T. Smith, *A filter bank for the directional decomposition of images: theory and design*, IEEE Trans. Signal Process. **40** (1992), 882–893.
3. E. J. Candés, L. Demanet, D. Donoho, L. Ying, *Fast discrete curvelet transforms*, Multiscale Model. Simul. **5** (2006), 861–899.
4. E. J. Candés and D. L. Donoho, *Curvelets – a suprisingly effective nonadaptive representation for objects with edges*, in Curve and Surface Fitting: Saint-Malo 1999, edited by A. Cohen, C. Rabut, and L. L. Schumaker, Vanderbilt University Press, Nashville, TN, 2000.
5. E. J. Candés and D. L. Donoho, *New tight frames of curvelets and optimal representations of objects with piecewise C^2 singularities*, Comm. Pure and Appl. Math. **56** (2004), 216–266.
6. S. Dahlke, G. Kutyniok, G. Steidl, and G. Teschke, *Shearlet coorbit spaces and associated Banach frames*, Appl. Comput. Harmon. Anal. **27** (2009), 195–214.

7. S. Dahlke, G. Kutyniok, P. Maass, C. Sagiv, H.-G. Stark, and G. Teschke, *The uncertainty principle associated with the continuous shearlet transform*, Int. J. Wavelets Multiresolut. Inf. Process. **6** (2008), 157–181.
8. S. Dahlke, G. Steidl, and G. Teschke, *The continuous shearlet transform in arbitrary space dimensions*, J. Fourier Anal. Appl. **16** (2010), 340–364.
9. S. Dahlke and G. Teschke, *The continuous shearlet transform in higher dimensions: variations of a theme*, in Group Theory: Classes, Representation and Connections, and Applications, edited by C. W. Danellis, Math. Res. Develop., Nova Publishers, 2010, 167–175.
10. M. N. Do and M. Vetterli, *The contourlet transform: an efficient directional multiresolution image representation*, IEEE Trans. Image Process. **14** (2005), 2091–2106.
11. D. L. Donoho, *Sparse components of images and optimal atomic decomposition*, Constr. Approx. **17** (2001), 353–382.
12. D. L. Donoho and G. Kutyniok, *Geometric separation using a wavelet-shearlet dictionary*, SampTA'09 (Marseille, France, 2009), Proc., 2009.
13. D. L. Donoho and G. Kutyniok, *Microlocal analysis of the geometric separation problem*, preprint.
14. D. L. Donoho, G. Kutyniok, M. Shahram, and X. Zhuang, *A rational design of a digital shearlet transform*, preprint.
15. D. L. Donoho, A. Maleki, M. Shahram, V. Stodden, and I. Ur-Rahman, *Fifteen years of reproducible research in computational harmonic analysis*, Comput. Sci. Engrg. **11** (2009), 8–18.
16. G. Easley, D. Labate, and W.-Q Lim, *Sparse directional image representations using the discrete shearlet transform*, Appl. Comput. Harmon. Anal. **25** (2008), 25–46.
17. R. Eslami and H. Radha, *A new family of nonredundant transforms using hybrid wavelets and directional filter banks*, IEEE Trans. Image Process. **16** (2007), 1152–1167.
18. P. Grohs, *Continuous shearlet frames and resolution of the wavefront set*, Monatsh. Math., to appear. DOI: 10.1007/s00605-010-0264-2.
19. P. Grohs, *Continuous shearlet tight frames*, J. Fourier Anal. Appl. **17** (2011), 506–518.
20. K. Guo, G. Kutyniok, and D. Labate, *Sparse multidimensional representations using anisotropic dilation and shear operators*, in Wavelets and Splines (Athens, GA, 2005), Nashboro Press, Nashville, TN, 2006, 189–201.
21. K. Guo and D. Labate, *Optimally sparse multidimensional representation using shearlets*, SIAM J. Math Anal. **39** (2007), 298–318.
22. K. Guo and D. Labate, *Analysis and detection of surface discontinuities using the 3D continuous shearlet transform*, Appl. Comput. Harmon. Anal. **30** (2011), 231–242.
23. K. Guo, D. Labate, W.-Q Lim, G. Weiss, and E. Wilson, *Wavelets with composite dilations*, Electron. Res. Announc. Amer. Math. Soc. **10** (2004), 78–87.
24. K. Guo, D. Labate, W.-Q Lim, G. Weiss, and E. Wilson, *The theory of wavelets with composite dilations*, Harmonic analysis and applications, Appl. Numer. Harmon. Anal., Birkhäuser Boston, Boston, MA, 2006, 231–250.
25. K. Guo, W.-Q Lim, D. Labate, G. Weiss, and E. Wilson, *Wavelets with composite dilations and their MRA properties*, Appl. Comput. Harmon. Anal. **20** (2006), 220–236.
26. B. Han, G. Kutyniok, and Z. Shen. *A unitary extension principle for shearlet systems*, SIAM J. Numer. Anal., to appear.
27. P. Kittipoom, G. Kutyniok, and W.-Q Lim, *Construction of compactly supported shearlet frames*, Constr. Approx., to appear.
28. P. Kittipoom, G. Kutyniok, and W.-Q Lim, *Irregular shearlet frames: Geometry and approximation properties*, J. Fourier Anal. Appl. **17** (2011), 604–639.
29. G. Kutyniok and D. Labate, *Construction of regular and irregular shearlets*, J. Wavelet Theory and Appl. **1** (2007), 1–10.
30. G. Kutyniok and D. Labate, *Resolution of the wavefront set using continuous shearlets*, Trans. Amer. Math. Soc. **361** (2009), 2719–2754.
31. G. Kutyniok, J. Lemvig, and W.-Q Lim, *Compactly supported shearlet frames and optimally sparse approximations of functions in* $L^2(\mathbb{R}^3)$ *with piecewise* C^α *singularities*, preprint.
32. G. Kutyniok and W.-Q Lim, *Compactly supported shearlets are optimally sparse*, J. Approx. Theory **163** (2011), 1564–1589.

33. G. Kutyniok and W.-Q Lim, *Image separation using shearlets*, in; Curves and Surfaces (Avignon, France, 2010), Lecture Notes in Computer Science, Springer, to appear.
34. G. Kutyniok and W.-Q Lim, *Shearlets on bounded domains*, in Approximation Theory XIII (San Antonio, TX, 2010), Springer, **13** (2012).
35. G. Kutyniok and T. Sauer, *Adaptive directional subdivision schemes and shearlet multiresolution analysis*, SIAM J. Math. Anal. **41** (2009), 1436–1471.
36. G. Kutyniok, M. Shahram, and D. L. Donoho, *Development of a digital shearlet transform based on pseudo-polar FFT*, in Wavelets XIII, edited by V. K. Goyal, M. Papadakis, D. Van De Ville, SPIE Proc. **7446**, SPIE, Bellingham, WA, 2009, 7446-12.
37. D. Labate, W.-Q Lim, G. Kutyniok, and G. Weiss. *Sparse multidimensional representation using shearlets*, in Wavelets XI, edited by M. Papadakis, A. F. Laine, and M. A. Unser, SPIE Proc. **5914**, SPIE, Bellingham, WA, 2005, 254–262,
38. W.-Q Lim, *The discrete shearlet transform: A new directional transform and compactly supported shearlet frames*, IEEE Trans. Image Process. **19** (2010), 1166–1180.
39. E. L. Pennec and S. Mallat, *Sparse geometric image representations with bandelets*, IEEE Trans. Image Process. **14** (2005), 423–438.
40. E. P. Simoncelli, W. T. Freeman, E. H. Adelson, D. J. Heeger, *Shiftable multiscale transforms*, IEEE Trans. Inform. Theory **38** (1992), 587–607.

Shearlets on Bounded Domains

Gitta Kutyniok and Wang-Q Lim

Abstract Shearlet systems have so far been only considered as a means to analyze L^2-functions defined on $\mathbb{R}^2$, which exhibit curvilinear singularities. However, in applications such as image processing or numerical solvers of partial differential equations the function to be analyzed or efficiently encoded is typically defined on a non-rectangular shaped bounded domain. Motivated by these applications, in this paper, we first introduce a novel model for cartoon-like images defined on a bounded domain. We then prove that compactly supported shearlet frames satisfying some weak decay and smoothness conditions, when orthogonally projected onto the bounded domain, do provide (almost) optimally sparse approximations of elements belonging to this model class.

1 Introduction

It is by now well accepted that L^2-functions supported on the unit square which are C^2 except for a C^2 discontinuity curve are a suitable model for images which are governed by edges. Of all directional representation systems which provide optimally sparse approximations of this model class, shearlet systems have distinguished themselves by the fact that they are the only system which provides a unified treatment of the continuum and digital setting, thereby making them particularly useful for both theoretical considerations as well as applications. However, most applications concern sparse approximations of functions on bounded domains, for instance, a numerical solver of a transport dominated equation could seek a solution

Gitta Kutyniok
Department of Mathematics, Technische Universitüt Berlin, 10623 Berlin
e-mail: kutyniok@math.tu-berlin.de

Wang-Q Lim
Department of Mathematics, Technische Universitüt Berlin, 10623 Berlin
e-mail: wlim@uni-osnabrueck.de

M. Neamtu and L. Schumaker (eds.), *Approximation Theory XIII: San Antonio 2010*,
Springer Proceedings in Mathematics 13, DOI 10.1007/978-1-4614-0772-0_11,

on a polygonal shaped area. This calls for shearlet systems which are adapted to bounded domains while still providing optimally sparse expansions.

In this paper, we therefore consider the following questions:

(I) What is a suitable model for a function on a bounded domain with curvilinear singularities?
(II) What is the "correct" definition of a shearlet system for a bounded domain?
(III) Do these shearlet systems provide optimally sparse approximations of the model functions introduced in (I)?

In the sequel we will indeed provide a complete answer to those questions. These results push the door open for the usability of shearlet systems in all areas where 2D functions on bounded domains require efficient encoding.

1.1 Optimally Sparse Approximations of Cartoon-like Images

The first complete model of cartoon-like images has been introduced in [1], the basic idea being that a closed C^2 curve separates smooth – in the sense of C^2 – functions. For the precise definition, we let $\rho : [0,2\pi] \to [0,1]$ be a C^2 function with $\rho(0) = \rho(2\pi)$ and define the set B by

$$B = \{x \in \mathbb{R}^2 : \|x\|_2 \leqslant \rho(\theta),\, x = (\|x\|_2, \theta) \text{ in polar coordinates}\}, \tag{1}$$

where

$$\sup|\rho''(\theta)| \leqslant \nu, \quad \rho \leq \rho_0 < 1. \tag{2}$$

This allows us to introduce $\mathrm{STAR}^2(\nu)$, a class of sets B with C^2 boundaries ∂B and curvature bounded by ν, as well as $\mathcal{E}^2(\nu)$, a class of cartoon-like images.

Definition 1 ([1]). For $\nu > 0$, the set $\mathrm{STAR}^2(\nu)$ is defined to be the set of all $B \subset [0,1]^2$ such that B is a translate of a set obeying (1) and (2). Further, $\mathcal{E}^2(\nu)$ denotes the set of functions f on $\mathbb{R}^2$ with compact support in $[0,1]^2$ of the form

$$f = f_0 + f_1 \chi_B,$$

where $B \in \mathrm{STAR}^2(\nu)$ and $f_0, f_1 \in C^2(\mathbb{R}^2)$ with compact support in $[0,1]^2$ as well as $\sum_{|\alpha|\leq 2} \|D^\alpha f_i\|_\infty \leq 1$ for each $i = 0,1$.

In [4], Donoho proved that for $f \in \mathcal{E}^2(\nu)$, the optimal rate which can be achieved under some restrictions on the representation system as well as on the selection procedure of the approximating coefficients is

$$\|f - f_N\|_2^2 \leqslant C \cdot N^{-2} \quad \text{as } N \to \infty,$$

where f_N is the best N-term approximation.

1.2 Shortcomings of this Cartoon-like Model Class

The first shortcoming of this model is the assumption that the discontinuity curve is C^2. Think, for instance, of an image, which pictures a building. Then the frames of the windows separate the dark interior of the windows from the presumably light color of the wall, however this frame is far from being C^2. Hence, a much more natural assumption would be to assume that the discontinuity curve is piecewise C^2.

The second shortcoming consists in the fact that the function is implicitly assumed to vanish on the boundary of $[0,1]^2$. More precisely, even if the function $f = f_0 + f_1\chi_B$ is non-zero on a section of positive measure of the boundary ∂B, this situation is not particularly treated at all. However, reminding ourselves of the very careful boundary treatment in the theory of partial differential equations, this situation should be paid close attention. Thus, a very natural approach to a careful handling of the boundary in a model for cartoon-like images seems to consist in regarding the boundary as a singularity curve itself.

The third and last shortcoming is the shape of the support $[0,1]^2$ of this model. Typically, in real-world situations the domain of 2D data can be very different from being a rectangle, and even a polygonal-shape model might not necessarily be sufficient. Examples to support this claim can be found, for instance, in fluid dynamics, where the flow can be supported on variously shaped domains. In this regard, a suitable model situation seems to be to allow the boundary to consist of any piecewise C^2 curve.

1.3 Our Model for Cartoon-like Images on Bounded Domains

The model for cartoon-like images on bounded domains, which we now define, will indeed take all considerations from the previous subsection into account. For an illustration, we refer to Fig. 1.

We first introduce $\mathrm{STAR}^2(\nu,L)$, a class of sets B with now piecewise C^2 boundaries ∂B consisting of C^2 smooth pieces whose curvature bounded by ν. This will serve us for both modeling the bounded domain as well as modeling the discontinuity curve.

Definition 2. Let $L > 0$ and let $\rho : [0,2\pi] \to [0,1]$ be a continuous function with $\rho(0) = \rho(2\pi)$. Further, let $[a_i,b_i) \subset [0,2\pi)$, $1 \leqslant i \leqslant L$ be disjoint intervals satisfying

$$\bigcup_{i=1}^{L} [a_i,b_i) = [0,2\pi),$$

and let $\rho_i : [a_i,b_i) \to [0,1]$, $1 \leqslant i \leqslant L$ be C^2 functions such that

$$\max_i \sup |\rho_i''(\theta)| \leq \nu, \quad \max_i \sup |\rho_i| \leq \rho_0 < 1, \quad \text{and} \quad \rho\,\big|_{[a_i,b_i)} = \rho_i.$$

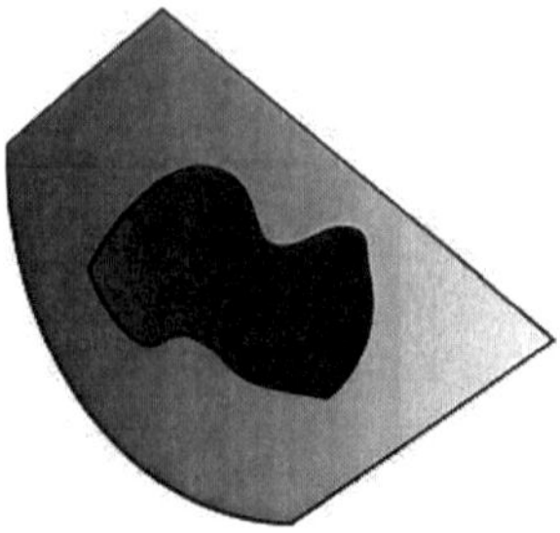

Fig. 1: Example of a function f belonging to our model class $\mathscr{E}^2_{\nu,L}(\Omega)$ of cartoon-like images on bounded domains

Then $B \in \mathrm{STAR}^2(\nu, L)$, if B is a bounded subset of $[0,1]^2$ and B is a translate of a set of the form

$$\{x \in \mathbb{R}^2 : \|x\|_2 \leq \rho(\theta),\, x = (\|x\|_2, \theta) \text{ in polar coordinates}\}.$$

The above definition allows us to introduce a model class of cartoon-like images in bounded domains. In accordance with modeling functions on bounded domains, we now consider functions defined on $[0,1]^2$; its "true" domain is brought into play by requiring these functions to be supported on $\Omega \subseteq (0,1)^2$, which we model as piecewise C^2 bounded. This ensures that we treat $\partial\Omega$ as a singularity curve, which would not have been possible when defining the model on Ω itself.

Definition 3. For $\nu > 0$ and $L \in \mathbb{Z}^+$, let $\Omega, B \in \mathrm{STAR}^2(\nu, L)$ be such that $B \subset \Omega^\circ$, where Ω° denotes the interior of the set Ω, and $\Omega \subset (0,1)^2$. Then, $\mathscr{E}^2_{\nu,L}(\Omega)$ denotes the set of functions f on $[0,1]^2$ with compact support in Ω of the form

$$f = f_0 + f_1 \chi_B,$$

where $f_0, f_1 \in C^2([0,1]^2)$ with compact support in Ω and $\sum_{|\alpha|\leq 2} \|D^\alpha f_i\|_\infty \leq 1$ for each $i = 0, 1$.

Later it will become important to analyze the points on boundaries of sets in $\mathrm{STAR}^2(\nu, L)$, in which the boundary is not C^2. For these points, we will employ the following customarily used notion.

Definition 4. For $\nu > 0$ and $L \in \mathbb{Z}^+$, let $B \in \mathrm{STAR}^2(\nu, L)$. Then a point $x_0 \in \partial B$ will be called a *corner point*, if ∂B is not C^2 at x_0.

Since the model $\mathscr{E}^2_{\nu,L}(\Omega)$, while containing the previous model $\mathscr{E}^2_\nu$ as a special case, is considerably more complicated, we would like to make the reader aware of the fact that it is now not clear at all whether the optimal approximation rate is still

$$\|f - f_N\|_2^2 \leq C \cdot N^{-2} \quad \text{as } N \to \infty.$$

1.4 Review of Shearlets

Inspired by works on generalized wavelets [16, 22], the directional representation system of *shearlets* has recently emerged – a first introduction dates back to 2005 in [20] – and rapidly gained attention due to the fact that, in contrast to other proposed directional representation systems, shearlets provide a unified treatment of the continuum and digital world similar to wavelets. We refer to, e.g., [9, 15] for the continuum theory, [8, 19, 21] for the digital theory, and [5, 10] for recent applications. Shearlets are scaled according to a parabolic scaling law and exhibit directionality by parameterizing slope by shearing, which allows the aforementioned unified treatment in contrast to rotation. Thus, shearlets are associated with three parameters: scale, orientation, and position. A precise definition will be given in Sect. 2.

A few months ago, the theory of shearlets focussed entirely on band-limited generators although precise spatial localization is evidently highly desirable for, e.g., edge detection. Recently, motivated by this desideratum, compactly supported shearlets were studied by Kittipoom and the two authors. It was shown that a large class of compactly supported shearlets generates a frame for $L^2(\mathbb{R}^2)$ with controllable frame bounds alongside with several explicit constructions [12]. By the two authors it was then proven in [18] that a large class of these compactly supported shearlet frames does in fact provide (almost) optimally sparse approximations of functions in $\mathcal{E}^2_\nu$ in the sense of

$$\|f - f_N\|_2^2 \leqslant C \cdot N^{-2} \cdot (\log N)^3 \quad \text{as } N \to \infty.$$

It should be mentioned that although the optimal rate is not completely achieved, the log-factor is typically considered negligible compared to the N^{-2}-factor, wherefore the term "almost optimal" has been adopted into the language.

1.5 Surprising Result

We now aim to discuss the ability of shearlets to sparsely approximate elements of the previously introduced model for cartoon-like images on bounded domains, $\mathcal{E}^2_{\nu,L}(\Omega)$. For this, we first need to define shearlet systems for functions in $L^2(\Omega)$. Assume we are given a (compactly supported) shearlet frame for $L^2(\mathbb{R}^2)$. The most crude approach to transform this into a shearlet system defined on $L^2(\Omega)$, where $\Omega \in \text{STAR}^2(\nu, L)$, is to just truncate each element at the boundary of Ω. Since it is well known in classical frame theory that the orthogonal projection of a frame onto a subspace does not change the frame bounds (cf. [2, Proposition 5.3.5]), this procedure will result in a (compactly supported) shearlet frame for $L^2(\Omega)$ with the same frame bounds as before.

We now apply this procedure to the family of compactly supported shearlet frames for $L^2(\mathbb{R}^2)$, which yield (almost) optimally sparse approximations of functions in $\mathcal{E}^2_\nu$ (see [18, Theorem 1.3]). The main result of this paper then proves that the resulting family of shearlet frames – now regarded as a system on $[0,1]^2$ with compact support in Ω – again provides (almost) optimally sparse approximations now of elements from our model of cartoon-like images on bounded domains $\mathcal{E}^2_{\nu,L}(\Omega)$ in the sense of

$$\|f - f_N\|_2^2 \leqslant C \cdot N^{-2} \cdot (\log N)^3 \quad \text{as } N \to \infty.$$

The precise statement is phrased in Theorem 1 in Sect. 3.

This result is quite surprising in two ways:

- *Surprise 1*. Regarding a log-factor as negligible – a customarily taken viewpoint –, the previous result shows that even for our much more sophisticated model of cartoon-like images on bounded domains the *same* optimal sparse approximation rate as for the simple model detailed in Sect. 1.1 can be achieved. This is even more surprising taking into account that our model contains point singularities at the corner points of the singularity curves. Naively, one would expect that these should worsen the approximation rate. However, observing that "not too many" shearlets intersect these "sparsely occurring" points unravels this mystery.
- *Surprise 2*. Orthogonally projecting a shearlet system onto the considered bounded domain, thereby merely truncating it, seems an exceptionally crude approach to derive shearlets for a bounded domain. However, these "modified" shearlet systems are indeed sufficient to achieve the optimal rate and no sophisticated adaptions are required, which is of significance for deriving fast algorithmic realizations.

1.6 Main Contributions

The main contributions of this paper are two-fold. Firstly, we introduce $\mathcal{E}^2_{\nu,L}(\Omega)$ as a suitable model for a function on a bounded domain with curvilinear singularities. Secondly, we show that the "crude" approach towards a shearlet system on a bounded domain by simply orthogonally projecting still provides optimally sparse approximations of elements belonging to our model class $\mathcal{E}^2_{\nu,L}(\Omega)$.

We should mention that although not formally stated the idea of one piecewise C^2 discontinuity curve in a model for functions on $\mathbb{R}^2$ as an extension of Definition 1 is already lurking in [1]. Also, a brief sketch of proof of (almost) optimally sparse approximations of curvelets is contained therein. These ideas are however very different from ours in two aspects. First of all, our goal is a suitable model for functions on bounded domains exhibiting discontinuity curves and also treating the boundary of the domain as a singularity curve. And secondly, in this paper we consider compactly supported shearlets – hence elements with superior spatial localization properties in contrast to the (band-limited) curvelets – which allows an elegant proof of the sparse approximation result in addition to a simplified treatment of the corner points.

1.7 Outline

In Sect. 2, after recalling the definition of shearlet systems, we introduce shearlet systems on bounded domains, thereby focussing in particular on compactly supported shearlet frames. The precise statement of our main result is presented in Sect. 3 together with a road map to its proof. The proof itself is then carried out in Sect. 4. Finally, in Sect. 5, we discuss our main result and possible extensions of it.

2 Compactly Supported Shearlets

We first review the main notions and definitions related to shearlet theory, focussing in particular on compactly supported generators. For more details we would like to refer the interested reader to the survey paper [17]. Then we present our definition of shearlet systems on a bounded domain $\Omega \in \mathrm{STAR}^2(\nu, L)$.

2.1 Compactly Supported Shearlet Frames for $L^2(\mathbb{R}^2)$

Shearlets are scaled according to a parabolic scaling law encoded in the *parabolic scaling matrices* A_{2^j} or $\tilde{A}_{2^j}$, $j \in \mathbb{Z}$, and exhibit directionality by parameterizing slope encoded in the *shear matrices* S_k, $k \in \mathbb{Z}$, defined by

$$A_{2^j} = \begin{pmatrix} 2^j & 0 \\ 0 & 2^{j/2} \end{pmatrix} \quad \text{or} \quad \tilde{A}_{2^j} = \begin{pmatrix} 2^{j/2} & 0 \\ 0 & 2^j \end{pmatrix}$$

and

$$S_k = \begin{pmatrix} 1 & k \\ 0 & 1 \end{pmatrix},$$

respectively.

We next partition the frequency plane into four cones $\mathscr{C}_1 - \mathscr{C}_4$. This allow the introduction of shearlet systems which treat different slopes equally in contrast to the shearlet group-based approach. We though wish to mention that historically the shearlet group-based approach was developed first due to its advantageous theoretical properties and it still often serves as a system for developing novel analysis strategies (see, for instance, [6, 7, 13]).

The four cones $\mathscr{C}_1 - \mathscr{C}_4$ are now defined by

$$\mathscr{C}_\iota = \begin{cases} \{(\xi_1,\xi_2) \in \mathbb{R}^2 : \xi_1 \geqslant 1, |\xi_2/\xi_1| \leqslant 1\} : \iota = 1, \\ \{(\xi_1,\xi_2) \in \mathbb{R}^2 : \xi_2 \geqslant 1, |\xi_1/\xi_2| \leqslant 1\} : \iota = 2, \\ \{(\xi_1,\xi_2) \in \mathbb{R}^2 : \xi_1 \leqslant -1, |\xi_2/\xi_1| \leqslant 1\} : \iota = 3, \\ \{(\xi_1,\xi_2) \in \mathbb{R}^2 : \xi_2 \leqslant -1, |\xi_1/\xi_2| \leqslant 1\} : \iota = 4, \end{cases}$$

and a centered rectangle

$$\mathscr{R} = \{(\xi_1,\xi_2) \in \mathbb{R}^2 : \|(\xi_1,\xi_2)\|_\infty < 1\}.$$

For an illustration, we refer to Fig. 2a.

The rectangle $\mathscr{R}$ corresponds to the low frequency content of a signal and is customarily represented by translations of some scaling function. Anisotropy comes into play when encoding the high frequency content of a signal which corresponds to the cones $\mathscr{C}_1 - \mathscr{C}_4$, where the cones $\mathscr{C}_1$ and $\mathscr{C}_3$ as well as $\mathscr{C}_2$ and $\mathscr{C}_4$ are treated separately as can be seen in the following

Definition 5. For some sampling constant $c > 0$, the *cone-adapted shearlet system* $SH(\phi,\psi,\tilde{\psi};c)$ generated by a *scaling function* $\phi \in L^2(\mathbb{R}^2)$ and *shearlets* $\psi,\tilde{\psi} \in L^2(\mathbb{R}^2)$ is defined by

$$SH(\phi,\psi,\tilde{\psi};c) = \Phi(\phi;c) \cup \Psi(\psi;c) \cup \tilde{\Psi}(\tilde{\psi};c),$$

where

$$\Phi(\phi;c) = \{\phi_m = \phi(\cdot - cm) : m \in \mathbb{Z}^2\},$$

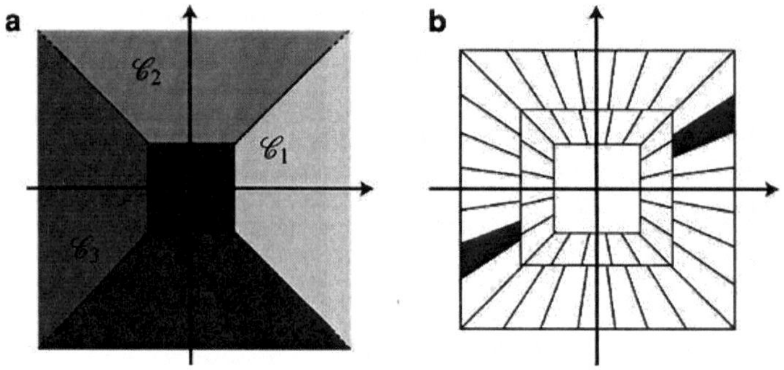

Fig. 2: (**a**) The cones $\mathscr{C}_1 - \mathscr{C}_4$ and the centered rectangle $\mathscr{R}$ in frequency domain. (**b**) The tiling of the frequency domain induced by a cone-adapted shearlet system, where the (essential) support of the Fourier transform of one shearlet generator is exemplary high-lighted

$$\Psi(\psi;c) = \{\psi_{j,k,m} = 2^{3j/4}\psi(S_k A_{2^j} \cdot -cm) : j \geqslant 0, |k| \leqslant \lceil 2^{j/2} \rceil, m \in \mathbb{Z}^2\},$$

and

$$\tilde{\Psi}(\tilde{\psi};c) = \{\tilde{\psi}_{j,k,m} = 2^{3j/4}\tilde{\psi}(S_k^T \tilde{A}_{2^j} \cdot -cm) : j \geqslant 0, |k| \leqslant \lceil 2^{j/2} \rceil, m \in \mathbb{Z}^2\}.$$

The tiling of frequency domain induced by $SH(\phi,\psi,\tilde{\psi};c)$ is illustrated in Fig. 2b. From this illustration, the anisotropic footprints of shearlets contained in $\Psi(\psi;c)$ and $\tilde{\Psi}(\tilde{\psi};c)$ can clearly be seen. The corresponding anisotropic footprints of shearlets *in spatial domain* are of size $2^{-j/2} \times 2^{-j}$.

The reader should keep in mind that although not indicated by the notation, the functions ϕ_m, $\psi_{j,k,m}$, and $\tilde{\psi}_{j,k,m}$ all depend on the sampling constant c. For the sake of brevity, we will often write ψ_λ and $\tilde{\psi}_\lambda$, where $\lambda = (j,k,m)$ index scale, shear, and position. For later use, we further let Λ_j and $\tilde{\Lambda}_j$ be the indexing sets of shearlets in $\Psi(\psi;c)$ and $\tilde{\Psi}(\tilde{\psi};c)$ at scale j, respectively, i.e.,

$$\Psi(\psi;c) = \{\psi_\lambda : \lambda \in \Lambda_j, j = 0,1,\ldots\} \text{ and } \tilde{\Psi}(\tilde{\psi};c) = \{\tilde{\psi}_\lambda : \lambda \in \tilde{\Lambda}_j, j = 0,1,\ldots\}.$$

Finally, we define

$$\Lambda = \bigcup_{j=0}^{\infty} \Lambda_j \quad \text{and} \quad \tilde{\Lambda} = \bigcup_{j=0}^{\infty} \tilde{\Lambda}_j.$$

The shearlet systems $SH(\phi,\psi,\tilde{\psi};c)$ have already been well studied with respect to their frame properties for $L^2(\mathbb{R}^2)$, and we would like to refer to results in [3, 9, 14]. It should be mentioned that those results typically concern frame properties of $\Psi(\psi;c)$, which immediately imply frame properties of $\tilde{\Psi}(\tilde{\psi};c)$ likewise, whereas numerous frame properties for the low-frequency part $\Phi(\phi;c)$ can be found in the wavelet literature. Combining those leads to frame properties of $SH(\phi,\psi,\tilde{\psi};c)$.

Recent results in [12] establish frame properties specifically for the case of spatially compactly supported shearlet systems, i.e., shearlet systems with compactly supported generators ϕ, ψ, and $\tilde{\psi}$ which lead to a shearlet system consisting of compactly supported elements. These results give sufficient conditions for the so-called t_q conditions to be satisfied. As one class of examples with "good" frame bounds, generating shearlets ψ and $\tilde{\psi}$ were chosen to be separable, i.e., of the form $\psi_1(x_1)\cdot\psi_2(x_2)$ and $\psi_1(x_2)\cdot\psi_2(x_1)$, respectively, where ψ_1 is a wavelet and ψ_2 a scaling function both associated with some carefully chosen (maximally flat) low pass filter. The separability has in addition the advantage to lead to fast accompanying algorithms.

We wish to mention that there is a trade-off between *compact support* of the shearlet generators, *tightness* of the associated frame, and *separability* of the shearlet generators. The known constructions of tight shearlet frames do not use separable generators, and these constructions can be shown to *not* be applicable to compactly supported generators. Tightness is difficult to obtain while allowing for compactly supported generators, but we can gain separability, hence fast algorithmic realizations. On the other hand, when allowing non-compactly supported generators, tightness is possible, but separability seems to be out of reach, which makes fast algorithmic realizations very difficult.

2.2 Compactly Supported Shearlet Frames for $L^2(\Omega)$

Let $\Omega \in \mathrm{STAR}^2(\nu, L)$ be a bounded domain as defined in Sect. 1.3. The main idea of constructing a shearlet frame for $L^2(\Omega)$, preferably with compactly supported elements, is to start with a compactly supported shearlet frame for $L^2(\mathbb{R}^2)$ and apply the orthogonal projection from $L^2(\mathbb{R}^2)$ onto $L^2(\Omega)$ to each element. To make this mathematically precise, we let $P_\Omega : L^2(\mathbb{R}^2) \to L^2(\Omega)$ denote the orthogonal projection from $L^2(\mathbb{R}^2)$ onto $L^2(\Omega)$. This allows us to state the following

Definition 6. Let $\Omega \in \mathrm{STAR}^2(\nu,L)$. For some sampling constant $c > 0$, the *cone-adapted shearlet system* $SH_\Omega(\phi,\psi,\tilde{\psi};c)$ for $L^2(\Omega)$ generated by a *scaling function* $\phi \in L^2(\mathbb{R}^2)$ and *shearlets* $\psi, \tilde{\psi} \in L^2(\mathbb{R}^2)$ is defined by

$$SH_\Omega(\phi,\psi,\tilde{\psi};c) = P_\Omega(\Phi(\phi;c) \cup \Psi(\psi;c) \cup \tilde{\Psi}(\tilde{\psi};c)),$$

where $\Phi(\phi;c)$, $\Psi(\psi;c)$, and $\tilde{\Psi}(\tilde{\psi};c)$ are defined as in Definition 5.

As a direct corollary from well known results in frame theory (see [2, Proposition 5.3.5]), we obtain the following result, which clarifies frame properties for systems $SH_\Omega(\phi,\psi,\tilde{\psi};c)$ to the extent to which they are known for systems $SH(\phi,\psi,\tilde{\psi};c)$. In the sequel, we will usually regard $SH_\Omega(\phi,\psi,\tilde{\psi};c)$ as a system defined on $[0,1]^2$ – in accordance with our model $\mathcal{E}^2_{\nu,L}(\Omega)$ – by which we simply mean extension by zero. This system will be sometimes referred to as the *extension of* $SH_\Omega(\phi,\psi,\tilde{\psi};c)$ *to* $[0,1]^2$. The following result also provides frame properties of these systems.

Proposition 1. *Let $c > 0$, let $\phi,\psi,\tilde{\psi} \in L^2(\mathbb{R}^2)$, and let $\Omega \in \mathrm{STAR}^2(\nu,L)$ with positive measure. Then the following statements are equivalent.*

(i) The shearlet system $SH(\phi,\psi,\tilde{\psi};c)$ is a frame for $L^2(\mathbb{R}^2)$ with frame bounds A and B.
(ii) The shearlet system $SH_\Omega(\phi,\psi,\tilde{\psi};c)$ is a frame for $L^2(\Omega)$ with frame bounds A and B.
(iii) The extension of the shearlet system $SH_\Omega(\phi,\psi,\tilde{\psi};c)$ to $[0,1]^2$ is a frame with frame bounds A and B for functions $L^2([0,1]^2)$ with compact support in Ω.

3 Optimal Sparsity of Shearlets on Bounded Domains

We now have all ingredients to formally state the result already announced in Sect. 1.5, which shows that even with the "crude" construction of shearlets on bounded domains *and* the significantly more sophisticated model for cartoon-like images on bounded domains we still obtain (almost) optimally sparse approximations.

3.1 Main Theorem 1

Theorem 1. *Let $c > 0$, and let $\phi,\psi,\tilde{\psi} \in L^2(\mathbb{R}^2)$ be compactly supported. Suppose that, in addition, for all $\xi = (\xi_1,\xi_2) \in \mathbb{R}^2$, the shearlet ψ satisfies*

(i) $|\hat{\psi}(\xi)| \leqslant C_1 \cdot \min(1,|\xi_1|^\alpha) \cdot \min(1,|\xi_1|^{-\gamma}) \cdot \min(1,|\xi_2|^{-\gamma})$, and

(ii) $\left|\frac{\partial}{\partial \xi_2}\hat{\psi}(\xi)\right| \leqslant |h(\xi_1)| \cdot \left(1+\frac{|\xi_2|}{|\xi_1|}\right)^{-\gamma}$,

where $\alpha > 5$, $\gamma \geqslant 4$, $h \in L^1(\mathbb{R})$, *and* C_1 *is a constant, and suppose that the shearlet* $\tilde{\psi}$ *satisfies (i) and (ii) with the roles of* ξ_1 *and* ξ_2 *reversed. Further, let* $\nu > 0, L \in \mathbb{Z}^+$ *and* $\Omega \in \mathrm{STAR}^2(\nu, L)$, *and suppose that* $SH_\Omega(\phi, \psi, \tilde{\psi}; c)$ *forms a frame for* $L^2(\Omega)$.

Then, the extension of the shearlet frame $SH_\Omega(\phi, \psi, \tilde{\psi}; c)$ *to* $[0,1]^2$ *provides (almost) optimally sparse approximations of functions* $f \in \mathscr{E}^2_{\nu,L}(\Omega)$ *in the sense that there exists some* $C > 0$ *such that*

$$\|f - f_N\|_2^2 \leq C \cdot N^{-2} \cdot (\log N)^3 \qquad \text{as } N \to \infty,$$

where f_N *is the nonlinear N-term approximation obtained by choosing the N largest shearlet coefficients of* f.

3.2 Architecture of the Proof of Theorem 1

Before delving into the proof in the following section, we present some preparation as well as describe the architecture of the proof for clarity purposes.

Let now $SH_\Omega(\phi, \psi, \tilde{\psi}; c)$ satisfy the hypotheses in Theorem 1, and let $f \in \mathscr{E}^2_{\nu,L}(\Omega)$. We first observe that, without loss of generality, we might assume the scaling index j to be sufficiently large, since f as well as all frame elements in the shearlet frame $SH_\Omega(\phi, \psi, \tilde{\psi}; c)$ are compactly supported in spatial domain, hence a finite number does not contribute to the asymptotic estimate we aim for. In particular, this means that we do not need to take frame elements from $\Phi(\phi; c)$ into account. Also, we are allowed to restrict our analysis to shearlets $\psi_{j,k,m}$, since the frame elements $\tilde{\psi}_{j,k,m}$ can be handled in a similar way.

We further observe that we can drive the analysis for the frame $SH(\phi, \psi, \tilde{\psi}; c)$ and for the domain $[0,1]^2$ instead, since, by hypothesis, Ω is contained in the interior of $[0,1]^2$, we treat the boundary of Ω as a singularity curve in $[0,1]^2$, and the frame properties are equal as shown in Proposition 1. In this viewpoint, the function to be sparsely approximated vanishes on $[0,1]^2 \setminus \Omega$.

Our main concern will now be to derive appropriate estimates for the shearlet coefficients $\{\langle f, \psi_\lambda \rangle : \lambda \in \Lambda\}$ of f. Letting $|\theta(f)|_n$ denote the nth largest shearlet coefficient $\langle f, \psi_\lambda \rangle$ in absolute value and exploring the frame property of $SH(\phi, \psi, \tilde{\psi}; c)$, we conclude that

$$\|f - f_N\|_2^2 \leq \frac{1}{A} \sum_{n>N} |\theta(f)|_n^2,$$

for any positive integer N, where A denotes the lower frame bound of the shearlet frame $SH(\phi, \psi, \tilde{\psi}; c)$. Thus, for the proof of Theorem 1, it suffices to show that

$$\sum_{n>N} |\theta(f)|_n^2 \leq C \cdot N^{-2} \cdot (\log N)^3 \qquad \text{as } N \to \infty. \tag{3}$$

To derive the anticipated estimate in (3), for any shearlet ψ_λ, we will study two separate cases:

- *Case 1 (The smooth part):* The compact support of the shearlet ψ_λ does not intersect the boundary of the set B (or $\partial\Omega$), i.e., $\text{supp}(\psi_\lambda) \cap (\partial B \cup \partial\Omega) = \emptyset$.
- *Case 2 (The non-smooth part):* The compact support of the shearlet ψ_λ intersects the boundary of the set B (or $\partial\Omega$), i.e., $\text{supp}(\psi_\lambda) \cap (\partial B \cup \partial\Omega) \neq \emptyset$.

Notice that this exact distinction is only possible due to the spatial compact support of all shearlets in the shearlet frame.

In contrast to Case 1, Case 2 will throughout the proof be further subdivided into the situations – which we now do not state precisely, but just give the reader the intuition behind them:

- *Case 2a*. The support of the shearlet intersects only one C^2 curve in $\partial B \cup \partial\Omega$.
- *Case 2b*. The support of the shearlet intersects at least two C^2 curves in $\partial B \cup \partial\Omega$.
  - *Case 2b-1*. The support of the shearlet intersects $\partial B \cup \partial\Omega$ in a corner point.
  - *Case 2b-2*. The support of the shearlet intersects two C^2 curves in $\partial B \cup \partial\Omega$ simultaneously, but does not intersect a corner point.

4 Proof of Theorem 1

In this section, we present the proof of Theorem 1, following the road map outlined in Sect. 3.2. We wish to mention that Case 1 and Case 2a are similar to the proof of (almost) optimally sparse approximations of the class $\mathcal{E}^2(\nu)$ using compactly supported shearlet frames in [18]. However, Case 2b differs significantly from it, since it, in particular, requires a careful handling of the corner points of ∂B and $\partial\Omega$.

In the sequel – since we are concerned with an asymptotic estimate – for simplicity we will often simply use C as a constant although it might differ for each estimate. Also all the results in the sequel are independent on the sampling constant $c > 0$, wherefore we now fix it once and for all.

4.1 Case 1: The Smooth Part

We start with Case 1 which deals with the smooth part of the function f. Without loss of generality, we can consider some $g \in C^2([0,1]^2)$ as a model of the smooth part of f and estimate its shearlet coefficients. The following proposition, which is taken from [18], implies the rate for optimal sparsity. Notice that the hypothesis on ψ of the following result is implied by condition (i) in Theorem 1.

Proposition 2 ([18]). *Let $g \in C^2([0,1]^2)$, and let $\psi \in L^2(\mathbb{R}^2)$ be compactly supported and satisfy*

$$|\hat{\psi}(\xi)| \leqslant C_1 \cdot \min(1,|\xi_1|^\alpha) \cdot \min(1,|\xi_1|^{-\gamma}) \cdot \min(1,|\xi_2|^{-\gamma}) \text{ for all } \xi = (\xi_1,\xi_2) \in \mathbb{R}^2,$$

where $\gamma > 3$, $\alpha > \gamma + 2$, *and* C_1 *is a constant. Then, there exists some* $C > 0$ *such that*

$$\sum_{n>N} |\theta(g)|_n^2 \leqslant C \cdot N^{-2} \qquad \text{as } N \to \infty.$$

This settles Theorem 1 for Case 1.

4.2 Case 2: The Non-Smooth Part

Next, we turn our attention to the non-smooth part, and aim to estimate the shearlet coefficients $\langle f, \psi_\lambda \rangle$ associated with those shearlets ψ_λ whose spatial support intersects the discontinuity curve ∂B or the boundary of the domain Ω. One of the main means of the proof will be the partitioning of the unit cube $[0,1]^2$ into dyadic cubes, picking those which contain such an intersection, and estimating the associated shearlet coefficients. For this, we first need to introduce the necessary notational concepts.

For any scale $j \geqslant 0$ and any grid point $p \in \mathbb{Z}^2$, we let $Q_{j,p}$ denote the dyadic cube defined by

$$Q_{j,p} = [-2^{-j/2}, 2^{-j/2}]^2 + 2^{-j/2} p.$$

Further, let Q_j be the collection of those dyadic cubes $Q_{j,p}$ which intersect $\partial B \cup \partial \Omega$, i.e.,

$$Q_j = \{Q_{j,p} : Q_{j,p} \cap (\partial B \cup \partial \Omega) \neq \emptyset, p \in \mathbb{Z}^2\}.$$

Of interest to us is also the set of shearlet indices, which are associated with shearlets intersecting the discontinuity curve inside some $Q_{j,p} \in Q_j$; hence, for $j \geqslant 0$ and $p \in \mathbb{Z}^2$ with $Q_{j,p} \in Q_j$, we will consider the index set

$$\Lambda_{j,p} = \{\lambda \in \Lambda_j : \operatorname{supp}(\psi_\lambda) \cap Q_{j,p} \cap (\partial B \cup \partial \Omega) \neq \emptyset\}.$$

Finally, for $j \geqslant 0$, $p \in \mathbb{Z}^2$, and $0 < \varepsilon < 1$, we define $\Lambda_{j,p}(\varepsilon)$ to be the index set of shearlets ψ_λ, $\lambda \in \Lambda_{j,p}$, such that the magnitude of the corresponding shearlet coefficient $\langle f, \psi_\lambda \rangle$ is larger than ε and the support of ψ_λ intersects $Q_{j,p}$ at the jth scale, i.e.,

$$\Lambda_{j,p}(\varepsilon) = \{\lambda \in \Lambda_{j,p} : |\langle f, \psi_\lambda \rangle| > \varepsilon\},$$

and we define $\Lambda(\varepsilon)$ to be the index set for shearlets so that $|\langle f, \psi_\lambda \rangle| > \varepsilon$ across all scales j, i.e.,

$$\Lambda(\varepsilon) = \bigcup_{j,p} \Lambda_{j,p}(\varepsilon).$$

The expert reader will have noticed that in contrast to the proofs in [1] and [11], which also split the domain into smaller scale boxes, we do not apply a weight function to obtain a smooth partition of unity. In our case, this is not necessary due to the spatial compact support of the frame elements. Finally, we set

$$S_{j,p} = \bigcup_{\lambda \in \Lambda_{j,p}} \operatorname{supp}(\psi_\lambda),$$

which is contained in a cubic window of size $C \cdot 2^{-j/2}$ by $C \cdot 2^{-j/2}$, hence, is of asymptotically the same size as $Q_{j,p}$. As mentioned earlier, we may assume that j is sufficiently large so that it is sufficient to consider the following two cases:

- *Case 2a.* There is only one edge curve $\Gamma_1 \subset \partial B$ (or $\partial\Omega$) which can be parameterized by $x_1 = E(x_2)$ (or $x_2 = E(x_1)$) with $E \in C^2$ inside $S_{j,p}$. For any $\lambda \in \Lambda_{j,p}$, there exists some $\hat{x} = (\hat{x}_1, \hat{x}_2) \in Q_{j,p} \cap \mathrm{supp}(\psi_\lambda) \cap \Gamma_1$.
- *Case 2b.* There are two edge curves $\Gamma_1, \Gamma_2 \subset \partial B$ (or $\partial\Omega$) which can be parameterized by $x_1 = E(x_2)$ (or $x_2 = E(x_1)$) with $E \in C^2$ inside $S_{j,p}$. For any $\lambda \in \Lambda_{j,p}$, there exist two distinct points $\hat{x} = (\hat{x}_1, \hat{x}_2)$ and $\hat{y} = (\hat{y}_1, \hat{y}_2)$ such that $\hat{x} \in Q_{j,p} \cap \mathrm{supp}(\psi_\lambda) \cap \Gamma_1$ and $\hat{y} \in Q_{j,p} \cap \mathrm{supp}(\psi_\lambda) \cap \Gamma_2$.

In the sequel, we only consider the edge curve ∂B to analyze shearlet coefficients associated with the non-smooth part, since the boundary of the domain Ω can be handled in a similar way; see also our elaboration on the fact that WLOG we can consider the approximation on $[0,1]^2$ rather than Ω in Sect. 3.2.

4.2.1 Case 2a: The Non-Smooth Part

This part was already studied in [18], where an (almost) optimally sparse approximation rate by the class of compactly supported shearlet frames $SH(\phi, \psi, \tilde{\psi}; c)$ under consideration was proven, and we refer to [18] for the precise argumentation. For intuition purposes as well as for later usage, we though state the key estimate, which implies (almost) optimally sparse approximation for Case 2a:

Proposition 3 ([18]). *Let $\psi \in L^2(\mathbb{R}^2)$ be compactly supported and satisfy the conditions (i), (ii) in Theorem 1 and assume that, for any $\lambda \in \Lambda_{j,p}$, there exists some $\hat{x} = (\hat{x}_1, \hat{x}_2) \in Q_{j,p} \cap supp(\psi_\lambda) \cap \partial B$. Let s be the slope[1] of the tangent to the edge curve ∂B at $(\hat{x}_1, \hat{x}_2)$, i.e.,*

- *$s = E'(\hat{x}_2)$, if ∂B is parameterized by $x_1 = E(x_2)$ with $E \in C^2$ in $S_{j,p}$,*
- *$s = (E'(\hat{x}_1))^{-1}$, if ∂B is parameterized by $x_2 = E(x_1)$ with $E \in C^2$ in $S_{j,p}$, and*
- *$s = \infty$, if ∂B is parameterized by $x_2 = E(x_1)$ with $E'(\hat{x}_1) = 0$ and $E \in C^2$ in $S_{j,p}$.*

Then, there exists some $C > 0$ such that

$$|\langle f, \psi_\lambda \rangle| \leq C \cdot 2^{-\frac{9}{4}j}, \qquad \text{if } |s| > \frac{3}{2} \text{ or } |s| = \infty, \tag{4}$$

and

$$|\langle f, \psi_\lambda \rangle| \leq C \cdot \frac{2^{-\frac{3}{4}j}}{|k + 2^{j/2} s|^3}, \qquad \text{if } |s| \leq 3. \tag{5}$$

Similar estimates with ∂B substituted by $\partial\Omega$ hold if, for any $\lambda \in \Lambda_{j,p}$, there exists some $\hat{x} = (\hat{x}_1, \hat{x}_2) \in Q_{j,p} \cap \mathrm{supp}(\psi_\lambda) \cap \partial\Omega$.

---

[1] Notice that here we regard the slope of the tangent to a curve $(E(x_2), x_2)$, i.e., we consider s of a curve $x_1 = sx_2 + b$, say. For analyzing shearlets $\tilde{\psi}_{j,k,m}$, the roles of x_1 and x_2 would need to be reversed.

4.2.2 Case 2b: The Non-Smooth Part

Letting $\varepsilon > 0$, our goal will now be to first estimate $|\Lambda_{j,p}(\varepsilon)|$ and, based on this, derive an estimate for $|\Lambda(\varepsilon)|$. WLOG we might assume $\|\psi\|_1 \leqslant 1$, which implies

$$|\langle f, \psi_\lambda \rangle| \leqslant 2^{-\frac{3}{4}j}.$$

Hence, for estimating $|\Lambda_{j,p}(\varepsilon)|$, it is sufficient to restrict our attention to scales $j \leqslant \frac{4}{3}\log_2(\varepsilon^{-1})$.

As already announced before, we now split Case 2b into the following two sub-cases:

- *Case 2b-1.* The shearlet ψ_λ intersects a corner point, in which two C^2 curves Γ_1 and Γ_2, say, meet (see Fig. 3a).
- *Case 2b-2.* The shearlet ψ_λ intersects two edge curves Γ_1 and Γ_2, say, simultaneously, but it does not intersect a corner point (see Fig. 3b).

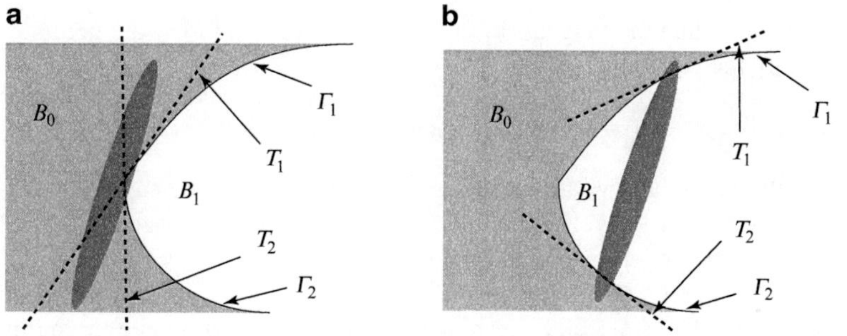

Fig. 3: (**a**) A shearlet ψ_λ intersecting a corner point where two edge curves Γ_1 and Γ_2 meet. T_1 and T_2 are tangents to the edge curves Γ_1 and Γ_2 in this corner point. (**b**) A shearlet ψ_λ intersecting two edge curves Γ_1 and Γ_2 which are a part of the boundary of sets B_0 and B_1. T_1 and T_2 are tangents to the edge curves Γ_1 and Γ_2 in points contained in the support of ψ_λ

Case 2b-1. We first consider *Case 2b-1*. In this case, by a counting argument, it follows that

$$|\Lambda_{j,p}(\varepsilon)| \leq C \cdot 2^{j/2}.$$

Since there are only finitely many corner points with its number not depending on scale $j \geqslant 0$, we have

$$|\Lambda(\varepsilon)| \leq C \cdot \sum_{j=0}^{\frac{4}{3}\log_2(\varepsilon^{-1})} 2^{j/2} \leq C \cdot \varepsilon^{-\frac{2}{3}}.$$

The value $\varepsilon > 0$ can be written as a function of the total number N of coefficients, which yields $\varepsilon(N) \leq C \cdot N^{-\frac{3}{2}}$. This implies that

$$\sum_{n>N} |\theta(f)|_n^2 \leq C \cdot N^{-2},$$

and the optimal sparse approximation rate is proven for *Case 2b-1*.

Case 2b-2. Next, we consider *Case 2b-2*. In this case, WLOG, we might assume that, for any $\lambda \in \Lambda_{j,p}$, there exist two distinct points $\hat{x} = (\hat{x}_1, \hat{x}_2), \hat{y} = (\hat{y}_1, \hat{y}_2)$ such that $\hat{x} \in Q_{j,p} \cap \mathrm{supp}(\psi_\lambda) \cap \Gamma_1$ and $\hat{y} \in Q_{j,p} \cap \mathrm{supp}(\psi_\lambda) \cap \Gamma_2$, and the two edge curves Γ_1 and Γ_2 are parameterized by $x_1 = E(x_2)$ (or $x_2 = E(x_1)$) with $E \in C^2$ inside $S_{j,p}$. We can then write the function $f \in \mathcal{E}^2_{\nu,L}(\Omega)$ as

$$f_0 \chi_{B_0} + f_1 \chi_{B_1} = (f_0 - f_1)\chi_{B_0} + f_1 \qquad \text{on } S_{j,p},$$

where $f_0, f_1 \in C^2([0,1]^2)$ and B_0, B_1 are two disjoint subsets of $[0,1]^2$ (see Fig. 3). By Proposition 2, the rate for optimal sparse approximation is achieved for the smooth part f_1. Thus, it is sufficient to consider $f = g_0 \chi_{B_0}$ with $g_0 = f_0 - f_1 \in C^2([0,1]^2)$.

Assume now that the tangents to the edge curves Γ_1 and Γ_2 at the points $\hat{x}$ and $\hat{y}$ are given by the equations

$$T_1 : x_1 - \hat{x}_1 = s_1(x_2 - \hat{x}_2) \quad \text{and} \quad T_2 : x_1 - \hat{y}_1 = s_2(x_2 - \hat{y}_2),$$

respectively, i.e., s_1 and s_2 are the slopes of the tangents to the edge curves Γ_1 and Γ_2 at $\hat{x}$ and $\hat{y}$, respectively. If the curve Γ_i, $i = 1, 2$, is parameterized by $x_2 = E(x_1)$ with $E'(\hat{x}_1) = 0$, we let $s_i = \infty$ and the tangent is given by $x_2 = \hat{x}_2$ (or $x_2 = \hat{y}_2$) in this case.

Next, for fixed scale j and shear index k, let $N^1_{j,k}(Q_{j,p})$ denote the number of shearlets ψ_λ intersecting Γ_1 in $Q_{j,p}$, i.e.,

$$N^1_{j,k}(Q_{j,p}) = |\{\lambda = (j,k,m) : Q_{j,p} \cap \mathrm{supp}(\psi_\lambda) \cap \Gamma_1 \neq \emptyset\}|,$$

let $N^2_{j,k}(Q_{j,p})$ denote the number of shearlets ψ_λ intersecting Γ_2 in $Q_{j,p}$, i.e.,

$$N^2_{j,k}(Q_{j,p}) = |\{\lambda = (j,k,m) : Q_{j,p} \cap \mathrm{supp}(\psi_\lambda) \cap \Gamma_2 \neq \emptyset\}|,$$

and let $N_{j,k}(Q_{j,p})$ denote the number of shearlets ψ_λ intersecting Γ_1 and Γ_2 in $Q_{j,p}$, i.e.,

$$N_{j,k}(Q_{j,p}) = |\{\lambda = (j,k,m) : Q_{j,p} \cap \mathrm{supp}(\psi_\lambda) \cap \Gamma_1 \neq \emptyset \text{ and } Q_{j,p} \cap \mathrm{supp}(\psi_\lambda) \cap \Gamma_2 \neq \emptyset\}|.$$

Then,

$$N_{j,k}(Q_{j,p}) \leq \min(N^1_{j,k}(Q_{j,p}), N^2_{j,k}(Q_{j,p})). \tag{6}$$

By a counting argument, there exists some $C > 0$ such that

$$N^i_{j,k}(Q_{j,p}) \leq C \cdot 2^{j/2} \qquad \text{for } i = 1, 2, \tag{7}$$

and the form of $\mathrm{supp}(\psi_\lambda)$ implies

$$N^i_{j,k}(Q_{j,p}) \le C \cdot (|2^{j/2}s_i + k| + 1) \qquad \text{for } i = 1,2. \tag{8}$$

We now subdivide into three subcases, namely, $|s_1|, |s_2| \le 2$, and $|s_1| \le 2, |s_2| > 2$ (or vice versa), and $|s_1|, |s_2| > 2$, and show in each case the (almost) optimal sparse approximation rate claimed in Theorem 1. This then finishes the proof.

Subcase $|s_1|, |s_2| \le 2$. In this case, (6) and (8) yield

$$N_{j,k}(Q_{j,p}) \le C \cdot \min(|2^{j/2}s_1 + k| + 1, |2^{j/2}s_2 + k| + 1).$$

We first show independence on the values of s_1 and s_2 within the interval $[-2,2]$. For this, let s and s' be the slopes of the tangents to the edge curve Γ_1 (or Γ_2) at $t \in Q_{j,p} \cap \mathrm{supp}(\psi_\lambda)$ and $t' \in Q_{j,p} \cap \mathrm{supp}(\psi_{\lambda'})$, respectively, with $s \in [-2,2]$. Since Γ_1 (or Γ_2) is C^2, we have $|s - s'| \le C \cdot 2^{-j/2}$, and hence

$$|2^{j/2}s' + k| \le C \cdot (|2^{j/2}s + k| + 1).$$

This implies that the estimate for $N_{j,k}(Q_{j,p})$ asymptotically remains the same, independent of the values of s_1 and s_2. Further, we may assume $s' \in [-3,3]$ for $s \in [-2,2]$, since a scaling index j can be chosen such that $|s - s'|$ is sufficiently small. Therefore, one can apply inequality (5) from Proposition 3 for both points t and t'. In fact, it can be easily checked that one can use (5) with the slope s instead of s' for the point t' (or vice versa); this replacement will not change the asymptotic estimates which we will derive. Thus, we might from now on use universal values for the slopes s_1 and s_2 at each point in $Q_{j,p}$.

Now using (5) from Proposition 3, we have

$$|\langle f, \psi_{j,k,m}\rangle| \le C \cdot \max\left(\frac{2^{-\frac{3}{4}j}}{|2^{j/2}s_1 + k|^3}, \frac{2^{-\frac{3}{4}j}}{|2^{j/2}s_2 + k|^3}\right). \tag{9}$$

Since $\frac{2^{-\frac{3}{4}j}}{|2^{j/2}s_i + k|^3} > \varepsilon$ implies

$$|2^{j/2}s_i + k| < \varepsilon^{-\frac{1}{3}} 2^{-\frac{1}{4}j} \qquad \text{for } i = 1,2,$$

the estimate (9) yields

$$\begin{aligned} |\Lambda_{j,p}(\varepsilon)| &\le C \cdot \sum_{k \in K^1_j(\varepsilon) \cup K^2_j(\varepsilon)} \min(|2^{j/2}s_1 + k| + 1, |2^{j/2}s_2 + k| + 1) \\ &\le C \cdot \sum_{i=1}^{2} \sum_{k \in K^i_j(\varepsilon)} (|2^{j/2}s_i + k| + 1) \\ &\le C \cdot (\varepsilon^{-\frac{1}{3}} 2^{-\frac{1}{4}j} + 1)^2, \end{aligned} \tag{10}$$

where

$$K_j^i(\varepsilon) = \{k \in \mathbb{Z} : |2^{j/2}s_i + k| < C \cdot \varepsilon^{-\frac{1}{3}} 2^{-\frac{1}{4}j}\} \quad \text{for } i = 1,2.$$

By the hypothesis for Case 2b-2, we have $|Q_j| \leqslant C$, where the constant C is independent of scale $j \geqslant 0$. Therefore, continuing (10),

$$|\Lambda(\varepsilon)| \leq C \cdot \sum_{j=0}^{\frac{4}{3}\log_2(\varepsilon^{-1})} |\Lambda_{j,p}(\varepsilon)| \leq C \cdot \varepsilon^{-\frac{2}{3}}.$$

This allows us to write $\varepsilon > 0$ as a function of the total number of coefficients N, which gives

$$\varepsilon(N) \leq C \cdot N^{-\frac{3}{2}}.$$

Thus,

$$\sum_{n>N} |\theta(f)|_n^2 \leq C \cdot N^{-2}, \tag{11}$$

which is the rate we sought.

Subcase $|s_1| \leq 2$ *and* $|s_2| > 2$ *or vice versa.* In this case, (6)–(8) yield

$$N_{j,k}(Q_{j,p}) \leq C \cdot \min(|2^{j/2}s_1 + k| + 1, 2^{j/2}).$$

Again utilizing the fact that the edge curves are C^2, and using similar arguments as in the first subcase, WLOG we can conclude that the slopes s_1, s_2 at each point in $Q_{j,p}$ are greater than $\frac{3}{2}$.

Now, exploiting inequalities (4) and (5) from Proposition 3, we have

$$|\langle f, \psi_{j,k,m}\rangle| \leq C \cdot \max\Big(\frac{2^{-\frac{3}{4}j}}{|2^{j/2}s_1 + k|^3}, 2^{-\frac{9}{4}j}\Big). \tag{12}$$

Since $\frac{2^{-\frac{3}{4}j}}{|2^{j/2}s_1+k|^3} > \varepsilon$ implies

$$|2^{j/2}s_1 + k| < \varepsilon^{-\frac{1}{3}} 2^{-\frac{1}{4}j},$$

and $2^{-\frac{9}{4}j} > \varepsilon$ implies

$$j \leq \frac{4}{9}\log_2(\varepsilon^{-1}),$$

it follows from (12), that

$$\begin{aligned}
|\Lambda(\varepsilon)| &\leq C \cdot \Big(\sum_{j=0}^{\frac{4}{3}\log_2(\varepsilon^{-1})} \sum_{k \in K_j^1(\varepsilon)} (|2^{j/2}s_1 + k| + 1) + \sum_{j=0}^{\frac{4}{9}\log_2(\varepsilon^{-1})} 2^{j/2} \Big) \\
&\leq C \cdot \Big(\sum_{j=0}^{\frac{4}{3}\log_2(\varepsilon^{-1})} (\varepsilon^{-\frac{1}{3}} 2^{-j/4} + 1)^2 + \sum_{j=0}^{\frac{4}{9}\log_2(\varepsilon^{-1})} 2^{j/2} \Big) \\
&\leq C \cdot \varepsilon^{-\frac{2}{3}}.
\end{aligned}$$

The value $\varepsilon > 0$ can now be written as a function of the total number of coefficients N, which gives

$$\varepsilon(N) \leq C \cdot N^{-\frac{3}{2}}.$$

Thus, we derive again the sought rate

$$\sum_{n>N} |\theta(f)|_n^2 \leq C \cdot N^{-2}.$$

Subcase $|s_1| > 2$ *and* $|s_2| > 2$. In this case, (6) and (7) yield

$$N_{j,k}(Q_{j,p}) \leq C \cdot 2^{j/2}.$$

Following similar arguments as before, we again derive the seeked rate (11).

5 Discussion

A variety of applications are concerned with efficient encoding of 2D functions defined on non-rectangular domains exhibiting curvilinear discontinuities, such as, e.g., a typical solution of a transport dominated partial differential equation. As an answer to this problem, our main result, Theorem 1, shows that compactly supported shearlets satisfying some weak decay and smoothness conditions, when orthogonally projected onto a given domain bounded by a piecewise C^2 curve, provide (almost) optimally sparse approximations of functions which are C^2 apart from a piecewise C^2 discontinuity curve. In this model the boundary curve is treated as a discontinuity curve.

Analyzing the proof of Theorem 1, it becomes evident that the presented optimal sparse approximation result for functions in $\mathcal{E}^2_{\nu,L}(\Omega)$ generalizes to an even more encompassing model, which does contain multiple piecewise C^2 possibly intersecting discontinuity curves separating C^2 regions in the bounded domain Ω.

In some applications, it is though of importance to avoid discontinuities at the boundary of the domain. Tackling this question requires further studies to carefully design shearlets near the boundary, and this will be one of our objective for the future.

Acknowledgements Both authors would like to thank Wolfgang Dahmen, Philipp Grohs, Chunyan Huang, Demetrio Labate, Christoph Schwab, and Gerrit Welper for various discussions on related topics, and Jakob Lemvig and Xiaosheng Zhuang for careful reading of an earlier version of this paper. They acknowledge support from DFG Grant SPP-1324, KU 1446/13. The first author also acknowledges support from DFG Grant KU 1446/14.

References

1. E. J. Candès and D. L. Donoho, *New tight frames of curvelets and optimal representations of objects with piecewise C^2 singularities*, Comm. Pure Appl. Math. **56** (2004), 219–266.
2. O. Christensen, *An Introduction to Frames and Riesz Bases*, Birkhauser, Boston (2003).
3. S. Dahlke, G. Kutyniok, G. Steidl, and G. Teschke, *Shearlet Coorbit Spaces and associated Banach Frames*, Appl. Comput. Harmon. Anal. **27** (2009), 195–214.
4. D. L. Donoho, *Sparse components of images and optimal atomic decomposition*, Constr. Approx. **17** (2001), 353–382.
5. D. L. Donoho and G. Kutyniok, *Microlocal Analysis of the Geometric Separation Problems*, preprint (2010).
6. S. Dahlke, G. Steidl, and G. Teschke, *The continuous shearlet transform in arbitrary space dimensions*, J. Fourier Anal. Appl. **16** (2010), 340–364.
7. S. Dahlke and G. Teschke, *The continuous shearlet transform in higher dimensions: variations of a theme*, Group Theory: Classes, Representation and Connections, and Applications (C. W. Danellis, ed.), Mathematics Research Developments, Nova Publishers, 2010.
8. G. Easley, D. Labate, and W.-Q Lim, *Sparse Directional Image Representations using the Discrete Shearlet Transform*, Appl. Comput. Harmon. Anal. **25** (2008), 25–46.
9. K. Guo, G. Kutyniok, and D. Labate, *Sparse Multidimensional Representations using Anisotropic Dilation and Shear Operators*, Wavelets and Splines (Athens, GA, 2005), Nashboro Press, Nashville, TN (2006), 189–201.
10. K. Guo, D. Labate, and W.-Q Lim, *Edge Analysis and identification using the Continuous Shearlet Transform*, Appl. Comput. Harmon. Anal. **27** (2009), 24–46.
11. K. Guo and D. Labate, *Optimally Sparse Multidimensional Representation using Shearlets*, SIAM J. Math Anal. **39** (2007), 298–318.
12. P. Kittipoom, G. Kutyniok, and W.-Q Lim, *Construction of Compactly Supported Shearlets*, Constr. Approx., to appear. (2010).
13. P. Kittipoom, G. Kutyniok, and W.-Q Lim, *Irregular Shearlet Frames: Geometry and Approximation Properties*, J. Fourier Anal. Appl. 17 (2011), 604–639.
14. G. Kutyniok and D. Labate, *Construction of Regular and Irregular Shearlet Frames*, J. Wavelet Theory and Appl. **1** (2007), 1–10.
15. G. Kutyniok and D. Labate, *Resolution of the Wavefront Set using Continuous Shearlets*, Trans. Amer. Math. Soc. **361** (2009), 2719–2754.
16. K. Guo, W.-Q Lim, D. Labate, G. Weiss, and E. Wilson, *Wavelets with composite dilations and their MRA properties*, Appl. Comput. Harmon. Anal. **20** (2006), 220–236.
17. G. Kutyniok, J. Lemvig, and W.-Q Lim, *Compactly Supported Shearlets*, Approximation Theory XIII (San Antonio, TX, 2010), Springer, to appear.
18. G. Kutyniok, and W.-Q Lim, *Compactly Supported Shearlets are Optimally Sparse*, J. Approx. Theory 163 (2011), 1564–1589.
19. G. Kutyniok, M. Shahram, and D. L. Donoho, *Development of a Digital Shearlet Transform Based on Pseudo-Polar FFT*, in Wavelets XIII (San Diego, CA, 2009), D. Van De Ville, V. K. Goyal und M. Papadakis, eds., 74460B-1 - 74460B-13, SPIE Proc. **7446**, SPIE, Bellingham, WA, 2009.
20. D. Labate, W-Q. Lim, G. Kutyniok, and G. Weiss. *Sparse multidimensional representation using shearlets*, Wavelets XI (San Diego, CA, 2005), 254-262, SPIE Proc. 5914, SPIE, Bellingham, WA, 2005.
21. W.-Q Lim, *The Discrete Shearlet Transform : A New Directional Transform and Compactly Supported Shearlet Frames*, IEEE Trans. Image Proc. **19** (2010), 1166–1180.
22. R. S. Laugesen, N. Weaver, G. L. Weiss and E. N. Wilson, *A characterization of the higher dimensional groups associated with continuous wavelets*, J. Geom. Anal. **12** (2002), 89–102.

On Christoffel Functions and Related Quantities for Compactly Supported Measures

D. S. Lubinsky

Abstract Let μ be a compactly supported positive measure on the real line, with associated orthogonal polynomials $\{p_n\}$. Without any global restrictions such as regularity, we discuss convergence in measure for

1. Ratio asymptotics for Christoffel functions
2. The Nevai operators (aka the Nevai condition)
3. Universality limits in the bulk

We also establish convergence a.e. for sufficiently sparse subsequences of Christoffel function ratios.

1 Introduction

Let μ be a positive measure on the real line, with compact support $\text{supp}[\mu]$, and infinitely many points in its support. Then we may define orthonormal polynomials

$$p_n(x) = \gamma_n x^n + \cdots, \quad \gamma_n > 0,$$

satisfying

$$\int p_n p_m \mathrm{d}\mu = \delta_{mn}.$$

The measure μ is said to be *regular* in the sense of Stahl, Totik, and Ullmann [29] if

$$\lim_{n\to\infty} \gamma_n^{1/n} = \frac{1}{cap(\text{supp}[\mu])}, \tag{1}$$

D. S. Lubinsky
School of Mathematics, Georgia Institute of Technology, Atlanta, GA 30332-0160, USA
e-mail: lubinsky@math.gatech.edu

M. Neamtu and L. Schumaker (eds.), *Approximation Theory XIII: San Antonio 2010*, Springer Proceedings in Mathematics 13, DOI 10.1007/978-1-4614-0772-0_12,

where $cap\,(\mathrm{supp}\,[\mu])$ is the logarithmic capacity of the support of μ. In particular, if the support is an interval $[a,b]$, the requirement is that

$$\lim_{n\to\infty} \gamma_n^{1/n} = \frac{4}{b-a}.$$

For definitions of logarithmic capacity, and the associated potential theory, see [22, 23, 29].

At first this particular definition seems technical and obscure – to the extent that one might doubt the utility of the concept. There are numerous equivalent definitions of regularity, but (1) is used because it is relatively direct. An important monograph by Stahl and Totik [29] comprehensively explores regular measures and the asymptotics in of their orthogonal polynomials. More recent analysis appears in [24].

Regularity of a measure is a very weak global requirement. Thus, the Erdős–Turán criterion asserts that if $\mu' > 0$ a.e. in $\mathrm{supp}[\mu]$, then μ is regular. But far less guarantees regularity, and there are pure jump, and pure singularly continuous measures, that are regular.

Here is a very useful equivalent formulation of regularity: μ is regular, iff for every sequence of polynomials $\{P_n\}$, where $\deg(P_n) \leq n$, we have

$$\limsup_{n\to\infty} \left[|P_n(x)| \Big/ \left(\int |P_n|^2 \,\mathrm{d}\mu \right)^{1/2} \right]^{1/n} \leq 1, \tag{2}$$

for quasi every $x \in \mathrm{supp}\,[\mu]$. Here quasi-every means except on a set of capacity 0. When $\mathbb{C} \backslash \mathrm{supp}\,[\mathrm{d}\mu]$ is regular for the Dirichlet problem, one can replace $|P_n(x)|$ by $\|P_n\|_{L_\infty(\mathrm{supp}[\mathrm{d}\mu])}$. Thus, in an nth root sense, the sup norms of polynomials are comparable to their $L_2(\mathrm{d}\mu)$ norms. Regularity of μ also permits asymptotics for $p_n(z)^{1/n}$ outside $\mathrm{supp}[\mu]$, and on $\mathrm{supp}[\mu]$. In particular, regularity is equivalent to

$$\limsup_{n\to\infty} |p_n(x)|^{1/n} = 1 \text{ for quasi every } x \in \mathrm{supp}\,[\mu].$$

Perhaps most surprising of all, is the appearance of this concept in so many orthogonal polynomial asymptotics that have nothing to do with nth root asymptotics. The reason for this often is that regularity of μ permits localization, allowing one to show that one can dispense with the behavior of polynomials outside a given neighborhood of a point (in an appropriate sense and setting of course). This is achieved by using polynomials that decay geometrically away from a given points, together with some version of (2).

This is particularly the case in studying asymptotics of Christoffel functions

$$\lambda_n(\mathrm{d}\mu, x) = \inf_{\deg(P)\leq n-1} \frac{\int P^2 \mathrm{d}\mu}{P^2(x)},$$

where the inf is taken over all polynomials P of degree $\leq n-1$. As is well known,

$$\lambda_n(\mathrm{d}\mu, x) = 1 \Big/ \sum_{j=0}^{n-1} p_j^2(x) = 1/K_n(x,x),$$

where

$$K_n(x,t) = \sum_{j=0}^{n-1} p_j(x)\, p_j(t)$$

is the nth reproducing kernel.

Vili Totik [31], [32] established the following result, which is the single most important asymptotic for Christoffel functions.

Theorem 1. *Let μ be a measure with compact support E. Assume that μ is regular in the sense of Stahl, Totik, and Ullmann. If I is an interval in the support for which*

$$\int_I \log \mu' > -\infty, \tag{3}$$

then for a.e. $x \in I$,

$$\lim_{n\to\infty} n\lambda_n(\mathrm{d}\mu, x) = \frac{\mu'(x)}{\nu_E'(x)}. \tag{4}$$

Here $\nu_E'(x)$ is the density of the *equilibrium measure* ν_E for E. Recall that if E is a compact set in the plane, with positive logarithmic capacity, it has an equilibrium measure ν_E. This is a probability measure with support in E such that the equilibrium potential

$$V^{\nu_E}(z) = \int \log \frac{1}{|t-z|} \mathrm{d}\nu_E(t)$$

satisfies

$$V^{\nu}(z) = -\log\, \mathrm{cap}(E)$$

quasi-everywhere on E. Moreover, this equation holds precisely at every point of E that is regular for the Dirichlet problem for $\mathbb{C}\backslash E$ – the so-called *regular points*. For further orientation, see [22, 23, 29].

In the special case $E = [-1,1]$, $\nu_E'(x) = \frac{1}{\pi\sqrt{1-x^2}}$, and Theorem 1 was established earlier by Maté, Nevai, and Totik [18]. Totik used regularity in localization, which permitted replacing difficult measures μ by locally "nicer" ones. Totik observes in [31] that some sort of global condition like regularity is necessary. He notes that, given any compact set E, properly containing an interval I, one can construct a non-regular measure that satisfies the local Szegő condition (3), but for which (4) fails at every point of I. However, this still leaves open the question as to what global condition is necessary, and whether the Szegő condition (3) is necessary.

Recently, Barry Simon [27] proved that if μ is regular, with compact support E, and $\mu' > 0$ a.e. on an interval I, then

$$\lim_{n\to\infty} \int_I \left| \nu_E' - \frac{\mu'}{n\lambda_n} \right| = 0.$$

An essentially weaker result than asymptotics for Christoffel functions are ratio asymptotics, for example, involving two closely related measures. This study goes back to a celebrated memoir of Nevai [20]. Typically, one might consider a non-negative function g that is integrable with respect to $\mathrm{d}\mu$, and try show that

$$\lim_{n\to\infty} \frac{\lambda_n(g\,\mathrm{d}\mu,x)}{\lambda_n(\mathrm{d}\mu,x)} = g(x), \tag{5}$$

in some sense. Note that this type of limit offers the hope of great generality, as its formulation does not involve equilibrium measures, or properties of the support. In particular, when μ is regular, and $g^{\pm 1}$ are bounded on supp$[\mu]$, while g is continuous at x, then methods pioneered by P. Nevai allow one to establish (5). This subject was further explored by Mate, Nevai, and Totik for orthogonal polynomials on the unit circle [17], and by Lopez [11] for measures on the whole real line.

A recent result of the author [15] shows that, at least for ratio asymptotics of Christoffel functions, it is possible to move beyond the class of regular measures. In fact, (5) holds in measure for arbitrary compactly supported measures:

Theorem 2. *Let μ be a compactly supported measure on the real line with infinitely many points in its support. Let $g:\mathbb{R}\to(0,\infty)$ be a $\mathrm{d}\mu$ measurable function such that $g^{\pm 1}$ are bounded on supp$[\mu]$. Let $\varepsilon>0$. Then, as $n\to\infty$,*

$$meas\left\{x\in\{\mu'>0\}:\left|\frac{\lambda_n(g\,\mathrm{d}\mu,x)}{\lambda_n(\mathrm{d}\mu,x)}-g(x)\right|>\varepsilon\right\}\to 0. \tag{6}$$

Moreover, for every $p>0$,

$$\lim_{n\to\infty}\int_{\{\mu'>0\}}\left|\frac{\lambda_n(g\,\mathrm{d}\mu,x)}{\lambda_n(\mathrm{d}\mu,x)}-g(x)\right|^p \mathrm{d}x=0. \tag{7}$$

Here, of course, $\{\mu'>0\}=\{x:\mu'(x)>0\}$ and *meas* denotes linear Lebesgue measure. The essential feature of this result is the absence of local and global restrictions on μ.

One important application of Totik's Theorem 1 is to universality limits for random matrices in the bulk of the spectrum. This much studied limit takes the form

$$\lim_{n\to\infty}\frac{\tilde{K}_n\left(\xi+\frac{a}{\tilde{K}_n(\xi,\xi)},\xi+\frac{b}{\tilde{K}_n(\xi,\xi)}\right)}{\tilde{K}_n(\xi,\xi)}=\frac{\sin\pi(a-b)}{\pi(a-b)}, \tag{8}$$

uniformly for a,b in compact subsets of the real line. Here, ξ lies in the interior of supp$[\mu]$, and

$$\tilde{K}_n(s,t)=\mu'(s)^{1/2}\mu'(t)^{1/2}K_n(s,t)$$

is a normalized form of the reproducing kernel. Quite often, we remove the normalization from the outer K_n, so that (8) takes the form

$$\lim_{n\to\infty}\frac{K_n\left(\xi+\frac{a}{\tilde{K}_n(\xi,\xi)},\xi+\frac{b}{\tilde{K}_n(\xi,\xi)}\right)}{K_n(\xi,\xi)}=\frac{\sin\pi(a-b)}{\pi(a-b)}, \tag{9}$$

with a,b now lying in compact subsets of the complex plane.

The limits (8) and (9) arise in describing the correlation of spacings of eigenvalues of $n \times n$ Hermitian matrices with random entries. A probability distribution is placed on the space of such matrices, with a probability density that is related to the measure μ above. There are many settings for universality limits. In the most important cases, the fixed measure μ is replaced by measures that change with n. See [2, 3, 5–7, 9, 10, 19, 26, 28, 30] for further orientation.

One of the biggest challenges is to determine the minimal conditions on μ that permit the universality limit (8). This has been intensively investigated in recent years, with important advances in [1, 8, 12–14, 25, 32]. To date, the most general result for fixed measures is due to Totik, and uses Theorem 1 above, as well as its method of proof:

Theorem 3. *Let μ be a measure with compact support. Assume that μ is regular. If I is an interval in the support for which*

$$\int_I \log \mu' > -\infty,$$

then for a.e. $\xi \in I$, (8) *holds uniformly for a, b in compact subsets of the real line.*

Totik established this theorem using asymptotics for Christoffel functions, an inequality of the author, and the method of polynomial pullbacks. That allows one to pass from supp$[\mu]$ consisting of a single interval to several intervals, and then to general compact sets. Simon proved related results using Jost functions [25].

The drawback of this theorem is the global assumption of regularity, even though this is a weak global assumption. The author [12] came up with an alternative method to establish (9) that avoids the assumption of regularity. Its basic hypothesis is that (9) holds for $b = a$, that is,

$$\lim_{n\to\infty} \frac{K_n\left(\xi + \frac{a}{\tilde{K}_n(\xi,\xi)}, \xi + \frac{a}{\tilde{K}_n(\xi,\xi)}\right)}{K_n(\xi,\xi)} = 1, \tag{10}$$

for all real a, together with some local hypothesis, such as μ' bounded above and below in some interval. Note that (10) can be reformulated as a ratio asymptotic for Christoffel functions,

$$\lim_{n\to\infty} \frac{\lambda_n(\mathrm{d}\mu,\xi)}{\lambda_n\left(\mathrm{d}\mu, \xi + \frac{a}{\tilde{K}_n(\xi,\xi)}\right)} = 1. \tag{11}$$

This ought to be easier to establish than (8), because $\lambda_n(\mathrm{d}\mu, x)$ (or $K_n(x,x)$ along the "diagonal") admits an extremal property. Unfortunately, there do not seem to be any techniques that establish (11) without first establishing the much stronger limit (4) in Totik's Theorem 1.

Recently, the author [16] has established that for arbitrary measures with compact support, universality holds in measure:

Theorem 4. *Let μ be a measure with compact support and with infinitely many points in the support. Let $\varepsilon > 0$ and $r > 0$. Then as $n \to \infty$,*

$$\operatorname{meas}\left\{ \xi \in \{\mu' > 0\} : \right.$$
$$\left. \sup_{|u|,|v| \le r} \left| \frac{K_n\left(\xi + \frac{u}{\tilde{K}_n(\xi,\xi)}, \xi + \frac{u}{\tilde{K}_n(\xi,\xi)}\right)}{K_n(\xi,\xi)} - \frac{\sin \pi (u-v)}{\pi (u-v)} \right| \ge \varepsilon \right\}$$
$$\to 0 \text{ as } n \to \infty. \tag{12}$$

Using the standard equivalence between convergence in measure, and subsequences that converge a.e., one deduces:

Corollary 1. *Assume the hypotheses of Theorem 4. Let $\mathscr{S}$ be an infinite sequence of positive integers. Then there is a subsequence $\mathscr{S}'$ of $\mathscr{S}$ such that for a.e. $\xi \in \{\mu' > 0\}$,*

$$\lim_{n \to \infty, n \in \mathscr{S}'} \frac{K_n\left(\xi + \frac{a}{\tilde{K}_n(\xi,\xi)}, \xi + \frac{b}{\tilde{K}_n(\xi,\xi)}\right)}{K_n(\xi,\xi)} = \frac{\sin \pi (a-b)}{\pi (a-b)}, \tag{13}$$

uniformly for a, b in compact subsets of the plane.

The proof of Theorem 4 is complicated. It depends on a uniqueness theorem for the sinc kernel, on maximal functions, and Hilbert transforms, and the theory of entire functions of exponential type.

It is no coincidence that convergence in measure is the conclusion in Theorems 2 and 4. Both depend heavily on upper bounds for the reproducing kernel K_n that are true outside sets of small measure. The latter depend on bounds on Green's functions associated with $\mathbb{C} \backslash E$, where E is an arbitrary compact subset of the real line.

Another key tool in both Theorems 2 and 4 is an estimate for the tail integral

$$\Psi_n(x,r) = \frac{\int_{|t-x| \ge \frac{r}{\tilde{K}_n(x,x)}} K_n(x,t)^2 \, d\mu(t)}{K_n(x,x)}, \quad r > 0. \tag{14}$$

Here, if $\mu'(x) = 0$, or does not exist, we set $\Psi_n(x,r) = 0$. Also, let

$$A_n(x) = p_{n-1}^2(x) + p_n^2(x) \tag{15}$$

and define the maximal function

$$\mathscr{M}[d\nu](x) = \sup_{h>0} \frac{1}{2h} \int_{x-h}^{x+h} d\nu$$

for positive measures ν on the real line. In [15], we showed that for a.e. $x \in \operatorname{supp}[\mu]$,

$$\Psi_n(x,r) \le \frac{8}{r} \left(\frac{\gamma_{n-1}}{\gamma_n} \mathscr{M}[A_n d\mu](x) \right)^2.$$

Using the classical weak (1,1) estimate for maximal functions readily yields, for $r, \varepsilon > 0$,

$$meas\{x \in \text{supp}[\mu] : \Psi_n(x,r) \geq \varepsilon\} \leq \frac{\gamma_{n-1}}{\gamma_n}\frac{17}{\sqrt{r\varepsilon}}. \tag{16}$$

This estimate has some applications to what Barry Simon calls the *Nevai condition*. One way to formulate this involves the Nevai operators $\{G_n\}$. Given a function f that is integrable with respect to $d\mu$, we define

$$G_n[d\mu, f](x) = \frac{\int K_n^2(x,t) f(t) d\mu(t)}{K_n(x,x)}.$$

The Nevai condition at x is that

$$\lim_{n\to\infty} G_n[d\mu, f](x) = f(x) \tag{17}$$

for every continuous f. Paul Nevai [20] introduced the operators $\{G_n\}$ as a means to establishing the ratio asymptotic (5) for Christoffel functions.

A very interesting recent result of Breuer, Last, and Simon [4], relates the Nevai condition to sub-exponential growth of orthogonal polynomials:

Theorem 5. *Assume that*

$$0 < \inf_n \frac{\gamma_{n-1}}{\gamma_n} \leq \sup_n \frac{\gamma_{n-1}}{\gamma_n} < \infty. \tag{18}$$

Then (17) *holds at x for every continuous compactly supportly function f iff*

$$\lim_{n\to\infty} \frac{p_n^2(x)}{\sum_{j=0}^n p_j^2(x)} = 0. \tag{19}$$

An equivalent formulation of (19) is that

$$\lim_{n\to\infty} \frac{\lambda_{n-1}(d\mu, x)}{\lambda_n(d\mu, x)} = 1.$$

Sub-exponential growth of orthogonal polynomials has been studied intensively over the years [20, 21]. It was Nevai and his collaborators who showed that when the measure μ has $[-1,1]$ as its essential support, and its recurrence coefficients have appropriate limits, then (19) is true throughout $[-1,1]$. In particular, this is true when $\mu' > 0$ a.e. in $[-1,1]$. More recently, Breuer, Last, and Simon [4] constructed an example of a regular measure with support $[-2,2]$ such that (19) fails at every point of (for example) $[1,2]$. Nevertheless, they formulated the following:

Conjecture 1. Let μ have compact support. The Nevai condition (19) holds for $d\mu$ a.e. $x \in \text{supp}[d\mu]$.

Here, we shall prove the following simple:

Theorem 6. *Let μ be compactly supported with infinitely many points in its support. Let $\{n_k\}_{k=1}^{\infty}$ be an increasing sequence of positive integers with*

$$\sum_{k=1}^{\infty} \frac{1}{n_k} < \infty. \tag{20}$$

(a) Then for Lebesgue a.e. $x \in \{\mu' > 0\}$,

$$\lim_{k\to\infty} A_{n_k}(x) / K_{n_k}(x,x) = 0. \tag{21}$$

(b) Let $f : \mathbb{R} \to \mathbb{R}$ be continuous and of compact support. Then for Lebesgue a.e. $x \in \{\mu' > 0\}$,

$$\lim_{k\to\infty} G_{n_k}[f](x) = f(x). \tag{22}$$

(c) Let $g : \mathbb{R} \to \mathbb{R}$ be continuous, of compact support, and positive on supp$[\mu]$. *Then for Lebesgue a.e. $x \in \{\mu' > 0\}$,*

$$\lim_{k\to\infty} \frac{\lambda_{n_k}(g\,\mathrm{d}\mu, x)}{\lambda_{n_k}(\mathrm{d}\mu, x)} = g(x). \tag{23}$$

Recall that A_n was defined at (15). Since a sequence of functions converges in measure iff every subsequence contains another subsequence that converges a.e., Theorem 6(c) has the following consequence: as $n \to \infty$, $\frac{\lambda_n(g\,\mathrm{d}\mu,\cdot)}{\lambda_n(\mathrm{d}\mu,\cdot)} \to g$ in measure in $\{\mu' > 0\}$. This provides an alternative, and simpler, proof of the special case of Theorem 2 in which g is continuous.

We shall discuss the application of (16) to estimates of $A_n(x)/K_n(x,x)$, and prove Theorem 6 in the next section.

2 Proof of Theorem 6

We may assume that supp$[\mu]$ is contained in $[-1,1]$. Maté, Nevai, and Totik [18] proved, without any further restrictions on μ, that

$$\begin{aligned}\limsup_{n\to\infty} n\lambda_n(\mathrm{d}\mu, x) &\le \mu'(x)/\nu_{[-1,1]}(x) \\ &= \pi\sqrt{1-x^2}\mu'(x)\end{aligned}$$

for a.e. $x \in \mathrm{supp}[\mu]$. It follows that if we let

$$\mathcal{S}_j = \{\mu' > 0\} \cap \{x : n\lambda_n(\mathrm{d}\mu, x) \le 4\mu'(x) \text{ for all } n \ge j\},$$

then

$$\mathcal{G} = \mathrm{supp}[\mu] \setminus \bigcup_{j=1}^{\infty} \mathcal{S}_j \text{ has meas}(\mathcal{G}) = 0. \tag{24}$$

Note that

$$x \in \mathscr{S}_j \Rightarrow \frac{n}{\tilde{K}_n(x,x)} \leq 4 \text{ for all } n \geq j. \tag{25}$$

We recall that $\Psi_n(x,r)$ was defined by (14). We also let

$$\Omega_n(x,r) = \frac{\int_{|t-x|\geq r} K_n^2(x,t)\,\mathrm{d}\mu(t)}{K_n(x,x)}, \tag{26}$$

and d denotes the diameter of $\mathrm{supp}[\mu]$.

Our first estimate is a consequence of (16):

Lemma 1. *Let $n \geq j \geq 1$, and $\varepsilon, r > 0$.*

(a)

$$\mathrm{meas}(\mathscr{S}_j \cap \{x : \Omega_n(x,r) \geq \varepsilon\}) \leq \frac{\gamma_{n-1}}{\gamma_n}\frac{34}{\sqrt{nr\varepsilon}}. \tag{27}$$

(b)

$$\mathrm{meas}\left(\mathscr{S}_j \cap \left\{x : \frac{A_n(x)}{K_n(x,x)} \geq \varepsilon\right\}\right) \leq \left(\frac{\gamma_{n-1}}{\gamma_n}\right)^{-1/2}\frac{60d}{\sqrt{n}\varepsilon^{3/4}}. \tag{28}$$

Proof. (a) For $x \in \mathscr{S}_j$ and $n \geq j$, we have by (25),

$$\Omega_n(x,r) \leq \frac{\int_{|t-x|\geq r\frac{n}{4\tilde{K}_n(x,x)}} K_n^2(x,t)\,\mathrm{d}\mu(t)}{K_n(x,x)} = \Psi_n\left(x, \frac{nr}{4}\right).$$

Thus,

$$\begin{aligned}\mathrm{meas}(\mathscr{S}_j \cap \{x : \Omega_n(x,r) \geq \varepsilon\}) &\leq \mathrm{meas}\left(\mathscr{S}_j \cap \left\{x : \Psi_n\left(x, \frac{nr}{4}\right) \geq \varepsilon\right\}\right)\\ &\leq \frac{\gamma_{n-1}}{\gamma_n}\frac{17(2)}{\sqrt{nr\varepsilon}},\end{aligned}$$

by (16).

(b) We use an idea of Breuer, Last, and Simon [4]: from the Christoffel–Darboux formula, and orthogonality,

$$\int (t-x)^2 K_n^2(x,t)\,\mathrm{d}\mu(t) = \left(\frac{\gamma_{n-1}}{\gamma_n}\right)^2 A_n(x). \tag{29}$$

Then, given $\eta > 0$, we see that

$$\begin{aligned}\left(\frac{\gamma_{n-1}}{\gamma_n}\right)^2 \frac{A_n(x)}{K_n(x,x)} &\leq \eta^2 \frac{\int_{|t-x|\leq\eta} K_n^2(x,t)\,\mathrm{d}\mu(t)}{K_n(x,x)} + d^2 \frac{\int_{|t-x|>\eta} K_n^2(x,t)\,\mathrm{d}\mu(t)}{K_n(x,x)}\\ &\leq \eta^2 + \mathrm{d}^2\Omega_n(x,\eta).\end{aligned}$$

Then,

$$\left(\frac{\gamma_{n-1}}{\gamma_n}\right)^2\frac{A_n(x)}{K_n(x,x)}\geq 2\eta^2\Rightarrow \mathrm{d}^2\Omega_n(x,\eta)\geq\eta^2$$

so

$$\begin{aligned}&\operatorname{meas}\left(\mathscr{S}_j\cap\left\{x:\frac{A_n(x)}{K_n(x,x)}\geq\left(\frac{\gamma_{n-1}}{\gamma_n}\right)^{-2}2\eta^2\right\}\right)\\&\leq\operatorname{meas}\left(\mathscr{S}_j\cap\left\{x:\Omega_n(x,\eta)\geq\frac{\eta^2}{d^2}\right\}\right)\leq\frac{\gamma_{n-1}}{\gamma_n}\frac{34d}{\sqrt{n}\eta^{3/2}},\end{aligned}$$

by (a). Now make the substitution

$$\varepsilon=\left(\frac{\gamma_{n-1}}{\gamma_n}\right)^{-2}2\eta^2$$

to obtain (28).

We can obtain an alternative estimate, by more elementary means. It has a larger, better power of n in the denominator, but also a worse power of ε:

Lemma 2. *Let $n\geq j\geq 1$, and $\varepsilon,r>0$. We have*
(a)

$$\operatorname{meas}\left(\mathscr{S}_j\cap\left\{x:\frac{A_n(x)}{K_n(x,x)}\geq\varepsilon\right\}\right)\leq\frac{8}{n\varepsilon}.\tag{30}$$

(b)

$$\operatorname{meas}\left(\mathscr{S}_j\cap\{x:\Omega_n(x,r)\geq\varepsilon\}\right)\leq\frac{8}{n\varepsilon}\left(\frac{1}{r}\frac{\gamma_{n-1}}{\gamma_n}\right)^2.\tag{31}$$

Proof. (a) For $x\in\mathscr{S}_j$, (25) shows that

$$\frac{A_n(x)}{K_n(x,x)}=\frac{A_n(x)\mu'(x)}{n}\frac{n}{\tilde{K}_n(x,x)}\leq\frac{4A_n(x)\mu'(x)}{n},$$

so,

$$\begin{aligned}\operatorname{meas}\left(\mathscr{S}_j\cap\left\{x:\frac{A_n(x)}{K_n(x,x)}\geq\varepsilon\right\}\right)&\leq\operatorname{meas}\left(\mathscr{S}_j\cap\left\{x:A_n(x)\mu'(x)\geq\frac{n\varepsilon}{4}\right\}\right)\\&\leq\frac{4}{n\varepsilon}\int A_n(x)\mu'(x)\,\mathrm{d}x\leq\frac{8}{n\varepsilon}.\end{aligned}$$

(b) From (29),

$$\int_{|t-x|\geq r}K_n^2(x,t)\,\mathrm{d}\mu(t)\leq\int\left(\frac{t-x}{r}\right)^2K_n^2(x,t)\,\mathrm{d}\mu(t)=\left(\frac{\gamma_{n-1}}{r\gamma_n}\right)^2A_n(x),$$

so

$$\Omega_n(x,r) \le \left(\frac{\gamma_{n-1}}{r\gamma_n}\right)^2 \frac{A_n(x)}{K_n(x,x)}.$$

Thus,

$$\begin{aligned}\operatorname{meas}(\mathscr{S}_j \cap \{x : \Omega_n(x,r) \ge \varepsilon\}) &\le \operatorname{meas}\left(\mathscr{S}_j \cap \left\{x : \frac{A_n(x)}{K_n(x,x)} \ge \varepsilon \left(\frac{\gamma_{n-1}}{r\gamma_n}\right)^{-2}\right\}\right) \\ &\le \frac{8}{n\varepsilon}\left(\frac{\gamma_{n-1}}{r\gamma_n}\right)^2.\end{aligned}$$

We turn to the

Proof of Theorem 6. (a) Fix $j \ge 1$, $\varepsilon > 0$, and let

$$\mathscr{E}_n(j,\varepsilon) = \mathscr{S}_j \cap \left\{x : \frac{A_n(x)}{K_n(x,x)} \ge \varepsilon\right\}.$$

For $n \ge j$, Lemma 2(a) gives

$$\operatorname{meas}(\mathscr{E}_n(j,\varepsilon)) \le \frac{8}{n\varepsilon}.$$

Let

$$\mathscr{E}(j,\varepsilon) = \limsup_{k\to\infty} \mathscr{E}_{n_k}(j,\varepsilon) = \bigcap_{\ell=1}^{\infty}\bigcup_{k=\ell}^{\infty} \mathscr{E}_{n_k}(j,\varepsilon).$$

Because of (20), $\mathscr{E}(j,\varepsilon)$ has linear Lebesgue measure 0. For $x \in \{\mu' > 0\} \setminus (\mathscr{E}(j,\varepsilon) \cup \mathscr{G})$, we have for large enough k,

$$\frac{A_{n_k}(x)}{K_{n_k}(x,x)} < \varepsilon.$$

Recall that $\mathscr{G}$ was defined at (24). Then, if

$$\mathscr{E} = \mathscr{G} \cup \bigcup_{j,\ell \ge 1} \mathscr{E}\left(j, \frac{1}{\ell}\right),$$

we see that $\mathscr{E}$ has linear Lebesgue measure 0, and for $x \in \{\mu' > 0\} \setminus \mathscr{E}$,

$$\lim_{k\to\infty} \frac{A_{n_k}(x)}{K_{n_k}(x,x)} = 0.$$

(b) Let

$$\mathscr{F}_n(j,r,\varepsilon) = \mathscr{S}_j \cap \{x : \Omega_n(x,r) \ge \varepsilon\},$$

so that by Lemma 2(b),

$$\text{meas}\left(\mathscr{F}_n(j,r,\varepsilon)\right) \leq \frac{8}{n\varepsilon}\left(\frac{\gamma_{n-1}}{r\gamma_n}\right)^2 \leq \frac{8}{n\varepsilon}\left(\frac{d}{r}\right)^2.$$

Recall that d is the diameter of supp$[\mu]$. Let

$$\mathscr{F}(j,r,\varepsilon) = \limsup_{k\to\infty} \mathscr{F}_{n_k}(j,r,\varepsilon),$$

so that $\mathscr{F}(j,r,\varepsilon)$ has Lebesgue measure 0, by (20) again. For $x \in \{\mu' > 0\} \setminus (\mathscr{G} \cup \mathscr{F}(j,r,\varepsilon))$, we have

$$\Omega_{n_k}(x,r) \leq \varepsilon \text{ for } k \text{ large enough.}$$

Finally, let

$$\mathscr{F} = \mathscr{G} \cup \bigcup_{j,\ell,m\geq 1} \mathscr{F}\left(j,\frac{1}{\ell},\frac{1}{m}\right).$$

Then $\mathscr{F}$ has Lebesgue measure 0, and for $x \in \{\mu' > 0\} \setminus \mathscr{F}$, we have, for each $r > 0$,

$$\lim_{k\to\infty} \Omega_{n_k}(x,r) = 0.$$

Now let f be continuous and of compact support. We see that

$$\begin{aligned} |G_n[\mathrm{d}\mu, f](x) - f(x)| &\leq \frac{1}{K_n(x,x)} \int_{|t-x|\leq r} |f(t) - f(x)| K_n^2(x,t)\,\mathrm{d}\mu(t) \\ &\quad + 2\|f\|_{L_\infty(\mathbb{R})} \Omega_n(x,r) \\ &\leq \sup_{|t-x|\leq r} |f(t) - f(x)| + 2\|f\|_{L_\infty(\mathbb{R})} \Omega_n(x,r). \end{aligned}$$

It follows that for $x \in \{\mu' > 0\} \setminus \mathscr{F}$,

$$\limsup_{k\to\infty} \left|G_{n_k}[\mathrm{d}\mu, f](x) - f(x)\right| \leq \sup_{|t-x|\leq r} |f(t) - f(x)|.$$

As $r > 0$ is arbitrary, continuity of f gives (22).
(c) We use the elementary inequality [20, p. 76]

$$\frac{\lambda_n(g\,\mathrm{d}\mu, x)}{\lambda_n(\mathrm{d}\mu, x)} \leq G_n[\mathrm{d}\mu, g](x).$$

Together with (b), this gives, for a.e. $x \in \{\mu' > 0\}$,

$$\limsup_{k\to\infty} \frac{\lambda_{n_k}(g\,\mathrm{d}\mu, x)}{\lambda_{n_k}(\mathrm{d}\mu, x)} \leq g(x).$$

Replacing $d\mu$ by $g\ d\mu$, and g by g^{-1}, gives for a.e. $x \in \{\mu' > 0\}$ (recall that g is bounded below on the compact set supp $[\mu]$),

$$\limsup_{k\to\infty} \frac{\lambda_{n_k}(d\mu,x)}{\lambda_{n_k}(g\ d\mu,x)} \leq g^{-1}(x).$$

Then, (23) follows. □

Acknowledgements Research supported by NSF grant DMS1001182 and US-Israel BSF grant 2008399.

References

1. A. Avila, Y. Last, and B. Simon, *Bulk universality and clock spacing of zeros for ergodic Jacobi matrices with a.c. spectrum,* Analysis & PDE, 3(2010), 81–108.
2. J. Baik, T. Kriecherbauer, K. T-R. McLaughlin, P.D. Miller, *Uniform Asymptotics for Polynomials Orthogonal with respect to a General Class of Discrete Weights and Universality Results for Associated Ensembles*, Princeton Annals of Mathematics Studies, 2006.
3. P. Bleher and A. Its, *Random matrix models and their applications,* Cambridge University Press, Cambridge, 2001.
4. J. Breuer, Y. Last, B. Simon, *The Nevai Condition*, Constr. Approx., 32 (2010), 221–254.
5. P. Deift, *Orthogonal Polynomials and Random Matrices: A Riemann-Hilbert Approach,* Courant Institute Lecture Notes, Vol. 3, New York University Press, New York, 1999.
6. P. Deift, D. Gioev, *Random Matrix Theory: Invariant Ensembles and Universality,* Courant Institute Lecture Notes, Vol. 18, New York University Press, New York, 2009.
7. P. Deift, T. Kriecherbauer, K. T-R. McLaughlin, S. Venakides and X. Zhou, *Uniform Asymptotics for Polynomials Orthogonal with respect to Varying Exponential Weights and Applications to Universality Questions in Random Matrix Theory*, Communications in Pure and Applied Maths., 52(1999), 1335–1425.
8. E. Findley, *Universality for Regular Measures satisfying Szegő's Condition,* J. Approx. Theory, 155 (2008), 136–154.
9. P. J. Forrester, *Log-gases and Random matrices*, Princeton University Press, Princeton, 2010.
10. Eli Levin and D.S. Lubinsky, *Universality Limits in the Bulk for Varying Measures,* Advances in Mathematics, 219(2008), 743–779.
11. G. Lopez, *Relative Asymptotics for Polynomials Orthogonal on the Real Axis*, Math. USSR. Sbornik, 65(1990), 505–529.
12. D.S. Lubinsky, *Universality limits in the bulk for arbitrary measures on compact sets*, J. d'Analyse Mathematique, 106 (2008), 373–394.
13. D.S. Lubinsky, *A New Approach to Universality Limits involving Orthogonal Polynomials*, Annals of Mathematics, 170(2009), 915–939.
14. D.S. Lubinsky, *Universality Limits for Random Matrices and de Branges Spaces of Entire Functions*, Journal of Functional Analysis, 256(2009), 3688–3729.
15. D.S. Lubinsky, *A Maximal Function Approach to Christoffel Functions and Nevai's Operators*, to appear in Constr. Approx.
16. D.S. Lubinsky, *Bulk Universality Holds in Measure for Compactly Supported Measures*, to appear in J. d'Analyse Mathematique.
17. A. Maté, P. Nevai, V. Totik, *Extensions of Szegő's Theory of Orthogonal Polynomials, II*, Constr. Approx., 3(1987), 51–72.

18. A. Mate, P. Nevai, V. Totik, *Szegő's Extremum Problem on the Unit Circle*, Annals of Mathematics, 134(1991), 433–453.
19. M.L. Mehta, *Random Matrices*, 2nd edn., Academic Press, Boston, 1991.
20. P. Nevai, *Orthogonal Polynomials*, Memoirs of the AMS no. 213 (1979).
21. P. Nevai, V.Totik, and J. Zhang, *Orthogonal Polynomials: their growth relative to their Sums*, J. Approx. Theory, 67(1991), 215–234.
22. T. Ransford, *Potential Theory in the Complex Plane*, Cambridge University Press, Cambridge, 1995.
23. E.B. Saff and V. Totik, *Logarithmic Potentials with External Fields*, Springer, New York, 1997.
24. B. Simon, *Orthogonal Polynomials on the Unit Circle*, Parts 1 and 2, American Mathematical Society, Providence, 2005.
25. B. Simon, *Two Extensions of Lubinsky's Universality Theorem*, J. d'Analyse Mathematique, 105 (2008), 345–362.
26. B. Simon, *The Christoffel-Darboux Kernel*, (in) Perspectives in PDE, Harmonic Analysis and Applications, a volume in honor of V.G. Maz'ya's 70th birthday, Proceedings of Symposia in Pure Mathematics, 79 (2008), 295–335.
27. B. Simon, *Weak Convergence of CD Kernels and Applications*, Duke Math. J., 146 (2009) 305–330.
28. A. Soshnikov, *Universality at the Edge of the Spectrum in Wigner Random Matrices,* Comm. Math. Phys., 207(1999), 697–733.
29. H. Stahl and V. Totik, *General Orthogonal Polynomials*, Cambridge University Press, Cambridge, 1992.
30. T. Tao, V. Vu, *From the Littlewood-Offord Problem to the Circular Law: Universality of the Spectral Distribution of Random Matrices*, Bull. Amer. Math. Soc., 46 (2009), 377–396.
31. V. Totik, *Asymptotics for Christoffel Functions for General Measures on the Real Line*, J. d' Analyse de Mathematique, 81(2000), 283–303.
32. V. Totik, *Universality and fine zero spacing on general sets*, Arkiv för Matematik, 47(2009), 361–391.

Exact Solutions of Some Extremal Problems of Approximation Theory

A.L. Lukashov

Dedicated to the memory of Franz Peherstorfer

Abstract F. Peherstorfer and R. Steinbauer introduced the complex T-polynomials. Recently, their rational analogues appear as Chebyshev – Markov rational functions on arcs with zeros on these arcs. Explicit representation and detailed proof for the particular case of one arc is given here. Author's reminiscences about Franz Peherstorfer are included.

1 Introduction

In this paper we would like to discuss complex T-polynomials which were introduced in paper [13] written by Franz Peherstorfer and his student Robert Steinbauer. These polynomials (and their generalizations) turned out to be very useful in other questions (Bernstein type inequalities, orthogonal polynomials on the unit circle with (quasi)periodic Verblunsky coefficients, compare [4, 5], [14, Chap. 11]).

Recently, they appeared in [8, 15] as Chebyshev polynomials on arcs with zeros on those arcs. It turned out that the case of one arc [13, Example 3.1 (a),(b)] has different applications [1, 10] and was considered there independently of [13, 15]. In that case the solution is closely connected with trigonometric polynomials which were introduced by Videnskii [16] as extremal polynomials for estimating derivatives of trigonometric polynomials on an interval shorter than the period. Rational analogues of the complex T-polynomials appeared firstly for the case of one arc (implicitly, trigonometric counterparts in different form) in [17], for two arcs in [7, Corollary 4](they were represented by elliptic functions) and in [4, Theorem 2] for the case of several arcs (with representations in terms of harmonic measures or in automorphic functions). They appear also (in the case of several arcs) in [9] as

A.L. Lukashov
Fatih University, Istanbul, Turkey, e-mail: alukashov@fatih.edu.tr
Saratov State University, Saratov, Russia, e-mail: LukashovAL@info.sgu.ru

M. Neamtu and L. Schumaker (eds.), *Approximation Theory XIII: San Antonio 2010*, Springer Proceedings in Mathematics 13, DOI 10.1007/978-1-4614-0772-0_13,

Chebyshev – Markov rational functions on arcs with zeros on those arcs. The case of one arc was not considered explicitely in [9] (in addition the paper is available in Russian only), so it seems useful to give detailed proof in that situation.

We consider the functions

$$R_N(z) = \frac{P_N(z)}{\sqrt{D(z)}}, \quad P_N(z) = \prod_{j=1}^{N}(z - z_j), \quad z_j \in \Gamma_{\mathscr{E}}, \quad j = \overline{1,N}, \tag{1}$$

on the set $\Gamma_{\mathscr{E}} = \{z \in \mathbb{C} : z = \mathrm{e}^{\mathrm{i}\varphi}, \varphi \in \mathscr{E}\}$, where $\mathscr{E} = [\alpha_1, \alpha_2] \cup \cdots \cup [\alpha_{2l-1}, \alpha_{2l}]$, $0 \leqslant \alpha_1 < \alpha_2 < \cdots < \alpha_{2l} < 2\pi$, and $D(z)$ is a polynomial of degree $2a$, $z^{-a}D(z) > 0$ for $z \in \Gamma_{\mathscr{E}}$; The branch of square root is chosen by $\sqrt{z^{-a}D(z)} > 0$ for $z \in \Gamma_{\mathscr{E}}$. The class of such functions will be denoted by $\mathscr{R}_N^D(\mathscr{E})$.

Let $\mathscr{T}_N^{(A,B,\mathscr{A})}$ be the class of rational trigonometric functions of the form

$$r_N(\varphi) = \frac{A\cos\frac{N}{2}\varphi + B\sin\frac{N}{2}\varphi + a_1\cos\left(\frac{N}{2} - 1\right)\varphi + \ldots + b_{\left[\frac{N}{2}\right]}\sin\left(\frac{N}{2} - \left[\frac{N}{2}\right]\right)\varphi}{\sqrt{\mathscr{A}(\varphi)}}, \tag{2}$$

where $N \in \mathbb{N}; A, B \in \mathbb{R}, A^2 + B^2 \neq 0$, are fixed numbers; $\mathscr{A}(\varphi)$ is a fixed real trigonometric polynomial of degree $a \leq N$, which is positive on the given finite system of intervals $\mathscr{E}$; $D(\mathrm{e}^{\mathrm{i}\varphi}) = \mathrm{e}^{\mathrm{i}a\varphi}\mathscr{A}(\varphi)$;and let $\mathscr{T}_N^{(A,B,\mathscr{A})}(\mathscr{E})$ be the class of rational trigonometric functions of that form with zeros in $\mathscr{E}$. Besides, we call deviation points of $r_N(\varphi)$ on $\mathscr{E}$ those points where $|r_N(\varphi)|$ attains its maximum value on $\mathscr{E}$. By $T_n(x)$ we shall denote the classical Chebyshev polynomials $T_n(x) = \cos n \arccos x$.

Theorem 1. *[9] Let $g_{\mathscr{E}}(\xi, z)$ be the Green function of the domain $C \setminus \Gamma_{\mathscr{E}}$,*

$$\Gamma_{\mathscr{E}_j} = \{\xi : \xi = \mathrm{e}^{\mathrm{i}\varphi}, \ \varphi \in \mathscr{E}_j = [\alpha_{2j-1}, \alpha_{2j}]\}.$$

If for any j, $j = 1, \ldots, l$, the sum of the harmonic measures of $\Gamma_{\mathscr{E}_j}$ with respect to the zeros of the polynomial $D(z) = \mathrm{e}^{\mathrm{i}a\varphi}\mathscr{A}(\varphi) = \prod_{j=1}^{m^*}(z - z_j)^{m_j}$ is a natural number, namely

$$(N - a)\omega_j(\infty) + \frac{1}{2}\sum_{k=1}^{m^*} m_k \omega_j(z_k) = q_{j-1}^{(N)}, \quad q_{j-1}^{(N)} \in \mathbb{N}, \quad j = 2, \ldots, l,$$

where

$$\varpi(z,x) = \frac{\partial}{\partial x}\omega\left(z, \ \Gamma_{\mathscr{E}} \cap \{\mathrm{e}^{\mathrm{i}\varphi} : \ b \leq \varphi \leq x\}, \mathbb{C} \setminus \Gamma_{\mathscr{E}}\right)$$

is the density of the harmonic measure, then the minimum of the extremal problem

$$\max_{z \in \Gamma_{\mathscr{E}}} |R_N^*(z)| = \min_{R_N \in \mathscr{R}_N^D(\mathscr{E})} \max_{z \in \Gamma_{\mathscr{E}}} |R_N(z)| \tag{3}$$

is attained on the functions $R_N^*(\mathrm{e}^{\mathrm{i}\varphi})$ *given by*

$$A_N^*\varepsilon \mathrm{e}^{\mathrm{i}\frac{N-a}{2}\varphi}\cos\left(\frac{\pi}{2}\int\limits_{\mathscr{E}\cap[b,\varphi]}\Big((N-a)(\varpi(\infty,\xi)+\varpi(0,\xi))+\sum_{j=1}^{m^*}m_j\varpi(z_j,\xi)\Big)\,\mathrm{d}\xi\right),$$

$|\varepsilon|=1$, *with a suitable constant* $A_N^*>0$.

The next theorem is a particular case of Theorem 1, but the calculations with harmonic measures of the arc with respect to finite points are not so easy, and we give here a direct proof using the same method.

Theorem 2. *For* $\mathscr{E}=[-\alpha,\alpha]$ *the minimum in the extremal problem (3) is attained on the functions*

$$\frac{\tau_N(\varphi)}{\sqrt{\mathscr{A}(\varphi)}}=A_N\cos\frac{1}{2}\left(\sum_{j=1}^{m^*}m_j\arccos\frac{\sin^2(\alpha/2)-\sin(\beta_j/2)\sin(\varphi/2)}{(\sin(\varphi/2)-\sin(\beta_j/2))\sin(\alpha/2)}\right.$$
$$\left.+\,(N-a)\arccos\frac{\sin(\varphi/2)}{\sin(\alpha/2)}\right),$$

where $z_j=\mathrm{e}^{\mathrm{i}\beta_j}$, $j=1,\ldots,m^*$, *and* $A_N^*>0$ *is a suitable constant.*

2 Proofs

The following lemmas are used for proving Theorem 2.

Lemma 1. *If*

$$r_N(\varphi)=\frac{A\cos\frac{N}{2}\varphi+B\sin\frac{N}{2}\varphi+a_1\cos\left(\frac{N}{2}-1\right)\varphi+\cdots+b_{[\frac{N}{2}]}\sin\left(\frac{N}{2}-[\frac{N}{2}]\right)\varphi}{\sqrt{\mathscr{A}(\varphi)}},$$

where $A^2+B^2=1$, *is least deviated from zero on* $\mathscr{E}$ *in the class* $\mathscr{T}_N^{(A,B,\mathscr{A})}$, *then there exists some* $\psi\in\mathbb{R}$ *such that*

$$\hat{r}_N(\varphi)=r_N(\varphi+\psi)=\frac{\cos\frac{N}{2}\varphi+\hat{a}_1\cos\left(\frac{N}{2}-1\right)\varphi+\cdots+\hat{b}_{[\frac{N}{2}]}\sin\left(\frac{N}{2}-[\frac{N}{2}]\right)\varphi}{\sqrt{\hat{\mathscr{A}}(\varphi)}},$$

with $\hat{\mathscr{A}}(\varphi)=\mathscr{A}(\varphi+\psi)$, $\hat{a}_i=a_i\cos\psi+b_i\sin\psi$, $\hat{b}_i=b_i\cos\psi-a_i\sin\psi$, $i=12,\ldots,N$, *is least deviated from zero on* $\mathscr{E}+\psi=[-\alpha+\psi,\alpha+\psi]$ *in the class* $\mathscr{T}_N^{(1,0,\mathscr{A})}$.

Proof. It is not difficult to see that for any $\psi \in \mathbb{R}$ the function $r_N(\varphi+\psi)$ can be written in the form

$$r_N(\varphi+\psi) = \frac{\tau_N(\varphi+\psi)}{\sqrt{\mathscr{A}(\varphi+\psi)}},$$

$$\begin{aligned}\tau_N(\varphi+\psi) = &\left(A\cos\frac{N}{2}\psi + B\sin\frac{N}{2}\psi\right)\cos\frac{N}{2}\varphi\\ &+\left(B\cos\frac{N}{2}\psi - A\sin\frac{N}{2}\psi\right)\sin\frac{N}{2}\varphi\\ &+\hat{a}_1\cos\left(\frac{N}{2}-1\right)\varphi + \hat{b}_1\sin\left(\frac{N}{2}-1\right)\varphi\\ &+\cdots+\hat{a}_{\left[\frac{N}{2}\right]}\cos\left(\frac{N}{2}-\left[\frac{N}{2}\right]\right)\varphi + \ \hat{b}_{\left[\frac{N}{2}\right]}\sin\left(\frac{N}{2}-\left[\frac{N}{2}\right]\right)\varphi,\end{aligned}$$

where

$$\hat{a}_k = a_k\cos\left(\frac{N}{2}-k\right)\psi + b_k\sin\left(\frac{N}{2}-k\right)\psi,$$

$$\hat{b}_k = b_k\cos\left(\frac{N}{2}-k\right)\psi - a_k\sin\left(\frac{N}{2}-k\right)\psi, \qquad k=\overline{1,\left[\frac{N}{2}\right]}.$$

It is obvious that $\hat{r}_N(\varphi) := r_N(\varphi+\psi)$ is least deviated from zero on $\mathscr{E}+\psi = [-\alpha+\psi, \alpha+\psi]$ in the class $\mathscr{T}_N^{(A_1,B_1,\mathscr{A})}$, where $A_1 = A\cos\frac{N}{2}\psi + B\sin\frac{N}{2}\psi$, $B_1 = B\cos\frac{N}{2}\psi - A\sin\frac{N}{2}\psi$. Now, choose a ψ such that $A\cos\frac{N}{2}\psi + B\sin\frac{N}{2}\psi = 1$ and $B\cos\frac{N}{2}\psi - A\sin\frac{N}{2}\psi = 0$. Then observe that these equalities are equivalent to $\cos\frac{N}{2}\psi = \frac{A}{\sqrt{A^2+B^2}}$, $\sin\frac{N}{2}\psi = \frac{B}{\sqrt{A^2+B^2}}$. Besides, under our choice of ψ the function

$$\hat{r}_N(\varphi) = \frac{\cos\frac{N}{2}\varphi + \hat{a}_1\cos\left(\frac{N}{2}-1\right)\varphi + \hat{b}_1\sin\left(\frac{N}{2}-1\right)\varphi + \cdots}{\sqrt{\mathscr{A}(\varphi+\psi)}}$$

is least deviated from zero on $\mathscr{E}+\psi$ in the class $\mathscr{T}_N^{(1,0,\mathscr{A})}$. $\square$

Lemma 2. *If on* $\mathscr{E}$

$$r_N^*(\varphi) = \frac{\cos\frac{N}{2}\varphi + a_1^*\cos\left(\frac{N}{2}-1\right)\varphi + \cdots + b^*_{\left[\frac{N}{2}\right]}\sin\left(\frac{N}{2}-\left[\frac{N}{2}\right]\right)\varphi}{\sqrt{\mathscr{A}(\varphi)}} \in \mathscr{T}_N^{(1,0,\mathscr{A})}$$

has the maximal number $N+1$ of deviation points, then it is least deviated from zero on the functions of the class $\mathscr{T}_N^{\mathscr{A}} = \bigcup\limits_{A,B:A^2+B^2=1} \mathscr{T}_N^{(A,B,\mathscr{A})}$.

Proof. To prove that contrary assumption is false, suppose that the function

$$r_N^*(\varphi) = \frac{\cos\frac{N}{2}\varphi + a_1^*\cos\left(\frac{N}{2}-1\right)\varphi + \cdots + b_{\left[\frac{N}{2}\right]}^*\sin\left(\frac{N}{2}-\left[\frac{N}{2}\right]\right)\varphi}{\sqrt{\mathscr{A}(\varphi)}}$$

has maximal number of deviation points on $\mathscr{E}$. Besides, suppose that for some ψ we have $\|r_N^*\| > \|r_{N,\psi}^*\|$, where

$$r_{N,\psi}^*(\varphi) = \mathscr{A}(\varphi)^{-\frac{1}{2}}\left(\cos\psi\cos\frac{N}{2}\varphi + \sin\psi\sin\frac{N}{2}\varphi + a_{1,\psi}^*\cos\left(\frac{N}{2}-1\right)\varphi\right.$$
$$\left. + \cdots + b_{\left[\frac{N}{2}\right],\psi}^*\sin\left(\frac{N}{2}-\left[\frac{N}{2}\right]\right)\varphi\right) \tag{4}$$

is a function in $\mathscr{T}_N^{(\cos\psi,\sin\psi,\mathscr{A})}$, least deviated from zero on $\mathscr{E}$. Then, $\|r_{N,\psi}^*\|$ continuously depends on ψ, and hence we can assume that $\frac{\psi}{2\pi}\in\mathbb{Q}$, i.e. $\psi = \frac{2\pi p}{q}$, $p\in\mathbb{Z}$, $q\in\mathbb{N}$.

Further,

$$\left(\cos\psi\cos\frac{N}{2}\varphi + \sin\psi\sin\frac{N}{2}\varphi\right)^q = \left(\cos\left(\frac{N}{2}\varphi-\psi\right)\right)^q$$
$$= \frac{1}{2^{q-1}}\cos\left(\frac{N}{2}\varphi-\psi\right)q + \cdots,$$

and hence the pairs of leading coefficients of the numerators of the functions $T_q\left(\frac{r_N^*(\varphi)}{\|r_N^*\|}\right)$ and $T_q\left(\frac{r_{N,\psi}^*(\varphi)}{\|r_{N,\psi}^*\|}\right)$ with the denominator $\sqrt{\mathscr{A}^q(\varphi)}$ are equal to $\left(\frac{1}{\|r_N^*\|^q},0\right)$ and $\left(\frac{1}{\|r_{N,\psi}^*\|^q},0\right)$ correspondingly, and their norms are equal to 1. Therefore, the functions

$$\|r_N^*\|^q T_q\left(\frac{r_N^*(\varphi)}{\|r_N^*\|}\right) = r_{Nq}^*(\varphi) \text{ and } \|r_{N,\psi}^*\|^q T_q\left(\frac{r_{N,\psi}^*(\varphi)}{\|r_{N,\psi}^*\|}\right) = r_{Nq,\psi}^*(\varphi)$$

have the same pairs of leading coefficients $(1,0)$ of the numerator, and the following inequality is true:

$$\|r_{Nq}^*\| > \|r_{Nq,\psi}^*\|.$$

On the other hand, in $\mathscr{E}$ the values of $\frac{r_N^*(\varphi)}{\|r_N^*\|}$ cover the interval $[-1,1]$ N times, and hence in $\mathscr{E}$ the values of $r_{Nq}^*(\varphi) = \|r_N^*\|T_q\left(\frac{r_N^*(\varphi)}{\|r_N^*\|}\right)$ cover the interval $[-\|r_{Nq}^*\|, \|r_{Nq}^*\|]$ Nq times, thus this function is least deviated from zero on $\mathscr{E}$ in the class $\mathscr{T}_{Nq}^{(1,0,\mathscr{A}^q)}$. This contradicts the last inequality. □

Proof of Theorem 2. The function

$$x = \frac{\sin \varphi/2}{\sin \alpha/2}$$

gives a one-to-one mapping of $\mathscr{E}$ onto $[-1,1]$. For any algebraic polynomial $P(x)$ of degree n that substitution gives a (half-order) trigonometric polynomial of order $n/2$. Hence, the same substitution into the Chebyshev–Markov rational function deviated least from zero on $[-1,1]$ with given denominator

$$\sqrt{\prod_{j=1}^{m^*}(x-a_j)^{m_j}},$$

where

$$a_j = \frac{\sin \beta_j/2}{\sin \alpha/2},$$

gives a trigonometric function of the type under consideration. Besides, it is deviated least from zero on $\mathscr{E}$ with maximal number of deviation points and all its zeros are on $\mathscr{E}$. Thus, the function $r_N^*(\varphi)$ is a solution of problem (3) from the class $\mathscr{T}_N^{(A,B,\mathscr{A})}(\mathscr{E})$.

Consequently, Lemma 1 implies that there exist some ψ such that the function $\hat{r}_N^*(\varphi) = r_N^*(\varphi+\psi)$ is least deviated from zero on $\mathscr{E}+\psi$ in the class of functions $\mathscr{T}_N^{(1,0,\mathscr{A})}$, and by Lemma 2 it also is least deviated from zero on $\mathscr{E}+\psi$ in the class

$$\mathscr{T}_N^{\mathscr{A}} = \bigcup_{A,B\in\mathbb{R}:\ A^2+B^2=1} \mathscr{T}_N^{(A,B,\mathscr{A})}.$$

Consequently, $r_N^*(\varphi)$ is least deviated from zero on $\mathscr{E}$ in the class $\mathscr{T}_N^{\mathscr{A}}(\mathscr{E})$ as well.

Consider now an arbitrary function of the class

$$\mathscr{T}_N^{\mathscr{A}}(\mathscr{E}) = \{r_N(\varphi):\ r_N(\varphi) = \frac{\tau_N(\varphi)}{\sqrt{\mathscr{A}(\varphi)}},\ \tau_N(\varphi) = A\cos\frac{N}{2}\varphi + B\sin\frac{N}{2}\varphi$$
$$+a_1\cos\left(\frac{N}{2}-1\right)\varphi + \cdots + b_{[\frac{N}{2}]}\sin\left(\frac{N}{2}-\left[\frac{N}{2}\right]\right)\varphi,$$
$$r_N(\varphi_j)=0,\ \varphi_j\in\mathscr{E},\ j=\overline{1,n}\}.$$

Evidently, its numerator can be expressed by its zeros φ_j, $j=\overline{1,N}$, in the form

$$\tau_N(\varphi) = c_\tau \prod_{j=1}^{N} \sin\frac{\varphi-\varphi_j}{2} = c_\tau \frac{1}{(2i)^N} e^{-\frac{i}{2}\sum_{j=1}^{N}\varphi_j} z^{-\frac{N}{2}} \prod_{j=1}^{N}(z-z_j),$$

where $z=\mathrm{e}^{\mathrm{i}\varphi}, z_j=\mathrm{e}^{\mathrm{i}\varphi_j}, j=1,\dots,N$, and $|c_\tau|=1$. Hence, there is some c_1, $|c_1|=1$, such that $\pi_N(z)=c_1\prod_{j=1}^{N}(z-z_j)$ is a self-reciprocal polynomial and $z_j\in\Gamma_{\mathscr{E}}$, $j=\overline{1,N}$.

Besides, the constant c_1 can be explicitly found by the formula $c_1=\frac{(2i)^N}{c_\tau}\mathrm{e}^{\frac{\mathrm{i}}{2}\sum_{j=1}^{N}\varphi_j}$.

Therefore, to the first (leading) and last coefficients of the numerator of the function $\pi_N(z)$ correspond a pair of the coefficients of the polynomial $\tau_N(\varphi)$. The following is the detailed description of the correspondence:

$$\begin{aligned}\frac{1}{(2i)^N}&\mathrm{e}^{-\frac{\mathrm{i}}{2}\sum_{j=1}^{N}\varphi_j}\mathrm{e}^{\mathrm{i}\frac{N}{2}\varphi}+\frac{(-1)^N\mathrm{e}^{\frac{\mathrm{i}}{2}\sum_{j=1}^{N}\varphi_j}}{(2i)^N}\mathrm{e}^{-\mathrm{i}\frac{N}{2}\varphi}+\cdots\\&=\frac{1}{(2i)^N}\left(\mathrm{e}^{-\frac{\mathrm{i}}{2}\sum_{j=1}^{N}\varphi_j}+(-1)^N\mathrm{e}^{\frac{\mathrm{i}}{2}\sum_{j=1}^{N}\varphi_j}\right)\cos\frac{N}{2}\varphi\\&+\frac{i}{(2i)^N}\left(\mathrm{e}^{-\frac{\mathrm{i}}{2}\sum_{j=1}^{N}\varphi_j}-(-1)^N\mathrm{e}^{\frac{\mathrm{i}}{2}\sum_{j=1}^{N}\varphi_j}\right)\sin\frac{N}{2}\varphi+\cdots\end{aligned}$$

$$=\begin{cases}\dfrac{\cos\frac{1}{2}\sum_{j=1}^{2m}\varphi_j}{(-1)^m2^{2m-1}}\cos m\varphi+\dfrac{\sin\frac{1}{2}\sum_{j=1}^{2m}\varphi_j}{(-1)^m2^{2m-1}}\sin m\varphi+\cdots, & N=2m,\\[2ex] \dfrac{-\sin\frac{1}{2}\sum_{j=1}^{2m-1}\varphi_j}{(-1)^{m-1}2^{2m-2}}\cos\frac{2m-1}{2}\varphi+\dfrac{\cos\frac{1}{2}\sum_{j=1}^{2m-1}\varphi_j}{(-1)^{m-1}2^{2m-2}}\sin\frac{2m-1}{2}\varphi+\cdots, & N=2m-1,\\ & m\in\mathbb{N}.\end{cases}\tag{5}$$

Conversely, to each polynomial $P_N(z)=\prod_{j=1}^{N}(z-z_j)$, $z_j\in\Gamma_{\mathscr{E}}$, $j=\overline{1,N}$, with zeros in $\Gamma_{\mathscr{E}}$ is put in correspondence a polynomial $\pi_N(z)=c_{P_N}P_N(z)$, $|c_{P_N}|=1$, to which, in its turn, the above procedure puts in correspondance a trigonometric polynomial $\tau_N(\varphi)$ with a pair of leading coefficients A,B, $A^2+B^2=1$, depending on zeros z_j.

Besides, $r_N^*(\varphi)$ is least deviated from zero on $\mathscr{E}$ in the class $\mathscr{T}_N^{\mathscr{A}}(\mathscr{E})$, and hence it corresponds to the function $R_N^*(z)$, least deviated from zero on $\Gamma_{\mathscr{E}}$ in the class $\mathscr{R}_N^D(\mathscr{E})$, and by (5),

$$\min_{R_N\in\mathscr{R}_N^D(\mathscr{E})}\max_{z\in\Gamma_{\mathscr{E}}}|R_N(z)|=2^{N-1}\min_{r_N\in\mathscr{T}_N^{\mathscr{A}}(\mathscr{E})}\max_{\varphi\in\mathscr{E}}|r_N(\varphi)|.\qquad\square$$

My Friend and Collaborator

One of the goals of this note is to pay a tribute to Franz Peherstorfer (of course reminiscences from below should be done for his obituary [2], but I was late). His role in my professional carrier and his influence on my mathematical interests were very significant.

He was a reviewer of my first serious paper which I submitted to JAT in 1993. By the influence of studies in Master class in geometry in the Netherlands I wrote a lot of words "obviously," "evidently," "clearly", and so on. Franz wrote a very gentle, clever, and patient report asking for clarifying any places of that kind and giving a counterexample to one of them. Stupid guy, I was so disappointed that for a long time I didn't try to revise the manuscript, but finally I rewrote it in a quite different form and it appeared in 1998 [3]. Almost simultaneously with the acceptance letter I got an e-mail from Franz with an invitation to Linz for a couple of weeks to give a short lecture course on automorphic functions and their application to extremal problems in approximation theory. Those two weeks were so beautiful: full of fruitful discussions, nice hiking in the Alps, and good Austrian beers! Franz posed a question: is it possible to generalize results of his seminal paper [11] to the case of several intervals using the automorphic functions approach? After a couple of trips to Linz and long discussions (the question was enlarged by including the results from [12] to the task), we wrote a paper [6].

These trips changed my visions of mathematics a lot, and also my position in home university in Saratov and so on. Unfortunately, our collaboration was not so fruitful after we finished our second joint paper [7], but we had some similar ideas and I hoped to realize them in a suitable time with Franz but alas...

References

1. Arestov V.V., Mendelev A.S.: On the trigonometric polynomials that deviate least from zero. Dokl. Akad. Nauk. **425**, 733–736 (2009)
2. Kroó A., Nevai P., Totik V.: Franz Peherstorfer July 26, 1950–November 27, 2009. J. Approx. Theory. (2010) doi:10.1016/j.jat.2010.03.008
3. Lukashov A.L.: On ChebyshevMarkov rational functions over several intervals. J. Approx. Theory. **95**, 333–352 (1998)
4. Lukashov A.L.: Inequalities for the derivatives of rational functions on several intervals. Izv. Math. **68**, 543–565 (2004)
5. Lukashov A.L.: Circular parameters of polynomials that are orthogonal on several arcs of the unit circle. Sb. Math. **195**, 1639–1663 (2004)
6. Lukashov A.L., Peherstorfer F.: Automorphic orthogonal and extremal polynomials. Canad. J. Math. **55**, 576–608 (2003)
7. Lukashov A.L., Peherstorfer F.: Zeros of polynomials orthogonal on two arcs of the unit circle. J. Approx. Theory. **132**, 42–71 (2005)
8. Lukashov A.L.,Tyshkevich S.V.: Extremal polynomials on arcs of the circle with zeros on these arcs. J. Contemp. Math. Anal. - Arm. Acad. Sci. **44**, 172–179 (2009)
9. Lukashov A.L.,Tyshkevich S.V.: Extremal rational functions on arcs of the circle with zeros on those arcs. Proceedings of Saratov University. Ser.Mathematics,mechanics,informatics. **9**(1), 8–14 (2009)
10. Maergoĭz L. S., Rybakova N. N.: Chebyshev polynomials with a zero set on a circular arc. Dokl. Akad. Nauk. **426**, 26–28 (2009)
11. Peherstorfer F.: Elliptic orthogonal and extremal polynomials. Proc. London Math. Soc. **70**, 605–624 (1995)
12. Peherstorfer F.: On the zeros of orthogonal polynomials: the elliptic case. Constr. Approx. **20**, 377–397 (2004)

13. Peherstorfer F., Steinbauer R.: Orthogonal polynomials on arcs of the unit circle.II. Orthogonal polynomials with periodic reflection coefficients. J. Approx. Theory. **87**, 60–102 (1996)
14. Simon B.: Orthogonal polynomials on the unit circle. Part 2. Spectral theory. American Mathematical Society, Providence, RI (2005)
15. Tyshkevich S. V.: On Chebyshev polynomials on arcs of a circle. Math. Notes. **81**, 851–853 (2007)
16. Videnskij V.S.: Extremal estimate for the derivative of a trigonometric polynomial on an interval shorter than its period. Sov. Math., Dokl. **1**, 5-8 (1960)
17. Videnskii V. S.: Extremal estimates of derivatives of a polynomial on a segment. In: Smirnov V.I. (ed.) Studies of Modern Problems of Constructive Theory of Functions, pp. 98-106. Fizmatgiz, Moscow (1961)

A Lagrange Interpolation Method by Trivariate Cubic C^1 Splines of Low Locality

G. Nürnberger and G. Schneider

Abstract We develop a local Lagrange interpolation method for trivariate cubic C^1 splines. The splines are constructed on a uniform partition consisting of octahedra (with one additional edge) and tetrahedra. The method is 2-local and stable and therefore yields optimal approximation order. The numerical results and visualizations confirm the efficiency of the method.

1 Introduction

Locality and stability are important properties for a Lagrange interpolation method, since they imply optimal approximation order. In recent years, a number of papers appeared where locality is achieved by decomposing the partition on which the spline space is defined into classes with common vertices and edges [3, 4, 6]. The locality of these methods depends on the number of classes in the decomposition. The method for cubic C^1 splines on Freudenthal partitions achieves a locality of 5, while the method for cubic C^1 splines on type-4 partitions is 4-local. For cubic C^1 splines on an arbitrary partition, the locality can be as high as 10.

In this paper, we investigate the problem of how to construct Lagrange interpolation methods of low locality. In other words, which tetrahedral partitions can be decomposed into only few classes. We consider a partition consisting of tetrahedra and octahedra alternately. In order to obtain a tetrahedral partition, we subdivide each octahedron into four tetrahedra by adding one additional edge. For this partition, we construct a local Lagrange interpolation method which is 2-local.

It is also desirable to construct a partition where all tetrahedra are almost the same size and shape. However, there is a basic problem. It is well known that the space

G. Nürnberger and G. Schneider
University of Mannheim, Mannheim, Germany
e-mail: nuern@staff.mail.uni-mannheim.de,
gschneid@rumms.uni-mannheim.de

M. Neamtu and L. Schumaker (eds.), *Approximation Theory XIII: San Antonio 2010*, Springer Proceedings in Mathematics 13, DOI 10.1007/978-1-4614-0772-0_14,

$\mathbb{R}^3$ cannot be decomposed into tetrahedra with edges of the same length [2]. The partition that is used in this paper consists of tetrahedra which are congruent. Moreover, the edges of these tetrahedra are nearly of the same length.
The paper is organized as follows. In Sect. 2, we recall some basic facts about spline spaces on tetrahedral partitions and the Bernstein–Bézier representation of splines. Section 3 deals with the partition on which the spline space is defined. We introduce some notation and describe the decomposition of the partition. We briefly recall some facts about spline interpolation on partial Worsey–Farin splits in Sect. 4, before we introduce a refinement of the partition. Based on this refinement and on the decomposition of the partition, we define a Lagrange interpolation set for the space of cubic C^1 splines in Sect. 5. Finally, we provide some numerical results and visualizations in Sect. 6.

2 Preliminaries

In this section, we recall some basic facts about spline spaces on tetrahedral partitions and the Bernstein–Bézier representation of these splines. For more detail on these subjects, see [5]. We also give a few lemmas which we need for the proof of our main result.

Given a tetrahedral partition Δ of some polygonal domain $\Omega \subset \mathbb{R}^3$ and non-negative integers d and r, the *space of trivariate splines of degree d and smoothness r* on Δ is defined by

$$\mathscr{S}_d^r(\Delta) := \left\{ s \in C^r(\Omega);\ s_{|T} \in \mathscr{P}_d \ \forall\, T \in \Delta \right\},$$

where

$$\mathscr{P}_d := \left\{ x^i y^j z^k;\ i+j+k \leq d,\ 0 \leq i,j,k \leq d \right\}$$

is the *space of trivariate polynomials of total degree d*. In this paper, we are only interested in cubic C^1 splines where $d = 3$ and $r = 1$. For the remainder of this section, let all indices i,j,k,l be non-negative integers. Let $T := \Delta(v_0, v_1, v_2, v_3)$ be a non-degenerate tetrahedron. The domain points associated with T are defined by

$$\mathscr{D}_T := \left\{ \xi^T_{i,j,k,l} := (iv_0 + jv_1 + kv_2 + lv_3)/3;\ i+j+k+l = 3 \right\}.$$

The *ball* of radius 1 around the vertex v_0 with respect to T is the subset of domain points defined by

$$D_1^T(v_0) := \left\{ \xi_{i,j,k,l} \in \mathscr{D}_T;\ i \geq 2 \right\},$$

with similar definitions for the other vertices. We further define

$$D_1(v) := \bigcup_{T \in \Delta;\ v \in T} D_1^T(v).$$

The *tube* of radius 1 with respect to T around the edge $e = \langle v_0, v_1 \rangle$ is defined by

$$E_1^T(e) := \left\{ \xi_{i,j,k,l}^T \in \mathscr{D}_T;\ i + j \geq 2 \right\}.$$

The definitions for the other edges are similar. Further, we define

$$E_1(e) := \bigcup_{T \in \Delta;\, e \in T} E_1^T(e).$$

The set of domain points of the tetrahedral partition Δ is defined by

$$\mathscr{D}(\Delta) := \bigcup_{T \in \Delta} \mathscr{D}_T.$$

We use the Bernstein–Bézier representation

$$s_{|T} := \sum_{i+j+k+l=3} c_{i,j,k,l}^T B_{i,j,k,l}^T,$$

for a continuous spline $s \in \mathscr{S}_3^0$, where the coefficients $c_{i,j,k,l}^T$ are the *Bernstein–Bézier coefficients* (*B-coefficients* for the remainder of this paper) with respect to T, and

$$B_{i,j,k,l}^T := \frac{3!}{i!j!k!l!} \phi_0^i \phi_1^j \phi_2^k \phi_3^l.$$

is the cubic *Bernstein polynomial* with respect to T. The linear polynomials $\phi_m \in \mathscr{P}_1$ are the *barycentric coordinates* with respect to T satisfying $\phi_m(v_n) := \delta_{m,n}$, $0 \leq m, n \leq 3$, where $\delta_{m,n}$ denotes Kronecker's delta. Note that if $l = 0$, the Bernstein polynomial $B_{i,j,k,l}^T$ degenerates to a bivariate polynomial on the face $\Delta(v_0, v_1, v_2)$. If $k = 0$ and $l = 0$, it degenerates to a univariate polynomial on the edge $\langle v_0, v_1 \rangle$. $s_{|T}$ is uniquely determined by its B-coefficients $c_{i,j,k,l}^T$, $i + j + k + l = 3$. We associate the B-coefficient $c_{i,j,k,l}^T$ and the Bernstein polynomial $B_{i,j,k,l}^T$ with the respective domain point $\xi_{i,j,k,l}^T$. For a domain point $\eta \in \mathscr{D}_T$, we denote the B-coefficient of $s_{|T}$ associated with η by c_η, and the Bernstein polynomial associated with η by B_η. Let $T = \Delta(v_0, v_1, v_2, v_3)$ and $\widetilde{T} = \Delta(v_0, v_1, v_2, \widetilde{v}_3)$ be two neighboring tetrahedra, and $s \in \mathscr{S}_3^0(T \cup \widetilde{T})$. Then, $s \in \mathscr{S}_3^1(T \cup \widetilde{T})$ if and only if the B-coefficients of $s_{|T}$ and $s_{|\widetilde{T}}$ satisfy

$$c_{i,j,k,1}^{\widetilde{T}} = \phi_0(\widetilde{v}_3) c_{i+1,j,k,0}^T + \phi_1(\widetilde{v}_3) c_{i,j+1,k,0}^T + \phi_2(\widetilde{v}_3) c_{i,j,k+1,0}^T + \phi_3(\widetilde{v}_3) c_{i,j,k,1}^T, \quad (1)$$

for $i + j + k = 2$, where the ϕ_m are the barycentric coordinates with respect to T. Note that if the B-coefficients associated with the ball $B_1^T(v)$ for some vertex v of T are known, then the remaining B-coefficients associated with $B_1(v)$ are determined by this equation. Likewise, if the B-coefficients associated with $E_1^T(e)$ are known for some edge e of T, the remaining B-coefficients associated with $E_1(e)$ are also determined by (1).

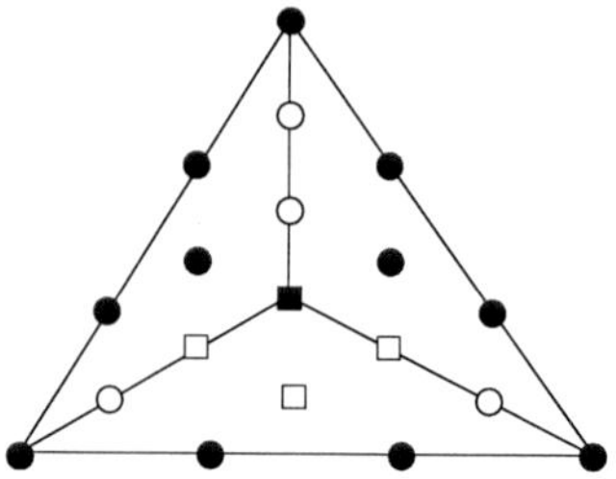

Fig. 1 The Clough–Tocher refinement of a face. *Black circles* are known B-coefficients, *black square* is a given data value

We now give some lemmas concerning Lagrange interpolation on faces and edges of tetrahedra.

Lemma 1. *Let $F := \Delta(v_0, v_1, v_2)$ be a face of some non-degenerate tetrahedron $T := \Delta(v_0, v_1, v_2, v_3)$, and let s be a spline in $\mathscr{S}_3^1(T)$. Given some real value f and all B-coefficients of $\mathscr{M} := (\mathscr{D}_T \cap F) \setminus \{\xi_{1110}^T\}$, the B-coefficient c_{1110}^T is uniquely and stably determined by $s(\xi_{1110}^T) = f$.*

Proof. Since $s_{|F} = \sum_{i+j+k=3} c_{i,j,k,0}^T B_{i,j,k,0}^T$, and since $B_{1110}^T(\xi_{1110}^T) = \frac{2}{3} \neq 0$, we have

$$c_{1110}^T = \frac{1}{B_{1110}^T(\xi_{1110}^T)} \left(f - \sum_{\eta \in \mathscr{M}} c_\eta^T B_\eta^T(\xi_{1110}^T) \right).$$

This is clearly stable. □

Lemma 2. *Let $T := \Delta(v_0, v_1, v_2, v_3)$ be a non-degenerate tetrahedron, and let v_F be a point in the interior of the face $F := \Delta(v_0, v_1, v_2)$. By applying the well-known Clough–Tocher split to F, we obtain the subtetrahedra $T_0 := \Delta(v_0, v_1, v_F, v_3)$, $T_1 := \Delta(v_1, v_2, v_F, v_3)$, and $T_2 := \Delta(v_2, v_0, v_F, v_3)$. We denote the resulting partition by $\Delta_{CT}(T)$. Let*

$$M := \{\xi_{3000}^{T_i}, \xi_{2100}^{T_i}, \xi_{1200}^{T_i}, \xi_{1110}^{T_j};\ i = 0, \ldots, 2, j = 1, 2\} \subset \mathscr{D}_3(\Delta_{CT}(T)),$$

and let the B-coefficients associated with the domain points in M be given (black circles in Fig. 1). Let s be a spline in $\mathscr{S}_3^1(\Delta_{CT}(T))$. Then, for any real value f, the B-coefficients of $s_{|F}$ are uniquely and stably determined by

$$s(v_F) = f.$$

Proof. Since $s(v_F) = c_{0030}^{T_0}$, it follows immediately that $c_{0030}^{T_0} = f$. The B-coefficients associated with $\xi_{2010}^{T_i}$, $i = 0, \ldots, 2$, (white circles in Fig. 1) can be computed from C^1 smoothness across the interior faces of $\Delta_{CT}(T)$. The remaining unknown B-coefficients are $c_{1020}^{T_0}, c_{0120}^{T_0}$ and $c_{1110}^{T_0}$ (white squares in Fig. 1). The C^1 smoothness conditions involving these B-coefficients lead to the linear system

$$\begin{pmatrix} -s & 1 & 0 \\ -r & 0 & 1 \\ 0 & -r & -s \end{pmatrix} \begin{pmatrix} c^{T_0}_{1110} \\ c^{T_0}_{1020} \\ c^{T_0}_{0120} \end{pmatrix} = \begin{pmatrix} d_1 \\ d_2 \\ d_3 \end{pmatrix},$$

where d_1, d_2 and d_3 are linear combinations of known B-coefficients and r, s are barycentric coordinates of v_F satisfying $v_F = rv_0 + sv_1 + (1-r-s)v_2$. Since v_F is in the interior of F, both r and s are greater than 0, and thus the determinant of the matrix $-2rs \neq 0$. □

Lemma 3. *Let $e := \langle v_0, v_1 \rangle$ be an edge of some non-degenerate tetrahedron $T := \Delta(v_0, v_1, v_2, v_3)$, and let s be a spline in $\mathscr{S}^1_3(T)$. Given real value $f_{i,j,0,0}$, $i+j=3$, the B-coefficients $c^T_{i,j,0,0}$, $i+j=3$, of a spline $s \in \mathscr{S}^1_3(T)$ are uniquely and stably determined by the linear system*

$$s(\xi^T_{i,j,0,0}) = f_{i,j,0,0}, \quad i+j=3.$$

Proof. Since $s_{|e}$ is a univariate cubic polynomial, the problem is equivalent to univariate Lagrange interpolation at equidistant interpolation points and therefore has a unique solution. □

Lemma 4. *Let $e := \langle v_0, v_1 \rangle$ be an edge of some non-degenerate tetrahedron $T := \Delta(v_0, v_1, v_2, v_3)$, and let s be a spline in $\mathscr{S}^1_3(T)$. Given the B-coefficients $c^T_{3,0,0,0}$ and $c^T_{2,1,0,0}$, and real values $f_{1,2,0,0}$ and $f_{0,3,0,0}$, the B-coefficients $c^T_{1,2,0,0}$ and $c^T_{0,3,0,0}$ of a spline $s \in \mathscr{S}^1_3(T)$ are uniquely and stably determined by the interpolation conditions*

$$s(\xi^T_{1,2,0,0}) = f_{1,2,0,0}, \qquad s(\xi^T_{0,3,0,0}) = f_{0,3,0,0}.$$

Proof. Since $s(\xi^T_{0,3,0,0}) = c^T_{0,3,0,0}$, this B-coefficient is immediately determined by the second equation. The first equation yields

$$s(\xi^T_{1,2,0,0}) = \sum_{i+j=3} c^T_{i,j,0,0} B^T_{i,j,0,0}(\xi^T_{1,2,0,0}) = f_{1,2,0,0}.$$

Since $B^T_{1,2,0,0}(\xi^T_{1,2,0,0}) \neq 0$, the B-coefficient $c^T_{1,2,0,0}$ is uniquely determined. □

3 A Uniform Partition Consisting of Tetrahedra and Octahedra

In this section, we describe a uniform partition over which the splines are constructed. Let N be an even integer and $h := 1/N$. Then for $i, j, l \in \mathbb{Z}$, let

$$v_{i,j,2l} := \left(\left(i + \frac{1}{2}\right)h, jh, lh\right)$$

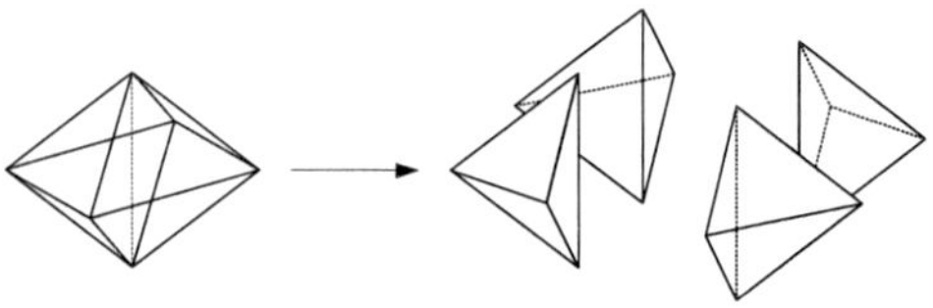

Fig. 2 An octahedron split into four tetrahedra

and

$$v_{i,j,2l+1} := \left(ih, \left(j+\frac{1}{2}\right)h, \left(l+\frac{1}{2}\right)h\right)$$

be the vertices of the body-centered cubic (BCC) grid. We denote the k-th *layer* of vertices by $V_k := \{v_{i,j,k},\ i,j \in \mathbb{Z}\}$. We call a layer V_k *even* or *odd* depending on the parity of the index k. To obtain a uniform partition of space, we connect the vertices of each layer V_k with their 12 nearest neighbors. The resulting partition consists of the following octahedra and tetrahedra. For even k, we denote the tetrahedra by

$$T^A_{i,j,k} := \Delta(v_{i,j,k}, v_{i,j+1,k}, v_{i,j,k+1}, v_{i+1,j,k+1}),$$
$$T^B_{i,j,k} := \Delta(v_{i,j,k}, v_{i+1,j,k}, v_{i+1,j,k+1}, v_{i+1,j-1,k+1}),$$

and the octahedra by

$$O_{i,j,k} := \diamondsuit(v_{i,j,k}, v_{i+1,j,k}, v_{i,j+1,k}, v_{i+1,j+1,k}, v_{i+1,j,k-1}, v_{i+1,j,k+1}).$$

Similarly, for odd k we have

$$T^A_{i,j,k} := \Delta(v_{i,j,k}, v_{i,j+1,k}, v_{i-1,j+1,k+1}, v_{i,j+1,k+1}),$$
$$T^B_{i,j,k} := \Delta(v_{i,j,k}, v_{i+1,j,k}, v_{i,j+1,k+1}, v_{i,j,k+1}),$$
$$O_{i,j,k} := \diamondsuit(v_{i,j,k}, v_{i+1,j,k}, v_{i,j+1,k}, v_{i+1,j+1,k}, v_{i,j+1,k-1}, v_{i,j+1,k+1}).$$

To obtain a tetrahedral partition, we add an additional edge to each octahedron $O_{i,j,k}$ which connects the vertex in the layer V_{k-1} with the vertex in the layer V_{k+1}. This splits the octahedron into four tetrahedra (Fig. 2). We denote these tetrahedra by $T^{(m)}_{i,j,k}$, $m = 1,\ldots,4$. We denote the resulting infinite partition by $\widehat{\diamondsuit}$.

We now define the finite partition $\diamondsuit := \bigcup_{k=0}^{2(N+1)} L_k$, where

$$\begin{aligned}
L_0 &:= \left\{T^A_{i,j,0};\ i=0,\ldots,N,\ j=0,\ldots,N\right\},\\
L_k &:= \left\{T^A_{i,j,k};\ i=0,\ldots,N,\ j=0,\ldots,N\right\}\\
&\quad \cup \left\{T^B_{i,j,k};\ i=0,\ldots,N-1,\ j=1,\ldots,N\right\}\\
&\quad \cup \left\{T^{(m)}_{i,j,k};\ i=0,\ldots,N-1,\ j=0,\ldots,N,\ m=1,\ldots,4\right\}, \quad 2 \le k \le 2N,\\
&\qquad\qquad k \text{ even},
\end{aligned}$$

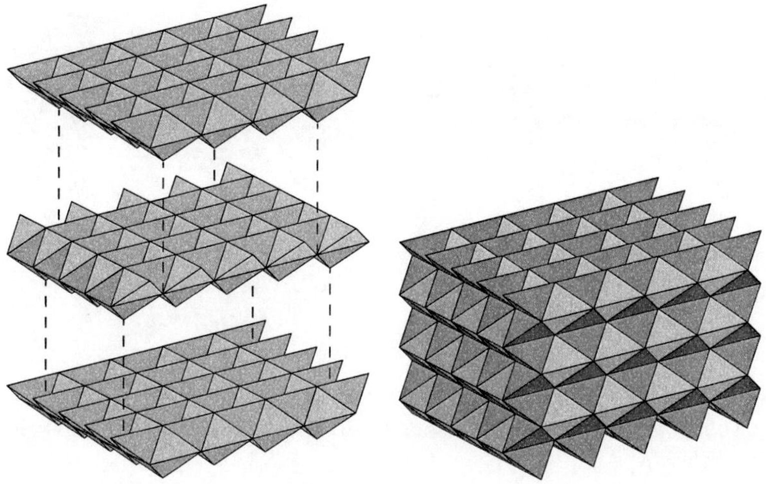

Fig. 3: *Left*: The layers of the partition. Tetrahedra $T^A_{i,j,k}$ are *light gray*, while tetrahedra $T^B_{i,j,k}$ are *white*.
Right: A partition consisting of tetrahedra and octahedra

$$
\begin{aligned}
L_k := &\left\{T^A_{i,j,k};\ i=1,\ldots,N,\ j=0,\ldots,N-1\right\}\\
&\cup\left\{T^B_{i,j,k};\ i=0,\ldots,N,\ j=0,\ldots,N\right\}\\
&\cup\left\{T^{(m)}_{i,j,k};\ i=0,\ldots,N,\ j=0,\ldots,N-1,\ m=1,\ldots,4\right\},\quad 1\le k\le 2N+1,\\
&\qquad\qquad\qquad\qquad\qquad\qquad\qquad\qquad\qquad\qquad\qquad\qquad\qquad k \text{ odd},
\end{aligned}
$$

$$
\begin{aligned}
L_{2(N+1)} := &\left\{T^A_{i,j,k};\ i=0,\ldots,N,\ j=0,\ldots,N\right\}\\
&\cup\left\{T^{(m)}_{i,j,k};\ i=0,\ldots,N-1,\ j=0,\ldots,N,\ m=1,\ldots,4\right\}.
\end{aligned}
$$

Figure 3 gives an impression of these layers and of the partition. Note that the tetrahedra of $\diamondsuit$ are congruent. Each tetrahedron has two edges of length h. These edges have no common vertices. The length of the remaining four shorter edges is $\sqrt{3}h/2$.

As usual, we say that two tetrahedra are *neighbors* if and only if they share a common face. A tetrahedron is said to be an *interior tetrahedron* if it has four neighbors. All other tetrahedra are called *boundary tetrahedra*. A face is called a *boundary face* if it is the face of exactly one tetrahedron. All other faces are called *interior faces*. A vertex and edge is called a *boundary vertex and edge*, respectively, if there exists some boundary face it belongs to. All other vertices and edges are *interior vertices and edges*, respectively.

This partition can be decomposed into the following few classes of tetrahedra. We first consider classes of tetrahedra with respect common vertices (Fig. 4, left).

$$
K_0 := \left\{T^A_{i,j,k} \in \diamondsuit;\ i+j \text{ even},\ k \text{ even}\right\},
$$
$$
K_1 := \diamondsuit \setminus K_0.
$$

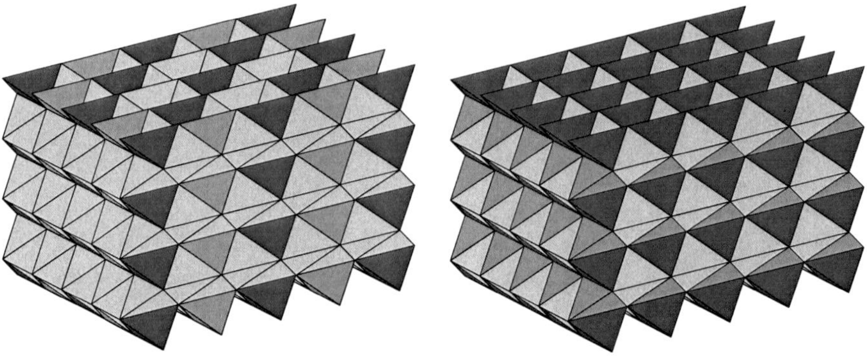

Fig. 4: *Left*: Classes with regard to common Vertices. *Black tetrahedra* belong to K_0, while *white and gray tetrahedra* belong to K_1, with *gray tetrahedra* also belonging to the subclass K_1'.
Right: Classes with regard to common Edges. *Black tetrahedra* belong to K_2, *gray tetrahedra* belong to K_3, and *white tetrahedra* belong to $K_4 \cup K_5$

We say a vertex v is a K_0*-vertex* if there exists some tetrahedron $T \in K_0$ with $v \in T$. All other vertices are K_1*-vertices*.

To deal with a special situation on the boundary of the partition, we define the following subclass of K_1.

$$K_1' := \left\{ T^A_{i,j,k} \in K_1;\ T^A_{i,j,k} \text{ has a } K_1\text{-vertex} \right\}.$$

Note that the tetrahedra in class K_0 are disjoint, and for each interior vertex of the partition, there is exactly one tetrahedron in class K_0 that the vertex belongs to. Moreover, each boundary vertex of the partition belongs to exactly one tetrahedron in $K_0 \cup K_1'$. This also means that each tetrahedron in class K_1' shares exactly three common vertices with the tetrahedra in class K_0, and each tetrahedron in class $K_1 \setminus K_1'$ shares exactly four common vertices with the tetrahedra in class $K_0 \cup K_1'$.

Next, starting new, we consider classes of tetrahedra with common edges (Fig. 4, right).

$$K_2 := \left\{ T^A_{i,j,k} \in \diamondsuit \right\},$$
$$K_3 := \left\{ T^B_{i,j,k} \in \diamondsuit \right\}.$$

Each tetrahedron that is not a subtetrahedron of one of the octahedra is either in class K_2 or in class K_3. For the next class, we choose exactly one tetrahedron $T^{(m)}_{i,j,k}$ for each octahedron $O_{i,j,k}$ of the partition. If there exists a tetrahedron T of $O_{i,j,k}$ which has exactly two boundary faces, we choose this tetrahedron. If there exists no such tetrahedron, we choose $T^{(1)}_{i,j,k}$.

$$K_4 := \left\{ T^{(m)}_{i,j,k} \in \diamondsuit \right\},$$
$$K_5 := \diamondsuit \setminus (K_2 \cup K_3 \cup K_4).$$

These last two classes contain all the subtetrahedra of the octahedra of $\diamondsuit$. Let e be an edge of the partition. Let $m \in \{2,\dots,5\}$ be the smallest index such that there exists a tetrahedron $T \in K_m$ sharing e. We call e a *K_m-edge*.

We need the following subclasses for some special situations on the boundary of the partition.

$$K_3' := \left\{T^B_{i,j,k} \in K_3;\; T^B_{i,j,k} \text{ has exactly two } K_3\text{-edges}\right\},$$

$$K_3'' := \left\{T^B_{i,j,k} \in K_3;\; T^B_{i,j,k} \text{ has exactly three } K_3\text{-edges}\right\},$$

$$K_4' := \{T \in K_4;\; T \text{ has exactly two } K_4\text{-edges}\}.$$

The tetrahedra in class K_3' have exactly one boundary face each, while the tetrahedra in classes K_3'' and K_4' have exactly two boundary faces each. Note that for each K_m-edge there exists exactly one tetrahedron in class K_m sharing that edge.

4 A (Partial) Worsey–Farin Refinement of $\diamondsuit$

To improve the locality of the interpolation method, we introduce a refinement of $\diamondsuit$ by splitting some of the tetrahedra in $\diamondsuit$ using partial Worsey–Farin splits. To fully describe these splits, we need some additional notation.

Let $T := \Delta(v_1, v_2, v_3, v_4)$ be some tetrahedron and $0 \le m \le 4$. For each vertex v_i, $i = 1,\dots,m$, let F_i be the face of T defined by the other three vertices of T. Let v_T be a point in the interior of T, and v_{F_i}, $i = 1,\dots,m$, a point in the interior of the F_i. Then the *m-th degree partial Worsey–Farin refinement of T* (Fig. 5), denoted by $\Delta^m_{WF}(T)$, is the set of tetrahedra that results from connecting v_{F_i} to the vertices of F_i, and v_T to the vertices v_i and v_{F_i}, $i = 1,\dots,m$. For $m = 0$, this is the well-known Alfeld split, while for $m = 4$ it is the Worsey–Farin split. We call v_T the *split point* of the tetrahedron, and v_{F_i} the *split point* of the face F_i.

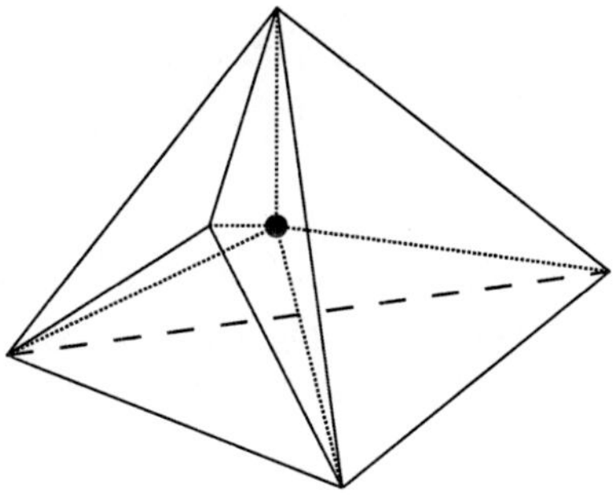

Fig. 5: A partial Worsey–Farin split of degree 1

Theorem 1. *Let $\Delta_{WF}^m(T)$ be the partial Worsey–Farin split of degree $0 \le m \le 4$ of some tetrahedron T. Let*

$$\begin{aligned}\mathcal{M} := &\{v_i;\ i=1,\dots,4\}\\ &\cup \left\{\frac{2}{3}v_i + \frac{1}{3}v_j, \frac{1}{3}v_i + \frac{1}{2}v_j;\ i,j=1,\dots,4, i \neq j\right\}\\ &\cup \left\{\frac{1}{3}(u+w+v_{F_i});\ \langle u,w\rangle \text{an edge of } F_i, i=1,\dots,m\right\}\\ &\cup \left\{\frac{1}{3}\sum_{\substack{j=1\\ j\neq i}}^{4} v_j;\ i=m+1,\dots,4\right\}.\end{aligned}$$

Then the set $\mathcal{M}$ is a minimal determining set for $\mathcal{S}_3^1(\Delta_{WF}^m(T))$.

For the proof of this theorem, see Theorem 6.3 in [3].

We now describe a refinement of the partition $\diamondsuit$. We first split certain faces of the partition using a Clough–Tocher split. Then we apply a (partial) Worsey–Farin split to all tetrahedra that have at least one of those faces. To define the Clough–Tocher splits, we make the following observation. Each face $F := \Delta(u,v,w)$ of the partition consists of one edge of the length h and two shorter edges of the lengths $\sqrt{\frac{3}{4}}h$. Let $\langle u,v\rangle$ be the longer edge. Then, when applying a Clough–Tocher split to F, we use $v_F := (3u+3v+2w)/8$ as the split point. This ensures that v_F and the barycenters of the two tetrahedra that share F are collinear. We split the following faces:

- For all $T \in K_3$, we split those faces of T which are interior faces of $\diamondsuit$. If T is an interior tetrahedron, then these faces have three K_2-edges each. For boundary tetrahedra, they have two K_2-edges and one K_3-edge.
- For all $T \in K_4'$, we split the two boundary faces of T. These faces have two K_2-edges and one K_4-edge.
- All interior faces of the octahedra are split. Again, these have two K_2-edges and one K_4-edge.

We then apply a (partial) Worsey–Farin split to each tetrahedron with at least one split face, using the barycenter as the split point. This results in the following refinement:

- The tetrahedra of class K_2 are not split.
- All interior tetrahedra $T \in K_3$ are refined with a (full) Worsey–Farin split.
- Each tetrahedron $T \in K_3'$ is refined using a partial Worsey–Farin split of degree 3. These tetrahedra have exactly one boundary face which is the only face that is not split.
- Each tetrahedron $T \in K_3''$ is refined using a partial Worsey–Farin split of degree 2, where a Clough–Tocher split is applied to the two interior faces.

- All interior tetrahedra $T \in K_4 \cup K_5$ are refined using a partial Worsey–Farin split of degree 3. Each of these tetrahedra has exactly one neighboring tetrahedron $\widetilde{T} \in K_2$. The common face of T and $\widetilde{T}$ is the only face that is not split.
- All boundary tetrahedra $T \in K_4'$ are refined using a (full) Worsey–Farin split.

We denote the resulting refined partition by $\widetilde{\diamondsuit}$.

5 Local Lagrange Interpolation by $\mathscr{S}_3^1(\widetilde{\diamondsuit})$

In this section, we state our main results. We provide a Lagrange interpolation set and an algorithm for Lagrange interpolation by $\mathscr{S}_3^1(\widetilde{\diamondsuit})$. We show that the interpolation set is both local and stable, and that the interpolation method therefore yields optimal approximation order 4.

We define the following subsets of the domain points.

$$\mathscr{M}_0 := \bigcup_{T \in K_0} \bigcup_{v \in T} D_1^T(v),$$

$$\mathscr{M}_1' := \bigcup_{T \in K_1'} D_1^T(v), \quad v \text{ a } K_1\text{-vertex of } T,$$

$$\mathscr{M}_2 := \bigcup_{T \in K_2} \{\xi_{1110}^T, \xi_{1101}^T, \xi_{1011}^T, \xi_{0111}^T\},$$

$$\mathscr{M}_3' := \bigcup_{T \in K_3'} \{v_F;\ F \text{ is a face of } T \text{ with exactly two } K_2\text{-edges}\},$$

$$\mathscr{M}_3'' := \bigcup_{T \in K_3''} \{v_F;\ F \text{ is a face of } T \text{ with exactly two } K_2\text{-edges}\},$$

$$\mathscr{M}_4 := \bigcup_{T \in K_4} \{v_F;\ F \text{ is a face of } T \text{ with exactly one } K_4\text{-edge}\},$$

We denote the union of these sets by

$$\mathscr{M} := \mathscr{M}_0 \cup \mathscr{M}_1' \cup \mathscr{M}_2 \cup \mathscr{M}_3' \cup \mathscr{M}_3'' \cup \mathscr{M}_4 \cup \mathscr{M}_4'.$$

The following theorem shows that $\mathscr{M}$ is a Lagrange interpolation set for $\mathscr{S}_3^1(\widetilde{\diamondsuit})$.

Theorem 2. *For each set $\{f_\eta;\ \eta \in \mathscr{M}\}$ of real values there exists a unique spline $s \in \mathscr{S}_3^1(\widetilde{\diamondsuit})$ which satisfies*

$$s(\eta) = f_\eta, \quad \forall\, \eta \in \mathscr{M}.$$

Before we proof this theorem, we give the following algorithm which provides an order in which the B-coefficients of s are computed from the data values $\{f_\eta;\ \eta \in \mathscr{M}\}$.

1. For each $T \in K_0$, use the data values associated with $\mathcal{M}_0$ to compute the B-coefficients on the edges of T.
2. For each $T \in K_1'$, compute the B-coefficients on the edges of T using C^1 smoothness at the vertices of T, and the data values associated with $\mathcal{M}_1'$.
3. For each $T \in K_1 \setminus K_1'$, compute the B-coefficients on the edges of T using C^1 smoothness at the vertices of T.
4. For each $T \in K_2$, compute the B-coefficients on the faces of T from the data values associated with $\mathcal{M}_2$, using Lemma 1.
5. For each $T \in K_3$, compute the B-coefficients on the faces of T using C^1 smoothness across the edges of T. If T is in class K_3' or K_3'', and F is a face of T which has interpolation points in $\mathcal{M}_3'$ or $\mathcal{M}_3''$, use the associated data values and Lemma 2 to compute the B-coefficients of F.
6. For each $T \in K_4$, and for each face F of T, compute the B-coefficients on F using C^1 smoothness across the edges of T. If F has interpolation points in $\mathcal{M}_4$ or $\mathcal{M}_4'$, use the associated data values and Lemma 2.
7. For each $T \in K_5$, compute the B-coefficients of the faces of T using C^1 smoothness.
8. For each $T \in \diamondsuit$, compute the B-coefficients associated with domain points in the interior of T by using C^1 smoothness.

Proof (of Theorem 2). Following the steps of algorithm in Sect. 5, we show how the B-coefficients of each tetrahedron can be computed using only smoothness conditions and data values located at the points of $\mathcal{M}$. We first show how the B-coefficients on the edges of $\diamondsuit$ can be computed.

Beginning with step (1), let e be the edge of some tetrahedron $T \in K_0$. Then all four domain points of that edge are in the set $\mathcal{M}_0$, and the B-coefficients associated with that edge can be computed from the data values located at these domain points by Lemma 3.

Now, moving to step (2), let T be some tetrahedron in class K_1'. Then one of the vertices of T is a K_1-vertex. Let v be that vertex. Then $\mathcal{M}_1'$ contains the domain points $D_1^T(v)$. Each of the other three vertices of T, u_1, u_2, u_3, is shared with some tetrahedron in class K_0. For each edge $e = \langle v, u_i \rangle$ of T, $i = 1,2,3$, the B-coefficients of e are determined by C^1 smoothness at u_i and the data values in $D_1^T(v)$ as in Lemma 4. The B-coefficients of the other edges of T are determined by C^1 smoothness only.

Following step (3), let $e = \langle u, v \rangle$ be one of the remaining edges of $\diamondsuit$. Then for both u and v there exist tetrahedra T_u and T_v, respectively, which are either in class K_0 or in class K_1'. Since the B-coefficients associated with the edges of T_u and T_v have already bean determined, the B-coefficients of e can be determined by C^1 smoothness at u and v.

We now show how the remaining B-coefficients can be computed, giving a brief outline of the process before we address steps (4)–(8) of the algorithm. We consider the tetrahedra of $\diamondsuit$ in the order given by algorithm in Sect. 5. Let T be a tetrahedron in class K_m. For each K_n-edge of T, $n < m$, there exists exactly one tetrahedron T' in class K_n that shares the edge. The B-coefficients associated with the faces of T' have already been computed in a previous step. Therefore, the B-coefficients associated with the tube around this edge are determined by C^1 smoothness. In all cases where the B-coefficients of a face cannot be computed by C^1 smoothness alone, there are data values associated with the face, and those B-coefficients are determined as in Lemma 2.

Following step (4) of algorithm in Sect. 5, let T be a tetrahedron in class K_2. Then T is not split, and the set $\mathcal{M}_2$ contains the face barycenters of T. Since the B-coefficients associated with the edges of T are already known, the B-coefficients associated with these barycenters can be computed using Lemma 1.

Moving to step (5), we consider the tetrahedra in class K_3. Let T be such a tetrahedron with no boundary faces. Then a full Worsey–Farin split has been applied to T, and each edge of T is a K_2-edge, i.e. the edge of some tetrahedron in class K_2. The B-coefficients associated with the faces of T are determined by C^1 smoothness across the edges of T.

Now let T be a tetrahedron in class K_3'. These tetrahedra have three interior faces and one boundary face. One of the interior faces, has three K_2-edges. The B-coefficients associated with that face are determined by C^1 smoothness across these edges. The boundary face, has one K_2-edge and two K_3-edges. This face is not split, and the B-coefficient associated with its barycenter is determined by C^1 smoothness across the K_2-edge. The remaining two faces of T are interior faces with two K_2-edges and one K_3-edge. These two faces are subdivided by a Clough–Tocher split, and the split point is contained in the set $\mathcal{M}_3'$. Therefore, the remaining B-coefficients of these faces are determined as in Lemma 2.2.

For the tetrahedra in class K_3'' the situation is similar. The two boundary faces are not split and have one K_2-edge and two K_3-edges each. The B-coefficient associated with the barycenter is determined by C^1 smoothness across the K_2-edge. The two interior faces are split with a Clough–Tocher split and have two K_2-edges and one K_3-edge. The split points are contained in $\mathcal{M}_3''$, and the remaining B-coefficients are determined as in Lemma 2.2.

At this point, only B-coefficients associated with domain points in the octahedra of the partition remain to be determined. These are dealt with in the last two steps of algorithm in Sect. 5. Following step (7), let T be a tetrahedron in class K_4 belonging to an octahedron O. If O is an interior octahedron, then two of the faces of T are shared with tetrahedra in the classes K_2 or K_3, and the B-coefficients associated with these faces are already determined. The remaining two faces are interior faces of the octahedron. They have two K_2-edges and one K_4-edge, and their split points are contained in the set $\mathcal{M}_4$. Thus, the remaining B-coefficients associated with these faces are determined as in Lemma 2.2. If O is a boundary octahedron, then none of the faces of T is shared with any of the previously considered tetrahedra. In this case, each face of T has two edges that are either K_2-edges or K_3-edges, and exactly one K_4-edge. The split point of each of these faces is contained in the set $\mathcal{M}_4$, and again Lemma 2.2 can be applied to determine the remaining B-coefficients.

Finally, moving to step (8), let T be a tetrahedron in class K_5. Each face of T has three K_m-edges with $m \leq 4$, and the B-coefficients associated with this face are determined by C^1 smoothness across the edges. Now that the B-coefficients associated with all the faces of the partition are already known, by Theorem 1 the remaining B-coefficients in the interior of the tetrahedra are also determined. Note that by the construction of the partition, the split point of each interior face F that is refined with a Clough–Tocher split lies on the line segment that connects the incenters of the two tetrahedra sharing F. This ensures that all C^1 smoothness conditions across that F are satisfied (cf. [8]). This concludes the proof. □

We now show that the Lagrange interpolation set $\mathcal{M}$ is *local* and *stable*.

Theorem 3. *The Lagrange interpolation method described in algorithm in Sect. 5 is 2-local, i.e. for each $T \in \diamondsuit$, the corresponding B-coefficients depend only on the data values*

$$\{f_\eta\}_{\eta \in \mathcal{M} \cap \operatorname{star}^2(T)},$$

where $\operatorname{star}^0(T) := T$ *and* $\operatorname{star}^\ell(T) := \bigcup\{T' \in \diamondsuit;\ T' \cap \operatorname{star}^{\ell-1}(T) \neq \emptyset\}$.
It is also stable, i.e. there exists a real constant $C > 0$ such that for each $s \in \mathcal{S}_3^1(\widetilde{\diamondsuit})$ and each B-coefficient c_η^T of s,

$$|c_\eta^T| \leq C \max_{\xi \in \mathcal{M} \cap \operatorname{star}^2(T)} |f_\xi|.$$

Proof. The B-coefficients of a spline $s \in \mathcal{S}_3^1(\widetilde{\diamondsuit})$ are computed either by one of the Lemmas 1–4, or by C^1 smoothness as in (1). Due to the uniformity of the partition $\diamondsuit$, these computations only depend on the data values $\{f_\eta;\ \eta \in \mathcal{M}\}$. Since each of these computations is stable, the interpolation method is also stable.

To establish the locality of the method, we observe that the B-coefficients of s are computed in a certain order prescribed by algorithm in Sect. 5. Let e be an edge. If e belongs to some tetrahedron $T \in K_0$, then the B-coefficients of $s_{|e}$ are computed locally from data values associated with e as in Lemma 3. If e belongs to some tetrahedron $T \in K_1$, then the B-coefficients of $s_{|e}$ are computed either as in Lemma 4 or by C^1 smoothness at the vertices of e. In either case, the computation of $s_{|e}$ is at most 1-local.

For the remaining B-coefficients of s, we have to consider the classes with respect to common edges. Let T be a tetrahedron in class K_2. Then T is not split, and the B-coefficients on the faces of T are computed as in Lemma 1. Since this depends on the B-coefficients on the edges of T, the computation of $s_{|T}$ is also at most 1-local.

Now let T be a tetrahedron in class K_3. Each edge of T is either the edge of some tetrahedron in class K_2, or it is a boundary edge. The computation of the B-coefficients on the faces of T depends either on C^1 smoothness across the K_2-edges, or on data values located on the faces using Lemma 2. In either case, the computation is at most 2-local.

Finally, let T be a tetrahedron in class K_4 or K_5. The tetrahedra in these classes form the octahedra of $\diamondsuit$. For an interior octahedron, the computation of its associated B-coefficients depends only on the surrounding K_2-tetrahedra (Fig. 6). Since these tetrahedra are in the 1-star of T, and the computation of their B-coefficients is at most 1-local, it follows that the computation of the B-coefficients of T is at most 2-local. We have to take a closer look at boundary octahedra, since the computation of their B-coefficients not only depends on the surrounding

Fig. 6 The K_2-tetrahedra surrounding an interior octahedron. For boundary octahedra, one of these tetrahedra is missing

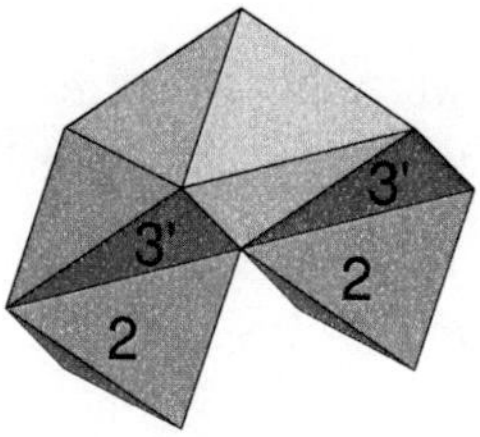

Fig. 7 A boundary octahedron with two adjacent tetrahedra in class K_3'

tetrahedra in class K_2, but also on the surrounding tetrahedra in classes K_3' and K_3''. Let O be a boundary octahedron. Of the 12 edges of O, nine are K_2-edges, two are K_3-edges, and one is a K_4-edge. The computation of the B-coefficients associated with the tubes around the K_2-edges is at most 2-local, since the computation of the corresponding K_2-tetrahedra is at most 1-local. This covers nine of the edges and four of the faces of O. Now consider the two faces of O which are shared with tetrahedra in classes K_3' or K_3'' (Fig. 7). These faces have two K_2-edges each, and their split point in contained in the sets $\mathscr{M}_3'$ or $\mathscr{M}_3''$, respectively. The B-coefficients associated with these faces are computed as in Lemma 2, using C^1 smoothness across the K_2-edges only. Thus, the computation is also 2-local. Finally, consider the two boundary faces of O. One of these faces has two K_2-edges and one K_4 edge. There is a data value associated with the split point of this face, since the split point is contained in the set $\mathscr{M}_4$. Using Lemma 2, the B-coefficients of this face can be computed 2-locally. The last face F of O belongs to a boundary tetrahedron T in class K_4'. It has two K_3-edges and one K_4-edge. It is coplanar with two boundary faces F_1 and F_2 of adjacent tetrahedra in classes K_3' or K_3'' (Fig. 7). These faces are not split and have a single K_2-edge each, and the B-coefficients associated with their barycenters are computed using C^1 smoothness across these K_2-edges. The B-coefficients of F are then computed as in Lemma 2, using C^1 smoothness across the K_3-edges and B-coefficients associated with F_1 and F_2. Since all relevant tetrahedra are within the 1-star of T, this computation is also 2-local. This shows that the computation of all B-coefficients associated with the faces of O is 2-local. Thus, the computation of the interior B-coefficients of O is also 2-local, completing the proof. □

The next theorem provides an error bound for $\|f-s\|_\Omega$ for sufficiently smooth functions f, where $\|\cdot\|_\Omega$ denotes the maximum norm on Ω. We define a linear operator $\mathfrak{I}$ mapping $C(\Omega)$ on $\mathscr{S}_3^1(\widetilde{\diamondsuit})$. Let $f \in C(\Omega)$. Then $\mathfrak{I}f \in \mathscr{S}_3^1(\widetilde{\diamondsuit})$ is defined as the unique spline that satisfies

$$\mathfrak{I}f(\eta) = f(\eta), \quad \forall\, \eta \in \mathscr{M}.$$

Let $W_\infty^m(B)$ denote the usual Sobolev space defined on some compact subset $B \subseteq \Omega$, equipped with the seminorm

$$|f|_{m,B} := \sum_{|\alpha|=m} \|D^\alpha f\|_B,$$

where $\|\cdot\|_B$ denotes the maximum norm on B, and $D^\alpha := D_x^{\alpha_1} D_y^{\alpha_2} D_z^{\alpha_3}$ the partial derivative operator with $\alpha := (\alpha_1, \alpha_2, \alpha_3)$ and $|\alpha| := \alpha_1 + \alpha_2 + \alpha_3$.

Theorem 4. *For $f \in W_\infty^{m+1}(\Omega)$, $0 \le m \le 3$, there exists a real constant $K > 0$ such that for $0 \le |\alpha| \le m$,*

$$\|D^\alpha(\mathfrak{I}f - f)\|_\Omega \le K|\Delta|^{m+1-|\alpha|}|f|_{m+1,\Omega}, \tag{2}$$

where $|\Delta|$ denotes the diameter of the tetrahedra of $\diamondsuit$.

Proof. Fix m, and let $f \in W_\infty^{m+1}(\Omega)$. Fix $T \in \diamondsuit$ and let $\Omega_T := \mathrm{star}^2(T)$. By Lemma 4.3.8 of [1], there exists a cubic polynomial p such that for all $0 \le |\beta| \le m$,

$$\|D^\beta(f-p)\|_{\Omega_T} \le C_1|\Omega_T|^{m+1-|\beta|}|f|_{m+1,\Omega_T}, \tag{3}$$

where C_1 is some constant and $|\Omega_T|$ is the diameter of Ω_T. Since $\mathfrak{I}p = p$, it follows that

$$\|D^\alpha(f-\mathfrak{I}f)\|_T \le \|D^\alpha(f-p)\|_T + \|D^\alpha \mathfrak{I}(f-p)\|_T.$$

We estimate the first term using (3) with $\beta = \alpha$. For the second term, using the Markov inequality [7] and Theorem 3, we obtain

$$\|D^\alpha \mathfrak{I}(f-p)\|_T \le C_2|\Delta|^{-|\alpha|}\|\mathfrak{I}(f-p)\|_T \le C_3|\Delta|^{-|\alpha|}\|f-p\|_{\Omega_T}.$$

Due to the uniformity of the tetrahedra, there exists a constant C_4 such that $|\Omega_T| \le C_4|\Delta|$. Using (3) with $\beta = 0$, we get

$$\|D^\alpha(f-\mathfrak{I}f)\|_T \le C_5|\Omega_T|^{m+1-|\alpha|}|f|_{m+1,\Omega_T}.$$

Taking the maximum over all tetrahedra, we obtain (2), thus completing the proof. □

6 Numerical Tests

In this section, we present our numerical results that confirm the optimal approximation order of our method. We constructed a spline interpolating the well-known Marschner–Lobb test function

$$p(x,y,z) := \frac{1-\sin(\pi\frac{z}{2})+\alpha(1+\rho_r(x^2+y^2))}{2(1+\alpha)},$$

where $\rho_r := \cos(2\pi f_M \cos(\pi\frac{r}{2}))$, for $f_M = 6$ and $\alpha = 0.25$. Table 1 shows the size parameter N of the partition, the maximum error $e_{\max}$, the decay exponent d, and the number of interpolation points $\#IP$. The maximum error was computed by sampling each tetrahedron at 120 points.

Figure 8 shows a rendering of a spline interpolating the Marschner–Lobb function. The spline was constructed on the partition described in Sect. 3 with the size parameter $N = 100$ using our method. The picture was obtained by ray tracing the volume, extracting the isosurface for the isovalue 0.5.

Table 1: Maximum errors of splines interpolating the Marschner–Lobb test function

N	$e_{\max}$	d	$\#IP$
82	$1.53\cdot10^{-4}$	–	17,890,456
164	$1.04\cdot10^{-5}$	3.88	142,127,672
330	$4.95\cdot10^{-7}$	4.39	1,143,476,372
660	$3.12\cdot10^{-8}$	3.99	9,173,788,772

Fig. 8: Isosurface of a spline interpolating the Marschner–Lobb test function

References

1. S. C. Brenner and L. R. Scott: The Mathematical Theory of Finite Element Methods. Texts in Applied Mathematics. Springer–Verlag, New York, 1994.
2. D. Eppstein, J. M. Sullivan, and A. Üngör, Tiling space and slabs with acute tetrahedra, Comput. Geom. **27** (3), 237–255 (2004).
3. G. Hecklin, G. Nürnberger, L. L. Schumaker, and F.Zeilfelder, A local lagrange interpolation method based on C^1 cubic splines on Freudenthal partitions, Math. Comp. **77** (1–2), 1017–1036 (2008).
4. G. Hecklin, G. Nürnberger, L. L. Schumaker, and F.Zeilfelder, Local lagrange interpolation with cubic C^1 splines on tetrahedral partitions, J. Approx. Theory **160** (1–2), 89–102 (2009).
5. M.-J. Lai and L.L. Schumaker, Spline Functions on Triangulations, Cambridge University Press, 2007.
6. M. A. Matt and G. Nürnberger, Local Lagrange interpolation using cubic C^1 splines on type-4 cube partitions, J. Approx. Theory **162** (3), 494–511 (2010).
7. D. R. Wilhelmsen, A Markov inequality in several dimensions, J. Approx. Theory **11**, 216–220 (1974).
8. A. Worsey and G. Farin, An *n*-dimensional Clough-Tocher element, Constr. Approx. **3**, 99–110 (1987).

Approximation of Besov Vectors by Paley–Wiener Vectors in Hilbert Spaces

Isaac Z. Pesenson and Meyer Z. Pesenson

Abstract We develop an approximation theory in Hilbert spaces that generalizes the classical theory of approximation by entire functions of exponential type. The results advance harmonic analysis on manifolds and graphs, thus facilitating data representation, compression, denoising and visualization. These tasks are of great importance to machine learning, complex data analysis and computer vision.

1 Introduction

One of the main themes in Analysis is correlation between frequency content of a function and its smoothness. In the classical approach, the frequency is understood in terms of the Fourier transform (or Fourier series) and smoothness is described in terms of the Sobolev or Lipshitz and Besov norms. For these notions it is well understood [1, 8] that there exists a perfect balance between the rate of approximation by bandlimited functions (trigonometric polynomials) and smoothness described by Besov norms. For more recent results of approximations by entire functions of exponential type we refer to [5]–[7].

The classical concepts and result were generalized to Riemannian manifolds, graphs, unitary representations of Lie groups and integral transforms in our work [11]–[21], [8].

Isaac Z. Pesenson
Department of Mathematics, Temple University, Philadelphia, PA 19122, USA
e-mail: pesenson@temple.edu

Meyer Z. Pesenson
Department of Computing and Mathematical Sciences, California Institute of Technology, Pasadena, CA 91125, USA
e-mail: mzp@cs.caltech.edu

M. Neamtu and L. Schumaker (eds.), *Approximation Theory XIII: San Antonio 2010*, Springer Proceedings in Mathematics 13, DOI 10.1007/978-1-4614-0772-0_15,

The goal of the present article is to develop a form of a Harmonic Analysis which holds true in general Hilbert spaces. In the introduction section, we formulate main results obtained in the paper. The exact definitions of all notions are given in the text.

We start with a self-adjoint positive definite operator L in a Hilbert space $\mathscr{H}$ and consider its positive root $D = L^{1/2}$. For the operator D one can introduce notion of the Spectral Transform $\mathscr{F}_D$ which is an isomorphism between $\mathscr{H}$ and a direct integral of Hilbert spaces over $\mathbb{R}$.

A Paley–Wiener space $PW_{\omega}(D), \omega > 0$, is introduced as the set of all $f \in \mathscr{H}$ whose image $\mathscr{F}_D f$ has support in $[0, \omega]$. In the case when $\mathscr{H} = L_2(\mathbb{R})^d$ and D is a positive square root from the Laplace operator, our definition produces regular Paley–Wiener spaces of spherical exponential type.

The domain $\mathscr{D}_s, s \in \mathbb{R}$, of the operator $D^s, s \in \mathbb{R}$, plays the role of the Sobolev space and we introduce Besov spaces $\mathbf{B}^{\alpha}_{2,q} = \mathbf{B}^{\alpha}_{2,q}(D), \alpha > 0, 1 \le q \le \infty$, by using Peetre's interpolation K-functor [2,4,8,9,23].

$$\mathbf{B}^{\alpha}_{2,q}(D) = \left(\mathscr{H}, \mathscr{D}_{r/2}\right)^K_{\alpha/r,q}, \tag{1}$$

where r can be any natural such that $0 < \alpha < r, 1 \le q < \infty$, or $0 \le \alpha \le r, q = \infty$. It is crucial for us that Besov norms can be described in terms of a modulus of continuity constructed in terms of the Schrodinger group $\mathrm{e}^{\mathrm{i}tD^2}$, wave semigroup $\mathrm{e}^{\mathrm{i}tD}$, or the heat semigroup e^{-tD^2}. In what follows the notation $\|\cdot\|$ below means $\|\cdot\|_{\mathscr{H}}$. We introduce a notion of best approximation

$$\mathscr{E}(f,\omega) = \inf_{g \in PW_{\omega}(D)} \|f - g\|, \; f \in \mathscr{H}. \tag{2}$$

We also consider the following family of functionals which describe a rate of decay of the Spectral transform $\mathscr{F}_D$

$$\mathscr{R}(f,\omega) = \left(\int_{\omega}^{\infty} \|\mathscr{F}_D(f)(\lambda)\|^2_{X(\lambda)} \mathrm{d}m(\lambda)\right)^{1/2}, \omega > 0. \tag{3}$$

The Plancherel Theorem for $\mathscr{F}_D$ implies that every such functional is exactly the best approximation of f by Paley–Wiener functions from $PW_{\omega}(D)$:

$$\mathscr{R}(f,\omega) = \mathscr{E}(f,\omega) = \inf_{g \in PW_{\omega}(D)} \|f - g\|. \tag{4}$$

Our main results are the following.

Theorem 1. *The norm of the Besov space $\mathbf{B}^{\alpha}_{2,q}(D), \alpha > 0, 1 \le q \le \infty$ is equivalent to the following norms*

$$\|f\| + \left(\int_0^{\infty} (s^{\alpha}\mathscr{E}(f,s))^q \frac{\mathrm{d}s}{s}\right)^{1/q}, \tag{5}$$

$$\|f\| + \left(\sum_{k=0}^{\infty}\left(a^{k\alpha}\mathscr{E}(f,a^k)\right)^q\right)^{1/q}, a > 1. \tag{6}$$

$$\|f\| + \left(\int_0^{\infty}(s^{\alpha}\mathscr{R}(f,s))^q \frac{\mathrm{d}s}{s}\right)^{1/q}, \tag{7}$$

and

$$\|f\| + \left(\sum_{k=0}^{\infty}\left(a^{k\alpha}\mathscr{R}(f,a^k)\right)^q\right)^{1/q}, a > 1. \tag{8}$$

Theorem 2. *A vector $f \in \mathscr{H}$ belongs to $\mathbf{B}_{2,q}^{\alpha}(D), \alpha > 0, 1 \le q \le \infty$, if and only if there exists a sequence of vectors $f_k = f_k(f) \in PW_{a^k}(D), a > 1, k \in \mathbb{N}$ such that the series $\sum_k f_k$ converges to f in $\mathscr{H}$ and the following inequalities hold for some $c_1 > 0, c_2 > 0$ which are independent on $f \in \mathbf{B}_{2,q}^{\alpha}(D)$*

$$c_1\|f\|_{\mathbf{B}_{2,q}^{\alpha}(D)} \le \left(\sum_{k=0}^{\infty}\left(a^{k\alpha}\|f_k\|\right)^q\right)^{1/q} \le c_2\|f\|_{\mathbf{B}_{2,q}^{\alpha}(D)}, a > 1. \tag{9}$$

In the case when $\alpha > 0, q = \infty$ one has to make appropriate modifications in the above formulas.

According to (4) the functional $\mathscr{E}(f,\omega)$ is a measure of decay of the Spectral Transform $\mathscr{F}_D$ and the Theorems 1 and 2 show that Besov spaces on a manifold M describe decay of the Spectral transform $\mathscr{F}_D$ associated with any appropriate operator D.

In the case $\mathscr{H} = L_2(\mathbb{R}^d)$, the Theorems 1 and 2 are classical and can be found in [1, 8] and [22]. In the case when $\mathscr{H}$ is L_2-space on a Riemannian manifold or a graph and D is the square root from the corresponding Laplace operator Theorems 1 and 2 were proved in our papers [11]–[20].

2 Paley–Wiener Subspaces Generated by a Self-adjoint Operator in a Hilbert Space

Now we describe Paley-Wiener functions for a self-adjoint positive definite operator D in $\mathscr{H}$. According to the spectral theory [4] for any self-adjoint operator D in a Hilbert space $\mathscr{H}$ there exist a direct integral of Hilbert spaces $X = \int X(\lambda)\mathrm{d}m(\lambda)$ and a unitary operator $\mathscr{F}_D$ from $\mathscr{H}$ onto X, which transforms domain of $D^k, k \in \mathbb{N}$, onto $X_k = \{x \in X | \lambda^k x \in X\}$ with norm

$$\|x(\lambda)\|_{X_k} = \left(\int_0^{\infty}\lambda^{2k}\|x(\lambda)\|_{X(\lambda)}^2 \mathrm{d}m(\lambda)\right)^{1/2} \tag{10}$$

besides $\mathscr{F}_D(D^k f) = \lambda^k(\mathscr{F}_D f)$, if f belongs to the domain of D^k. As it is known, X is the set of all m-measurable functions $\lambda \to x(\lambda) \in X(\lambda)$, for which the norm

$$\|x\|_X = \left(\int_0^\infty \|x(\lambda)\|^2_{X(\lambda)} \mathrm{d}m(\lambda)\right)^{1/2}$$

is finite.

Definition 1. We will say that a vector f from $\mathscr{H}$ belongs to the Paley–Wiener space $PW_\omega(D)$ if the support of the Spectral transform $\mathscr{F}_D f$ belong to $[0,\omega]$. For a vector $f \in PW_\omega(D)$ the notation ω_f will be used for a positive number such that $[0,\omega_f]$ is the smallest interval which contains the support of the Spectral transform $\mathscr{F}_D f$.

Using the spectral resolution of identity P_λ we define the unitary group of operators by the formula

$$\mathrm{e}^{\mathrm{i}tD} f = \int_0^\infty \mathrm{e}^{\mathrm{i}t\tau} \mathrm{d}P_\tau f, f \in \mathscr{H}, t \in \mathbb{R}.$$

Let us introduce the operator

$$\mathbf{R}_D^\omega f = \frac{\omega}{\pi^2} \sum_{k\in\mathbb{Z}} \frac{(-1)^{k-1}}{(k-1/2)^2} \mathrm{e}^{\mathrm{i}\left(\frac{\pi}{\omega}(k-1/2)\right)D} f, f \in \mathscr{H}, \omega > 0. \tag{11}$$

Since $\left\|\mathrm{e}^{\mathrm{i}t\mathscr{L}} f\right\| = \|f\|$ and

$$\frac{\omega}{\pi^2} \sum_{k\in\mathbb{Z}} \frac{1}{(k-1/2)^2} = \omega, \tag{12}$$

the series in (11) is convergent and it shows that $\mathbf{R}_D^\omega$ is a bounded operator in $\mathscr{H}$ with the norm ω:

$$\|\mathbf{R}_D^\omega f\| \le \omega \|f\|, f \in \mathscr{H}. \tag{13}$$

The next theorem contains generalizations of several results from the classical harmonic analysis (in particular the Paley–Wiener theorem) and it follows essentially from our results in [13, 14, 19].

Theorem 3. *The following statements hold*

1. *The set* $\bigcup_{\omega>0} PW_\omega(D)$ *is dense in* $\mathscr{H}$;
2. *The space* $PW_\omega(D)$ *is a linear closed subspace in* $\mathscr{H}$;
3. *A function* $f \in \mathscr{H}$ *belongs to* $PW_\omega(D)$ *if and only if it belongs to the set*

$$\mathscr{D}_\infty = \bigcap_{k=1}^\infty \mathscr{D}_k(D),$$

and for all $s \in \mathbb{R}_+$ *the following Bernstein inequality takes place*

$$\|D^s f\| \le \omega^s \|f\|; \tag{14}$$

4. *A vector* $f \in \mathscr{H}$ *belongs to the space* $PW_{\omega_f}(D), 0 < \omega_f < \infty$, *if and only if* f *belongs to the set* $\mathscr{D}_\infty$, *the limit*

$$\lim_{k\to\infty} \|D^k f\|^{1/k}$$

exists and

$$\lim_{k\to\infty} \|D^k f\|^{1/k} = \omega_f. \tag{15}$$

5. *A vector* $f \in \mathscr{H}$ *belongs to* $PW_\omega(D)$ *if and only if* $f \in \mathscr{D}_\infty$ *and the upper bound*

$$\sup_{k\in N} \left(\omega^{-k} \|D^k f\| \right) < \infty \tag{16}$$

is finite,

6. *A vector* $f \in \mathscr{H}$ *belongs to* $PW_\omega(D)$ *if and only if* $f \in \mathscr{D}_\infty$ *and*

$$\underline{\lim}_{k\to\infty} \|D^k f\|^{1/k} = \omega < \infty. \tag{17}$$

In this case, $\omega = \omega_f$.

7. *A vector* $f \in \mathscr{H}$ *belongs to* $PW_\omega(D)$ *if and only if it belongs to the to the set* $\mathscr{D}_\infty$ *and the following Riesz interpolation formula holds*

$$(\mathrm{i}D)^n f = (\boldsymbol{R}_D^\omega)^n f, n \in \mathbb{N}; \tag{18}$$

8. $f \in PW_\omega(D)$ *if and only if for every* $g \in \mathscr{H}$ *the scalar-valued function of the real variable* $\left\langle \mathrm{e}^{\mathrm{i}tD} f, g \right\rangle, t \in \mathbb{R}^1$, *is bounded on the real line and has an extension to the complex plane as an entire function of the exponential type* ω;

9. $f \in PW_\omega(D)$ *if and only if the abstract-valued function* $\mathrm{e}^{\mathrm{i}tD} f$ *is bounded on the real line and has an extension to the complex plane as an entire function of the exponential type* ω;

10. $f \in PW_\omega(D)$ *if and only if the solution* $u(t), t \in \mathbb{R}^1$ *of the Cauchy problem for the corresponding abstract Schrodinger equation*

$$\mathrm{i}\frac{\partial u(t)}{\partial t} = Du(t), u(0) = f, \mathrm{i} = \sqrt{-1},$$

has analytic extension $u(z)$ *to the complex plane* $\mathbb{C}$ *as an entire function and satisfies the estimate*

$$\|u(z)\|_{\mathscr{H}} \leq \mathrm{e}^{\omega|\Im z|} \|f\|_{\mathscr{H}}.$$

3 Direct and Inverse Approximation Theorems

Now we are going to use the notion of the best approximation (2) to introduce Approximation spaces $E_{2,q}^\alpha(D), 0 < \alpha < r, r \in \mathbb{N}, 1 \leq q \leq \infty$, as spaces for which the following norm is finite

$$\|f\|_{E^{\alpha}_{2,q}(D)} = \|f\| + \left(\int_0^{\infty} (s^{\alpha}\mathscr{E}(f,s))^q \frac{\mathrm{d}s}{s}\right)^{1/q}, \tag{19}$$

where $0 < \alpha < r, 1 \leq q < \infty$, or $0 \leq \alpha \leq r, q = \infty$. It is easy to verify that this norm is equivalent to the following "discrete" norm

$$\|f\| + \left(\sum_{j \in \mathbb{N}} \left(a^{j\alpha}\mathscr{E}(f,a^j)\right)^q\right)^{1/q}, a > 1, \tag{20}$$

The Plancherel Theorem for $\mathscr{F}_D$ also gives the following inequality

$$\mathscr{E}(f,\omega) \leq \omega^{-k}\left(\int_{\omega}^{\infty} \|\mathscr{F}_D(D^k f)(\lambda)\|^2_{X(\lambda)}\mathrm{d}m(\lambda)\right)^{1/2} \leq \omega^{-k}\|D^k f\|. \tag{21}$$

In the classical Approximation theory the Direct and Inverse Theorems give equivalence of the Approximation and Besov spaces. Our goal is to extend these results to a more general setting.

For any $f \in \mathscr{H}$ we introduce a difference operator of order $m \in \mathbb{N}$ as

$$\Delta^m_{\tau} f = (-1)^{m+1}\sum_{j=0}^{m}(-1)^{j-1}C^j_m \mathrm{e}^{j\tau(\mathrm{i}D)}f, \tau \in \mathbb{R}. \tag{22}$$

and the modulus of continuity is defined as

$$\Omega_m(f,s) = \sup_{|\tau| \leq s} \|\Delta^m_{\tau} f\| \tag{23}$$

The following theorem is a generalization of the classical Direct Approximation Theorem by entire functions of exponential type [8].

Theorem 4. *There exists a constant $C > 0$ such that for all $\omega > 0$ and all f*

$$\mathscr{E}(f,\omega) \leq \frac{C}{\omega^k}\Omega_{m-k}\left(D^k f, 1/\omega\right), 0 \leq k \leq m. \tag{24}$$

In particular, the following embeddings hold true

$$\mathbf{B}^{\alpha}_{2,q}(D) \subset E^{\alpha}_q(D), 1 \leq q \leq \infty. \tag{25}$$

Proof. If $h \in L_1(\mathbb{R})$ is an entire function of exponential type ω then for any $f \in \mathscr{H}$ the vector

$$g = \int_{-\infty}^{\infty} h(t)\mathrm{e}^{\mathrm{i}tD} f \mathrm{d}t$$

belongs to $PW_{\omega}(D)$. Indeed, for every real τ we have

$$\mathrm{e}^{\mathrm{i}\tau D}g=\int_{-\infty}^{\infty}h(t)\mathrm{e}^{\mathrm{i}(t+\tau)D}f\mathrm{d}t=\int_{-\infty}^{\infty}h(t-\tau)\mathrm{e}^{\mathrm{i}tD}f\mathrm{d}t.$$

Using this formula we can extend the abstract function $\mathrm{e}^{\mathrm{i}\tau D}g$ to the complex plane as

$$\mathrm{e}^{\mathrm{i}zD}g=\int_{-\infty}^{\infty}h(t-z)\mathrm{e}^{\mathrm{i}tD}f\mathrm{d}t.$$

Since by assumption $h\in L_1(\mathbb{R})$ is an entire function of exponential type ω we have

$$\|\mathrm{e}^{\mathrm{i}zD}g\|\leq\|f\|\int_{-\infty}^{\infty}|h(t-z)|\mathrm{d}t\leq\|f\|\mathrm{e}^{\omega|z|}\int_{-\infty}^{\infty}|h(t)|\mathrm{d}t.$$

It shows that for every functional $g^*\in\mathscr{H}$ the function $\left\langle \mathrm{e}^{\mathrm{i}zD}g,g^*\right\rangle$ is an entire function and

$$\left|\left\langle \mathrm{e}^{\mathrm{i}zD}g,g^*\right\rangle\right|\leq\|g^*\|\|f\|\mathrm{e}^{\omega|z|}\int_{-\infty}^{\infty}|h(t)|\mathrm{d}t.$$

In other words, $\left\langle \mathrm{e}^{\mathrm{i}zD}g,g^*\right\rangle$ is an entire function of the exponential type ω which is bounded on the real line and application of the classical Bernstein theorem gives the following inequality

$$\left|\left(\frac{\mathrm{d}}{\mathrm{d}t}\right)^k\left\langle \mathrm{e}^{\mathrm{i}tD}g,g^*\right\rangle\right|\leq\omega^k\sup_{t\in\mathbb{R}}\left|\left\langle \mathrm{e}^{\mathrm{i}tD}g,g^*\right\rangle\right|.$$

Since

$$\left(\frac{\mathrm{d}}{\mathrm{d}t}\right)^k\left\langle \mathrm{e}^{\mathrm{i}tD}g,g^*\right\rangle=\left\langle \mathrm{e}^{\mathrm{i}tD}(\mathrm{i}D)^k g,g^*\right\rangle$$

we obtain for $t=0$

$$\left|\left\langle D^k g,g^*\right\rangle\right|\leq\omega^k\|g^*\|\|f\|\int_{-\infty}^{\infty}|h(\tau)|\mathrm{d}\tau.$$

Choosing g^* such that $\|g^*\|=1$ and $\langle D^k g,g^*\rangle=\|D^k g\|$ we obtain the following inequality

$$\|D^k g\|\leq\omega^k\|f\|\int_{-\infty}^{\infty}|h(\tau)|\mathrm{d}\tau$$

which implies that g belongs to $PW_{\omega}(D)$.

Let

$$h(t)=a\left(\frac{\sin(t/n)}{t}\right)^n \tag{26}$$

where $n\geq m+3$ is an even integer and

$$a=\left(\int_{-\infty}^{\infty}\left(\frac{\sin(t/n)}{t}\right)^n\mathrm{d}t\right)^{-1}.$$

With such choice of a and n the function h will have the following properties:

1. h is an even nonnegative entire function of exponential type one
2. h belongs to $L_1(\mathbb{R})$ and its $L_1(\mathbb{R})$-norm is 1
3. the integral

$$\int_{-\infty}^{\infty} h(t)|t|^m \mathrm{d}t \tag{27}$$

is finite.

Consider the following vector

$$\mathscr{Q}_h^{\omega,m}(f) = \int_{-\infty}^{\infty} h(t)\left\{(-1)^{m-1}\Delta_{t/\omega}^m f + f\right\}\mathrm{d}t, \tag{28}$$

where

$$(-1)^{m+1}\Delta_s^m f = (-1)^{m+1}\sum_{j=0}^{m}(-1)^{j-1}C_m^j \mathrm{e}^{js(\mathrm{i}D)} f = \sum_{j=1}^{m} b_j \mathrm{e}^{js(\mathrm{i}D)} f - f, \tag{29}$$

and

$$b_1 + b_2 + \cdots + b_m = 1. \tag{30}$$

The formulas (28) and (29) imply the following formula

$$\mathscr{Q}_h^{\omega,m}(f) = \int_{-\infty}^{\infty} h(t)\sum_{j=1}^{m} b_j \mathrm{e}^{j\frac{t}{\omega}(\mathrm{i}D)} f \mathrm{d}t = \int_{-\infty}^{\infty} \Phi(t)\mathrm{e}^{t(\mathrm{i}D)} f \mathrm{d}t.$$

where

$$\Phi(t) = \sum_{j=1}^{m} b_j\left(\frac{\omega}{j}\right) h\left(t\frac{\omega}{j}\right).$$

Since the function $h(t)$ is of the exponential type one every function $h(t\omega/j)$ is of the type ω/j. It also shows that the function $\Phi(t)$ is of the exponential type ω as well.

Now we estimate the error of approximation of f by $\mathscr{Q}_h^{\omega,m}(f)$. If the modulus of continuity is defined as

$$\Omega_m(f,s) = \sup_{|\tau|\le s} \|\Delta_\tau^m f\| \tag{31}$$

then since by (28)

$$f - \mathscr{Q}_h^{\omega,m}(f) = \int_{-\infty}^{\infty} h(t)\Delta_{t/\omega}^m f \mathrm{d}t$$

we obtain

$$\mathscr{E}(f,\omega) \le \|f - \mathscr{Q}_h^{\omega,m}(f)\| \le \int_{-\infty}^{\infty} h(t)\left\|\Delta_{t/\omega}^m f\right\| \mathrm{d}t \le \int_{-\infty}^{\infty} h(t)\Omega_m(f,t/\omega)\,\mathrm{d}t.$$

Now we are going to use the following inequalities

$$\Omega_m(f,s) \leq s^k \Omega_{m-k}(D^k f, s) \tag{32}$$

$$\Omega_m(f,as) \leq (1+a)^m \Omega_m(f,s), a \in \mathbb{R}_+. \tag{33}$$

The first one follows from the identity

$$\Delta_t^k f = \left(e^{itD} - I\right)^k f = \int_0^t \ldots \int_0^t e^{i(\tau_1+\cdots+\tau_k)D} D^k f \mathrm{d}\tau_1 \ldots \mathrm{d}\tau_k, \tag{34}$$

where I is the identity operator and $k \in \mathbb{N}$. The second one follows from the property

$$\Omega_1(f, s_1 + s_2) \leq \Omega_1(f, s_1) + \Omega_1(f, s_2)$$

which is easy to verify. We can continue our estimation of $E(f,\omega)$.

$$\mathscr{E}(f,\omega) \leq \int_{-\infty}^{\infty} h(t) \Omega_m(f,t/\omega)\,\mathrm{d}t \leq \frac{\Omega_{m-k}\left(D^k f, 1/\omega\right)}{\omega^k} \int_{-\infty}^{\infty} h(t)|t|^k(1+|t|)^{m-k}\mathrm{d}t$$

$$\leq \frac{C_{m,k}^h}{\omega^k} \Omega_{m-k}\left(D^k f, 1/\omega\right),$$

where the integral

$$C_{m,k}^h = \int_{-\infty}^{\infty} h(t)|t|^k(1+|t|)^{m-k}\mathrm{d}t$$

is finite by the choice of h. The inequality (24) is proved and it implies the second part of the theorem.

In fact, we proved a little bit more. Namely, for the same choice of the function h the following holds.

Corollary 1. *For any $0 \leq k \leq m, k, m \in \mathbb{N}$, here exists a constant $C_{m,k}^h$ such that for all $0 < \omega < \infty$ and all $f \in \mathscr{H}$ the following inequality holds*

$$\mathscr{E}(f,\omega) \leq \|\mathscr{Q}_h^{\omega,m}(f) - f\| \leq \frac{C_{m,k}^h}{\omega^k} \Omega_{m-k}\left(D^k f, 1/\omega\right), \tag{35}$$

where

$$C_{m,k}^h = \int_{-\infty}^{\infty} h(t)|t|^k(1+|t|)^m \mathrm{d}t, 0 \leq k \leq m,$$

and the operator

$$\mathscr{Q}_h^{\omega,m} : \mathscr{H} \to PW_\omega(D)$$

is defined in (28).

Next, we are going to obtain the Inverse Approximation Theorem in the case $q = \infty$.

Lemma 1. *If there exist $r > \alpha - n > 0, \alpha > 0, r, n \in \mathbb{N}$, such that the quantity*

$$\mathbf{b}_{\infty,n,r}^{\alpha}(f) = \sup_{s>0} \left(s^{n-\alpha} \Omega_r(D^n f, s)\right) \tag{36}$$

is finite, then there exists a constant $A = A(n,r)$ for which

$$\sup_{s>0} s^{\alpha}\mathscr{E}(f,s) \le A(n,r)\mathbf{b}^{\alpha}_{\infty,n,r}(f). \tag{37}$$

Proof. Assume that (5) holds, then

$$\Omega_r(D^n f, s) \le \mathbf{b}^{\alpha}_{\infty,n,r}(f)s^{\alpha-n}$$

and (35) implies

$$\begin{aligned}\mathscr{E}(f,s) &\le C^h_{n+r,n}s^{-n}\mathbf{b}^{\alpha}_{\infty,n,r}(f)s^{n-\alpha}\\ &= A(n,r)\mathbf{b}^{\alpha}_{\infty,n,r}(f)s^{-\alpha}.\end{aligned} \tag{38}$$

Lemma is proved.

Lemma 2. *If for an $f \in \mathscr{H}$ and for an $\alpha > 0$ the following upper bound is finite*

$$\sup_{s>0} s^{\alpha}\mathscr{E}(f,s) = T(f,\alpha) < \infty, \tag{39}$$

then for every $r > \alpha - n > 0, \alpha > 0, r, n \in \mathbb{N}$, there exists a constant $C(\alpha,n,r)$ such that the next inequality holds

$$\mathbf{b}^{\alpha}_{\infty,n,r}(f) \le C(\alpha,n,r)\left(\|f\| + T(f,\alpha)\right). \tag{40}$$

Proof. The assumption implies that for a given $f \in \mathscr{H}$ and a sequence of numbers $a^j, a > 1, j = 0,1,2,\ldots$ one can find a sequence $g_j \in PW_{a^j}(D)$ such that

$$\|f - g_j\| \le T(f,\alpha)a^{-j\alpha}, a > 1.$$

Then for

$$f_0 = g_0, f_j = g_j - g_{j-1} \in PW_{a^j}(D), \tag{41}$$

the series

$$f = f_0 + f_1 + f_2 + \cdots \tag{42}$$

converges in $\mathscr{H}$. Moreover, we have the following estimates

$$\begin{aligned}&\|f_0\| = \|g_0\| \le \|g_0 - f\| + \|f\| \le \|f\| + T(f,\alpha),\\ &\|f_j\| \le \|f - g_j\| + \|f - g_{j-1}\|\\ &\quad \le T(f,\alpha)a^{-j\alpha} + T(f,\alpha)a^{-(j-1)\alpha} = T(f,\alpha)(1+a^{\alpha})a^{-j\alpha},\end{aligned} \tag{43}$$

which imply the following inequality

$$\|f_j\| \le C(a,\alpha)a^{-j\alpha}\left(\|f\| + T(f,\alpha)\right), j \in \mathbb{N}. \tag{44}$$

Since $f_j \in PW_{a^j}(D)$ we have for any $n \in \mathbb{N}$

$$\|D^n f_j\| \le a^{jn}\|f_j\|, a > 1, \tag{45}$$

we obtain

$$\|D^n f_j\| \le C(a,\alpha)a^{-j(\alpha-n)}\left(\|f\| + T(f,\alpha)\right)$$

which shows that the series

$$\sum_{j\in\mathbb{N}} D^n f_j$$

converges in $\mathscr{H}$ and because the operator D^n is closed the sum f of this series belongs to the domain of D^n and

$$D^n f = \sum_{j\in\mathbb{N}} D^n f_j.$$

Next, let $F_j = D^n f_j$ then we have that $D^n f = \sum_j F_j$, where $F_j \in PW_{a^j}(D)$ and according to (44) and (45)

$$\|F_j\| = \|D^n f_j\| \le a^{jn}\|f_j\| \le C(a,\alpha)a^{-j(\alpha-n)}(\|f\| + T(f,\alpha)). \tag{46}$$

Pick a positive t and a natural N such that

$$a^{-N} \le t < a^{-N+1}, a > 1, \tag{47}$$

then we obviously have the following formula for any natural r

$$\Delta_t^r D^n f = \sum_{j=0}^{N-1} \Delta_t^r F_j + \sum_{j=N}^{\infty} \Delta_t^r F_j, \tag{48}$$

where Δ_t^r is defined in (22). Note, that the Bernstein inequality and the formula (34) imply that if $f \in PW_\omega(D)$, then

$$\|\Delta_t^r f\| \le (t\omega)^r\|f\|. \tag{49}$$

Since (19) and (47) hold we obtain for $j \le N-1$ the following inequalities

$$\|\Delta_t^r F_j\| \le (a^j t)^r\|F_j\| \le C(a,\alpha)(\|f\| + T(f,\alpha))a^{j(n+r-\alpha)-(N-1)r}, a > 1.$$

These inequalities imply

$$\begin{aligned}\left\|\sum_{j=0}^{N-1} \Delta_t^r F_j\right\| &\le C(a,\alpha)(\|f\| + T(f,\alpha))a^{-r(N-1)}\sum_{j=0}^{N-1} a^{(n+r-\alpha)j} \\ &= C(a,\alpha)(\|f\| + T(f,\alpha))a^{-r(N-1)}\frac{1-a^{(n+r-\alpha)N}}{1-a^{(n+r-\alpha)}} \\ &\le C(a,\alpha,n,r)(\|f\| + T(f,\alpha))t^{\alpha-n}. \end{aligned} \tag{50}$$

By applying the following inequality

$$\|\Delta_t^r F_j\| \le 2^r \|F_j\|$$

to terms with $j \ge N$ we can continue our estimation as

$$\begin{aligned}\left\|\sum_{j=N}^{\infty} \Delta_t^r F_j\right\| &\le 2^r C(a,\alpha)(\|f\| + T(f,\alpha)) \sum_{j=N}^{\infty} a^{-(\alpha-n)j} \\ &= C(a,\alpha) 2^r (\|f\| + T(f,\alpha)) a^{-N(\alpha-n)} (1 - a^{(n-\alpha)})^{-1} \\ &\le C(a,\alpha,n,r)(\|f\| + T(f,\alpha)) t^{\alpha-n}. \end{aligned} \quad (51)$$

It gives the following inequality

$$\|\Delta_t^r D^n f\| \le C(a,\alpha,n,r) t^{\alpha-n} (\|f\| + T(f,\alpha)),$$

from which one has

$$\Omega_r(D^n f, s) \le C(a,\alpha,n,r)(\|f\| + T(f,\alpha)) s^{\alpha-n}, s > 0,$$

and

$$\mathbf{b}^{\alpha}_{\infty,n,r}(f) \le C(a,\alpha,n,r)(\|f\| + T(f,\alpha)).$$

The Lemma is proved.

Our main result concerning spaces $\mathbf{B}^{\alpha}_{2,\infty}(D), \alpha > 0$, is the following.

Theorem 5. *The norm of the space* $\mathbf{B}^{\alpha}_{2,\infty}D), \alpha > 0$, *is equivalent to the following norms*

$$\|f\| + \sup_{s>0} (s^{\alpha} \mathscr{E}(f,s)), \quad (52)$$

$$\|f\| + \sup_{s>0} (s^{\alpha} \mathscr{R}(f,s))), \quad (53)$$

$$\|f\| + \sup_{k \in \mathbb{N}} \left(a^{k\alpha} \mathscr{E}(f, a^k)\right), a > 1, \quad (54)$$

$$\|f\| + \sup_{k \in \mathbb{N}} \left(a^{k\alpha} \mathscr{R}(f, a^k))\right), a > 1. \quad (55)$$

Moreover, a vector $f \in \mathscr{H}$ *belongs to* $\mathbf{B}^{\alpha}_{2,\infty}(D), \alpha > 0$, *if and only if there exists a sequence of vectors* $f_k = f_k(f) \in PW_{a^k}(D), a > 1$, *such that the series* $\sum f_k$ *converges to* f *in* $\mathscr{H}$ *and*

$$c_1 \|f\|_{\mathbf{B}^{\alpha}_{2,\infty}(D)} \le \sup_{k \in \mathbb{N}} \left(a^{k\alpha} \|f_k\|\right) \le c_2 \|f\|_{\mathbf{B}^{\alpha}_{2,\infty}(D)}, a > 1, \quad (56)$$

for certain $c_1 = c_1(D,\alpha), c_2 = c_2(D,\alpha)$ *which are independent of* $f \in \mathbf{B}^{\alpha}_{2,\infty}(D)$.

Proof. That the norm of $\mathbf{B}^{\alpha}_{2,\infty}(D), \alpha > 0$, is equivalent to any of the norms (52)–(55) follows from the last two Lemmas and (4).

Next, if the norm (52) is finite then it was shown in the proof of the last Lemma that there exists a sequence of vectors $f_k = f_k(f) \in PW_{a^k}(D), a > 1$, such that the series $\sum f_k$ converges to f in $\mathscr{H}$. Moreover, the inequality (44) shows existence of constant c which is independent of $f \in \mathbf{B}^{\alpha}_{2,\infty}(D)$ for which the following inequality holds

$$\sup_{k\in\mathbb{N}} \left(a^{k\alpha}\|f_k\|\right) \le c\|f\|_{\mathbf{B}^{\alpha}_{2,\infty}(D)}, a > 1,$$

Conversely, let us assume that there exists a sequence of vectors $f_k = f_k(f) \in PW_{a^k}(D), a > 1$, such that the series $\sum f_k$ converges to f in $\mathscr{H}$ and

$$\sup_{k\in\mathbb{N}} \left(a^{k\alpha}\|f_k\|\right) < \infty.$$

We have

$$\mathscr{E}(f,a^N) \le \left\| f - \sum_{k=0}^{N-1} f_k \right\| = \sum_{k=N}^{\infty} \|f_k\| \le \sup_{k\in\mathbb{N}} \left(a^{k\alpha}\|f_k\|\right) \sum_{k=N}^{\infty} a^{-\alpha j}$$

$$\le C \sup_{k\in\mathbb{N}} \left(a^{k\alpha}\|f_k\|\right) a^{-N\alpha},$$

or

$$\sup_N a^{N\alpha} \mathscr{E}(f,a^N) \le C \sup_{k\in\mathbb{N}} \left(a^{k\alpha}\|f_k\|\right).$$

Since we also have

$$\|f\| \le \sum_k \|f_k\| \le \sup_{k\in\mathbb{N}} \left(a^{k\alpha}\|f_k\|\right) \sum_k a^{-\alpha k}, a > 1,$$

the theorem is proved.

Theorems 1 and 2 from the Introduction are extensions of the Theorem 5 to all indices $1 \le q \le \infty$. Their proofs go essentially along the same lines as the proof of the last theorem and are omitted.

Acknowledgements This paper was supported in part by the National Geospatial-Intelligence Agency University Research Initiative (NURI), grant HM1582-08-1-0019.

References

1. J. Akhiezer, *Theory of approximation*, Ungar, NY, 1956.
2. J. Bergh, J. Lofstrom, *Interpolation spaces*, Springer-Verlag, 1976.
3. M. Birman and M. Solomyak, *Spectral thory of selfadjoint operators in Hilbert space*, D.Reidel Publishing Co., Dordrecht, 1987.
4. P. Butzer, H. Berens, *Semi-Groups of operators and approximation*, Springer, Berlin, 1967.
5. M. Ganzburg, *Best constants of harmonic approximation on classes associated with the Laplace operator*, J. Approx. Theory 150 (2008), no. 2, 199–213.

6. M. Ganzburg, *Limit theorems in spline approximation*, J.Math. Anal. Appl. 318 (2006), no. 1, 15–31.
7. M. Ganzburg, *Limit theorems in approximation theory*, Anal. Math. 18 (1992), no. 1, 37–57.
8. S. Krein, I. Pesenson, *Interpolation Spaces and Approximation on Lie Groups*, The Voronezh State University, Voronezh, 1990,
9. S. Krein, Y. Petunin, E. Semenov, *Interpolation of linear operators*, Translations of Mathematical Monographs, 54. AMS, Providence, R.I., 1982.
10. S. M. Nikolskii, *Approximation of functions of several variables and imbedding theorems*, Springer, Berlin, 1975.
11. I. Pesenson, *Interpolation spaces on Lie groups*, (Russian) Dokl. Akad. Nauk SSSR 246 (1979), no. 6, 1298–1303.
12. I. Pesenson, *Nikolskii- Besov spaces connected with representations of Lie groups*, (Russian) Dokl. Akad. Nauk SSSR 273 (1983), no. 1, 45–49.
13. I. Pesenson, *The Best Approximation in a Representation Space of a Lie Group*, Dokl. Acad. Nauk USSR, v. 302, No 5, pp. 1055-1059, (1988) (Engl. Transl. in Soviet Math. Dokl., v.38, No 2, pp. 384-388, 1989.)
14. I. Pesenson, *The Bernstein Inequality in the Space of Representation of Lie group*, Dokl. Acad. Nauk USSR **313** (1990), 86–90; English transl. in Soviet Math. Dokl. **42** (1991).
15. I. Pesenson, *Approximations in the representation space of a Lie group*, (Russian) Izv. Vyssh. Uchebn. Zaved. Mat. 1990, no. 7, 43–50; translation in Soviet Math. (Iz. VUZ) 34 (1990), no. 7, 49–57
16. I. Pesenson, *On the abstract theory of Nikolskii-Besov spaces*, (Russian) Izv. Vyssh. Uchebn. Zaved. Mat. 1988, no. 6, 59–68; translation in Soviet Math. (Iz. VUZ) 32 (1988), no. 6, 80–92
17. I. Pesenson, *Bernstein-Nikolskii inequalities and Riesz interpolation formula on compact homogeneous manifolds*, J. of Approx. Theory 150, (2008), no. 2, 175–198.
18. I. Pesenson, *Paley-Wiener approximations and multiscale approximations in Sobolev and Besov spaces on manifolds*, J. Geom. Anal. 19 (2009), no. 2, 390–419.
19. I. Pesenson, *A discrete Helgason-Fourier transform for Sobolev and Besov functions on noncompact symmetric spaces*, Radon transforms, geometry, and wavelets, 231–247, Contemp. Math., 464, Amer. Math. Soc., Providence, RI, 2008.
20. I. Pesenson, A. Zayed, *Paley-Wiener subspace of vectors in a Hilbert space with applications to integral transforms*, J. Math. Anal. Appl. 353 (2009), no. 2, 566582.
21. I.Z. Pesenson and M.Z. Pesenson, *Sampling, filtering and sparse approximation on combinatorial graphs*, J. Fourier Anal. Appl. 16 (2010), 921–942.
22. E. Titchmarsh, *Theory of Fourier Integrals*, Oxford University Press, 1948.
23. H. Triebel, *Theory of function spaces II,* Monographs in Mathematics, 84. Birkhuser Verlag, Basel, 1992.

A Subclass of the Length 12 Parameterized Wavelets

David W. Roach

Abstract In this paper, a subclass of the length 12 parameterized wavelets is given. This subclass is a parameterization of the coefficients of a subset of the trigonometric polynomials, $m(\omega)$, that satisfy the necessary conditions for orthogonality, that is $m(0) = 1$ and $|m(\omega)|^2 + |m(\omega + \pi)|^2 = 1$, but is not sufficient to represent all possible trigonometric polynomials satisfying these constraints. This parameterization has three free parameters whereas the general parameterization would have five free parameters. Finally, we graph some example scaling functions from the parameterization and conclude with a numerical experiment.

1 Introduction

Since the discovery of wavelets in the 1980s, the variety of uses for these functions has been demonstrated including image compression, de-noising, image recognition, and the solution of numerical PDE's to name a few. One restraint with the development of these applications is the accessibility to a large variety of wavelets with varying properties. Typically, a researcher has a handful of wavelets to experiment with which include the standard Daubechies orthogonal wavelets as well as a few biorthogonal wavelets. The Daubechies standard wavelets have minimal phase and a maximum number of vanishing moments for a variety of lengths. A list of the trigonometric polynomial coefficients were published in the seminal work [2] on wavelets. Although the standard wavelets were highlighted because of their minimal phase and vanishing moments, a whole continuum of wavelets with varying properties were identified theoretically in [2] and would be accessible through a spectral factorization technique. In this paper, we give a simple approach to finding a closed representation for the coefficients of a subset of the trigonometric poly-

David W. Roach
Murray State University, Murray, KY 42071, USA
e-mail: david.roach@murraystate.edu

M. Neamtu and L. Schumaker (eds.), *Approximation Theory XIII: San Antonio 2010*, Springer Proceedings in Mathematics 13, DOI 10.1007/978-1-4614-0772-0_16,

nomials, $m(\omega)$, which satisfy the necessary conditions for the orthogonality of the associated scaling functions, i.e. $m(0) = 1$ and $|m(\omega)|^2 + |m(\omega+\pi)|^2 = 1$, for the specific length of 12. In earlier papers, [4, 8, 9], the complete necessary and sufficient parameterizations were given for all trigonometric polynomials up to length ten which satisfy the necessary conditions for orthogonality. Here, we extend the results to a subclass of the length 12 parameterization which is sufficient to satisfy the conditions for orthogonality but do not characterize the entire class of length 12 trigonometric polynomials.

Other researchers have investigated the parameterization of orthogonal wavelets (see [13]). It appears that Schneid and Pittner [11] were the first to give formulas that would lead to the explicit parameterizations for the class of finite length orthogonal scaling functions after finding the Kronecker product of some matrices for wavelet lengths of two through ten. Colella and Heil investigated the length four parameterization in [1]. Others have constructed parameterizations for biorthogonal wavelets as well as multiwavelets (see [3] and [7]). Regensburger, in [6], constructed the explicit parameterizations for the orthogonal scaling functions with multiple vanishing moments up to length ten by first solving the linear system of equations that result from the vanishing moment conditions and then solving the necessary condition for orthogonality.

In this current work, we give an explicit parameterization for a subclass of the length 12 wavelets by forcing an extra nonlinear condition that simplifies the solution. We then give examples from this continuum of length 12 wavelets that have a varying number of vanishing moments and conclude with a numerical experiment.

2 The Nonlinear Equations

The necessary conditions for orthogonality are well known in the literature (see [2, 5], and others). A scaling function ϕ that satisfies the dilation equation

$$\phi(x) = \sum_{k=0}^{N} h_k \phi(2x-k)$$

has an associated trigonometric polynomial m of degree N which can be expressed as

$$m(\omega) = \sum_{k=0}^{N} h_k e^{ik\omega}.$$

Moreover, it is well known that m can be written as an infinite product. In order for this product to converge, m must not vanish at the origin, i.e. $m(0) = c \neq 0$. This condition immediately implies that

$$\sum_{k=0}^{N} h_k = 1. \tag{1}$$

We have chosen a nonstandard normalization of one in order to avoid carrying around a $\sqrt{2}$ in our construction. When implementing the coefficients, we rescale the coefficients to sum to $\sqrt{2}$ in order to maintain orthonormality.

Moreover, the necessary condition for the orthogonality of ϕ with its integer shifts is given by

$$|m(\omega)|^2 + |m(\omega+\pi)|^2 = 1. \tag{2}$$

This condition is equivalent to the dilation coefficients satisfying a system of non-linear equations, specifically

$$\sum_{k=0}^{N-2j} h_k h_{k+2j} = \frac{1}{2}\delta(j), \quad j = 0, \ldots, \frac{N-1}{2}$$

where $\delta(0) = 1$ and $\delta(j) = 0$ for $j \neq 0$.

For the length 12 case (i.e. $N = 11$) that we are currently considering, we have the following underdetermined nonlinear system:

$$\begin{aligned}
h_0^2 + h_1^2 + h_2^2 + h_3^2 + h_4^2 + h_5^2 + h_6^2 + h_7^2 + h_8^2 + h_9^2 + h_{10}^2 + h_{11}^2 &= \frac{1}{2} \\
h_0h_2 + h_1h_3 + h_2h_4 + h_3h_5 + h_4h_6 + h_5h_7 + h_6h_8 + h_7h_9 + h_8h_{10} + h_9h_{11} &= 0 \\
h_0h_4 + h_1h_5 + h_2h_6 + h_3h_7 + h_4h_8 + h_5h_9 + h_6h_{10} + h_7h_{11} &= 0 \\
h_0h_6 + h_1h_7 + h_2h_8 + h_3h_9 + h_4h_{10} + h_5h_{11} &= 0 \\
h_0h_8 + h_1h_9 + h_2h_{10} + h_3h_{11} &= 0 \\
h_0h_{10} + h_1h_{11} &= 0.
\end{aligned}$$

Additionally, these two conditions (1) and (2) imply the zeroth vanishing moment condition $m(\pi) = 0$ or equivalently the linear equations

$$\sum_{k=0}^{(N-1)/2} h_{2k} = \sum_{k=0}^{(N-1)/2} h_{2k+1} = \frac{1}{2}.$$

Because the products of the coefficients in the system of nonlinear equations have the pattern that the odd indices multiply the other odd indices and similarly for the even indices, it is convenient to separate the even indices from the odd indices in the following fashion

$$m(\omega) = \sum_{k=0}^{n} a_k e^{2ki\omega} + b_k e^{(2k+1)i\omega}$$

where we let $n = (N-1)/2$. Note that, since there are no odd length scaling functions satisfying the necessary condition for orthogonality, N will always be an odd integer.

As a means of summary with our new notation, we conclude with the following statements. Given a scaling function ϕ and its associated trigonometric polynomial

$$m(\omega) = \sum_{k=0}^{n} a_k e^{2ki\omega} + b_k e^{(2k+1)i\omega}$$

of degree $2n+1$, the necessary condition for orthogonality,

$$|m(\omega)|^2 + |m(\omega+\pi)|^2 = 1,$$

is equivalent to the following system of nonlinear equations:

$$\sum_{k=0}^{n-j} a_k a_{k+j} + b_k b_{k+j} = \frac{1}{2}\delta(j), \quad j = 0,\ldots,n-1$$

where $\delta(0) = 1$ and $\delta(j) = 0$ for $j \neq 0$.

3 Length Four

Although the length four parameterization is well known (see [8, 13]), it is used in the construction of the length 12 parameterization and is presented here for completeness.

For length four ($N = 3$ and $n = 1$), the nonlinear system of equations is

$$a_0 + a_1 = \frac{1}{2} \tag{3}$$

$$b_0 + b_1 = \frac{1}{2} \tag{4}$$

$$a_0^2 + a_1^2 + b_0^2 + b_1^2 = \frac{1}{2} \tag{5}$$

$$a_0 a_1 + b_0 b_1 = 0. \tag{6}$$

Subtracting twice (6) from (5) gives

$$(a_0 - a_1)^2 + (b_0 - b_1)^2 = \frac{1}{2}.$$

This equation allows the introduction of a free parameter, that is,

$$a_0 - a_1 = \frac{1}{\sqrt{2}}\sin\theta \tag{7}$$

$$b_0 - b_1 = \frac{1}{\sqrt{2}}\cos\theta. \tag{8}$$

Combining (3) and (4) with (7) and (8) gives the length four parameterization

$$a_0 = \frac{1}{4} + \frac{1}{2\sqrt{2}}\sin\theta, \quad b_0 = \frac{1}{4} + \frac{1}{2\sqrt{2}}\cos\theta,$$

$$a_1 = \frac{1}{4} - \frac{1}{2\sqrt{2}}\sin\theta, \quad b_1 = \frac{1}{4} - \frac{1}{2\sqrt{2}}\cos\theta.$$

These formulas are well known (see [1,8,13], and others). To aid in the construction of the longer parameterizations, a different period and phase shift are chosen for the length four solution, that is $\theta = 2\alpha - \pi/4$. With this substitution and some simplification, the length four solution can be written as

$$\begin{aligned}
a_0 &= \frac{1}{4}(1 - \cos 2\alpha + \sin 2\alpha) \\
b_0 &= \frac{1}{4}(1 + \cos 2\alpha + \sin 2\alpha) \\
a_1 &= \frac{1}{4}(1 + \cos 2\alpha - \sin 2\alpha) \\
b_1 &= \frac{1}{4}(1 - \cos 2\alpha - \sin 2\alpha)
\end{aligned}$$

where this form will simplify future computations. It should be noted that this parameterization is a necessary representation for the coefficients and upon substituting them back into the system of (3)–(6), we see that they are also sufficient.

4 Length 12 Subclass

For the construction of the complete parameterizations for lengths six, eight, and ten see [8, 9]. In those constructions, the parameterization describes the complete set of all trigonometric polynomials that satisfy the necessary conditions for the orthogonality of the associated scaling functions. For the length 12 case ($N = 11$ and $n = 5$), we will impose an additional nonlinear equation that will simplify the parameterization but will relegate the parameterization to a subclass rather than a complete characterization.

For the length 12 case, the nonlinear system of equations is given by

$$a_0 + a_1 + a_2 + a_3 + a_4 + a_5 = \frac{1}{2} \quad (9)$$

$$b_0 + b_1 + b_2 + b_3 + b_4 + b_5 = \frac{1}{2} \quad (10)$$

$$a_0^2 + a_1^2 + a_2^2 + a_3^2 + a_4^2 + a_5^2 + b_0^2 + b_1^2 + b_2^2 + b_3^2 + b_4^2 + b_5^2 = \frac{1}{2} \quad (11)$$

$$a_0a_1 + a_1a_2 + a_2a_3 + a_3a_4 + a_4a_5 + b_0b_1 + b_1b_2 + b_2b_3 + b_3b_4 + b_4b_5 = 0 \quad (12)$$

$$a_0a_2 + a_1a_3 + a_2a_4 + a_3a_5 + b_0b_2 + b_1b_3 + b_2b_4 + b_3b_5 = 0 \quad (13)$$

$$a_0a_3 + a_1a_4 + a_2a_5 + b_0b_3 + b_1b_4 + b_2b_5 = 0 \quad (14)$$

$$a_0a_4 + a_1a_5 + b_0b_4 + b_1b_5 = 0 \quad (15)$$

$$a_0a_5 + b_0b_5 = 0 \quad (16)$$

An important step in the construction is establishing the connection between the sums of the even and odd indexed coefficients back to the length four parameterization. More specifically, the sums $a_0+a_2+a_4$, $a_1+a_3+a_5$, $b_0+b_2+b_4$, and $b_1+b_3+b_5$ satisfy the system of equations associated with the length four parameterization, i.e.

$$\begin{aligned}(a_0+a_2+a_4)+(a_1+a_3+a_5) &= \frac{1}{2}\\(b_0+b_2+b_4)+(b_1+b_3+b_5) &= \frac{1}{2}\\(a_0+a_2+a_4)^2+(a_1+a_3+b_5)^2+(b_0+b_2+b_4)^2+(b_1+b_3+b_5)^2 &= \frac{1}{2}\\(a_0+a_2+a_4)(a_1+a_3+b_5)+(b_0+b_2+b_4)(b_1+b_3+b_5) &= 0.\end{aligned}$$

The third equation is equivalent to the sum of (11), (13), and (15), and the last one is equivalent to the sum of (12), (14), and (16). Therefore, we can use the length four parameterization for these sums, i.e.

$$\begin{aligned}a_0+a_2+a_4 &= \frac{1}{4}(1-\cos 2\alpha+\sin 2\alpha)\\b_0+b_2+b_4 &= \frac{1}{4}(1+\cos 2\alpha+\sin 2\alpha)\\a_1+a_3+a_5 &= \frac{1}{4}(1+\cos 2\alpha-\sin 2\alpha)\\b_1+b_3+b_5 &= \frac{1}{4}(1-\cos 2\alpha-\sin 2\alpha).\end{aligned}$$

In an effort to linearize the system of equations, note that the sum and difference of (11) and twice (16) give the two equations:

$$(a_0+a_5)^2+(b_0+b_5)^2 = \frac{1}{2}-a_1^2-a_2^2-a_3^2-a_4^2-b_1^2-b_2^2-b_3^2-b_4^2 \tag{17}$$

$$(a_0-a_5)^2+(b_0-b_5)^2 = \frac{1}{2}-a_1^2-a_2^2-a_3^2-a_4^2-b_1^2-b_2^2-b_3^2-b_4^2 := p^2 \tag{18}$$

Although the right hand side, p^2, has not yet been determined, we use the fact that the right-hand sides of (17) and (18) are equivalent and introduce two new free parameters β and γ in the following fashion:

$$\begin{aligned}a_0+a_5 &= p\cos\beta\\b_0+b_5 &= p\sin\beta\\a_0-a_5 &= p\cos\gamma\\b_0-b_5 &= p\sin\gamma,\end{aligned}$$

which can be solved directly for a_0, a_5, b_0, and b_5. There are now eight linear equations that are each a necessary constraint for all trigonometric polynomials of length 12 which satisfy the necessary conditions for orthogonality. It should be noted that the nonlinear equation (16) is satisfied.

Upon examination of (15), we can see that a_1, a_4, b_1, and b_4 have an orthogonality condition with the known coefficients a_0, a_5, b_0, and b_5. Using this observation, we impose an additional constraint which is not part of the original set of nonlinear equations, i.e.

$$a_1 a_4 + b_1 b_4 = 0.$$

This additional constraint allows us to solve for a_1, a_4, b_1, and b_4 in the same way as a_0, a_5, b_0, and b_5 since

$$\begin{aligned}(a_1 + a_4)^2 + (b_1 + b_4)^2 &= \frac{1}{2} - a_0^2 - a_2^2 - a_3^2 - a_5^2 - b_0^2 - b_2^2 - b_3^2 - b_5^2 \\ (a_1 - a_4)^2 + (b_1 - b_4)^2 &= \frac{1}{2} - a_0^2 - a_2^2 - a_3^2 - a_5^2 - b_0^2 - b_2^2 - b_3^2 - b_5^2 := q^2\end{aligned}$$

giving us

$$\begin{aligned}a_1 + a_4 &= q\cos\delta \\ b_1 + b_4 &= q\sin\delta \\ a_1 - a_4 &= q\cos\sigma \\ b_1 - b_4 &= q\sin\sigma,\end{aligned}$$

which readily gives the solutions for a_1, a_4, b_1, and b_4. Plugging these solutions and the ones for a_0, a_5, b_0, and b_5 into (15) yields

$$\frac{1}{2}pq(\cos(\beta - \delta) - \cos(\gamma - \sigma)) = 0$$

which implies $p = 0$, $q = 0$ or $\cos(\beta - \delta) = \cos(\gamma - \sigma)$. The first two possibilities give us a restricted parameterization of shorter length polynomials. The third possibility has a continuum of solutions, but for the sake of simplifying the equations, we choose $\delta = \beta$ and $\sigma = \gamma$. This choice reduces (14) to

$$\frac{p}{4}(-4q + \cos\beta + \cos(2\alpha + \gamma) + \sin\beta - \sin(2\alpha + \gamma)) = 0$$

which can be solved for q, i.e.

$$q = \frac{1}{4}(\cos\beta + \cos(2\alpha + \gamma) + \sin\beta - \sin(2\alpha + \gamma)).$$

Using this parameterization for q and (13) gives us the parameterization for p, i.e,

$$p = \frac{1}{4}(\cos\beta - \cos(2\alpha + \gamma) + \sin\beta + \sin(2\alpha + \gamma)).$$

Although these choices for p and q are not necessary, they are sufficient to solve the nonlinear system of equations and give a three parameter subclass of the length 12 trigonometric polynomials that satisfy the necessary conditions for orthogonality.

Theorem 1. *For any real numbers α, β, and γ, the trigonometric polynomial $m(\omega)$ of the form*

$$m(\omega) = \sum_{k=0}^{5} a_k e^{2ki\omega} + b_k e^{(2k+1)i\omega}$$

with coefficients defined as

$$\begin{aligned}
p &= \frac{1}{4}(\cos\beta - \cos(2\alpha+\gamma) + \sin\beta + \sin(2\alpha+\gamma)) \\
q &= \frac{1}{4}(\cos\beta + \cos(2\alpha+\gamma) + \sin\beta - \sin(2\alpha+\gamma)) \\
a_0 &= \frac{p}{2}(\cos\beta + \cos\gamma) \\
b_0 &= \frac{p}{2}(\sin\beta + \sin\gamma) \\
a_1 &= \frac{q}{2}(\cos\beta + \cos\gamma) \\
b_1 &= \frac{q}{2}(\sin\beta + \sin\gamma) \\
a_2 &= \frac{1}{4}(1 - \cos 2\alpha + \sin 2\alpha) - \frac{p}{2}(\cos\beta + \cos\gamma) - \frac{q}{2}(\cos\beta - \cos\gamma) \\
b_2 &= \frac{1}{4}(1 + \cos 2\alpha + \sin 2\alpha) - \frac{p}{2}(\sin\beta + \sin\gamma) - \frac{q}{2}(\sin\beta - \sin\gamma), \\
a_3 &= \frac{1}{4}(1 + \cos 2\alpha - \sin 2\alpha) - \frac{p}{2}(\cos\beta - \cos\gamma) - \frac{q}{2}(\cos\beta + \cos\gamma) \\
b_3 &= \frac{1}{4}(1 - \cos 2\alpha - \sin 2\alpha) - \frac{p}{2}(\sin\beta - \sin\gamma) - \frac{q}{2}(\sin\beta + \sin\gamma) \\
a_4 &= \frac{q}{2}(\cos\beta - \cos\gamma) \\
b_4 &= \frac{q}{2}(\sin\beta - \sin\gamma) \\
a_5 &= \frac{p}{2}(\cos\beta - \cos\gamma) \\
b_5 &= \frac{p}{2}(\sin\beta - \sin\gamma),
\end{aligned}$$

satisfies

$$m(0) = 1 \text{ and } |m(\omega)|^2 + |m(\omega+\pi)|^2 = 1.$$

Proof. A simple verification that these coefficients satisfy the nonlinear system completes the proof.

A few example parameterized wavelets were selected where the parameters are given in Table 1. and the graphs of their associated scaling function are given

Table 1: Parameters associated with some example length 12 parameterized scaling functions

Wavelet	α	β	γ
B12	-2.24199155163278	-0.963122858225026	-0.720149522024697
C12	-2.07373764689361	-1.00906384228338	-0.826372187328246
S12	1.3	1.9793277207032078	1.9580372879121213

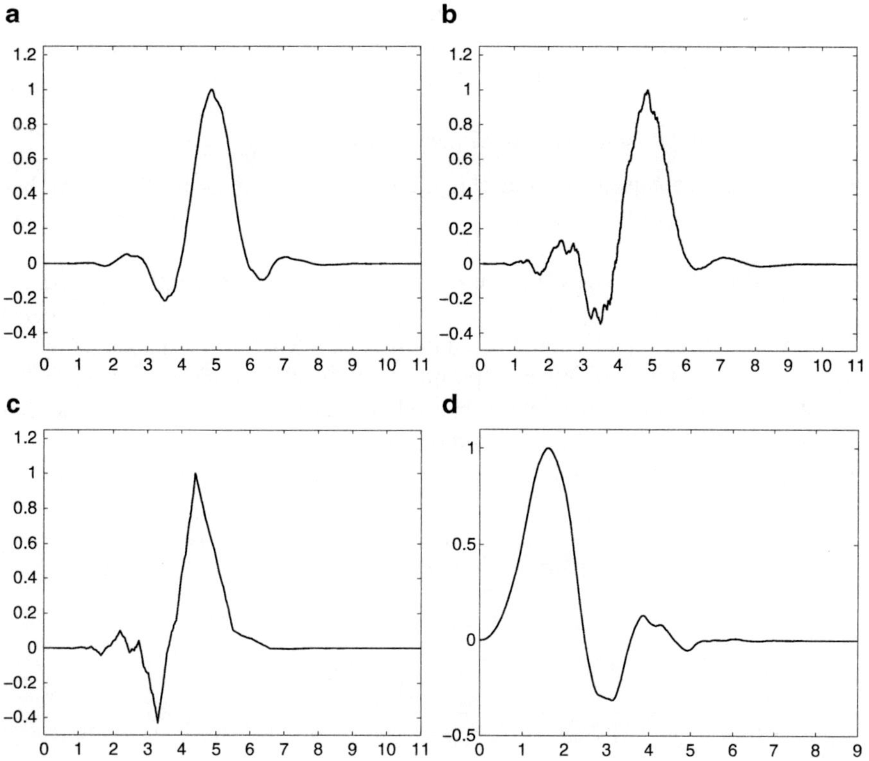

Fig. 1: Graphs for the parameterized length 12 scaling functions given in Table 1. (**a**) B12, (**b**) C12, (**c**) S12, and (**d**) the standard Daubechies scaling function D10 of length 10

in Fig. 1 along with the standard Daubechies scaling function of length ten. The examples B12 and C12 each have only one vanishing moment, but were chosen because of their comparable performance in the image compression scheme. The example S12 has three vanishing moments and the Daubechies standard wavelet D10 has five vanishing moments.

5 A Numerical Experiment

In this section, we present a numerical experiment using image compression as a comparison for the parameterized wavelets B12, C12, and S12, with the standard length ten Daubechies wavelet D10 all with periodic boundary extensions and the FBI biorthogonal 9/7 wavelet with symmetric boundary extensions. Because of its common use as an industry standard and similar length, we chose to include the biorthogonal 9/7 wavelet in the comparison. The 9/7 biorthogonal wavelet has one advantage over the orthogonal parameterized wavelets in that it is symmetric. This symmetry can be used to improve the performance of image compression at the image boundaries. We have included this advantage in our numerical results (Table 2).

The details of the numerical experiment are as follows:

- Eight level decomposition with periodic boundaries (except for 9/7 which has symmetric extensions) using D10, B12, C12, S12, and 9/7 FBI for the seven images in Fig. 2.
- Embedded Zero-tree (EZW) compression (see [12] and [10]) with a file size ratio of 32:1. For this experiment, all of the images are 512×512 with a PGM file-size of 256 Kb and a compressed file-size of 8 Kb. This particular EZW implementation is not completely optimized and would not necessarily yield the maximum PSNR possible but serves well as a comparative measure of the true compressibility of the wavelet decomposition.
- Eight level reconstruction followed by a Peak Signal to Noise Ratio (PSNR), i.e.

$$\text{RMSE} = \sqrt{\frac{1}{512^2} \sum_{i=1}^{512} \sum_{i=1}^{512} |A_{i,j} - \tilde{A}_{i,j}|^2}$$

$$\text{PSNR} = 20 \log_{10} \left(\frac{255}{\text{RMSE}} \right)$$

where $A_{i,j}$ is the original matrix of grayscale values and $\tilde{A}_{i,j}$ is the compressed version.

The results from the experiment are given in Table 2.

Table 2: PSNR results for the seven images Barb, Boat, Lena, Marm, Bark, Fing, and Sand using the wavelets D10, B12, C12, S12, and 9/7 FBI

Wavelet	Barb	Boat	Lena	Marm	Bark	Fing	Sand
D10	26.29	28.74	32.30	34.78	21.27	30.14	23.27
B12	**26.38**	28.74	32.32	**35.37**	**21.45**	**30.25**	23.44
C12	26.22	28.52	31.89	33.92	21.34	30.22	23.33
S12	25.74	28.34	31.58	32.40	21.02	29.66	23.03
FBI9/7	26.30	**29.23**	**32.82**	35.12	21.43	30.14	**23.46**

The best PSNR for each image is boldfaced

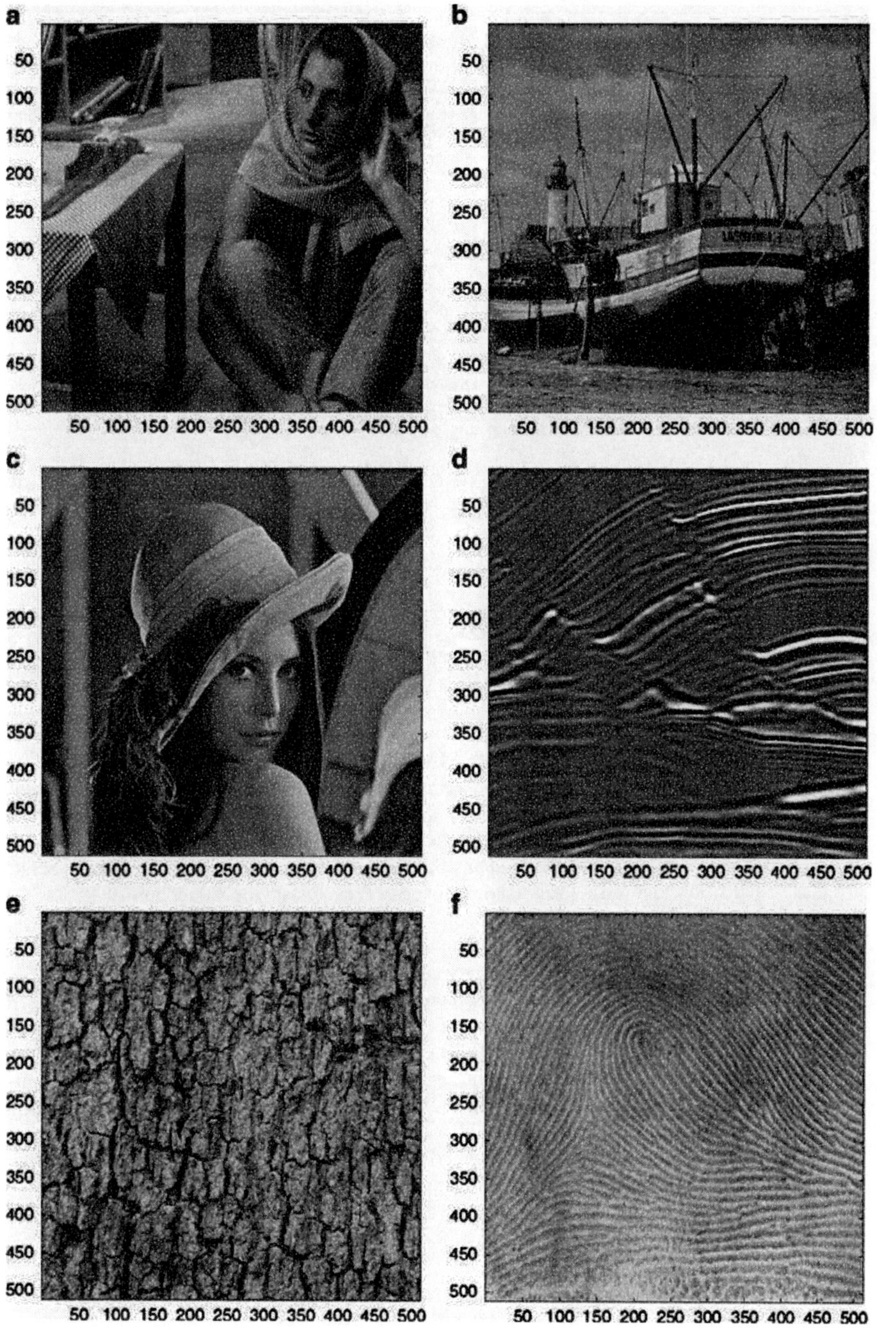
a
b
c
d
e
f
50
100
150
200
250
300
350
400
450
500
50 100 150 200 250 300 350 400 450 500

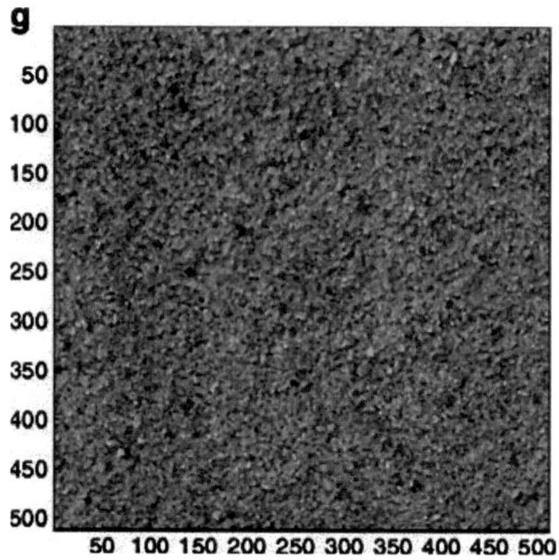

Fig. 2: The seven test images used in the compression scheme (512×512 grayscale images): (**a**) Barb, (**b**) Boat, (**c**) Lena, (**d**) Marm, (**e**) Bark, (**f**) Fing, and (**g**) Sand

References

1. D. Colella and C. Heil, The characterization of continuous, four-coefficient scaling functions and wavelets, IEEE Trans. Inf. Th., Special Issue on Wavelet Transforms and Multiresolution Signal Analysis, 38 (1992), pp. 876-881.
2. I. Daubechies, *Ten Lectures on Wavelets*, SIAM, Philadelphia, 1992.
3. Q.T. Jiang, Paramterization of m-channel orthogonal multifilter banks, *Advances in Computational Mathematics* **12** (2000), 189–211.
4. M. J. Lai and D. W. Roach, Parameterizations of univariate orthogonal wavelets with short support,*Approximation Theory X: Splines, Wavelets, and Applications*, C. K. Chui, L. L. Schumaker, and J. Stockler (eds.), Vanderbilt University Press, Nashville, 2002, 369–384.
5. W. Lawton, Necessary and sufficient conditions for constructing orthonormal wavelet bases, J. Math. Phys.**32** (1991), 57–61.
6. G. Regensburger, Parametrizing compactly supported orthonormal wavelets by discrete moments. Appl. Algebra Eng., Commun. Comput. 18, 6 (Nov. 2007), 583-601.
7. H. L. Resnikoff, J. Tian, R. O. Wells, Jr., Biorthogonal wavelet space: parametrization and factorization, SIAM J. Math. Anal. **33** (2001), no. 1, 194–215.
8. D. W. Roach, The Parameterization of the Length Eight Orthogonal Wavelets with No Parameter Constraints, *Approximation Theory XII: San Antonio 2007*, M. Neamtu and L. Schumaker (eds.), Nashboro Press, pp. 332-347, 2008.
9. D. W. Roach, Frequency selective parameterized wavelets of length ten, *Journal of Concrete and Applicable Mathematics*, vol. 8, no. 1, pp. 1675-179, 2010.
10. A. Said and W. A. Pearlman, A new fast and efficient image codec based on set partitioning in hierarchical trees, *IEEE Transactions on Circuits and Systems for Video Technology* **6** (1996), 243–250.
11. J. Schneid and S. Pittner, On the parametrization of the coefficients of dilation equations for compactly supported wavelets, *Computing* **51** (1993), 165–173.

12. J. M. Shapiro, Embedded image coding using zerotrees of wavelet coefficients, IEEE Transactions Signal Processing **41** (1993), 3445–3462.
13. R. O. Wells, Jr., Parameterizing smooth compactly supported wavelets, Trans. Amer. Math. Soc. **338** (1993), 919–931.

Geometric Properties of Inverse Polynomial Images

Klaus Schiefermayr

Abstract Given a polynomial $\mathscr{T}_n$ of degree n, consider the inverse image of $\mathbb{R}$ and $[-1,1]$, denoted by $\mathscr{T}_n^{-1}(\mathbb{R})$ and $\mathscr{T}_n^{-1}([-1,1])$, respectively. It is well known that $\mathscr{T}_n^{-1}(\mathbb{R})$ consists of n analytic Jordan arcs moving from ∞ to ∞. In this paper, we give a necessary and sufficient condition such that (1) $\mathscr{T}_n^{-1}([-1,1])$ consists of ν analytic Jordan arcs and (2) $\mathscr{T}_n^{-1}([-1,1])$ is connected, respectively.

1 Introduction

Let $\mathbb{P}_n$ be the set of all polynomials of degree n with complex coefficients. For a polynomial $\mathscr{T}_n \in \mathbb{P}_n$, consider the inverse images $\mathscr{T}_n^{-1}(\mathbb{R})$ and $\mathscr{T}_n^{-1}([-1,1])$, defined by

$$\mathscr{T}_n^{-1}(\mathbb{R}) := \{z \in \mathbb{C} : \mathscr{T}_n(z) \in \mathbb{R}\} \tag{1}$$

and

$$\mathscr{T}_n^{-1}([-1,1]) := \{z \in \mathbb{C} : \mathscr{T}_n(z) \in [-1,1]\}, \tag{2}$$

respectively. It is well known that $\mathscr{T}_n^{-1}(\mathbb{R})$ consists of n analytic Jordan arcs moving from ∞ to ∞ which cross each other at points which are zeros of the derivative $\mathscr{T}_n'$. In [11], Peherstorfer proved that $\mathscr{T}_n^{-1}(\mathbb{R})$ may be split up into n Jordan arcs (not necessarily analytic) moving from ∞ to ∞ with the additional property that $\mathscr{T}_n$ is strictly monotone decreasing from $+\infty$ to $-\infty$ on each of the n Jordan arcs. Thus,

Klaus Schiefermayr
University of Applied Sciences Upper Austria, School of Engineering and Environmental Sciences, Stelzhamerstr. 23, 4600 Wels, Austria
e-mail: klaus.schiefermayr@fh-wels.at

M. Neamtu and L. Schumaker (eds.), *Approximation Theory XIII: San Antonio 2010*, Springer Proceedings in Mathematics 13, DOI 10.1007/978-1-4614-0772-0_17,

$\mathcal{T}_n^{-1}([-1,1])$ is the union of n (analytic) Jordan arcs and is obtained from $\mathcal{T}_n^{-1}(\mathbb{R})$ by cutting off the n arcs of $\mathcal{T}_n^{-1}(\mathbb{R})$. In [14, Theorem 3], we gave a necessary and sufficient condition such that $\mathcal{T}_n^{-1}([-1,1])$ consists of 2 Jordan arcs, compare also [5], where the proof can easily be extended to the case of ℓ arcs, see also [11, Remark after Corollary 2.2]. In the present paper, we will give a necessary and sufficient condition such that (1) $\mathcal{T}_n^{-1}([-1,1])$ consists of ν (but not less than ν) *analytic* Jordan arcs (in Sect. 2) and (2) $\mathcal{T}_n^{-1}([-1,1])$ is connected (in Sect. 3), respectively. From a different point of view as in this paper, inverse polynomial images are considered, e.g., in [6, 7, 15], and [8].

Inverse polynomial images are interesting for instance in approximation theory, since each polynomial (suitable normed) of degree n is the minimal polynomial with respect to the maximum norm on its inverse image, see [2, 4, 10], and [3].

2 The Number of (Analytic) Jordan Arcs of an Inverse Polynomial Image

Let us start with a collection of important properties of the inverse images $\mathcal{T}_n^{-1}(\mathbb{R})$ and $\mathcal{T}_n^{-1}([-1,1])$. Most of them are due to Peherstorfer [11] or classical well known results. Let us point out that $\mathcal{T}_n^{-1}(\mathbb{R})$ (and also $\mathcal{T}_n^{-1}([-1,1])$), on the one hand side, may be characterized by n analytic Jordan arcs and, on the other side, by n (not necessarily analytic) Jordan arcs, on which $\mathcal{T}_n$ is strictly monotone.

Let $C := \{\gamma(t) : t \in [0,1]\}$ be an analytic Jordan arc in $\mathbb{C}$ and let $\mathcal{T}_n \in \mathbb{P}_n$ be a polynomial such that $\mathcal{T}_n(\gamma(t)) \in \mathbb{R}$ for all $t \in [0,1]$. We call a point $z_0 = \gamma(t_0)$ a *saddle point* of $\mathcal{T}_n$ on C if $\mathcal{T}_n'(z_0) = 0$ and z_0 is no extremum of $\mathcal{T}_n$ on C.

Lemma 1. *Let $\mathcal{T}_n \in \mathbb{P}_n$ be a polynomial of degree n.*

(i) $\mathcal{T}_n^{-1}(\mathbb{R})$ consists of n analytic Jordan arcs, denoted by $\tilde{C}_1, \tilde{C}_2, \ldots, \tilde{C}_n$, in the complex plane running from ∞ to ∞.

(ii) $\mathcal{T}_n^{-1}(\mathbb{R})$ consists of n Jordan arcs, denoted by $\tilde{\Gamma}_1, \tilde{\Gamma}_2, \ldots, \tilde{\Gamma}_n$, in the complex plane running from ∞ to ∞, where on each $\tilde{\Gamma}_j$, $j = 1, 2, \ldots, n$, $\mathcal{T}_n(z)$ is strictly monotone decreasing from $+\infty$ to $-\infty$.

(iii) A point $z_0 \in \mathcal{T}_n^{-1}(\mathbb{R})$ is a crossing point of exactly m, $m \geq 2$, analytic Jordan arcs $\tilde{C}_{i_1}, \tilde{C}_{i_2}, \ldots, \tilde{C}_{i_m}$, $1 \leq i_1 < i_2 < \cdots < i_m \leq n$, if and only if z_0 is a zero of $\mathcal{T}_n'$ with multiplicity $m-1$. In this case, the m arcs are cutting each other at z_0 in successive angles of π/m. If m is odd then z_0 is a saddle point of $\operatorname{Re}\{\mathcal{T}_n(z)\}$ *on each of the m arcs. If m is even then, on $m/2$ arcs, z_0 is a minimum of* $\operatorname{Re}\{\mathcal{T}_n(z)\}$ *and on the other $m/2$ arcs, z_0 is a maximum of* $\operatorname{Re}\{\mathcal{T}_n(z)\}$.

(iv) A point $z_0 \in \mathcal{T}_n^{-1}(\mathbb{R})$ is a crossing point of exactly m, $m \geq 2$, Jordan arcs $\tilde{\Gamma}_{i_1}$, $\tilde{\Gamma}_{i_2}, \ldots, \tilde{\Gamma}_{i_m}$, $1 \leq i_1 < i_2 < \cdots < i_m \leq n$, if and only if z_0 is a zero of $\mathcal{T}_n'$ with multiplicity $m-1$.

(v) $\mathcal{T}_n^{-1}([-1,1])$ consists of n analytic Jordan arcs, denoted by $C_1, C_2, \ldots, C_n$, where the $2n$ zeros of $\mathcal{T}_n^2 - 1$ are the endpoints of the n arcs. If $z_0 \in \mathbb{C}$ is a zero of $\mathcal{T}_n^2 - 1$ of multiplicity m, then exactly m analytic Jordan arcs

$C_{i_1}, C_{i_2}, \ldots, C_{i_m}$ of $\mathscr{T}_n^{-1}([-1,1])$, $1 \le i_1 < i_2 < \cdots < i_m \le n$, have z_0 as common endpoint.

(vi) *$\mathscr{T}_n^{-1}([-1,1])$ consists of n Jordan arcs, denoted by $\Gamma_1, \Gamma_2, \ldots, \Gamma_n$, with $\Gamma_j \subset \tilde{\Gamma}_j$, $j = 1,2,\ldots,n$, where on each Γ_j, $\mathscr{T}_n(z)$ is strictly monotone decreasing from $+1$ to -1. If $z_0 \in \mathbb{C}$ is a zero of $\mathscr{T}_n^2 - 1$ of multiplicity m then exactly m Jordan arcs $\Gamma_{i_1}, \ldots, \Gamma_{i_m}$ of $\mathscr{T}_n^{-1}([-1,1])$, $1 \le i_1 < i_2 < \cdots < i_m \le n$, have z_0 as common endpoint.*

(vii) *Two arcs C_j, C_k, $j \ne k$, cross each other at most once (the same holds for Γ_j, Γ_k).*

(viii) *Let $S := \mathscr{T}_n^{-1}([-1,1])$, then the complement $\mathbb{C} \setminus S$ is connected.*

(ix) *Let $S := \mathscr{T}_n^{-1}([-1,1])$ then, for $P_n(z) := \mathscr{T}_n((z-b)/a)$, $a, b \in \mathbb{C}$, $a \ne 0$, the inverse image is $P_n^{-1}([-1,1]) = aS + b$.*

(x) *$\mathscr{T}_n^{-1}([-1,1]) \subseteq \mathbb{R}$ if and only if the coefficients of $\mathscr{T}_n$ are real, $\mathscr{T}_n$ has n simple real zeros and $\min\{|\mathscr{T}_n(z)| : \mathscr{T}_n'(z) = 0\} \ge 1$.*

(xi) *$\mathscr{T}_n^{-1}(\mathbb{R})$ is symmetric with respect to the real line if and only if $\mathscr{T}_n(z)$ or $\mathrm{i}\mathscr{T}_n(z)$ has real coefficients only.*

Proof. (i), (iii), (iv), and (xi) are well known.
For (ii), see [11, Theorem 2.2].
Concerning the connection between (iii),(iv) and (v),(vi) note that each zero z_0 of $Q_{2n}(z) = \mathscr{T}_n^2(z) - 1 \in \mathbb{P}_{2n}$ with multiplicity m is a zero of $Q_{2n}'(z) = 2\mathscr{T}_n(z)\,\mathscr{T}_n'(z)$ with multiplicity $m-1$, hence a zero of $\mathscr{T}_n'(z)$ with multiplicity $m-1$. Thus, (v) and (vi) follow immediately from (i) and (iii) and (ii) and (iv), respectively.
(vii) follows immediately from (viii).
Concerning (viii), suppose that there exists a simple connected domain B, which is surrounded by a subset of $\mathscr{T}_n^{-1}([-1,1])$. Then the harmonic function $v(x,y) := \mathrm{Im}\{\mathscr{T}_n(x+\mathrm{i}y)\}$ is zero on ∂B thus, by the maximum principle, $v(x,y)$ is zero on B, which is a contradiction.
(ix) follows from the definition of $\mathscr{T}_n^{-1}([-1,1])$.
For (x), see [11, Corrolary 2.3].

Example 1. Consider the polynomial $\mathscr{T}_n(z) := 1 + z^2(z-1)^3(z-2)^4$ of degree $n = 9$. Figure 1 shows the inverse images $\mathscr{T}_n^{-1}([-1,1])$ (solid line) and $\mathscr{T}_n^{-1}(\mathbb{R})$ (dotted and solid line). The zeros of $\mathscr{T}_n + 1$ and $\mathscr{T}_n - 1$ are marked with a circle and a disk, respectively. One can easily identify the $n = 9$ analytic Jordan arcs $\tilde{C}_1, \tilde{C}_2, \ldots, \tilde{C}_n$ which $\mathscr{T}_n^{-1}(\mathbb{R})$ consists of, compare Lemma 1 (i), and the $n = 9$ analytic Jordan arcs $C_1, C_2, \ldots, C_n$ which $\mathscr{T}_n^{-1}([-1,1])$ consists of, compare Lemma 1 (v), where the endpoints of the arcs are exactly the circles and disks, i.e., the zeros of $\mathscr{T}_n^2 - 1$. Note that $\tilde{C}_1 = \mathbb{R}$, $C_1 = [-0.215\ldots, 0]$ and $C_2 = [0,1]$.

Before we state the result concerning the minimal number of analytic Jordan arcs $\mathscr{T}_n^{-1}([-1,1])$ consists of, let us do some preparations. Let $\mathscr{T}_n \in \mathbb{P}_n$ and consider the zeros of the polynomial $\mathscr{T}_n^2 - 1 \in \mathbb{P}_{2n}$. Let $\{a_1, a_2, \ldots, a_{2\ell}\}$ be the set of all zeros of $\mathscr{T}_n^2 - 1$ with *odd* multiplicity, where $a_1, a_2, \ldots, a_{2\ell}$ are pairwise distinct and each a_j

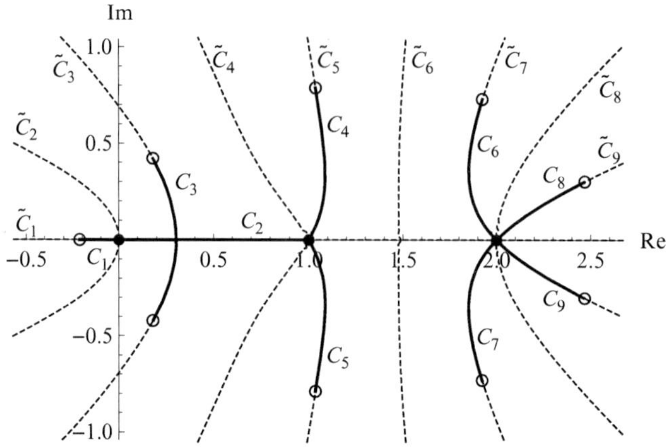

Fig. 1: Inverse images $\mathscr{T}_9^{-1}([-1,1])$ (*solid line*) and $\mathscr{T}_9^{-1}(\mathbb{R})$ (*dotted and solid line*) for the polynomial $\mathscr{T}_9(z) := 1 + z^2(z-1)^3(z-2)^4$

has multiplicity $2\beta_j - 1$, $j = 1, \ldots, 2\ell$. Further, let

$$(b_1, b_2, \ldots, b_{2\nu}) := (\underbrace{a_1, \ldots, a_1}_{(2\beta_1-1)-\text{times}}, \underbrace{a_2, \ldots, a_2}_{(2\beta_2-1)-\text{times}}, \ldots, \underbrace{a_{2\ell}, \ldots, a_{2\ell}}_{(2\beta_{2\ell}-1)-\text{times}}), \tag{3}$$

thus

$$2\nu = \sum_{j=1}^{2\ell} (2\beta_j - 1), \tag{4}$$

i.e., $b_1, b_2, \ldots, b_{2\nu}$ are the zeros of odd multiplicity *written according to their multiplicity*.

Theorem 1. *Let $\mathscr{T}_n \in \mathbb{P}_n$ be any polynomial of degree n. Then, $\mathscr{T}_n^{-1}([-1,1])$ consists of ν (but not less than ν) analytic Jordan arcs with endpoints $b_1, b_2, \ldots, b_{2\nu}$ if and only if $\mathscr{T}_n^2 - 1$ has exactly 2ν zeros $b_1, b_2, \ldots, b_{2\nu}$ (written according to their multiplicity) of odd multiplicity.*

Proof. By Lemma 1 (v), $\mathscr{T}_n^{-1}([-1,1])$ consists of n analytic Jordan arcs $C_1, C_2, \ldots, C_n$, which can be combined into ν analytic Jordan arcs in the following way. Clearly, two analytic Jordan arcs C_{i_1} and C_{i_2} can be joined together into one analytic Jordan arc if they have the same endpoint, which is a zero of $\mathscr{T}_n^2 - 1$, and if they lie on the same analytic Jordan arc $\tilde{C}_{i_3}$ of Lemma 1 (i). By Lemma 1 (iii) and (v), such combinations are possible only at the zeros of $\mathscr{T}_n^2 - 1$ of *even* multiplicity. More precisely, let $d_1, d_2, \ldots, d_k$ be the zeros of $\mathscr{T}_n^2 - 1$ with even multiplicities $2\alpha_1, 2\alpha_2, \ldots, 2\alpha_k$, where, by assumption,

$$2\alpha_1 + 2\alpha_2 + \cdots + 2\alpha_k = 2n - 2\nu.$$

By Lemma 1 (iii) and (v), at each point d_j, the $2\alpha_j$ analytic Jordan arcs of $\mathscr{T}_n^{-1}([-1,1])$ can be combined into α_j analytic arcs, $j=1,2,\ldots,k$. Altogether, the number of such combinations is $\alpha_1+\alpha_2+\cdots+\alpha_k=n-\nu$, thus the total number of n analytic Jordan arcs is reduced by $n-\nu$, hence ν analytic Jordan arcs remain and the sufficiency part is proved. Since, for each polynomial $\mathscr{T}_n\in\mathbb{P}_n$, there is a unique $\nu\in\{1,2,\ldots,n\}$ such that $\mathscr{T}_n^2-1$ has exactly 2ν zeros of odd multiplicity (counted with multiplicity), the necessity part follows.

Example 2. For a better understanding of the combination of two analytic Jordan arcs into one analytic Jordan arc, as done in the proof of Theorem 1, let us again consider the inverse image of the polynomial of Example 1.

- The point $d_1=0$ is a zero of $\mathscr{T}_n-1$ with multiplicity $2\alpha_1=2$, thus 2 analytic Jordan arcs, here C_1 and C_2, have d_1 as endpoint, compare Lemma 1 (v). Along the arc $\tilde{C}_1$, d_1 is a maximum, along the arc $\tilde{C}_2$, d_1 is a minimum, compare Lemma 1 (iii), thus the 2 analytic Jordan arcs C_1 and C_2 can be joined together into one analytic Jordan arc $C_1\cup C_2$.
- The point $d_2=2$ is a zero of $\mathscr{T}_n-1$ with multiplicity $2\alpha_2=4$, thus 4 analytic Jordan arcs, here C_6, C_7, C_8 and C_9, have d_2 as endpoint. Along the arc $\tilde{C}_7$ or $\tilde{C}_9$, d_3 is a maximum, along the arc $\tilde{C}_8$ or $\tilde{C}_1$, d_3 is a minimum, compare Lemma 1 (iii). Hence, the analytic Jordan arcs C_6 and C_9 can be combined into one analytic Jordan arc $C_6\cup C_9$, analogously C_7 and C_8 can be combined into $C_7\cup C_8$.
- The point $a_1=1$ is a zero of $\mathscr{T}_n-1$ with multiplicity 3, thus 3 analytic Jordan arcs, here C_2, C_4 and C_5, have a_1 as endpoint. Since a_1 is a saddle point along each of the three analytic Jordan arcs $\tilde{C}_1,\tilde{C}_4,\tilde{C}_5$, compare Lemma 1 (iii), no combination of arcs can be done.

Altogether, we get $\alpha_1+\alpha_2=3=n-\nu$ combinations and therefore $\mathscr{T}_n^{-1}([-1,1])$ consists of $\nu=6$ analytic Jordan arcs, which are given by $C_1\cup C_2$, C_3, C_4, C_5, $C_6\cup C_9$ and $C_7\cup C_8$.

Lemma 2. *For any polynomial $\mathscr{T}_n(z)=c_nz^n+\cdots\in\mathbb{P}_n$, $c_n\in\mathbb{C}\setminus\{0\}$, there exists a unique $\ell\in\{1,2,\ldots,n\}$, a unique monic polynomial $\mathscr{H}_{2\ell}(z)=z^{2\ell}+\cdots\in\mathbb{P}_{2\ell}$ with pairwise distinct zeros $a_1,a_2,\ldots,a_{2\ell}$, i.e.,*

$$\mathscr{H}_{2\ell}(z)=\prod_{j=1}^{2\ell}(z-a_j), \tag{5}$$

and a unique polynomial $\mathscr{U}_{n-\ell}(z)=c_nz^{n-\ell}+\cdots\in\mathbb{P}_{n-\ell}$ with the same leading coefficient c_n such that the polynomial equation

$$\mathscr{T}_n^2(z)-1=\mathscr{H}_{2\ell}(z)\,\mathscr{U}_{n-\ell}^2(z) \tag{6}$$

holds. Note that the points $a_1,a_2,\ldots,a_{2\ell}$ are exactly those zeros of $\mathscr{T}_n^2-1$ which have odd multiplicity.

Proof. The assertion follows immediately by the fundamental theorem of algebra for the polynomial $Q_{2n}(z):=\mathscr{T}_n^2(z)-1=c_n^2z^{2n}+\cdots\in\mathbb{P}_{2n}$, where 2ℓ is the number of distinct zeros of Q_{2n} with odd multiplicity. It only remains to show that the case $\ell = 0$ is not possible. If $\ell = 0$, then all zeros of Q_{2n} are of even multiplicity. Thus there are at least n zeros (counted with multiplicity) of Q_{2n}' which are also zeros of Q_{2n} but not zeros of $\mathscr{T}_n$. Since $Q_{2n}'(z) = 2\,\mathscr{T}_n(z)\,\mathscr{T}_n'(z)$, there are at least n zeros (counted with multiplicity) of $\mathscr{T}_n'$, which is a contradiction.

Let us point out that the polynomial equation (6) (sometimes called Pell equation) is the starting point for investigations concerning minimal or orthogonal polynomials on several intervals, see, e.g., [7, 9, 10, 12, 13, 18], and [19].

In [14, Theorem 3], we proved that the polynomial equation (6) (for $\ell = 2$) is equivalent to the fact that $\mathscr{T}_n^{-1}([-1,1])$ consists of two Jordan arcs (not necessarily analytic), compare also [5]. The condition and the proof can be easily extended to the general case of ℓ arcs, compare also [11, Remark after Corollary 2.2]. In addition, we give an alternative proof similar to that of Theorem 1.

Theorem 2. *Let $\mathscr{T}_n \in \mathbb{P}_n$ be any polynomial of degree n. Then $\mathscr{T}_n^{-1}([-1,1])$ consists of ℓ (but not less than ℓ) Jordan arcs with endpoints $a_1, a_2, \ldots, a_{2\ell}$ if and only if $\mathscr{T}_n^2 - 1$ has exactly 2ℓ pairwise distinct zeros $a_1, a_2, \ldots, a_{2\ell}$, $1 \le \ell \le n$, of odd multiplicity, i.e., if and only if $\mathscr{T}_n$ satisfies a polynomial equation of the form* (6) *with $\mathscr{H}_{2\ell}$ given in* (5).

Proof. By Lemma 1 (vi), $\mathscr{T}_n^{-1}([-1,1])$ consists of n Jordan arcs $\Gamma_1, \Gamma_2, \ldots, \Gamma_n$, which can be combined into ℓ Jordan arcs in the following way: Let $d_1, d_2, \ldots, d_k$ be those zeros of $\mathscr{T}_n^2 - 1$ with *even* multiplicities $2\alpha_1, 2\alpha_2, \ldots, 2\alpha_k$ and let, as assumed in the Theorem, $a_1, a_2, \ldots, a_{2\ell}$ be those zeros of $\mathscr{T}_n^2 - 1$ with *odd* multiplicities $2\beta_1 - 1, 2\beta_2 - 1, \ldots, 2\beta_{2\ell} - 1$, where

$$2\alpha_1 + 2\alpha_2 + \cdots + 2\alpha_k + (2\beta_1 - 1) + (2\beta_2 - 1) + \cdots + (2\beta_{2\ell} - 1) = 2n \qquad (7)$$

holds. By Lemma 1 (vi), at each point d_j, the $2\alpha_j$ Jordan arcs can be combined into α_j Jordan arcs, $j = 1, 2, \ldots, \nu$, and at each point a_j, the $2\beta_j - 1$ Jordan arcs can be combined into β_j Jordan arcs, $j = 1, 2, \ldots, 2\ell$. Altogether, the number of such combinations, using (7), is

$$\alpha_1 + \alpha_2 + \cdots + \alpha_\nu + (\beta_1 - 1) + (\beta_2 - 1) + \cdots + (\beta_{2\ell} - 1) = (n + \ell) - 2\ell = n - \ell,$$

i.e., the total number n of Jordan arcs is reduced by $n - \ell$, thus ℓ Jordan arcs remain and the sufficiency part is proved. Since, by Lemma 2, for each polynomial $\mathscr{T}_n \in \mathbb{P}_n$ there is a unique $\ell \in \{1, 2, \ldots, n\}$ such that $\mathscr{T}_n^2 - 1$ has exactly 2ℓ distinct zeros of odd multiplicity, the necessity part is clear.

Example 3. Similar as after the proof of Theorem 1, let us illustrate the combination of Jordan arcs by the polynomial of Example 1. Taking a look at Fig. 1, one can easily identify the $n = 9$ Jordan arcs $\Gamma_1, \Gamma_2, \ldots, \Gamma_n \in \mathscr{T}_n^{-1}([-1,1])$, where each arc Γ_j runs from a disk to a circle. Note that the two arcs, which cross at $z \approx 0.3$, may be chosen in two different ways. Now, $\mathscr{T}_n^2 - 1$ has the zero $d_1 = 0$ with multiplicity $2\alpha_1 = 2$, the zero $d_2 = 2$ with multiplicity $2\alpha_2 = 4$, and a zero $a_1 = 1$

with multiplicity $2\beta_1 - 1 = 3$, all other zeros a_j have multiplicity $2\beta_j - 1 = 1$, $j = 2,3,\ldots,2\ell$. Thus, it is possible to have one combination at $d_1 = 0$, two combinations at $d_2 = 2$ and one combination of Jordan arcs at $a_1 = 1$. Altogether, we obtain $\alpha_1 + \alpha_2 + (\beta_1 - 1) = 4 = n - \ell$ combinations and the number of Jordan arcs is $\ell = 5$.

For the sake of completeness, let us mention two simple special cases, first the case $\ell = 1$, see, e.g., [14, Remark 4], and second, the case when all endpoints $a_1, a_2, \ldots, a_{2\ell}$ of the arcs are real, see [9].

Corollary 1. *Let $\mathcal{T}_n \in \mathbb{P}_n$.*

(i) $\mathcal{T}_n^{-1}([-1,1])$ consists of $\ell = 1$ Jordan arc with endpoints $a_1, a_2 \in \mathbb{C}$, $a_1 \neq a_2$, if and only if $\mathcal{T}_n$ is the classical Chebyshev polynomial of the first kind (suitable normed), i.e., $\mathcal{T}_n(z) = T_n((2z - a_1 - a_2)/(a_2 - a_1))$, where $T_n(z) := \cos(n \arccos z)$. In this case, $\mathcal{T}_n^{-1}([-1,1])$ is the complex interval $[a_1, a_2]$.

(ii) $\mathcal{T}_n^{-1}([-1,1]) = [a_1,a_2] \cup [a_3,a_4] \cup \ldots \cup [a_{2\ell-1}, a_{2\ell}]$, $a_1, a_2, \ldots, a_{2\ell} \in \mathbb{R}$, $a_1 < a_2 < \cdots < a_{2\ell}$, if and only if $\mathcal{T}_n$ satisfies the polynomial equation (6) with $\mathcal{H}_{2\ell}$ as in (5) and $a_1, a_2, \ldots, a_{2\ell} \in \mathbb{R}$, $a_1 < a_2 < \cdots < a_{2\ell}$.

Let us consider the case of $\ell = 2$ Jordan arcs in more detail. Given four pairwise distinct points $a_1, a_2, a_3, a_4 \in \mathbb{C}$ in the complex plane, define

$$\mathcal{H}_4(z) := (z - a_1)(z - a_2)(z - a_3)(z - a_4), \tag{8}$$

and suppose that $\mathcal{T}_n(z) = c_n z^n + \cdots \in \mathbb{P}_n$ satisfies a polynomial equation of the form

$$\mathcal{T}_n^2(z) - 1 = \mathcal{H}_4(z)\, \mathcal{U}_{n-2}^2(z) \tag{9}$$

with $\mathcal{U}_{n-2}(z) = c_n z^{n-2} + \cdots \in \mathbb{P}_{n-2}$. Then, by (9), there exists a $z^* \in \mathbb{C}$ such that the derivative of $\mathcal{T}_n$ is given by

$$\mathcal{T}_n'(z) = n(z - z^*)\, \mathcal{U}_{n-2}(z). \tag{10}$$

By Theorem 2, $\mathcal{T}_n^{-1}([-1,1])$ consists of two Jordan arcs. Moreover, it is proved in [14, Theorem 3] that the two Jordan arcs are crossing each other if and only if $z^* \in \mathcal{T}_n^{-1}([-1,1])$ (compare also Theorem 4). In this case, z^* is the only crossing point. Interestingly, the minimum number of analytic Jordan arcs is not always two, as the next theorem says. In order to prove this result, we need the following lemma [14, Lemma 1].

Lemma 3. *Suppose that $\mathcal{T}_n \in \mathbb{P}_n$ satisfies a polynomial equation of the form (9), where $\mathcal{H}_4$ is given by (8), and let z^* be given by (10).*

(i) If z^* is a zero of $\mathcal{U}_{n-2}$ then it is either a double zero of $\mathcal{U}_{n-2}$ or a zero of $\mathcal{H}$.

(ii) If z^* is a zero of $\mathcal{H}$ then z^* is a simple zero of $\mathcal{U}_{n-2}$.

(iii) The point z^* is the only possible common zero of $\mathcal{H}$ and $\mathcal{U}_{n-2}$.

(iv) If $\mathcal{U}_{n-2}$ has a zero y^* of order greater than one then $y^* = z^*$ and z^* is a double zero of $\mathcal{U}_{n-2}$.

Theorem 3. *Suppose that $\mathcal{T}_n \in \mathbb{P}_n$ satisfies a polynomial equation of the form* (9)*, where $\mathcal{H}_4$ is given by* (8)*, and let z^* be given by* (10)*. If $z^* \notin \{a_1,a_2,a_3,a_4\}$ then $\mathcal{T}_n^{-1}([-1,1])$ consists of two analytic Jordan arcs. If $z^* \in \{a_1,a_2,a_3,a_4\}$ then $\mathcal{T}_n^{-1}([-1,1])$ consists of three analytic Jordan arcs, all with one endpoint at z^*, and an angle of $2\pi/3$ between two arcs at z^*.*

Proof. We distinguish two cases:

1. $\mathcal{T}_n(z^*) \notin \{-1,1\}$: By Lemma 3, $\mathcal{T}_n^2 - 1$ has 4 simple zeros $\{a_1,a_2,a_3,a_4\}$ and $n-2$ double zeros. Thus, by Theorem 1, $\mathcal{T}_n^{-1}([-1,1])$ consists of two analytic Jordan arcs.
2. $\mathcal{T}_n(z^*) \in \{-1,1\}$:

   2.1 If $z^* \in \{a_1,a_2,a_3,a_4\}$ then, by Lemma 3, $\mathcal{T}_n^2 - 1$ has 3 simple zeros given by $\{a_1,a_2,a_3,a_4\} \setminus \{z^*\}$, $n-3$ double zeros and one zero of multiplicity 3 (that is z^*). Thus, by Theorem 1, $\mathcal{T}_n^{-1}([-1,1])$ consists of three analytic Jordan arcs.

   2.2 If $z^* \notin \{a_1,a_2,a_3,a_4\}$ then, by Lemma 3, z^* is a double zero of $\mathcal{U}_{n-2}$. Thus $\mathcal{T}_n^2 - 1$ has 4 simple zeros $\{a_1,a_2,a_3,a_4\}$, $n-4$ double zeros and one zero of multiplicity 4 (that is z^*). Thus, by Theorem 1, $\mathcal{T}_n^{-1}([-1,1])$ consists of two analytic Jordan arcs.

The very last statement of the theorem follows immediately by Lemma 1 (iii).

Let us mention that in [14], see also [16] and [17], necessary and sufficient conditions for four points $a_1,a_2,a_3,a_4 \in \mathbb{C}$ are given with the help of Jacobian elliptic functions such that there exists a polynomial of degree n whose inverse image consists of two Jordan arcs with the four points as endpoints. Concluding this section, let us give two simple examples of inverse polynomial images.

Example 4.

(i) Let $a_1 = -1$, $a_2 = -a$, $a_3 = a$ and $a_4 = 1$ with $0 < a < 1$ and

$$\mathcal{H}_4(z) = (z-a_1)(z-a_2)(z-a_3)(z-a_4) = (z^2-1)(z^2-a^2).$$

If

$$\mathcal{T}_2(z) := \frac{2z^2-a^2-1}{1-a^2}, \quad \mathcal{U}_0(z) := \frac{2}{1-a^2},$$

then

$$\mathcal{T}_2^2(z) - \mathcal{H}_4(z)\mathcal{U}_0^2(z) = 1.$$

Thus, by Theorem 2, $\mathcal{T}_2^{-1}([-1,1])$ consists of two Jordan arcs with endpoints a_1, a_2, a_3, a_4, more precisely $\mathcal{T}_2^{-1}([-1,1]) = [-1,-a] \cup [a,1]$.

(ii) Let $a_1 = \mathrm{i}$, $a_2 = -\mathrm{i}$, $a_3 = a - \mathrm{i}$ and $a_4 = a + \mathrm{i}$ with $a > 0$ and

$$\mathcal{H}_4(z) = (z-a_1)(z-a_2)(z-a_3)(z-a_4) = (z^2+1)((z-a)^2+1).$$

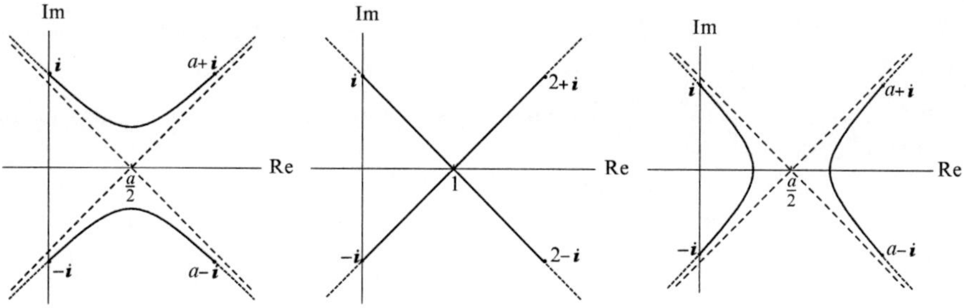

Fig. 2: The inverse image $\mathscr{T}_2^{-1}([-1,1])$ for $0<a<2$ (*left plot*), for $a=2$ (*middle plot*) and for $a>2$ (*right plot*)

If

$$\mathscr{T}_2(z) := \frac{\mathrm{i}}{a}\big(z^2-az+1\big), \quad \mathscr{U}_0(z) := \frac{\mathrm{i}}{a},$$

then

$$\mathscr{T}_2^2(z)-\mathscr{H}_4(z)\mathscr{U}_0^2(z)=1.$$

Thus, by Theorem 2, $\mathscr{T}_2^{-1}([-1,1])$ consists of two Jordan arcs with endpoints a_1, a_2, a_3, a_4. More precisely, if $0<a<2$,

$$\mathscr{T}_2^{-1}([-1,1]) = \big\{x+\mathrm{i}y\in\mathbb{C} : -\frac{(x-a/2)^2}{1-a^2/4}+\frac{y^2}{1-a^2/4}=1\big\},$$

i.e., $\mathscr{T}_2^{-1}([-1,1])$ is an equilateral hyperbola (not crossing the real line) with center at $z_0=a/2$ and asymptotes $y=\pm(x-a/2)$.
If $a=2$, $\mathscr{T}_2^{-1}([-1,1]) = [\mathrm{i},a-\mathrm{i}]\cup[-\mathrm{i},a+\mathrm{i}]$, i.e., the union of two complex intervals.
If $2<a<\infty$,

$$\mathscr{T}_2^{-1}([-1,1]) = \big\{x+\mathrm{i}y\in\mathbb{C} : \frac{(x-a/2)^2}{a^2/4-1}-\frac{y^2}{a^2/4-1}=1\big\},$$

i.e., $\mathscr{T}_2^{-1}([-1,1])$ is an equilateral hyperbola with center at $z_0=a/2$, crossing the real line at $a/2\pm\sqrt{a^2/4-1}$ and asymptotes $y=\pm(x-a/2)$.
In Fig. 2, the sets $\mathscr{T}_2^{-1}([-1,1])$ including the asymptotes are plotted for the three cases discussed above.

3 The Connectedness of an Inverse Polynomial Image

In the next theorem, we give a necessary and sufficient condition such that the inverse image is connected.

Theorem 4. *Let $\mathscr{T}_n \in \mathbb{P}_n$. The inverse image $\mathscr{T}_n^{-1}([-1,1])$ is connected if and only if all zeros of the derivative $\mathscr{T}_n'$ lie in $\mathscr{T}_n^{-1}([-1,1])$.*

Proof. Let $\Gamma := \{\Gamma_1, \Gamma_2, \ldots, \Gamma_n\}$ denote the set of arcs of $\mathscr{T}_n^{-1}([-1,1])$ as in Lemma 1 (vi).

"$\Longleftarrow$": Suppose that all zeros of $\mathscr{T}_n'$ lie in Γ. Let $A_1 \in \Gamma$ be such that it contains at least one zero z_1 of $\mathscr{T}_n'$ with multiplicity $m_1 \geq 1$. By Lemma 1 (ii), (iv) and (vi), there are m_1 additional arcs $A_2, A_3, \ldots, A_{m_1+1} \in \Gamma$ containing z_1. By Lemma 1 (vii),

$$A_j \cap A_k = \{z_1\} \text{ for } j,k \in \{1,2,\ldots,m_1+1\},\ j \neq k.$$

Now assume that there is another zero z_2 of $\mathscr{T}_n'$, $z_2 \neq z_1$, with multiplicity m_2, on A_{j^*}, $j^* \in \{1,2,\ldots,m_1+1\}$. Since no arc A_j, $j \in \{1,2,\ldots,m_1+1\} \setminus \{j^*\}$ contains z_2, there are m_2 curves $A_{m_1+1+j} \in \Gamma$, $j = 1,2,\ldots,m_2$, which cross each other at z_2 and for which, by Lemma 1 (vii),

$$\begin{aligned}
A_j \cap A_k &= \{z_2\} && \text{for} \quad j,k \in \{m_1+2,\ldots,m_1+m_2+1\}, j \neq k,\\
A_j \cap A_k &= \emptyset && \text{for} \quad j \in \{1,2,\ldots,m_1+1\} \setminus \{j^*\},\\
& && \phantom{\text{for}} \quad k \in \{m_1+2,\ldots,m_1+m_2+1\}\\
A_{j^*} \cap A_k &= \{z_2\} && \text{for} \quad k \in \{m_1+2,\ldots,m_1+m_2+1\}.
\end{aligned}$$

If there is another zero z_3 of $\mathscr{T}_n'$, $z_3 \notin \{z_1, z_2\}$, on $A_{j^{**}}$, $j^{**} \in \{1,2,\ldots,m_1+m_2+1\}$, of multiplicity m_3, we proceed as before.

We proceed like this until we have considered all zeros of $\mathscr{T}_n'$ lying on the constructed set of arcs. Thus, we get a connected set of k^*+1 curves

$$A^* := A_1 \cup A_2 \cup \ldots \cup A_{k^*+1}$$

with k^* zeros of $\mathscr{T}_n'$, counted with multiplicity, on A^*.

Next, we claim that $k^* = n-1$. Assume that $k^* < n-1$, then, by assumption, there exists a curve $A_{k^*+2} \in \Gamma$, for which

$$A_{k^*+2} \cap A^* = \{\}$$

and on which there is another zero of $\mathscr{T}_n'$. By the same procedure as before, we get a set A^{**} of $k^{**}+1$ arcs of Γ for which $A^* \cap A^{**} = \{\}$ and k^{**} zeros of $\mathscr{T}_n'$, counted with multiplicity. If $k^* + k^{**} = n-1$, then we would get a set of $k^* + k^{**} + 2 = n+1$ arcs, which is a contradiction to Lemma 1 (i). If $k^* + k^{**} < n-1$, we proceed analogously and again, we get too many arcs, i.e., a contradiction to Lemma 1 (vi). Thus, $k^* = n-1$ must hold and thus Γ is connected.

"$\Longrightarrow$": Suppose that Γ is connected. Thus, it is possible to reorder $\Gamma_1, \Gamma_2, \ldots, \Gamma_n$ into $\Gamma_{k_1}, \Gamma_{k_2}, \ldots, \Gamma_{k_n}$ such that $\Gamma_{k_1} \cup \ldots \cup \Gamma_{k_j}$ is connected for each $j \in \{2,\ldots,n\}$. Now we will count the crossing points (common points) of the arcs in the following way: If there are $m+1$ arcs $A_1, A_2, \ldots, A_{m+1} \in \Gamma$ such that $z_0 \in A_j$, $j = 1,2,\ldots,A_{m+1}$, then we will count the crossing point z_0 m-times, i.e., we say $A_1, \ldots, A_{m+1}$ has m crossing points. Hence, $\Gamma_{k_1} \cup \Gamma_{k_2}$ has one crossing point, $\Gamma_{k_1} \cup \Gamma_{k_2} \cup \Gamma_{k_3}$ has two crossing points, $\Gamma_{k_1} \cup \Gamma_{k_2} \cup \Gamma_{k_3} \cup \Gamma_{k_4}$ has 3 crossing points, and so on. Summing up, we arrive at $n-1$ crossing points which are, by Lemma 1 (iv) the zeros of $\mathscr{T}_n'$.

Theorem 4 may be generalized to the question how many connected sets $\mathcal{T}_n^{-1}([-1,1])$ consists of. The proof runs along the same lines as that of Theorem 4.

Theorem 5. *Let $\mathcal{T}_n \in \mathbb{P}_n$. The inverse image $\mathcal{T}_n^{-1}([-1,1])$ consists of k, $k \in \{1,2,\ldots,n\}$, connected components $B_1, B_2, \ldots, B_k$ with $B_1 \cup B_2 \cup \ldots \cup B_k = \mathcal{T}_n^{-1}([-1,1])$ and $B_i \cap B_j = \{\}$, $i \neq j$, if and only if $n-k$ zeros of the derivative $\mathcal{T}_n'$ lie in $\mathcal{T}_n^{-1}([-1,1])$.*

References

1. Bogatyrëv, A.B., On the efficient computation of Chebyshev polynomials for several intervals, Sb. Math. **190**, 1571–1605 (1999).
2. Kamo, S.O. and Borodin, P.A., Chebyshev polynomials for Julia sets, Moscow Univ. Math. Bull. **49**, 44–45 (1994).
3. Fischer, B., Chebyshev polynomials for disjoint compact sets, Constr. Approx. **8**, 309–329 (1992).
4. Fischer, B. and Peherstorfer, F., Chebyshev approximation via polynomial mappings and the convergence behaviour of Krylov subspace methods, Electron. Trans. Numer. Anal. **12**, 205–215 (electronic) (2001).
5. Pakovich, F., Elliptic polynomials, Russian Math. Surveys **50**, 1292–1294 (1995).
6. Pakovich, F., Combinatoire des arbres planaires et arithmétique des courbes hyperelliptiques, Ann. Inst. Fourier (Grenoble) **48**, 323–351 (1998).
7. Pakovich, F., On trees that cover chains or stars, Fundam. Prikl. Mat. **13**, 207–215 (2007).
8. Pakovich, F., On polynomials sharing preimages of compact sets, and related questions, Geom. Funct. Anal. **18**, 163–183 (2008).
9. Peherstorfer, F., Orthogonal and extremal polynomials on several intervals, J. Comput. Appl. Math. **48**, 187–205 (1993).
10. Peherstorfer, F., Minimal polynomials for compact sets of the complex plane, Constr. Approx. **12**, 481–488 (1996).
11. Peherstorfer, F., Inverse images of polynomial mappings and polynomials orthogonal on them, J. Comput. Appl. Math. **153**, 371–385 (2003).
12. Peherstorfer, F., Deformation of minimal polynomials and approximation of several intervals by an inverse polynomial mapping, J. Approx. Theory **111**, 180–195 (2001).
13. Peherstorfer, F. and Schiefermayr, K., Description of extremal polynomials on several intervals and their computation. I, II, Acta Math. Hungar. **83**, 27–58, 59–83 (1999).
14. Peherstorfer, F. and Schiefermayr, K., Description of inverse polynomial images which consist of two Jordan arcs with the help of Jacobi's elliptic functions, Comput. Methods Funct. Theory **4**, 355–390 (2004).
15. Peherstorfer, F. and Steinbauer, R., Orthogonal and L_q-extremal polynomials on inverse images of polynomial mappings, J. Comput. Appl. Math. **127**, 297–315 (2001).
16. Schiefermayr, K., Inverse polynomial images which consists of two Jordan arcs – An algebraic solution, J. Approx. Theory **148**, 148–157 (2007).
17. Schiefermayr, K., Inverse polynomial images consisting of an interval and an arc, Comput. Methods Funct. Theory **9**, 407–420 (2009).
18. Sodin, M.L. and Yuditskiĭ, P.M., Algebraic solution of a problem of E.I. Zolotarev and N.I. Akhiezer on polynomials with smallest deviation from zero, J. Math. Sci. **76**, 2486–2492 (1995).
19. Totik, V., Polynomial inverse images and polynomial inequalities, Acta Math. **187**, 139–160 (2001).

On Symbolic Computation of Ideal Projectors and Inverse Systems

Boris Shekhtman

Abstract A zero-dimensional ideal J in the ring $\mathbb{k}[\mathbf{x}]$ of polynomials in d variables is often given in terms of its "border basis"; that is a particular finite set of polynomials that generate the ideal. We produce a convenient formula for symbolic computation of the space of functionals on $\mathbb{k}[\mathbf{x}]$ that annihilate J. The formula is particularly useful for computing an explicit form of an ideal projector from its values on a certain finite set of polynomials.

1 Introduction

Throughout, $\mathbb{k}$ will stand for the field of complex numbers or the field of real numbers, $\mathbb{k}[\mathbf{x}] := \mathbb{k}[x_1, \ldots, x_d]$ will denote the space (algebra, ring) of polynomials in d indeterminants with coefficients in the field $\mathbb{k}$ and $(\mathbb{k}[\mathbf{x}])'$ is the algebraic dual of $\mathbb{k}[\mathbf{x}]$, i.e., the space of all linear functionals on $\mathbb{k}[\mathbf{x}]$.

Definition 1 ([1]). A linear idempotent operator $P : \mathbb{k}[\mathbf{x}] \to \mathbb{k}[\mathbf{x}]$ is called an ideal projector if $\ker P$ is an ideal in $\mathbb{k}[\mathbf{x}]$.

Lagrange interpolation projectors, Taylor projectors and, in one variable, Hermite interpolation projectors are all examples of ideal projectors. Thus, the study of ideal projectors holds a promise of an elegant extension of operators, traditionally used in approximation theory, to multivariate setting. The theory was initiated by Birkhoff [1], Carl de Boor [2], de Boor and Ron [1], Mőller [7], and Sauer [10].

Boris Shekhtman
Department of Mathematics and Statistics, University of South Florida, Tampa, FL 33620, USA
e-mail: boris@math.usf.edu

M. Neamtu and L. Schumaker (eds.), *Approximation Theory XIII: San Antonio 2010*, Springer Proceedings in Mathematics 13, DOI 10.1007/978-1-4614-0772-0_18,

Any finite-dimensional projector, ideal or not, can be written as

$$Pf = \sum \lambda_j(f) g_j \tag{1}$$

where $(g_j) \subset \mathbb{k}[\mathbf{x}]$ is a (linear) basis for the range of P and $(\lambda_j) \subset (\mathbb{k}[\mathbf{x}])'$ are dual functionals:

$$\lambda_k(g_j) = \delta_{k,j} \tag{2}$$

forming a basis in the space

$$\operatorname{ran} P^* = (\ker P)^{\perp}$$

Thus, the functionals (λ_j) determine the kernel of P

$$\ker P = \{f \in \mathbb{k}[\mathbf{x}] : \lambda_j(f) = 0 \text{ for all } j\}.$$

Conversely, the kernel of P determines the span of (λ_j) since

$$\operatorname{span}\{\lambda_j\} = (\ker P)^{\perp}.$$

When the projector P is ideal, its kernel is often given by the ideal basis, a finite set of polynomials that generate the ideal $\ker P$ and thus the projector P is define by its values on a finite subsets of polynomials. The purpose of this note is to present a convenient formula (9) for symbolic computation of the functionals (λ_j) from these values.

To expand on this point, recall

Theorem 1 ([2]). *A linear operator $P : \mathbb{k}[\mathbf{x}] \to \mathbb{k}[\mathbf{x}]$ is an ideal projector if and only if*

$$P(fg) = P(f \cdot P(g)) \tag{3}$$

for all $f, g \in \mathbb{k}[\mathbf{x}]$.

In terms of the quotient algebra $\mathbb{k}[\mathbf{x}]/\ker P$, (3) says that $[f[g]] = [fg] \in \mathbb{k}[\mathbf{x}]/J$, for all $f, g \in \mathbb{k}[\mathbf{x}]$.

Let $G \subset \mathbb{k}[\mathbf{x}]$ stand for a finite-dimensional range of the projector P, and let $\mathfrak{g} = (g_1, \ldots, g_N)$ be a linear basis for G. We define the border for $\mathfrak{g}$ to be

$$\partial\mathfrak{g} := \{1, x_i g_k, i = 1, \ldots, d, k = 1, \ldots, N\} \backslash G.$$

By the de Boor's formula (3), the ideal projector is completely determined by its finitely many values (cf. [2])

$$\{Pf, f \in \partial\mathfrak{g}\}. \tag{4}$$

Equivalently, the polynomials $\{f - Pf, f \in \partial\mathfrak{g}\}$ form an ideal basis, called the border basis (cf. [5, 9, 11]) for the ideal $\ker P$.

The formula (9) computes the functionals λ_j from the finitely many values of the ideal projector given by (4).

2 Preliminaries

2.1 Multiplication Operators

Using the polynomials (4) one can define a sequence of multiplication operators on G: $\mathbf{M}_{P,\mathfrak{g}} = (M_1, \ldots, M_d)$ where

$$M_i(g) = P(x_i g). \tag{5}$$

With the aid of (3) it is easy to see (cf. [2, 11]) that this is a sequence of pairwise commuting operators with the cyclic vector $P1$:

$$\{p(\mathbf{M}_{P,\mathfrak{g}})(P1), p \in \mathbb{k}[\mathbf{x}]\} = G.$$

The sequence $\mathbf{M}_{P,\mathfrak{g}}$ is similar (literally and figuratively) to the operators of multiplication by x_i on $\mathbb{k}[\mathbf{x}]/J$.

There is a partial converse to this statement ([4, 8, 11]): every cyclic sequence $\mathbf{L} = (L_1, \ldots, L_d)$ of commuting operators on a finite-dimensional subspace $G \subset \mathbb{k}[\mathbf{x}]$ defines the (unique) ideal projector. The sequence of multiplication operators for this projector is similar to $\mathbf{L}$.

2.2 Duality

The space of all formal power series in $\mathbf{x}$ is denoted by $\mathbb{k}[[\mathbf{x}]]$. For $\hat{\lambda} \in \mathbb{k}[[\mathbf{x}]]$, we use $\hat{\lambda}(D)$ to denote the differential operator on $\mathbb{k}[\mathbf{x}]$ obtained by formally replacing the indeterminants with the corresponding partial derivatives with respect to these indeterminants. Every $\hat{\lambda} \in \mathbb{k}[[\mathbf{x}]]$ defines a linear functional λ on $\mathbb{k}[\mathbf{x}]$ by

$$\lambda(f) := \left(\hat{\lambda}(D)f\right)(0) \text{ for every } f \in \mathbb{k}[\mathbf{x}].$$

It is well-known (cf. [1] and [6] in its original form) that the map

$$\hat{\lambda} \longmapsto \lambda$$

defined by the display above is a linear isomorphism between $\mathbb{k}[[\mathbf{x}]]$ and $(\mathbb{k}[\mathbf{x}])'$. Thus, every functional λ is identified with the power series $\hat{\lambda} \in \mathbb{k}[[\mathbf{x}]]$ and, when there is no possibility for confusion, we will denote both by the same letter. For instance, the point evaluation functional $(\mathbb{k}[\mathbf{x}])' \ni \lambda : \lambda(f) := f(\mathbf{z})$ is identified with the power series for the exponential function $\lambda(\mathbf{x}) = e^{\mathbf{x}\cdot\mathbf{z}}$.

For a set $J \subset \mathbb{k}[\mathbf{x}]$, we define

$$J^{\perp} := \{\lambda \in \mathbb{k}[[\mathbf{x}]] : \lambda(f) = 0 \text{ for every } f \in J\}.$$

For every $f \in \Bbbk[\mathbf{x}]$ we have (cf. [1,6])

$$(D_i\lambda)(f) = \lambda(x_i f). \tag{6}$$

That is, the operator D_i is the adjoint to the operator of multiplication by independent variable on $\Bbbk[\mathbf{x}]$.

A linear subspace $\Lambda \subset \Bbbk[[\mathbf{x}]]$ is D-invariant if $D_i\lambda \in \Lambda$ for every $\lambda \in \Lambda$ and every $i = 1,\dots,d$. The next theorem (cf. [1]) is an easy consequence of (6):

Theorem 2 ([6]). *A subspace $J \subset \Bbbk[\mathbf{x}]$ is an ideal if and only if $J^\perp \subset \Bbbk[[\mathbf{x}]]$ is D-invariant.*

3 The Main Result

Theorem 3. *Let $P : Pf = \sum_{j=1}^{N} \lambda_j(f) g_j$ be a N-dimensional ideal projector on $\Bbbk[\mathbf{x}]$. Let $\mathfrak{g} = (g_1,\dots,g_N)$ be a basis for ran P and let $\tilde{\mathbf{M}}_{P,\mathfrak{g}} = \left(\tilde{M}_1,\dots,\tilde{M}_d\right)$ be the matrices representing the operators M_i defined by (5) in the basis $\mathfrak{g}$. Then*

$$\lambda := (\lambda_1,\dots,\lambda_N)^t = \mathrm{e}^{\left(\sum_{i=1}^{\mathrm{d}} x_i\tilde{M}_i\right)}\lambda(0). \tag{7}$$

Proof. First, we claim that

$$M_i^t = D_i\,|_G\,.$$

Indeed, for every $g \in G = \operatorname{ran} P$ and every $\lambda \in \operatorname{ran} P^* = (\ker P)^\perp$ we have $\lambda(M_i g) = (M_i^t\lambda)(g)$. On the other hand

$$\lambda(M_i g) = \lambda(P(x_i g)) = (P^*\lambda)(x_i g) = \lambda(x_i g) = (D_i\lambda)(g),$$

where the second equality follows from $\lambda \in (\ker P^*)^\perp = \operatorname{ran} P^*$ and the last from (6).

This means that

$$(D_i\lambda_k) = \sum_{j=1}^{N} m_{j,k}^{(i)}\lambda_j, \tag{8}$$

where $m_{j,k}^{(i)}$ is the j,k-th entry in the matrix $\tilde{M}_i^t$. By D-invariance, $D_i\lambda_k = \sum_{j=1}^{N} a_{j,k}^{(i)}\lambda_j$ for some coefficients $a_{j,k}^{(i)}$. Since (λ_j) is a basis in ran P^* it follows that $a_{j,k}^{(i)} = m_{j,k}^{(i)}$.

The display (8) means that, as a vector-valued function of x_1, λ is the solution of the initial value problem

$$D_1\mathbf{u} = \tilde{M}_1\mathbf{u},\ \mathbf{u}(0) = \lambda(0,x_2,\dots,x_d).$$

Thus,

$$\lambda = \mathrm{e}^{x_1\tilde{M}_1}\mathbf{u}(0) = \mathrm{e}^{x_1\tilde{M}_1}\lambda(0,x_2,\dots,x_d).$$

Next, we observe that, as a vector-valued function of x_2, $e^{x_1\tilde{M}_1}\lambda(0,x_2,\ldots,x_d)$ solves the initial value problem

$$D_2\mathbf{u} = \tilde{M}_2\mathbf{u},\ \mathbf{u}(0) = \lambda(x_1,0,x_3,\ldots,x_d) = e^{x_1\tilde{M}_1}\lambda(0,0,x_3,\ldots,x_d).$$

Hence,

$$\lambda = e^{x_1\tilde{M}_1}e^{x_2\tilde{M}_2}\lambda(0,0,x_3,\ldots,x_d) = e^{x_1\tilde{M}_1+x_2\tilde{M}_2}\lambda(0,0,x_3,\ldots,x_d).$$

Repeating this process $d-2$ more times we obtain (7).

Corollary 1. *Suppose that* $g_1 = 1 \in \operatorname{ran} G$. *Then*

$$\lambda = (\lambda_1,\ldots,\lambda_N)^t = e^{\left(\sum_{i=1}^{d} x_i\tilde{M}_i\right)}e_1, \tag{9}$$

where $e_1 = (1,0,\ldots,0)^t \in \mathbb{k}^N$.

Proof. Observe that for every $\lambda \in \mathbb{k}[[\mathbf{x}]]$ we have $\lambda(D)1 = \lambda(0)$. Since, by (2), $(\lambda_j(D)1)(0) = \lambda_j(g_1) = \delta_{j,1}$ it follows that $\lambda_j(0) = \delta_{j,1}$.

Remark 1. The formula (9) can be interpreted algebraically in terms of an inverse systems for the quotient ring of a zero-dimensional ideal. For a zero-dimensional ideal $J \subset \mathbb{k}[\mathbf{x}]$, we define multiplication operators $\hat{M}_i$ on $\mathbb{k}[\mathbf{x}]/J$ by

$$\hat{M}_i[f] = [x_i f]$$

and the matrices $\tilde{M}_i$ as the matrices of the operators $\hat{M}_i$ in any linear basis for $\mathbb{k}[\mathbf{x}]/J$. Then the formula (8) gives a linear basis for the Macaulay inverse systems (cf. [6]) for $\mathbb{k}[\mathbf{x}]/J$.

The formula can also be interpreted as follows: Any N-dimensional D-invariant subspace of $\mathbb{k}[[\mathbf{x}]]$ has a basis of the form (9) for some cyclic sequence of commuting matrices $(\tilde{M}_1,\ldots,\tilde{M}_d)$. Conversely, every cyclic sequence of commuting matrices $(\tilde{M}_1,\ldots,\tilde{M}_d)$ generate an N-dimensional D-invariant subspace of $\mathbb{k}[[\mathbf{x}]]$ via (9).

4 A Couple of Examples

Both example will deal with the ideal projector onto the span of $\mathfrak{g} = (1,x,y)$; hence, $\partial\mathfrak{g} = \{x^2,xy,y^2\}$.

Example 1. Define $Px^2 = Pxy = Py^2 = 0$. Then the matrices of multiplication operators in the basis $\mathfrak{g}$ are

$$M_1 = \begin{bmatrix} 0&0&0\\ 1&0&0\\ 0&0&0 \end{bmatrix},\ M_2 = \begin{bmatrix} 0&0&0\\ 0&0&0\\ 1&0&0 \end{bmatrix}.$$

It is easy to verify that these matrices commute, hence define an ideal projector P. Maple computations give

$$\left(e^{xM_1+yM_2}\right)\begin{bmatrix}1\\0\\0\end{bmatrix}=\begin{bmatrix}1\\x\\y\end{bmatrix}$$

and

$$Pf=f(0)1+(D_xf)(0)x+(D_yf)(0)y$$

is the Taylor projector onto first degree polynomials.

Example 2. Define $Px^2=y$, $Pxy=Py^2=0$. Then the matrices of multiplication operators in the basis $\mathfrak{g}$ are

$$M_1=\begin{bmatrix}0&0&0\\1&0&0\\0&1&0\end{bmatrix},M_2=\begin{bmatrix}0&0&0\\0&0&0\\1&0&0\end{bmatrix}.$$

Again, it is easy to verify that these matrices commute, hence define an ideal projector P. Maple computations give

$$\left(e^{xM_1+yM_2}\right)\begin{bmatrix}1\\0\\0\end{bmatrix}=\begin{bmatrix}1\\x\\\frac{1}{2}x^2+y\end{bmatrix}$$

and the ideal projector given by

$$Pf=f(0)1+(D_xf)(0)x+\left(\frac{1}{2}D_x^2+D_y\right)f(0)y$$

is the ideal projector in question.

References

1. Birkhoff, Garrett, The algebra of multivariate interpolation, Constructive approaches to mathematical models (Proc. Conf. in honor of R. J. Duffin, Pittsburgh, Pa., 1978), pp. 345–363, Academic Press, New York, 1979.
2. de Boor, Carl, Ideal interpolation, Approximation theory XI: Gatlinburg 2004, Mod. Methods Math., pp. 59–91, Nashboro Press, Brentwood, TN, 2005.
3. de Boor, Carl and Ron, Amos, On polynomial ideals of finite codimension with applications to box spline theory, J. Math. Anal. Appl. **158** (1), 168–193 (1991).
4. de Boor, C. and Shekhtman, B., On the pointwise limits of bivariate Lagrange projectors, Linear Algebra Appl. **429** (1), 311–325 (2008).
5. Kreuzer, Martin and Robbiano, Lorenzo, Computational commutative algebra. 2, Springer-Verlag, Berlin, 2005.

6. Macaulay, F. S., The algebraic theory of modular systems, Cambridge Mathematical Library, Revised reprint of the 1916 original, With an introduction by Paul Roberts, Cambridge University Press, Cambridge, 1994.
7. Möller, Hans Michael, Hermite interpolation in several variables using ideal-theoretic methods, Constructive theory of functions of several variables (Proc. Conf., Math. Res. Inst., Oberwolfach, 1976), pp. 155–163, Lecture Notes in Math., Vol. 571, Springer, Berlin, 1977.
8. Nakajima, Hiraku, Lectures on Hilbert schemes of points on surfaces, University Lecture Series, Vol. 18, American Mathematical Society, Providence, RI, 1999.
9. Robbiano, L., On border basis and Gröbner basis schemes, Collect. Math. **60** (1), 11–25 (2009).
10. Sauer, Tomas, Polynomial interpolation in several variables: lattices, differences, and ideals, Topics in multivariate approximation and interpolation, Stud. Comput. Math. **12**, 191–230 (2006).
11. Shekhtman, B., Some tidbits on ideal projectors, commuting matrices and their applications, Elec. Trans. Numer. Anal. **36**, 17–26 (2009).

The Dimension of the Space of Smooth Splines of Degree 8 on Tetrahedral Partitions

Xiquan Shi, Ben Kamau, Fengshan Liu, and Baocai Yin

Abstract Let $\Omega \subset \mathbb{R}^3$ be a connected polyhedral domain that is allowed to contain polyhedral holes and Δ be a tetrahedral partition of Ω. Given $0 \leq r \leq d$, we define

$$S_d^r(\Delta) = \{s \in C^r(\Omega);\ s|_\sigma \in P_d \text{ for any tetrahedron} \sigma \in \Delta)\},$$

the spline space of degree d and smoothness r, where P_d is the trivariate polynomial space of total degree not exceeding d.

In this paper, we obtained the following result.

Theorem

$$\dim S_8^1(\Delta) = \sum_{\mathbf{v}\in V} \dim S_3^1(\mathrm{Star}(\mathbf{v})) + 5|E| + 9|F| + |T| + 3|E_b| + 3|E_\delta|,$$

where V, E, F, T, E_b, and E_δ are the sets of vertices, edges, triangles, tetrahedra, boundary edges, and singular edges of Δ, respectively.

Xiquan Shi • Fengshan Liu
Department of Mathematical Sciences, Delaware State University, Dover, DE 19901, USA
e-mail: xshi@desu.edu, fliu@desu.edu

Ben Kamau
Mathematics Department, Columbus State University, Columbus, GA 31907, USA
e-mail: kamau_ben@colstate.edu

Baocai Yin
Beijing Key Laboratory of Multimedia and Intelligent Software, College of Computer Science and Technology, Beijing University of Technology, Beijing 100022, China
e-mail: ybc@bjut.edu.cn

M. Neamtu and L. Schumaker (eds.), *Approximation Theory XIII: San Antonio 2010*, Springer Proceedings in Mathematics 13, DOI 10.1007/978-1-4614-0772-0_19,

1 Introduction

We first introduce some terminology. Let $V = \{\mathbf{v}_1, \ldots, \mathbf{v}_{n+1}\} \subset \mathbb{R}^n$ be a set of $n+1$ points which are in general position, i.e., $\{\mathbf{v}_1 - \mathbf{v}_{n+1}, \ldots, \mathbf{v}_n - \mathbf{v}_{n+1}\}$ is a basis of $\mathbb{R}^n$. The convex hull $[V]$ of V is called a n-simplex. The convex hull of m+1 points of V is also a simplex, called an m-face. A 0-face is called a vertex, a 1-face is called an edge, a 2-face is called a triangle, a 3-face is called a tetrahedron, and an $(n-1)$-face is called a facet. For an n-simplex σ, we denote by $\mathrm{Face}_i(\sigma)$ the collection of all i-faces of σ and $\mathrm{Face}(\sigma) = \bigcup_{i=0}^{n} \mathrm{Face}_i(\sigma)$.

Definition 1. A simplicial complex Δ in $\mathbb{R}^n$ is a collection of simplices in $\mathbb{R}^n$ such that

- Every simplex of Δ is a face of a n-simplex of Δ or itself is a n-simplex,
- Every face of a simplex of Δ is still in Δ, and
- The intersection of any two simplices of Δ is empty or a face of each other.

Similar to a simplex, for a simplicial complex Δ we denote by $\mathrm{Face}_i(\Delta)$ the collection of all i-simplices of Δ. We denote by $\Omega = \bigcup_{\sigma \in \mathrm{Face}_n(\Delta)} \sigma$ the region covered by the simplicial complex Δ. A simplex $\delta \in \Delta$ is called boundary if $\delta \subset \partial\Omega$, the boundary of Ω; otherwise, it is called inner (or interior).

For a simplex σ of Δ, we denote by

$$\mathrm{Star}(\sigma) = \{\delta \in \Delta;\ \sigma \text{ is a face of } \delta\}$$

the simplicial complex formed by the collection of all the simplices (together with their faces) with σ as a common face. $\mathrm{Star}(\sigma)$ is called the σ-star of σ.

For an m-simplex $\sigma = [\mathbf{v}_0, \mathbf{v}_1, \ldots, \mathbf{v}_m]$, we denote by

$$D^{\mathbf{k}}_{\mathbf{v}_i,\sigma} = \prod_{j=0, j\neq i}^{m} D^{k_j}_{\mathbf{v}_j - \mathbf{v}_i}$$

the mixed directional derivative of order $\mathbf{k} = (k_0, \ldots, \widehat{k_i}, \ldots, k_m) \in Z_+^m$ (Z_+ is the set of all nonnegative integers), where $\widehat{k_i}$ means that the component k_i in $\mathbf{k}$ is missing, $D_{\mathbf{v}_j - \mathbf{v}_i} = (\mathbf{v}_j - \mathbf{v}_i) \cdot \bigtriangledown$ the directional derivative of $\mathbf{v}_j - \mathbf{v}_i$, and $D^{k_j}_{\mathbf{v}_j - \mathbf{v}_i} = ((\mathbf{v}_j - \mathbf{v}_i) \cdot \bigtriangledown)^{k_j}$ ($\bigtriangledown$ is the gradient vector).

Similarly, for an i-simplex $\sigma = [\mathbf{v}_0, \mathbf{v}_1, \ldots, \mathbf{v}_i]$ ($i \geq 1$), if $[\mathbf{v}_0, \mathbf{v}_1, \ldots, \mathbf{v}_i, \mathbf{w}]$ is an $(i+1)$-simplex, we denote by

$$D^k_{\sigma,\mathbf{w}} = \frac{\partial^k}{\partial \mathbf{n}^k_{\sigma,\mathbf{w}}}$$

the directional derivative of order k along $\mathbf{n}_{\sigma,\mathbf{w}}$, where $\mathbf{n}_{\sigma,\mathbf{w}}$ is the unit inner normal vector of $[\mathbf{v}_0, \mathbf{v}_1, \ldots, \mathbf{v}_i, \mathbf{w}]$ to σ.

More generally, if $W = \{\mathbf{w}_1,\ldots,\mathbf{w}_m\} (m \geq 1)$ such that $\delta = [\mathbf{v}_0,\mathbf{v}_1,\ldots,\mathbf{v}_i,\mathbf{w}_1,\ldots,\mathbf{w}_m]$ is an $(m+i)$-simplex, we denote

$$D^{\mathbf{k}}_{\sigma,W} = \prod_{\mathbf{w}\in W} D^{k_{\mathbf{w}}}_{\sigma,\mathbf{w}} \text{ or } D^{\mathbf{k}}_{\sigma,\delta} = \prod_{\mathbf{w}\in W} D^{k_{\mathbf{w}}}_{\sigma,\mathbf{w}},$$

where $\sigma = [\mathbf{v}_0,\mathbf{v}_1,\ldots,\mathbf{v}_i]$. We also denote by

$$\delta/\sigma = [\mathbf{0},\mathbf{n}_{\sigma,\mathbf{w}_1},\mathbf{n}_{\sigma,\mathbf{w}_2},\ldots,\mathbf{n}_{\sigma,\mathbf{w}_m}]$$

the simplex composed of the original and the normal vectors $\mathbf{n}_{\sigma,\mathbf{w}_1},\mathbf{n}_{\sigma,\mathbf{w}_2},\ldots,\mathbf{n}_{\sigma,\mathbf{w}_m}$, where we identify a vector as a point which has the same components with this vector.

For an i-simplex σ of a simplicial complex Δ in $\mathbb{R}^n$, we define the simplicial complex

$$\mathrm{TStar}(\sigma) = \{\delta/\sigma;\ \delta \in \mathrm{Face}_n(\mathrm{Star}(\sigma))\}$$

the transversal star of σ, where we assume that $\mathrm{TStar}(\sigma)$ also contains all faces of δ/σ and ask $i \leq n-1$. Clearly, $\mathrm{TStar}(\sigma)$ is a simplicial complex in $\mathbb{R}^{n-i}$. $\mathrm{TStar}(\sigma)$ is called the σ-TStar of σ.

For convenience, we call a simplicial complex in $\mathbb{R}^3$ a tetrahedral partition. For a tetrahedral partition Δ, we introduce the following definition.

Definition 2. For an edge $e \in \mathrm{Face}_1(\Delta)$, the degree of e, denoted by degree(e), is the number of triangles in $\mathrm{Star}(e)$ sharing e as a common edge. An inner edge e of Δ is called odd (even) if degree(e) is odd (even) and it is called singular if degree(e)=4 and these four triangles are pairwise coplanar.

In addition, we denote

$$|\mathbf{v}| = v_1 + \cdots + v_m$$

for a vector $\mathbf{v} = (v_1,\ldots,v_m) \in \mathbb{R}^m$ and $|V|$ the number of elements of a set V.

Let $\Omega \subset \mathbb{R}^3$ be a connected polyhedral domain which is allowed to contain polyhedral holes and Δ be a tetrahedral partition of Ω, i.e., $\Omega = \bigcup_{\sigma\in\Delta}\sigma$. Given $0 \leq r \leq d$, we define

$$S^r_d(\Delta) = \{s \in C^r(\Omega);\ s|_\sigma \in P_d \text{ for any } \sigma \in \mathrm{Face}_3(\Delta)\}$$

the spline space of degree d and smoothness r, where P_d is the trivariate polynomial space of total degree not exceeding d. For convenience, we define $S^{-1}_d(\Delta)$ the collection of all functions whose restrictions to a tetrahedron $\sigma \in \mathrm{Face}_3(\Delta)$ belong to P_d.

Clearly, $S^r_d(\Delta)$ is a linear space with finite dimension. In particular, if we denote by $\dim S^r_d(\Delta)$ the dimension of $S^r_d(\Delta)$, then $\dim S^r_d(\Delta) \leq \dim S^{-1}_d(\Delta) = \frac{1}{6}(d+3)(d+2)(d+1)|\mathrm{Face}_3(\Delta)|$. Owing to the importance in a variety of areas, including finite element method, wavelets, data fitting, and computer aided geometric design, spline spaces are extensively studied. Among others, the dimension problem of spline spaces attracts considerable attention in the study. This paper is devoted to the dimension problem. In this paper, we obtain the dimension of $S_8^1(\Delta)$, where Δ is a tetrahedral partition.

For Δ being a triangulation, i.e., a simplicial complex in $\mathbb{R}^2$, the dimensions of the spline spaces $S_d^r(\Delta)$ are obtained in the following cases: $\dim S_d^1(\Delta)(d \geq 5)$ is obtained by Morgan and Scott ([5]); $\dim S_d^r(\Delta)(d \geq 4r+1)$ is respectively obtained by Wang and Lu ([11]) and Alfeld and Schumaker ([2]); Dong ([4]) solved the case $d \geq 3r+2$; and $\dim S_4^1(\Delta)$ is obtained by Alfeld et al. ([1]). In the case where Δ is a tetrahedral partition, X. Shi ([9] and [10]) found $\dim S_d^r(\Delta)$ for $d \geq 8r+1$ for a general tetrahedral partition and $\dim S_d^1(\Delta)$ $(d \geq 7)$ ([9] and [10]) for an odd tetrahedral partition. A tetrahedral partition Δ is odd if it has only odd inner edges or singular edges. For a generic tetrahedral partition Δ, Alfeld/Schumaker/Whiteley ([3]) obtained $\dim S_d^1(\Delta)(d \geq 8)$ and X. Shi ([6]) obtained $\dim S_7^1(\Delta)$. A tetrahedral partition Δ is called generic provided that for a sufficiently small perturbation of the location of the vertices of Δ, the resulting tetrahedral partition $\widetilde{\Delta}$ satisfies $\dim S_d^r(\widetilde{\Delta})=\dim S_d^r(\Delta)$. For a general simplicial complex Δ in $\mathbb{R}^n$, $\dim S_d^r(\Delta)$ was obtained by Shi ([10]) if $d \geq 2^n r+1$.

2 Main Results

In this section, Δ represents a tetrahedral partition unless stated otherwise. Similar to Alfeld/Schumaker/Whiteley ([3]), we use the following notations.

$$
\begin{aligned}
V_b &= \text{the set of boundary vertices of } \Delta,\\
V_I &= \text{the set of inner vertices of } \Delta,\\
E_b &= \text{the set of boundary edges of } \Delta,\\
E_I &= \text{the set of inner edges of } \Delta,\\
E_e &= \text{the set of even edges of } \Delta,\\
E_o &= \text{the set of odd edges of } \Delta,\\
E_\delta &= \text{the set of singular edges of } \Delta,\\
F_b &= \text{the set of boundary triangular faces of } \Delta,\\
F_I &= \text{the set of inner triangular faces of } \Delta,\\
V &= V_I \bigcup V_b,\ E = E_I \bigcup E_b,\ F = F_I \bigcup F_b,\\
T &= \text{the set of tetrahedra of } \Delta,\\
E_{\mathbf{v}} &= \{e \in E;\ \mathbf{v} \text{ is a vertex of } e\},\\
F_\delta &= \{f \in F;\ \delta \text{ is a face of } f\},\\
T_\delta &= \{t \in T;\ \delta \text{ is a face of } t\},\\
E_{\mathbf{v}}^{\partial} &= E_{\mathbf{v}} \bigcap E_b, \qquad E_{\mathbf{v}}^{\delta} = E_{\mathbf{v}} \bigcap E_\delta.
\end{aligned}
\tag{1}
$$

In this paper, we will prove the following main result.

Theorem 1. *Let Δ be a tetrahedral partition. Then*

$$dimS_8^1(\Delta) = \Sigma_{\mathbf{v}\in V}\left(dimS_3^1(\text{Star}(\mathbf{v})) + N_{\mathbf{v}}\right) + \Sigma_{e\in E}\left(dimS_2^1(TStar(e)) - 3\right) + |E| + 3|F| + |T|, \tag{2}$$

where $N_{\mathbf{v}} = 2|E_{\mathbf{v}}| + |E_{\mathbf{v}}^{\partial}| + |E_{\mathbf{v}}^{\delta}| + |F_{\mathbf{v}}|$.

Remark 1. For Δ being a triangulation, $\dim S_d^1(\Delta)$ $(d \geq 5)$ is obtained by Morgan/Scott in 1975 ([5]), but $\dim S_4^1(\Delta)$ is obtained by Alfeld/Piper/Schumaker in 1987 ([1]). This is because that the method of deriving $\dim S_4^1(\Delta)$ is very different and more difficult than that of deriving $\dim S_d^1(\Delta)$ $(d \geq 5)$.

Remark 2. Similarly, for Δ being a tetrahedral partition, the method of obtaining $\dim S_8^1(\Delta)$ is also very different from and much difficult than the method of obtaining $\dim S_d^1(\Delta)(d \geq 9)$.

Remark 3. Obtaining $\dim S_3^1(\text{Star}(\mathbf{v}))$ is even more difficult than obtaining both $\dim S_2^1(\Delta)$ and $\dim S_3^1(\Delta)$, the still famous open problems, where Δ is a general triangulation. In fact, let Δ be a $(n-1)$-simplicial complex embedded in the coordinate superplane x_n=0 of $\mathbb{R}^n$ and $\mathbf{v} \in \mathbb{R}^n$ be a point out of the superplane x_n=0. Denote by Star($\mathbf{v}$) the star formed by respectively joining $\mathbf{v}$ to all vertices of Δ. Then, it holds that ([7–9])

$$\dim S_d^r(\text{Star}(\mathbf{v})) = \sum_{k=0}^{d} \dim S_k^r(\Delta).$$

Especially, let $n = 3, r = 1, d = 3$, then

$$\dim S_3^1(\text{Star}(\mathbf{v})) = 4 + \dim S_2^1(\Delta) + \dim S_3^1(\Delta).$$

This means that to obtain $\dim S_3^1(\text{Star}(\mathbf{v}))$, one has to obtain both $\dim S_2^1(\Delta)$ and $\dim S_3^1(\Delta)$.

It is easy to obtain

$$\dim S_2^1(\text{TStar}(e)) = \text{degree}(e) + 3 + \delta_e + \partial_e,$$

where $\delta_e = 1$ if e is a singular edge; otherwise $\delta_e = 0$, and $\partial_e = 1$ if e is a boundary edge; otherwise $\partial_e = 0$. From the above equality, we have

$$\sum_{e\in E}\left(\dim S_2^1(\text{TStar}(e)) - 3\right) = \sum_{e\in E}(\text{degree}(e) + \delta_e + \partial_e) = 3|F| + |E_b| + |E_\delta|. \tag{3}$$

Similarly, it holds

$$\Sigma_{\mathbf{v}\in V}|E_{\mathbf{v}}| = 2|E|,\ \Sigma_{\mathbf{v}\in V}|E_{\mathbf{v}}^{\partial}| = 2|E_b|,\ \Sigma_{\mathbf{v}\in V}|E_{\mathbf{v}}^{\delta}| = 2|E_\delta|,\ \Sigma_{\mathbf{v}\in V}|F_{\mathbf{v}}| = 3|F|. \tag{4}$$

According to (3) and (4), Theorem 1 can be rewrite as

Theorem 2.

$$dimS_8^1(\Delta) = \sum_{\mathbf{v}\in V} dimS_3^1(\mathrm{Star}(\mathbf{v})) + 5|E| + 9|F| + |T| + 3|E_b| + 3|E_\delta|. \quad (5)$$

For convenience, we denote $\mathbf{v}_1$, $\mathbf{v}_2$, $\cdots$, $\mathbf{v}_{n_0}$ ($n_0 = |V|$) are all vertices of Δ, e_1, e_2, $\cdots$, e_{n_1} ($n_1 = |E|$) are all edges of Δ, and δ_1, δ_2, $\cdots$, δ_{m_2}, $\delta_{m_2+1}, \ldots, \delta_{n_2}(n_2 = |F|)$ are all vertices of Δ with first m_2 triangles are inner.

The remainder of this paper is dedicated to prove Theorem 1. The key to prove Theorem 1 is the analysis of how to obtain the numbers $N_\mathbf{v}$'s.

For a polynomial p of degree 8 defined on a tetrahedron $\sigma = [\mathbf{v}_0, \mathbf{v}_1, \mathbf{v}_2, \mathbf{v}_3]$, it has the following Bézier form:

$$p(\mathbf{x}) = \sum_{|\mathbf{i}|=8} c_\mathbf{i} \frac{8!}{\mathbf{i}!} \mathbf{u}^\mathbf{i}, \quad (6)$$

where $\mathbf{i} = (i_0, i_1, i_2, i_3) \in Z_+^4$, $\mathbf{i}! = i_0!\ i_1!\ i_2!\ i_3!$, and $\mathbf{u} = (u_0, u_1, u_2, u_3)$ are the barycentric coordinates of $\mathbf{x}$ with respect to σ, i.e., $\mathbf{x} = u_0\mathbf{v}_0 + u_1\mathbf{v}_1 + u_2\mathbf{v}_2 + u_3\mathbf{v}_3$, and $\mathbf{u}^\mathbf{i} = u_0^{i_0} u_1^{i_1} u_2^{i_2} u_3^{i_3}$. $c_\mathbf{i}$'s are called Bézier coefficients of p.

Lemma 1. *A polynomial $p(\mathbf{x})$ of degree 8 defined on σ is uniquely determined by its following values and Bézier coefficients:*

1. $C_j = \{c_\mathbf{i};\ i_j \geq 4\}$, $0 \leq j \leq 3$, *where* $\mathbf{i} = (i_0, i_1, i_2, i_3)$. *We should note that* $C_j \bigcap C_k$ *is not empty even for* $j \neq k$. *They have the common element* $c_{(\bar{i}_0, \bar{i}_1, \bar{i}_2, \bar{i}_3)}$*; where* $\bar{i}_t = 4$ *if* $t = j$ *or* k*; otherwise* $\bar{i}_t = 0$.
2. *For any edge e of σ, the values are*
$$\{D_{e,\sigma}^\mathbf{k} p(\mathbf{m}_e);\ |\mathbf{k}| = 2\},$$
where $\mathbf{m}_e$ is the centroid of the edge e.
3. *For any triangle face δ of σ, we select a unit normal vector $\mathbf{n}_\delta$ of δ. Assume that $\mathbf{n}_{1\delta}$ and $\mathbf{n}_{2\delta}$ are two unit vectors such that $\mathbf{n}_\delta$, $\mathbf{n}_{1\delta}$ and $\mathbf{n}_{2\delta}$ are perpendicular to each other. Then, the values are*
$$\{\frac{\partial p(\mathbf{m}_\delta)}{\partial \mathbf{n}_\delta}, \frac{\partial^2 p(\mathbf{m}_\delta)}{\partial \mathbf{n}_\delta \partial \mathbf{n}_{1\delta}}, \frac{\partial^2 p(\mathbf{m}_\delta)}{\partial \mathbf{n}_\delta \partial \mathbf{n}_{2\delta}}\},$$
where $\mathbf{m}_\delta$ is the centroid of the triangle δ.
4. *For σ itself, the values is*
$$\{p(\mathbf{m}_\sigma)\},$$
where $\mathbf{m}_\sigma$ is the centroid of σ.

The proof of Lemma 1 is straightforward.

For a spline $s \in S_8^{-1}(\Delta)$, its restriction $p_\sigma = s|_\sigma$ to a tetrahedron $\sigma \in \Delta$ has the following Bézier form

$$p_\sigma(\mathbf{x}) = \sum_{|\mathbf{i}|=8} c_{\mathbf{i}}^\sigma \frac{8!}{\mathbf{i}!} \mathbf{u}_\sigma^{\mathbf{i}}, \tag{7}$$

where the notations are similar to that in (6). $\{c_{\mathbf{i}}^\sigma;\ \sigma \in T \text{ and } |\mathbf{i}| = 8\}$ are called Bézier coefficients of s. Then, similar to Lemma 1, we have the following lemma.

Lemma 2. *$s \in S_8^{-1}(\Delta)$ is uniquely determined by its following values and Bézier coefficients:*

1. *For any vertex $\mathbf{v} \in V$, the coefficients are*

$$C_{\mathbf{v}}^k = \{c_{\mathbf{i}}^\sigma;\ \forall\ \sigma = [\mathbf{v}, \mathbf{v}_{\sigma 1}, \mathbf{v}_{\sigma 2}, \mathbf{v}_{\sigma 3}] \in T,\ \mathbf{i} = (i_0, i_1, i_2, i_3) \in Z_+^4,\ |\mathbf{i}| = 8,\ i_0 \geq k\},$$

   where $k = 4$.
2. *For any edge $e \in E$, the values are*

$$E_e' = \{D_{e,\sigma}^{\mathbf{k}} p_\sigma(\mathbf{m}_e);\ \forall \sigma \in T_e \text{ and } |\mathbf{k}| = 2\},$$

   where $\mathbf{m}_e$ is the centroid of the edge e.
3. *For any triangle face $\delta \in F$, we select a unit normal vector $\mathbf{n}_\delta$ of δ. Assume that $\mathbf{n}_{1\delta}$ and $\mathbf{n}_{2\delta}$ are two unit vectors such that $\mathbf{n}_\delta$, $\mathbf{n}_{1\delta}$ and $\mathbf{n}_{2\delta}$ are perpendicular to each other. Then, the values are*

$$T_\delta' = \{\frac{\partial p_\sigma(\mathbf{m}_\delta)}{\partial \mathbf{n}_\delta},\ \frac{\partial^2 p_\sigma(\mathbf{m}_\delta)}{\partial \mathbf{n}_\delta \partial \mathbf{n}_{1\delta}},\ \frac{\partial^2 p_\sigma(\mathbf{m}_\delta)}{\partial \mathbf{n}_\delta \partial \mathbf{n}_{2\delta}};\quad \forall \sigma \in T_\delta\},$$

   where $\mathbf{m}_\delta$ is the centroid of the triangle δ.
4. *For a tetrahedron $\sigma \in T$, the values is*

$$\{p_\sigma(\mathbf{m}_\sigma)\},$$

   where $\mathbf{m}_\sigma$ is the centroid of the edge σ. We denote

$$\mathbf{x}_T = (p_{\sigma_1}(\mathbf{m}_{\sigma_1}), p_{\sigma_2}(\mathbf{m}_{\sigma_2}), \cdots, p_{\sigma_{n_3}}(\mathbf{m}_{\sigma_{n_3}}))', \tag{8}$$

   where σ_i, $1 \leq i \leq n_3 = |T|$ are all tetrahedra in Δ and $\mathbf{x}'$ means the transpose of vector $\mathbf{x}$.

Two tetrahedra is called i-face adjacent if they have a common i-face. For two 2-face adjacent tetrahedra, say $\sigma = [\mathbf{v}_0, \mathbf{v}_1, \mathbf{v}_2, \mathbf{v}_3]$ and $\bar{\sigma} = [\mathbf{v}_0, \mathbf{v}_1, \mathbf{v}_2, \bar{\mathbf{v}}_3]$, we denote $\delta = [\mathbf{v}_0, \mathbf{v}_1, \mathbf{v}_2]$ and $e_i = [\mathbf{v}_j, \mathbf{v}_k]$, where $\{i, j, k\} = \{0, 1, 2\}$.

C^0 conditions. If $s \in S_8^{-1}(\Delta)$ is determined by its Bézier coefficients and values given in Lemma 2, the necessary and sufficient condition of $s \in S_d^0(\Delta)$ is that for any two 2-face adjacent tetrahedra of Δ, say $\sigma = [\mathbf{v}_0, \mathbf{v}_1, \mathbf{v}_2, \mathbf{v}_3]$ and $\bar{\sigma} = [\mathbf{v}_0, \mathbf{v}_1, \mathbf{v}_2, \bar{\mathbf{v}}_3]$, it holds the following conditions:

$$\begin{cases} c^{\sigma}_{(i_0,i_1,i_2,0)} = c^{\bar{\sigma}}_{(i_0,i_1,i_2,0)} & \text{where } i_0+i_1+i_2 = 8 \text{ and } \max\{i_0,i_1,i_2\} \geq 4 \\ D^2_{e_i,\mathbf{v}_i} p_\sigma(\mathbf{m}_{e_i}) = D^2_{e_i,\mathbf{v}_i} p_{\bar{\sigma}}(\mathbf{m}_{e_i}),\ i = 0,1,2. \end{cases} \tag{9}$$

In fact, the restriction $p_\sigma|_\delta$ of p_σ to δ is uniquely determined by $\{c^{\sigma}_{(i_0,i_1,i_2,0)};\ i_0+i_1+i_2 = 8, \max\{i_0,i_1,i_2\} \geq 4\} \bigcup \{D^2_{e_i,\mathbf{v}_i} p_\sigma(\mathbf{m}_{e_i});\ i = 0,1,2\}$ and the restriction $p_{\bar{\sigma}}|_\delta$ of $p_{\bar{\sigma}}$ to δ is uniquely determined by $\{c^{\bar{\sigma}}_{(i_0,i_1,i_2,0)};\ i_0+i_1+i_2 = 8, \max\{i_0,i_1,i_2\} \geq 4\} \bigcup \{D^2_{\delta_i,\mathbf{v}_i} p_{\bar{\sigma}}(\mathbf{m}_{\delta_i});\ i = 0,1,2\}$, respectively. Therefore, $p_\sigma|_\delta = p_{\bar{\sigma}}|_\delta$ iff (9) holds, i.e., $s \in S^0_8(\Delta)$ if (9) holds.

C^1 **conditions.** For convenience, we will divide Bézier coefficients and the function values in Lemma 2 into different classes. Correspondingly, the C^1 conditions of $s \in S^0_8(\Delta)$ are also divided into different classes.

1. The first class is composed of Bézier coefficients of $C^5_{\mathbf{v}}$ as defined in Lemma 2. Correspondingly, C^1 conditions among $C^5_{\mathbf{v}}$ are called vertex-C^1 conditions with respect to $\mathbf{v}$.
2. For a vertex $\mathbf{v} \in V$ and an edge $e = [\mathbf{v}, \mathbf{w}] \in E$, the second class Bézier coefficients is a subset of $C^4_{\mathbf{v}} \backslash C^5_{\mathbf{v}}$ given by

   $$C_{\mathbf{v},e} = \{c^{\sigma}_{\mathbf{i}};\ \forall \sigma = [\mathbf{v}, \mathbf{w}, \mathbf{v}_1, \mathbf{v}_2] \in T,\ \mathbf{i} = (4, i_{\mathbf{w}}, i_1, i_2),\ i_{\mathbf{w}} \geq 2,\ |\mathbf{i}| = 8\}. \tag{10}$$

   Any C^1 condition of s involves $C_{\mathbf{v},e}$ is called a vertex-edge-C^1 condition with respect to $\mathbf{v}$ and e.
3. The third class is E'_e (defined in Lemma 2). Any C^1 condition of s involves E'_e for is called an edge-C^1 condition with respect to e .
4. The fourth class is T'_δ (defined in Lemma 2). Any C^1 condition of s related to T'_δ for some δ is called a triangle-C^1 condition.

By the classification as above, we have

Lemma 3. *If $s \in S^0_8(\Delta)$ is determined by its Bézier coefficients and values given in Lemma 2, then the necessary and sufficient condition of $s \in S^1_8(\Delta)$ is that for any two 2-face adjacent tetrahedra of Δ (we use the same notations as them in C^0 case), it holds the following equalities:*

1. The vertex-C^1 conditions at $\mathbf{v}_j$, $j = 0,1,2$, are

$$\begin{aligned} c^{\bar{\sigma}}_{(i_0,i_1,i_2,1)} = {} & b_0 c^{\sigma}_{(i_0+1,i_1,i_2,0)} + b_1 c^{\sigma}_{(i_0,i_1+1,i_2,0)} + b_2 c^{\sigma}_{(i_0,i_1,i_2+1,0)} + \\ & b_3 c^{\sigma}_{(i_0,i_1,i_2,1)}, \quad i_j \geq 5,\ i_0+i_1+i_2 = 7, \end{aligned} \tag{11}$$

where (b_0,b_1,b_2,b_3) are the barycentric coordinates of $\bar{\mathbf{v}}_3$ with respect to σ.

2. *vertex-edge-C^1 conditions corresponding to vertex $\mathbf{v}_j$ and edge $e_k = [\mathbf{v}_j, \mathbf{v}_l]$ are*

$$\begin{aligned} c^{\bar\sigma}_{(i_0,i_1,i_2,1)} = b_0 c^{\sigma}_{(i_0+1,i_1,i_2,0)} + b_1 c^{\sigma}_{(i_0,i_1+1,i_2,0)} + b_2 c^{\sigma}_{(i_0,i_1,i_2+1,0)} + \\ b_3 c^{\sigma}_{(i_0,i_1,i_2,1)}, \quad i_j = 4,\ i_l = 2,3,\ i_k = 3 - i_l, \end{aligned} \tag{12}$$

where $\{j,k,l\} = \{0,1,2\}$.

3. *The edge-C^1 condition at edge $e_k = [\mathbf{v}_j, \mathbf{v}_l]$ is*

$$D_{e_k,\bar{\mathbf{v}}_3} D_{e_k,\mathbf{v}_k} p_{\bar\sigma}(\mathbf{m}_{e_k}) = (c_1 D_{e_k,\mathbf{v}_k} + c_2 D_{e_k,\mathbf{v}_3}) D_{e_k,\mathbf{v}_k} p_{\sigma}(\mathbf{m}_{e_k}), \tag{13}$$

where $\mathbf{n}_{e_k,\bar{\mathbf{v}}_3} = c_1 \mathbf{n}_{e_k,\mathbf{v}_k} + c_2 \mathbf{n}_{e_k,\mathbf{v}_3}$.

4. *The triangle-C^1 conditions on $\delta = [\mathbf{v}_0, \mathbf{v}_1, \mathbf{v}_2]$ are*

$$\frac{\partial p_{\bar\sigma}(\mathbf{m}_\delta)}{\partial \mathbf{n}_\delta} = \frac{\partial p_{\sigma}(\mathbf{m}_\delta)}{\partial \mathbf{n}_\delta}, \quad \frac{\partial^2 p_{\bar\sigma}(\mathbf{m}_\delta)}{\partial \mathbf{n}_\delta \partial \mathbf{n}_{1\delta}} = \frac{\partial^2 p_{\sigma}(\mathbf{m}_\delta)}{\partial \mathbf{n}_\delta \partial \mathbf{n}_{1\delta}}, \quad \frac{\partial^2 p_{\bar\sigma}(\mathbf{m}_\delta)}{\partial \mathbf{n}_\delta \partial \mathbf{n}_{2\delta}} = \frac{\partial^2 p_{\sigma}(\mathbf{m}_\delta)}{\partial \mathbf{n}_\delta \partial \mathbf{n}_{2\delta}}. \tag{14}$$

The matrix form of (14) is as follows.

$$W_\delta \mathbf{x}_\delta = 0, \tag{15}$$

where

$$\mathbf{x}_\delta = \left(\frac{\partial p_{\bar\sigma}(\mathbf{m}_\delta)}{\partial \mathbf{n}_\delta}, \frac{\partial^2 p_{\bar\sigma}(\mathbf{m}_\delta)}{\partial \mathbf{n}_\delta \partial \mathbf{n}_{1\delta}}, \frac{\partial^2 p_{\bar\sigma}(\mathbf{m}_\delta)}{\partial \mathbf{n}_\delta \partial \mathbf{n}_{2\delta}}, \frac{\partial p_{\sigma}(\mathbf{m}_\delta)}{\partial \mathbf{n}_\delta}, \frac{\partial^2 p_{\sigma}(\mathbf{m}_\delta)}{\partial \mathbf{n}_\delta \partial \mathbf{n}_{1\delta}}, \frac{\partial^2 p_{\sigma}(\mathbf{m}_\delta)}{\partial \mathbf{n}_\delta \partial \mathbf{n}_{2\delta}}\right)'$$

and

$$W_\delta = \begin{bmatrix} 1 & 0 & 0 & -1 & 0 & 0 \\ 0 & 1 & 0 & 0 & -1 & 0 \\ 0 & 0 & 1 & 0 & 0 & -1 \end{bmatrix}.$$

For a boundary triangle δ, we denote by

$$\mathbf{x}_\delta = \left(\frac{\partial p_{\sigma}(\mathbf{m}_\delta)}{\partial \mathbf{n}_\delta}, \frac{\partial^2 p_{\sigma}(\mathbf{m}_\delta)}{\partial \mathbf{n}_\delta \partial \mathbf{n}_{1\delta}}, \frac{\partial^2 p_{\sigma}(\mathbf{m}_\delta)}{\partial \mathbf{n}_\delta \partial \mathbf{n}_{2\delta}}\right)'.$$

and

$$\mathbf{x}_{\partial\delta} = (\mathbf{x}'_{\delta_{m_2+1}}, \mathbf{x}'_{\delta_{m_2+2}}, \ldots, \mathbf{x}'_{\delta_{n_2}})'.$$

Thus, the triangle-C^1 conditions for Δ have the following matrix form

$$W_\sigma \mathbf{x}_\sigma = 0, \tag{16}$$

where $\mathbf{x}_\sigma = (\mathbf{x}'_{\delta_1}, \mathbf{x}'_{\delta_2}, \ldots, \mathbf{x}'_{\delta_{m_2}}, \mathbf{x}'_{\partial\delta})'$ and

$$W_\sigma = \begin{bmatrix} W_\delta & 0 & \cdots & 0 & 0 \\ 0 & W_\delta & \cdots & 0 & 0 \\ \vdots & \vdots & \ddots & \vdots & \vdots \\ 0 & 0 & \cdots & W_\delta & 0 \end{bmatrix}.$$

Proof of Lemma 3. Clearly, we only need to prove $s|_{\sigma\bigcup\bar{\sigma}} \in C^1(\sigma\bigcup\bar{\sigma})$. This is equivalent to prove

$$\left.\frac{\partial p_{\bar{\sigma}}}{\partial \mathbf{n}_\delta}\right|_\delta = \left.\frac{\partial p_\sigma}{\partial \mathbf{n}_\delta}\right|_\delta, \tag{17}$$

where $\delta = \sigma\bigcap\bar{\sigma}$. Since $\frac{\partial p_\sigma}{\partial \mathbf{n}_\delta}|_\delta$ is a polynomial of degree 7 in two variables. It is easy to check that p_σ is determined by its values

$$\begin{aligned}&\{\tfrac{\partial}{\partial \mathbf{n}_\delta} D^{\mathbf{i}}_{\mathbf{v}_j,\delta} p_\sigma(\mathbf{v}_j);\ |\mathbf{i}| \le 3,\ j=0,1,2\}\bigcup\{\tfrac{\partial}{\partial \mathbf{n}_\delta} D_{e_k,v_k} p_\sigma(\mathbf{m}_{e_k});\ k=0,1,2\}\bigcup\\ &\{\tfrac{\partial p_\sigma(\mathbf{m}_\delta)}{\partial \mathbf{n}_\delta}, \tfrac{\partial^2 p_\sigma(\mathbf{m}_\delta)}{\partial \mathbf{n}_\delta \partial \mathbf{n}_{1\delta}}, \tfrac{\partial^2 p_\sigma(\mathbf{m}_\delta)}{\partial \mathbf{n}_\delta \partial \mathbf{n}_{2\delta}}\}\end{aligned} \tag{18}$$

Correspondingly, $\frac{\partial p_{\bar{\sigma}}}{\partial \mathbf{n}_\delta}|_\delta$ is determined by

$$\begin{aligned}&\{\tfrac{\partial}{\partial \mathbf{n}_\delta} D^{\mathbf{i}}_{\mathbf{v}_j,\delta} p_{\bar{\sigma}}(\mathbf{v}_j);\ |\mathbf{i}| \le 3,\ j=0,1,2\}\bigcup\{\tfrac{\partial}{\partial \mathbf{n}_\delta} D_{e_k,v_k} p_{\bar{\sigma}}(\mathbf{m}_{e_k});\ k=0,1,2\}\bigcup\\ &\{\tfrac{\partial p_{\bar{\sigma}}(\mathbf{m}_\delta)}{\partial \mathbf{n}_\delta}, \tfrac{\partial^2 p_{\bar{\sigma}}(\mathbf{m}_\delta)}{\partial \mathbf{n}_\delta \partial \mathbf{n}_{1\delta}}, \tfrac{\partial^2 p_{\bar{\sigma}}(\mathbf{m}_\delta)}{\partial \mathbf{n}_\delta \partial \mathbf{n}_{2\delta}}\}.\end{aligned} \tag{19}$$

Therefore, the proof of (17) is equivalent to prove the corresponding values in (18) and (19) are equal to each other.

Since (11) and (12) are directly from the well-known Bézier net C^1 conditions, it is not difficult to prove that (11) and (12) are equivalent to

$$\frac{\partial}{\partial \mathbf{n}_\delta} D^{\mathbf{i}}_{\mathbf{v}_j,\delta} p_{\bar{\sigma}}(\mathbf{v}_j) = \frac{\partial}{\partial \mathbf{n}_\delta} D^{\mathbf{i}}_{\mathbf{v}_j,\delta} p_\sigma(\mathbf{v}_j), \quad |\mathbf{i}| \le 3,\ j=0,1,2. \tag{20}$$

For edge $e_k = [\mathbf{v}_j, \mathbf{v}_l]$, since all the vectors $\mathbf{n}_\delta, \mathbf{n}_{e_k,\ \bar{v}_3}, \mathbf{n}_{e_k,\ v_k}$ are perpendicular to edge e_k, there exist constants b_1 and b_2 such that $\mathbf{n}_\delta = b_1\mathbf{n}_{e_k,\ \bar{v}_3} + b_2\mathbf{n}_{e_k,\ v_k}$. According to (13), we have

$$\begin{aligned}&\tfrac{\partial}{\partial \mathbf{n}_\delta} D_{e_k,\ v_k} p_{\bar{\sigma}}(\mathbf{m}_{e_k}) = b_1 \tfrac{\partial}{\partial \mathbf{n}_{e_k,\ \bar{v}_3}} D_{e_k,\ v_k} p_{\bar{\sigma}}(\mathbf{m}_{e_k}) + b_2 \tfrac{\partial}{\partial \mathbf{n}_{e_k,\ v_k}} D_{e_k,\ v_k} p_{\bar{\sigma}}(\mathbf{m}_{\delta_k})\\ &= b_1(c_1 D_{e_k,\ v_k} + c_2 D_{e_k,\ v_3}) \tfrac{\partial}{\partial \mathbf{n}_{e_k,\ \bar{v}_3}} D_{e_k,\ v_k} p_\sigma(\mathbf{m}_{e_k}) + b_2 D^2_{e_k,\ v_k} p_\sigma(\mathbf{m}_{e_k})\\ &= b_1 \tfrac{\partial}{\partial \mathbf{n}_{e_k,\ \bar{v}_3}} D_{e_k,v_k} p_\sigma(\mathbf{m}_{e_k}) + b_2 \tfrac{\partial}{\partial \mathbf{n}_{e_k,\ v_k}} D_{e_k,v_k} p_\sigma(\mathbf{m}_{e_k})\\ &= \tfrac{\partial}{\partial \mathbf{n}_\delta} D_{e_k,\ v_k} p_\sigma(\mathbf{m}_{e_k}).\end{aligned} \tag{21}$$

From (20), (21), and (14), we conclude that (17) holds, i.e., $s \in S^1_8(\Delta)$.

For convenience, we re-formula the equations (11)–(13) as follows.

- The vertex-C^1 conditions corresponding to $\mathbf{v}$ are, for any 2-face adjacent tetrahedra $\sigma = [\mathbf{v}, \mathbf{v}_1, \mathbf{v}_2, \mathbf{w}]$, $\bar{\sigma} = [\mathbf{v}, \mathbf{v}_1, \mathbf{v}_2, \bar{\mathbf{w}}] \in \mathrm{Star}(\mathbf{v})$,

$$c^{\bar{\sigma}}_{\mathbf{i}+\varepsilon_{\mathbf{w}}} = \sum_{\mathbf{u}\in V_\sigma} b_{\mathbf{u}} c^\sigma_{\mathbf{i}+\varepsilon_{\mathbf{u}}},\ \mathbf{i} = (i_{\mathbf{v}}, i_1, i_2, 0), i_{\mathbf{v}} \ge 5,\ |\mathbf{i}| = 7, \tag{22}$$

where V_σ is the set of vertices of σ, $(b_{\mathbf{u}})_{\mathbf{u}\in V_\sigma}$ are the barycentric coordinates of $\bar{\mathbf{w}}$ with respect to σ and $\varepsilon_{\mathbf{v}} = (1,0,0,0), \varepsilon_{\mathbf{v}_1} = (0,1,0,0), \varepsilon_{\mathbf{v}_2} = (0,0,1,0), \varepsilon_{\mathbf{w}} = (0,0,0,1)$. We denote (22) as the following matrix form.

$$W_{\mathbf{v}}\mathbf{x}_{\mathbf{v}} = 0, \tag{23}$$

where $\mathbf{x}_{\mathbf{v}}$ is the vector formed by all Bézier coefficients of $C^5_{\mathbf{v}}$ and $W_{\mathbf{v}}$ is the corresponding coefficient matrix.
The matrix form of vertex-C^1 conditions for Δ is

$$W_V\mathbf{x}_V = 0, \tag{24}$$

where $\mathbf{x}_V = (\mathbf{x}'_{\mathbf{v}_1}, \mathbf{x}'_{\mathbf{v}_2}, \ldots, \mathbf{x}'_{\mathbf{v}_{n_0}})'$ and

$$W_V = \begin{bmatrix} W_{\mathbf{v}_1} & 0 & \cdots & 0 \\ 0 & W_{\mathbf{v}_2} & \cdots & 0 \\ \vdots & \vdots & \ddots & \vdots \\ 0 & 0 & \cdots & W_{\mathbf{v}_{n_0}} \end{bmatrix}.$$

- For any 2-face adjacent tetrahedra $\sigma = [\mathbf{v}, \mathbf{v}_1, \mathbf{v}_2, \mathbf{w}]$, $\bar{\sigma} = [\mathbf{v}, \mathbf{v}_1, \mathbf{v}_2, \bar{\mathbf{w}}] \in \mathrm{Star}(e)$, the vertex-edge-$C^1$ conditions corresponding to vertex $\mathbf{v}$ and edge $e = [\mathbf{v}, \mathbf{v}_1]$ are

$$c^{\bar{\sigma}}_{\mathbf{i}+\varepsilon_{\mathbf{w}}} = \sum_{\mathbf{u}\in V_\sigma} b_{\mathbf{u}} c^{\sigma}_{\mathbf{i}+\varepsilon_{\mathbf{u}}},\ \mathbf{i} = (4, i_1, 3-i_1, 0),\ i_1 = 2,3. \tag{25}$$

We will write out the matrix form of (25) later.
- For any 2-face adjacent tetrahedra $\sigma = [\mathbf{v}, \mathbf{v}_1, \mathbf{v}_2, \mathbf{w}]$, $\bar{\sigma} = [\mathbf{v}, \mathbf{v}_1, \mathbf{v}_2, \bar{\mathbf{w}}] \in \mathrm{Star}(e)$, the edge-$C^1$ conditions at edge $e = [\mathbf{v}, \mathbf{v}_1]$ are

$$D_{e,\bar{\mathbf{w}}} D_{e,\mathbf{v}_2} p_{\bar{\sigma}}(\mathbf{m}_e) = (c_1 D_{e,\mathbf{v}_2} + c_2 D_{e,\mathbf{w}}) D_{e,\mathbf{v}_2} p_\sigma(\mathbf{m}_e), \tag{26}$$

where $\mathbf{n}_{e,\bar{\mathbf{w}}} = c_1\mathbf{n}_{e,\mathbf{v}_2} + c_2\mathbf{n}_{e,\mathbf{w}}$.

For $e = [\mathbf{v}, \mathbf{w}] \in E$, we assume that $\sigma_i = [\mathbf{v}, \mathbf{w}, \mathbf{v}_i, \mathbf{v}_{i+1}]$, $0 \le i \le m-1$, are all tetrahedra in $\mathrm{Star}(e)$, where $\mathbf{v}_m = \mathbf{v}_0$ if e is an inner edge. Then, the matrix form of (26) is as follows.

$$W_e\mathbf{x}_e = 0, \tag{27}$$

where W_e is the corresponding coefficient matrix and

$$\mathbf{x}_e = (D^2_{e,\mathbf{v}_0} p_{\sigma_0}(\mathbf{m}_e), D_{e,\mathbf{v}_0} D_{e,\mathbf{v}_1} p_{\sigma_0}(\mathbf{m}_e), \cdots, D^2_{e,\mathbf{v}_{m-1}} p_{\sigma_{m-1}}(\mathbf{m}_e), D_{e,\mathbf{v}_{m-1}} D_{e,\mathbf{v}_0} p_{\sigma_{m-1}}(\mathbf{m}_e))'$$

if e is an inner edge and

$$\mathbf{x}_e = (D^2_{e,\mathbf{v}_0} p_{\sigma_0}(\mathbf{m}_e), \ldots, D_{e,\mathbf{v}_{m-1}} D_{e,\mathbf{v}_m} p_{\sigma_{m-1}}(\mathbf{m}_e), D^2_{e,\mathbf{v}_m} p_{\sigma_{m-1}}(\mathbf{m}_e))'$$

if e is a boundary edge (comparing to the inner edge case, it has one more component $D^2_{e,\mathbf{v}_m} p_{\sigma_{m-1}}(\mathbf{m}_e)$).

The matrix form of vertex-C^1 conditions for Δ is

$$W_E \mathbf{x}_E = 0, \tag{28}$$

where $\mathbf{x}_E = (\mathbf{x}'_{e_1}, \mathbf{x}'_{e_2}, \dots, \mathbf{x}'_{e_{n_1}})'$ and

$$W_E = \begin{bmatrix} W_{e_1} & 0 & \cdots & 0 \\ 0 & W_{e_2} & \cdots & 0 \\ \vdots & \vdots & \ddots & \vdots \\ 0 & 0 & \cdots & W_{e_{n_1}} \end{bmatrix}.$$

To obtain the matrix form of (25), we assume that $e = [\mathbf{v}, \mathbf{w}]$ and $\sigma_i = [\mathbf{v}, \mathbf{w}, \mathbf{v}_i, \mathbf{v}_{i+1}]$, $0 \le i \le m-1$, are all tetrahedra in Star(e), where $\mathbf{v}_m = \mathbf{v}_0$ if e is an inner edge. We denote by

$$p_i(\mathbf{x}) = \sum_{i_0+i_1+i_2+i_3=8} c^i_{i_0,i_1,i_2,i_3} \frac{8!}{i_0!\,i_1!\,i_2!\,i_3!} u_0^{i_0} u_1^{i_1} u_2^{i_2} u_3^{i_3} \tag{29}$$

the restriction of a spline $s_e \in S^1_8(\text{Star}(e))$ to the tetrahedron σ_i and (u_0, u_1, u_2, u_3) are the barycentric coordinates of $\mathbf{x}$ with respect to σ_i, i.e., $\mathbf{x} = u_0\mathbf{v} + u_1\mathbf{w} + u_2\mathbf{v}_i + u_3\mathbf{v}_{i+1}$. We also denote $\det[\mathbf{a}, \mathbf{b}, \mathbf{c}, \mathbf{d}] = ((\mathbf{b}-\mathbf{a}) \times (\mathbf{c}-\mathbf{a})) \cdot (\mathbf{d}-\mathbf{a})$ the mixed product of vectors $\mathbf{b}-\mathbf{a}$, $\mathbf{c}-\mathbf{a}$, $\mathbf{d}-\mathbf{a}$. Then, (25) is equivalent to

$$\begin{cases} c^i_{4,3,0,1} = -\frac{W_i}{V_{i-1}} c^{i-1}_{5,3,0,0} + \frac{U_i}{V_{i-1}} c^{i-1}_{4,4,0,0} + \frac{T_i}{V_{i-1}} c^{i-1}_{4,3,0,1} - \frac{V_i}{V_{i-1}} c^{i-1}_{4,3,1,0}, \\ c^i_{4,2,1,1} = -\frac{W_i}{V_{i-1}} c^{i-1}_{5,2,0,1} + \frac{U_i}{V_{i-1}} c^{i-1}_{4,3,0,1} + \frac{T_i}{V_{i-1}} c^{i-1}_{4,2,0,2} - \frac{V_i}{V_{i-1}} c^{i-1}_{4,2,1,1}, \\ \qquad 1 \le i \le m - b_e, \end{cases} \tag{30}$$

where $c^m_{i,j,k,l} = c^0_{i,j,k,l}$ if e is inner, $W_i = \det[\mathbf{w}, \mathbf{v}_{i-1}, \mathbf{v}_i, \mathbf{v}_{i+1}]$,$U_i = \det[\mathbf{v}, \mathbf{v}_{i-1}, \mathbf{v}_i, \mathbf{v}_{i+1}]$, $T_i = \det[\mathbf{v}, \mathbf{w}, \mathbf{v}_{i-1}, \mathbf{v}_{i+1}]$, $V_i = \det[\mathbf{v}, \mathbf{w}, \mathbf{v}_i, \mathbf{v}_{i+1}]$, and $(-\frac{W_i}{V_{i-1}}, \frac{U_i}{V_{i-1}}, -\frac{V_i}{V_{i-1}}, \frac{T_i}{V_{i-1}})$ are the barycentric coordinates of $\mathbf{v}_{i+1}$ with respect to σ_{i-1}, i.e., $\mathbf{v}_{i+1} = -\frac{W_i}{V_{i-1}}\mathbf{v} + \frac{U_i}{V_{i-1}}\mathbf{w} - \frac{V_i}{V_{i-1}}\mathbf{v}_{i-1} + \frac{T_i}{V_{i-1}}\mathbf{v}_i$. In addition, since $p_i|_{\delta_i} = p_{i-1}|_{\delta_i}$ ($\delta_i = [\mathbf{v}, \mathbf{w}, \mathbf{v}_i]$), it holds

$$c^i_{j,k,l,0} = c^{i-1}_{j,k,0,l}, \quad j+k+l = 8,\ 1 \le i \le m - b_e. \tag{31}$$

Especially, we set $l = 0$ in (31), we have

$$c^i_{j,k,0,0} = c^0_{j,k,0,0}, \quad j+k = 8, 1 \le i \le m-1. \tag{32}$$

Remark 4. Since $\{c^0_{5,3,0,0}, c^0_{4,4,0,0}, c^i_{4,3,1,0}, c^i_{4,3,0,1}\}$ determine uniquely a linear polynomial and the first equation of (30) shows that those linear polynomials are joint smoothly, those linear polynomials have to be the same. Thus, there are exact four free variables among these unknowns, say $\{c^0_{5,3,0,0}, c^0_{4,4,0,0}, c^0_{4,3,1,0}, c^0_{4,3,0,1}\}$. Noting that $c^0_{5,3,0,0} \in C^5_{\mathbf{v}}$ (so it is already fixed in the previous steps) and $c^0_{4,4,0,0}$ (it will be fixed separately) will appear also in the vertex-edge-C^1 conditions corresponding to the edge $e=[\mathbf{v},\mathbf{w}]$ and the vertex $\mathbf{w}$, only $c^0_{4,3,1,0}$ and $c^0_{4,3,0,1}=c^1_{4,3,1,0}$ are real free variables. We denote

$$c_e=c^0_{4,4,0,0},\; c^0_{\mathbf{v},e}=c^0_{4,3,1,0},\; c^1_{\mathbf{v},e}=c^1_{4,3,1,0}.$$

Thus, (30) has the following matrix form

$$W_{\mathbf{v},e}\mathbf{x}_{\mathbf{v},e}+W'_{\mathbf{v},e}\mathbf{x}_{\mathbf{v}}+W''_{\mathbf{v},e}c_e=0, \tag{33}$$

where

$$\mathbf{x}_{\mathbf{v},e}=\begin{cases}(c^0_{4,2,2,0},c^0_{4,2,1,1},\dots,c^{m-1}_{4,2,2,0},c^{m-1}_{4,2,1,1},c^0_{\mathbf{v},e},c^1_{\mathbf{v},e})', & e \text{ is an inner edge},\\ (c^0_{4,2,2,0},c^0_{4,2,1,1},\cdots,c^{m-1}_{4,2,2,0},c^{m-1}_{4,2,1,1},c^{m-1}_{4,2,0,2},c^0_{\mathbf{v},e},c^1_{\mathbf{v},e})', & e \text{ is a boundary edge},\end{cases}$$

and $W_{\mathbf{v},e}$, $W'_{\mathbf{v},e}$ and $W''_{\mathbf{v},e}$ are the corresponding coefficient matrices.

The matrix form of vertex-edge-C^1 conditions for the edges of Star(v) is

$$W_{\mathbf{v},\mathrm{Star}}\mathbf{x}_{\mathbf{v},\mathrm{Star}}+W'_{\mathbf{v},\mathrm{Star}}\mathbf{x}_{\mathbf{v}}+W''_{\mathbf{v},\mathrm{Star}}\mathbf{x}_E=0, \tag{34}$$

where $\mathbf{x}_{\mathbf{v},\mathrm{Star}}=(\mathbf{x}'_{\mathbf{v},\tilde{e}_1},\mathbf{x}'_{\mathbf{v},\tilde{e}_2},\dots,\mathbf{x}'_{\mathbf{v},\tilde{e}_{n_s}})'$ $(\tilde{e}_i\in \mathrm{Star}(v),\ 1\le i\le n_s=|E_{\mathbf{v}}|)$, $\mathbf{x}_E=(c_{e_1},c_{e_2},\dots,c_{e_{n_1}})'$, and $W_{\mathbf{v},\mathrm{Star}}$, $W'_{\mathbf{v},\mathrm{Star}}$, $W''_{\mathbf{v},\mathrm{Star}}$ are the corresponding coefficient matrices. We will discuss the structure of $W_{\mathbf{v},\mathrm{Star}}$ later.

The matrix form of vertex-edge-C^1 conditions for $S^1_8(\Delta)$ is

$$W_{V,E}\mathbf{x}_{V,E}+W'_{V,E}\mathbf{x}_V+W''_{V,E}\mathbf{x}_E=0, \tag{35}$$

where $\mathbf{x}_{V,E}=(\mathbf{x}'_{\mathbf{v}_1,\mathrm{Star}},\dots,\mathbf{x}'_{\mathbf{v}_{n_0},\mathrm{Star}})'$,

$$W_{V,E}=\begin{bmatrix}W_{\mathbf{v}_1,\mathrm{Star}} & 0 & \cdots & 0\\ 0 & W_{\mathbf{v}_2,\mathrm{Star}} & \cdots & 0\\ \vdots & \vdots & \ddots & \vdots\\ 0 & 0 & \cdots & W_{\mathbf{v}_{n_0},\mathrm{Star}}\end{bmatrix},$$

and $W'_{V,E}, W''_{V,E}$ are the corresponding coefficient matrices.

Finally, the matrix form for the C^1 conditions for $S^1_8(\Delta)$ is

$$\begin{bmatrix}W_{V,E} & W'_{V,E} & 0 & 0 & 0 & W''_{V,E}\\ 0 & W_V & 0 & 0 & 0 & 0\\ 0 & 0 & W_E & 0 & 0 & 0\\ 0 & 0 & 0 & W_\sigma & 0 & 0\end{bmatrix}\begin{bmatrix}\mathbf{x}_{V,E}\\ \mathbf{x}_V\\ \mathbf{x}_E\\ \mathbf{x}_\sigma\\ \mathbf{x}_T\\ \mathbf{x}_E\end{bmatrix}=0. \tag{36}$$

Lemma 4. *If for any vertex* $\mathbf{v} \in \Delta$, rank$(W_{\mathbf{v},\text{Star}})$=rank$([W_{\mathbf{v},\text{Star}}\ W'_{\mathbf{v},\text{Star}}\ W''_{\mathbf{v},\text{Star}}])$*, then*

$$\begin{aligned}\dim S_8^1(\Delta) &= \dim(\text{null}(W_{V,E})) + \dim(\text{null}(W_V)) + \dim(\text{null}(W_E)) \\ &\quad + \dim(\text{null}(W_\sigma)) + |E| + |T|,\end{aligned}$$

where rank(M) *is the rank of matrix* M *and* null(M) *is the* null *space of* M.

The proof of Lemma 4 is straightforward. Note that rank$(W_{V,E})$ = rank$([W_{V,E}\ W'_{V,E}\ W''_{V,E}])$ if rank$(W_{\mathbf{v},\text{Star}})$ = rank$([W_{\mathbf{v},\text{Star}}\ W'_{\mathbf{v},\text{Star}}\ W''_{\mathbf{v},\text{Star}}])$ holds for all vertex $\mathbf{v}$ of Δ. Thus, the matrix $[W_{V,E}\ W'_{V,E}\ W''_{V,E}]$ can be simplified to a version $[\tilde{W}_{V,E}\ \tilde{W}'_{V,E}\ \tilde{W}''_{V,E}]$ such that rank$(\tilde{W}_{V,E})$= rank$([\tilde{W}_{V,E}\ \tilde{W}'_{V,E}\ \tilde{W}''_{V,E}])$=$\tilde{n}$, the number of rows of $\tilde{W}_{V,E}$. Thus, if we denote W the coefficient matrix of (36), it holds

$$\begin{aligned}\dim S_8^1(\Delta) &= \dim(\text{null}(W)) = \dim(\text{null}(\tilde{W}_{V,E})) + \dim(\text{null}(W_V)) \\ &\quad + \dim(\text{null}(W_E)) + \dim(\text{null}(W_\sigma)) + |E| + |T| \\ &= \dim(\text{null}(W_{V,E})) + \dim(\text{null}(W_V)) + \dim(\text{null}(W_E)) \\ &\quad + \dim(\text{null}(W_\sigma)) + |E| + |T|.\end{aligned}$$

Next, we analyze $\dim S_8^1(\Delta)$ in Lemma 4 term by term. We first consider $\dim(\text{null}(W_\sigma))$ and have the following lemma.

Lemma 5. $\dim(\text{null}(W_\sigma))= 3|F|$.

The proof of Lemma 5 is straightforward.

For $\dim(\text{null}(W_E))$, we chose a tetrahedron $\tilde{\sigma} \in T_e$. Then for any $\sigma \in T_e$, the values $\{D^{\mathbf{k}}_{e,\sigma} p_\sigma(\mathbf{m}_e);\ |\mathbf{k}| = 2\} \bigcup \{D^{\mathbf{k}}_{e,\tilde{\sigma}} p_{\tilde{\sigma}}(\mathbf{m}_e);\ |\mathbf{k}| \le 1\}$ determine uniquely a bivariate quadratic polynomial on triangle σ/e. If those polynomials satisfy (26), they define a spline $s_e \in S_2^1(\text{TStar}(e)))$. Thus,

$$\dim(\text{null}(W_e)) = \dim S_2^1(\text{TStar}(e))) - 3.$$

Therefore, we have the following lemma.

Lemma 6. $\dim(\text{null}(W_E))=\Sigma_{e\in E}\left(\dim S_2^1(\mathit{TStar}(e)) - 3\right)$.

Next, we consider $\dim(\text{null}(W_V))$. For a vertex $\mathbf{v} \in V$ and a tetrahedron $\sigma \in T_{\mathbf{v}}$, the Bézier coefficients $\{c^\sigma_{\mathbf{i}};\ \mathbf{i} = (i_{\mathbf{w}})_{\mathbf{w}\in V_\sigma} \in Z^4_+,\ |\mathbf{i}| = 8,\ i_{\mathbf{v}} \ge 5\}$ determine uniquely a trivariate cubic polynomial on σ. The vertex-C^1 conditions (22) insure that those cubic polynomials are C^1 joint between any two 2-face adjacent tetrahedra. It concludes that the Bézier coefficients $C^5_{\mathbf{v}}$ determine uniquely a spline $s_{\mathbf{v}} \in S_3^1(\text{Star}(\mathbf{v}))$ if they satisfy (22). Thus,

$$\dim(\text{null}(W_{\mathbf{v}})) = \dim S_3^1(\text{Star}(\mathbf{v})))$$

and

$$\dim(\text{null}(W_V)) = \Sigma_{\mathbf{v}\in V}\dim(\text{null}(W_{\mathbf{v}})) = \Sigma_{\mathbf{v}\in V}\dim S_3^1(\text{Star}(\mathbf{v})),$$

i.e., we have

Lemma 7. $\dim(\mathrm{null}(W_V)) = \sum_{\mathbf{v}\in V} \dim S_3^1(\mathrm{Star}(\mathbf{v}))$.

Next, we discuss the most difficult part of Lemma 4, i.e., how to obtain $\dim(\mathrm{null}(W_{\mathbf{v},\mathrm{Star}}))$ in the following equations.

$$\dim(\mathrm{null}(W_{V,E})) = \sum_{\mathbf{v}\in V} \dim(\mathrm{null}(W_{\mathbf{v},\mathrm{Star}})). \tag{37}$$

To obtain $\dim(\mathrm{null}(W_{V,E}))$, we only need to analyze the vertex-edge-C^1 conditions (25) which takes the Bézier coefficients of $C_{\mathbf{v},e}$ (defined in (10)) as unknowns. We need to re-formula (25) again. For $e = [\mathbf{v},\mathbf{w}] \in E$, we assume that $\sigma_i = [\mathbf{v},\mathbf{w},\mathbf{v}_i,\mathbf{v}_{i+1}]$, $0 \le i \le m-1$, are all tetrahedra in Star(e), where $\mathbf{v}_m = \mathbf{v}_0$ if e is inner. Next, we generalize a terminology in [1].

Definition 3. Let $A_{\mathbf{v},e}$ be a subset of $C_{\mathbf{v},e}$ (defined in (10)). $A_{\mathbf{v},e}$ is called determining $C_{\mathbf{v},e}$ if the Bézier coefficients in $A_{\mathbf{v},e}$, together with all other involved Bézier coefficients in the vertex-edge-C^1 conditions (30) that are not contained in $C_{\mathbf{v},e}$, are zeros, then all the Bézier coefficients in $C_{\mathbf{v},e}$ have to be zeros if they satisfy the vertex-edge-C^1 conditions (30). An edge $e \in E_{\mathbf{v}}$ is called confinable if there exists a subset $A_{\mathbf{v},e}$ of $C_{\mathbf{v},e}$ determining $C_{\mathbf{v},e}$ with $|A_{\mathbf{v},e}| = m+3+\varepsilon_e+2b_e$ and containing

$$M_{\mathbf{v},e} = \{c^0_{4,4,0,0},\ c^i_{4,2,2,0};\quad 0 \le i \le m-1+b_e\}, \tag{38}$$

where $\varepsilon_e = 1$ if e is a singular edge; $\varepsilon_e = 0$ otherwise.

Next, we analyze the second equation of (30) and, after applying (31), rewrite it as

$$\begin{aligned}\tfrac{1}{V_i}c^i_{4,2,1,1} &= -\tfrac{W_i}{V_{i-1}V_i}c^i_{5,2,1,0} + \tfrac{U_i}{V_{i-1}V_i}c^i_{4,3,1,0} + \tfrac{T_i}{V_{i-1}V_i}c^i_{4,2,2,0} - \tfrac{1}{V_{i-1}}c^{i-1}_{4,2,1,1},\\ &1 \le i \le m-b_e.\end{aligned} \tag{39}$$

We will discuss (39) case by case.

Case 1. $e = [\mathbf{v},\mathbf{w}]$ is a boundary edge, then (30) is the same as

$$\begin{aligned}\tfrac{1}{V_i}c^i_{4,2,1,1} &= -\tfrac{W_i}{V_{i-1}V_i}c^i_{5,2,1,0} + \tfrac{U_i}{V_iV_i}c^{i-1}_{4,3,1,0} + \tfrac{T_i}{V_{i-1}V_i}c^i_{4,2,2,0} - \tfrac{1}{V_{i-1}}c^{i-1}_{4,2,1,1},\\ &1 \le i \le m-1.\end{aligned} \tag{40}$$

The matrix form of (40) (i.e., (30)) is

$$B_{\mathbf{v},e}\mathbf{x}'_{\mathbf{v},e} + B'_{\mathbf{v},e}\mathbf{x}''_{\mathbf{v},e} + F_{\mathbf{v},e}\mathbf{x}_{\mathbf{v}} + G_{\mathbf{v},e}\mathbf{x}_E = 0, \tag{41}$$

where

$$\mathbf{x}'_{\mathbf{v},e} = (c^0_{4,2,1,1},\ldots,c^{m-2}_{4,2,1,1})',$$

$$\mathbf{x}''_{\mathbf{v},e} = (c^{m-1}_{4,2,1,1},c^0_{4,3,1,0},c^1_{4,3,1,0},c^0_{4,2,2,0},\ldots,c^{m-1}_{4,2,2,0})',$$

$$B_{\mathbf{v},e} = \begin{bmatrix} \frac{1}{V_0} & \frac{1}{V_1} & 0 & \cdots & 0 & 0 \\ 0 & \frac{1}{V_1} & \frac{1}{V_2} & \cdots & 0 & 0 \\ 0 & 0 & \frac{1}{V_2} & \cdots & 0 & 0 \\ \vdots & \vdots & \vdots & \ddots & \vdots & \vdots \\ 0 & 0 & 0 & \cdots & \frac{1}{V_{m-3}} & \frac{1}{V_{m-2}} \\ 0 & 0 & 0 & \cdots & 0 & \frac{1}{V_{m-2}} \end{bmatrix},$$

$B'_{\mathbf{v},e}, F_{\mathbf{v},e}, G_{\mathbf{v},e}$ are the coefficient matrices (but they are not important). If we treat a vector $\mathbf{x}$ as a set of its components, then $\mathbf{x}_{\mathbf{v},e} = \mathbf{x}'_{\mathbf{v},e} \bigcup \mathbf{x}''_{\mathbf{v},e}$.

It is obvious that

$$\operatorname{rank}(B_{\mathbf{v},e}) = m - 1 = \operatorname{degree}(e) - 2. \tag{42}$$

Let $A_{\mathbf{v},e} = \{c^0_{4,3,1,0},\ c^1_{4,3,1,0},\ c^{m-1}_{4,2,1,1}\} \bigcup M_{\mathbf{v},e}$. We set all the elements in $A_{\mathbf{v},e}$ are zeros and set $c^i_{5,2,1,0} = 0,\ 0 \le i \le m-1$, at the same time. Then, according to Remark 4 and (40), all the Bézier coefficients of $C_{\mathbf{v},e}$ are zeros. Therefore, we have

Lemma 8. *A boundary edge $e \in E^{\partial}_{\mathbf{v}}$ is confinable.*

Case 2. $e = [\mathbf{v}, \mathbf{w}]$ is an inner edge and m is an odd number. In this case, we re-write (39)(i.e., (30)) as the following matrix form

$$B_{\mathbf{v},e}\mathbf{x}'_{\mathbf{v},e} + B'_{\mathbf{v},e}\mathbf{x}''_{\mathbf{v},e} + F_{\mathbf{v},e}\mathbf{x}_{\mathbf{v}} + G_{\mathbf{v},e}\mathbf{x}_E = 0, \tag{43}$$

where $\mathbf{x}'_{\mathbf{v},e} = (c^0_{4,2,1,1}, \dots, c^{m-1}_{4,2,1,1})'$, $\mathbf{x}''_{\mathbf{v},e} = (c^0_{4,3,1,0}, c^1_{4,3,1,0}, c^0_{4,2,2,0}, \dots, c^{m-1}_{4,2,2,0})'$,

$$B_{\mathbf{v},e} = \begin{bmatrix} \frac{1}{V_0} & \frac{1}{V_1} & 0 & \cdots & 0 & 0 \\ 0 & \frac{1}{V_1} & \frac{1}{V_2} & \cdots & 0 & 0 \\ 0 & 0 & \frac{1}{V_2} & \cdots & 0 & 0 \\ \vdots & \vdots & \vdots & \ddots & \vdots & \vdots \\ 0 & 0 & 0 & \cdots & \frac{1}{V_{m-2}} & \frac{1}{V_{m-1}} \\ \frac{1}{V_0} & 0 & 0 & \cdots & 0 & \frac{1}{V_{m-1}} \end{bmatrix},$$

and again, $B'_{\mathbf{v},e}, F_{\mathbf{v},e}, G_{\mathbf{v},e}$ are the coefficient matrices. We still have $\mathbf{x}_{\mathbf{v},e} = \mathbf{x}'_{\mathbf{v},e} \bigcup \mathbf{x}''_{\mathbf{v},e}$.

Since $\det B_{\mathbf{v},e} = \frac{2}{\prod_{i=0}^{m-1} V_i} \neq 0$, it yields

$$\operatorname{rank}(B_{\mathbf{v},e}) = m = \operatorname{degree}(e). \tag{44}$$

Let $A_{\mathbf{v},e} = \{c^0_{4,3,1,0},\ c^1_{4,3,1,0}\} \bigcup M_{\mathbf{v},e}$. If we set all the elements in $A_{\mathbf{v},e}$ are zeros and set $c^i_{5,2,1,0} = 0,\ 0 \le i \le m-1$, at the same time. Then, according to (43) and (44), all the Bézier coefficients of $C_{\mathbf{v},e}$ are zeros. Therefore, we have

Lemma 9. *An odd inner edge $e \in E_{\mathbf{v}}$ is confinable.*

If e is an even inner edge, (39) is equivalent to

$$\frac{1}{V_i}c^i_{4,2,1,1} = -\frac{W_i}{V_{i-1}V_i}c^i_{5,2,1,0} + \frac{U_i}{V_{i-1}V_i}c^i_{4,3,1,0} + \frac{T_i}{V_{i-1}V_i}c^i_{4,2,2,0} - \frac{1}{V_{i-1}}c^{i-1}_{4,2,1,1}, \quad 1 \le i \le m-1 \tag{45}$$

and

$$-\sum_{i=0}^{m-1}(-1)^i\frac{W_i}{V_{i-1}V_i}c^i_{5,2,1,0} + \sum_{i=0}^{m-1}(-1)^i\frac{U_i}{V_{i-1}V_i}c^i_{4,3,1,0} + \sum_{i=0}^{m-1}(-1)^i\frac{T_i}{V_{i-1}V_i}c^i_{4,2,2,0} = 0. \tag{46}$$

Clearly, we only need to discuss (46).

Case 3. $e = [\mathbf{v}, \mathbf{w}]$ is an inner edge with $m = 4$, but e is not singular. According to (31) and (32), we re-write the first equation of (30) as follows.

$$\frac{1}{V_i}c^{i+1}_{4,3,1,0} = -\frac{W_i}{V_{i-1}V_i}c^0_{5,3,0,0} + \frac{U_i}{V_{i-1}V_i}c^0_{4,4,0,0} + \frac{T_i}{V_{i-1}V_i}c^i_{4,3,1,0} - \frac{1}{V_{i-1}}c^{i-1}_{4,3,1,0}, \quad 1 \le i \le 4 \tag{47}$$

Setting $i = 1$ and $i = 4$ in (47), respectively, we obtain

$$\begin{cases} \frac{1}{V_1}c^2_{4,3,1,0} = -\frac{W_1}{V_0V_1}c^0_{5,3,0,0} + \frac{U_1}{V_0V_1}c^0_{4,4,0,0} + \frac{T_1}{V_0V_1}c^1_{4,3,1,0} - \frac{1}{V_0}c^0_{4,3,1,0}, \\ \frac{1}{V_3}c^3_{4,3,1,0} = -\frac{W_0}{V_3V_0}c^0_{5,3,0,0} + \frac{U_0}{V_3V_0}c^0_{4,4,0,0} + \frac{T_0}{V_3V_0}c^0_{4,3,1,0} - \frac{1}{V_0}c^1_{4,3,1,0}. \end{cases} \tag{48}$$

Substituting (48) into (46) (m=4), it yields

$$\left(\frac{U_0}{V_3V_0} - \frac{U_2}{V_0V_2} - \frac{U_3T_0}{V_0V_2V_3}\right)c^0_{4,3,1,0} + \left(-\frac{U_1}{V_0V_1} + \frac{U_3}{V_0V_2} + \frac{U_2T_1}{V_0V_1V_2}\right)c^1_{4,3,1,0} + B = 0, \tag{49}$$

where

$$B = -\textstyle\sum_{i=0}^{m-1}(-1)^i\frac{W_i}{V_{i-1}V_i}c^i_{5,2,1,0} + \sum_{i=0}^{m-1}(-1)^i\frac{T_i}{V_{i-1}V_i}c^i_{4,2,2,0} + \left(\frac{U_3W_0}{V_0V_2V_3} - \frac{U_2W_1}{V_0V_1V_2}\right)c^0_{5,3,0,0} + \left(\frac{U_1U_2}{V_0V_1V_2} - \frac{U_0U_3}{V_0V_2V_3}\right)c^0_{4,4,0,0}.$$

Since $\mathbf{v}_0 - \mathbf{v}$, $\mathbf{v}_1 - \mathbf{v}$ and $\mathbf{w} - \mathbf{v}$ are linearly independent, there exist constants $a_\mathbf{w}, a_0, a_1$ and $b_\mathbf{w}, b_0, b_1$ such that

$$\begin{cases} \mathbf{v}_2 - \mathbf{v} = a_\mathbf{w}(\mathbf{w} - \mathbf{v}) + a_0(\mathbf{v}_0 - \mathbf{v}) + a_1(\mathbf{v}_1 - \mathbf{v}), \\ \mathbf{v}_3 - \mathbf{v} = b_\mathbf{w}(\mathbf{w} - \mathbf{v}) + b_0(\mathbf{v}_0 - \mathbf{v}) + b_1(\mathbf{v}_1 - \mathbf{v}). \end{cases} \tag{50}$$

According to the definitions of V_i, U_i and T_i in (30), the following equalities can be checked directly.

$$\begin{cases} V_1 = -a_0V_0, \quad V_2 = (a_0b_1 - a_1b_0)V_0, \quad V_3 = -b_1V_0, \\ U_0 = b_\mathbf{w}V_0, \quad U_1 = a_\mathbf{w}V_0, \quad U_2 = (a_\mathbf{w}b_0 - a_0b_\mathbf{w})V_0, \quad U_3 = (a_1b_\mathbf{w} - a_\mathbf{w}b_1)V_0, \\ T_0 = b_0V_0, \quad T_1 = a_1V_0, \end{cases} \tag{51}$$

From (51), it holds

$$V_2U_0 - V_3U_2 = ((a_0b_1 - a_1b_0)b_{\mathbf{w}} - (-b_1)(a_{\mathbf{w}}b_0 - a_0b_{\mathbf{w}}))V_0^2 = b_0(a_{\mathbf{w}}b_1 - a_1b_{\mathbf{w}})V_0^2.$$

Therefore,

$$V_2U_0 - V_3U_2 = -T_0U_3 \tag{52}$$

Similarly,

$$V_1U_3 - V_2U_1 = T_1U_2 \tag{53}$$

Substituting (52) and (53) into (49), it yields

$$-\frac{2U_3T_0}{V_0V_2V_3}c^0_{4,3,1,0} + \frac{2U_2T_1}{V_0V_1V_2}c^1_{4,3,1,0} + B = 0, \tag{54}$$

Clearly, $U_i \neq 0,\ 0 \leq i \leq 3$. Since e is not singular, $\{\mathbf{v},\ \mathbf{w},\ \mathbf{v}_1,\ \mathbf{v}_3\}$ or $\{\mathbf{v},\ \mathbf{w},\ \mathbf{v}_0,\ \mathbf{v}_2\}$ (or both) is a set of non-coplanar points, i.e., $T_0 \neq 0$ or $T_1 \neq 0$ (or both). Without loss of generality, we assume $T_0 \neq 0$.

Therefore, (45) and (54) (i.e., (30)) have the following matrix form.

$$B_{\mathbf{v},e}\mathbf{x}'_{\mathbf{v},e} + B'_{\mathbf{v},e}\mathbf{x}''_{\mathbf{v},e} + F_{\mathbf{v},e}\mathbf{x}_{\mathbf{v}} + G_{\mathbf{v},e}\mathbf{x}_E = 0, \tag{55}$$

where

$$\mathbf{x}'_{\mathbf{v},e} = (c^0_{4,2,1,1}, c^1_{4,2,1,1}, c^2_{4,2,1,1}, c^0_{4,3,1,0})',$$

$$\mathbf{x}''_{\mathbf{v},e} = (c^3_{4,2,1,1}, c^1_{4,3,1,0}, c^0_{4,2,2,0}, c^1_{4,2,2,0}, c^2_{4,2,2,0}, c^3_{4,2,2,0})',$$

$$B_{\mathbf{v},e} = \begin{bmatrix} \frac{1}{V_0} & \frac{1}{V_1} & 0 & \times \\ 0 & \frac{1}{V_1} & \frac{1}{V_2} & \times \\ 0 & 0 & \frac{1}{V_2} & \times \\ 0 & 0 & 0 & -\frac{2U_3T_0}{V_0V_2V_3} \end{bmatrix},$$

and again, $B'_{\mathbf{v},e}, F_{\mathbf{v},e}, G_{\mathbf{v},e}$ are the coefficient matrices and $\times$'s are some numbers. We still have $\mathbf{x}_{\mathbf{v},e} = \mathbf{x}'_{\mathbf{v},e} \bigcup \mathbf{x}''_{\mathbf{v},e}$.

It is obvious that

$$\text{rank}(B_{\mathbf{v},e}) = 4 = \text{degree}(e). \tag{56}$$

Let $A_{\mathbf{v},e} = \{c^3_{4,2,1,1},\ c^1_{4,3,1,0}\} \bigcup M_{\mathbf{v},e}$. We set all the elements in $A_{\mathbf{v},e}$ are zeros and set $c^0_{5,3,0,0} = c^i_{5,2,1,0} = 0,\ 0 \leq i \leq m-1$, at the same time. Then, according to Remark 4 and (54), all the Bézier coefficients of $C_{\mathbf{v},e}$ are zeros. Therefore, we have

Lemma 10. *A nonsingular inner edge $e \in E_{\mathbf{v}}$ with $m = 4$ is confinable.*

Case 4. $e = [\mathbf{v}, \mathbf{w}]$ is a singular edge. In this case, we can prove $B \equiv 0$, i.e., (54) is an identity. First of all,

$$T_i = 0, \qquad i = 0, 1, 2, 3, \tag{57}$$

since $\{\mathbf{v},\ \mathbf{w},\ \mathbf{v}_{i-1},\ \mathbf{v}_{i+1}\}$ are coplanar. According to (51), we have

$$a_1 = b_0 = 0. \tag{58}$$

Using (51) and (58), it yields

$$U_1U_2V_3 - U_0U_3V_1 = (a_{\mathbf{w}}(a_{\mathbf{w}}b_0 - a_0b_{\mathbf{w}})(-b_1) - b_{\mathbf{w}}(a_1b_{\mathbf{w}} - a_{\mathbf{w}}b_1)(-a_0))V_0^3 = 0. \tag{59}$$

Therefore, (54) is simplified as

$$-\sum_{i=0}^{m-1}(-1)^i\frac{W_i}{V_{i-1}V_i}c^i_{5,2,1,0} + \left(\frac{U_3W_0}{V_0V_2V_3} - \frac{U_2W_1}{V_0V_1V_2}\right)c^0_{5,3,0,0} = 0. \tag{60}$$

Similar to (47), since $T_i = 0$, for $c^i_{5,2,1,0}$, we have the following equations.

$$\frac{1}{V_i}c^{i+1}_{5,2,1,0} = -\frac{W_i}{V_{i-1}V_i}c^0_{6,2,0,0} + \frac{U_i}{V_{i-1}V_i}c^0_{5,3,0,0} - \frac{1}{V_{i-1}}c^{i-1}_{5,2,1,0},\ 1 \le i \le 4 \tag{61}$$

Setting $i = 1$ and $i = 4$ in (61), respectively, we obtain

$$\begin{cases} \frac{1}{V_1}c^2_{5,2,1,0} = -\frac{W_1}{V_0V_1}c^0_{6,2,0,0} + \frac{U_1}{V_0V_1}c^0_{5,3,0,0} - \frac{1}{V_0}c^0_{5,2,1,0}, \\ \frac{1}{V_3}c^3_{5,2,1,0} = -\frac{W_0}{V_3V_0}c^0_{6,2,0,0} + \frac{U_0}{V_3V_0}c^0_{5,3,0,0} - \frac{1}{V_0}c^1_{5,2,1,0}. \end{cases} \tag{62}$$

Substituting (62) into (60), it yields

$$\begin{aligned} \left(-\frac{W_0}{V_3V_0} + \frac{W_2}{V_0V_2}\right)c^0_{5,2,1,0} + \left(\frac{W_1}{V_0V_1} - \frac{W_3}{V_0V_2}\right)c^1_{5,2,1,0} + \left(\frac{W_1W_2}{V_0V_1V_2} - \frac{W_0W_3}{V_0V_2V_3}\right)c^0_{6,2,0,0} \\ + \left(\frac{U_3W_0}{V_0V_2V_3} - \frac{U_2W_1}{V_0V_1V_2} + \frac{W_3U_0}{V_0V_2V_3} - \frac{W_2U_1}{V_0V_1V_2}\right)c^0_{5,3,0,0} = 0. \end{aligned} \tag{63}$$

According to (52) and (53), since $T_i = 0$, it holds

$$V_2U_0 = V_3U_2, \qquad V_1U_3 = V_2U_1. \tag{64}$$

Symmetrically,

$$V_2W_0 = V_3W_2, \qquad V_1W_3 = V_2W_1. \tag{65}$$

(65) shows that the coefficients of $c^0_{5,2,1,0}$ and $c^1_{5,2,1,0}$ in (63) are zeros. According to (65),

$$W_1W_2V_3 - W_0W_3V_1 = W_1W_0V_2 - W_0W_1V_2 = 0.$$

Thus, the coefficient of $c^0_{6,2,0,0}$ in (63) is zero. Similarly, according to (64) and (65)

$$U_3W_0V_1 - U_2W_1V_3 + U_0W_3V_1 - U_1W_2V_3 = W_0V_2U_1 - W_1U_0V_2 + U_0W_1V_2 - U_1W_0V_2 = 0.$$

Thus, the coefficient of $c^0_{5,3,0,0}$ in (63) is also zero. Therefore, (63) is an identity if e is a singular edge.

Therefore, (45) (i.e., (30)), has the following matrix form.

$$B_{\mathbf{v},e}\mathbf{x}'_{\mathbf{v},e} + B'_{\mathbf{v},e}\mathbf{x}''_{\mathbf{v},e} + F_{\mathbf{v},e}\mathbf{x}_{\mathbf{v}} + G_{\mathbf{v},e}\mathbf{x}_E = 0, \tag{66}$$

where

$$\mathbf{x}'_{\mathbf{v},e} = (c^0_{4,2,1,1}, c^1_{4,2,1,1}, c^2_{4,2,1,1})',$$

$$\mathbf{x}''_{\mathbf{v},e} = (c^3_{4,2,1,1}, c^0_{4,3,1,0}, c^1_{4,3,1,0}, c^0_{4,2,2,0}, c^1_{4,2,2,0}, c^2_{4,2,2,0}, c^3_{4,2,2,0})',$$

$$B_{\mathbf{v},e} = \begin{bmatrix} \frac{1}{V_0} & \frac{1}{V_1} & 0 \\ 0 & \frac{1}{V_1} & \frac{1}{V_2} \\ 0 & 0 & \frac{1}{V_2} \end{bmatrix},$$

and again, $B'_{\mathbf{v},e}, F_{\mathbf{v},e}, G_{\mathbf{v},e}$ are the coefficient matrices. We still have $\mathbf{x}_{\mathbf{v},e} = \mathbf{x}'_{\mathbf{v},e} \bigcup \mathbf{x}''_{\mathbf{v},e}$. It is obvious that

$$\operatorname{rank}(B_{\mathbf{v},e}) = 3 = \operatorname{degree}(e) - 1. \tag{67}$$

Let $A_{\mathbf{v},e} = \{c^3_{4,2,1,1},\ c^0_{4,3,1,0},\ c^1_{4,3,1,0}\} \bigcup M_{\mathbf{v},e}$. We set all the elements in $A_{\mathbf{v},e}$ are zeros and set $c^0_{5,3,0,0} = c^i_{5,2,1,0} = 0,\ 0 \le i \le m-1$, at the same time. Then, according to Remark 4 and (45), all the Bézier coefficients of $C_{\mathbf{v},e}$ are zeros. Therefore, we have

Lemma 11. *A singular edge $e \in E_{\mathbf{v}}$ is confinable.*

Definition 4. We denote $\Gamma_{\mathbf{v}}$ the collection of all triangles, edges, and vertices in Star($\mathbf{v}$) which do not contain the vertex $\mathbf{v}$. $\Gamma_{\mathbf{v}}$ is called the surface triangulation of Star($\mathbf{v}$) or the link of vertex $\mathbf{v}$.

We have the following lemma

Lemma 12. *In* Star($\mathbf{v}$), *there are at least four confinable edges with degrees smaller than six if* $\mathbf{v}$ *is inner.*

Proof. According to Euler formula, it holds

$$|\Gamma_{\mathbf{v},0}| - |\Gamma_{\mathbf{v},1}| + |\Gamma_{\mathbf{v},2}| = 2, \tag{68}$$

where $\Gamma_{\mathbf{v},0}, \Gamma_{\mathbf{v},1}, \Gamma_{\mathbf{v},2}$ are the sets of vertices, edges, triangles of $\Gamma_{\mathbf{v}}$, respectively. It is clear that

$$2|\Gamma_{\mathbf{v},1}| = 3|\Gamma_{\mathbf{v},2}| \tag{69}$$

According to (68) and (69), it holds

$$|\Gamma_{\mathbf{v},1}| = 3(|\Gamma_{\mathbf{v},0}| - 2), \quad |\Gamma_{\mathbf{v},2})| = 2(|\Gamma_{\mathbf{v},0}| - 2). \tag{70}$$

Thus, we have

$$\sum_{\mathbf{w} \in \Gamma_{\mathbf{v},0}} \operatorname{degree}(\mathbf{w}) = 2|\Gamma_{\mathbf{v},1}| = 6(|\Gamma_{\mathbf{v},0}| - 2),$$

where degree($\mathbf{w}$) is the number of edges of $\Gamma_{\mathbf{v},1}$ sharing $\mathbf{w}$ as a common vertex. If there are at most three vertices with degrees smaller than six, then

$$\sum_{\mathbf{w}\in\Gamma_{\mathbf{v},0}} \text{degree}(\mathbf{w}) \geq 6(|\Gamma_{\mathbf{v},0}|-3)+3\times 3 > 6(|\Gamma_{\mathbf{v},0}|-2) = \sum_{\mathbf{w}\in\Gamma_{\mathbf{v},0}} \text{degree}(\mathbf{w}).$$

This is a contradiction. This shows that there exists at least four vertices $\mathbf{w} \in \Gamma_{\mathbf{v},0}$ such that their degrees degree($\mathbf{w}$) satisfy $3 \leq \text{degree}(\mathbf{w}) \leq 5$. Noting that $\text{degree}(\mathbf{w}) = \text{degree}(e)(e = [\mathbf{v},\mathbf{w}])$, Lemma 12 is proved.

Similar to [1], we introduce the following definition.

Definition 5. For an edge $e = [\mathbf{v},\mathbf{w}] \in E$, we assume that $\delta = [\mathbf{v},\mathbf{w},\mathbf{u}] \in F$ is an inner triangle, and that $\sigma' = [\mathbf{v},\mathbf{w},\mathbf{u},\mathbf{u}']$, $\sigma'' = [\mathbf{v},\mathbf{w},\mathbf{u},\mathbf{u}''] \in T$ are the two consecutive tetrahedra sharing δ as a common 2-face. Then δ is called degenerate (at edge e) whenever $\{\mathbf{v},\mathbf{w},\mathbf{u}',\mathbf{u}''\}$ are coplanar. Otherwise, δ is called nondegenerate (at edge e).

From Remark 4, (45) and (46), we have the following lemma.

Lemma 13. *If $e = [\mathbf{v},\mathbf{w}]$ is an inner edge and $[\mathbf{v},\mathbf{w},\mathbf{v}_j]$ is nondegenerate at $[\mathbf{v},\mathbf{w}]$, then*

$$A_{\mathbf{v},e} = \{c^0_{4,4,0,0},\, c^0_{4,3,1,0},\, c^1_{4,3,1,0}, c^{m-1}_{4,2,1,1}\} \bigcup \{c^i_{4,2,2,0};\ 0 \leq i \leq m-1,\ i \neq j\}$$

determines $C_{\mathbf{v},e}$, since in (46) $T_j \neq 0$.

In fact, in this case, (45) and (46) (i.e., (30)), has the following matrix form

$$B_{\mathbf{v},e}\mathbf{x}'_{\mathbf{v},e} + B'_{\mathbf{v},e}\mathbf{x}''_{\mathbf{v},e} + F_{\mathbf{v},e}\mathbf{x}_{\mathbf{v}} + G_{\mathbf{v},e}\mathbf{x}_E = 0, \tag{71}$$

where

$$\mathbf{x}'_{\mathbf{v},e} = (c^0_{4,2,1,1},\ldots,c^{m-2}_{4,2,1,1},c^j_{4,2,2,0})'$$

and

$$\mathbf{x}''_{\mathbf{v},e} = (c^{m-1}_{4,2,1,1},c^0_{4,3,1,0},c^1_{4,3,1,0},c^0_{4,2,2,0},\ldots,\widehat{c^j_{4,2,2,0}},\ldots,c^{m-1}_{4,2,2,0})',$$

where $\widehat{c^j_{4,2,2,0}}$ means that the term $c^j_{4,2,2,0}$ in $\mathbf{x}''_{\mathbf{v},e}$ is dropped,

$$B_{\mathbf{v},e} = \begin{bmatrix} \frac{1}{V_0} & \frac{1}{V_1} & 0 & \cdots & 0 & \times \\ 0 & \frac{1}{V_1} & \frac{1}{V_2} & \cdots & 0 & \times \\ 0 & 0 & \frac{1}{V_2} & \cdots & 0 & \times \\ \vdots & \vdots & \vdots & \ddots & \vdots & \vdots \\ 0 & 0 & 0 & \cdots & \frac{1}{V_{m-2}} & \times \\ 0 & 0 & 0 & \cdots & 0 & \frac{T_j}{V_{j-1}V_j} \end{bmatrix},$$

and again, $B'_{\mathbf{v},e}, F_{\mathbf{v},e}, G_{\mathbf{v},e}$ are the coefficient matrices and the $\times$s are some numbers. We still have $\mathbf{x}_{\mathbf{v},e} = \mathbf{x}'_{\mathbf{v},e} \bigcup \mathbf{x}''_{\mathbf{v},e}$.

It is obvious that

$$\mathrm{rank}(B_{\mathbf{v},e}) = m = \mathrm{degree}(e), \tag{72}$$

since $T_j \neq 0$.

According to (42), (44), (56), (67) and (72), for each $e \in E_{\mathbf{v}}$, there exist degree$(e) - \delta_e - 2\partial_e$ linear independent equations, where $\delta_e = 1$ if e is a singular edge and $\delta_e = 0$ otherwise, and $\partial_e = 1$ if e is a boundary edge and $\partial_e = 0$ otherwise. The total number of those equations is

$$\sum_{e \in E_{\mathbf{v}}} \mathrm{degree}(e) - |E_{\mathbf{v}}^{\delta}| - 2|E_{\mathbf{v}}^{\partial}|.$$

So it concludes that

$$\mathrm{rank}(W_{\mathbf{v},\mathrm{Star}}) \leq \sum_{e \in E_{\mathbf{v}}} \mathrm{degree}(e) - |E_{\mathbf{v}}^{\delta}| - 2|E_{\mathbf{v}}^{\partial}|. \tag{73}$$

To prove

$$\mathrm{rank}(W_{\mathbf{v},\mathrm{Star}}) = \sum_{e \in E_{\mathbf{v}}} \mathrm{degree}(e) - |E_{\mathbf{v}}^{\delta}| - 2|E_{\mathbf{v}}^{\partial}|, \tag{74}$$

we need some definitions and results in [1]. If (74) is right, then rank$(W_{\mathbf{v},\mathrm{Star}})$ equals its row numbers. This means that Lemma 4 holds. Thus, in the rest, we need only to prove (74).

Definition 6. A tree τ is a connected set of edges in $\Gamma_{\mathbf{v}}$ containing no loops. Precisely one vertex in τ is identified as the root of τ. The vertex of an edge e in τ that is closer to the root of τ is called the parent of the other vertex which is called its child. A chain is a tree such that no vertex has more than one child. Those vertices in a tree τ that have no children are called the leaves of τ. Two trees are called disjoint if their respective sets of vertices are disjoint. A forest is a set of disjoint trees. A descendant of a vertex $\mathbf{w} \in \Gamma_{\mathbf{v}}$ is any vertex $\mathbf{u} \in \Gamma_{\mathbf{v}}$ whose path from the root contains $\mathbf{w}$, and $\mathbf{w}$ is called an ancestor of $\mathbf{u}$

We divide the vertices $\mathbf{w} \in \Gamma_{\mathbf{v}}$ into two classes

Definition 7. $\mathbf{w}$ is called a terminating vertex if the edge $[\mathbf{v}, \mathbf{w}]$ is confinable. Any other vertex $\mathbf{w} \in \Gamma_{\mathbf{v}}$ is called propagating.

Definition 8. A proper tree τ is a tree such that

1. The root of τ is a terminating vertex.
2. Every vertex of τ besides its root is a propagating vertex.
3. Every edge $[\mathbf{v}_{\mathrm{parent}}, \mathbf{v}_{\mathrm{child}}]$ in τ is nondegenerate at the child vertex $\mathbf{v}_{\mathrm{child}}$, i.e., $[\mathbf{v}, \mathbf{v}_{\mathrm{parent}}, \mathbf{v}_{\mathrm{child}}]$ is nondegenerate at $[\mathbf{v}, \mathbf{v}_{\mathrm{child}}]$.

It holds the following lemma.

Lemma 14. *For $\Gamma_{\mathbf{v}}$, there exists a forest of disjoint proper trees that contains all propagating vertices.*

Proof. The proof of Lemma 14 is very similar to the proof of Theorem 3 of [1]. We omit the proof here.

We now consider the dimension of the null space of

$$W_{\mathbf{v},\text{Star}}\mathbf{x}_{\mathbf{v},\text{Star}} = 0. \tag{75}$$

Let Υ be a forest of disjoint proper trees that contains all vertices of $\Gamma_{\mathbf{v}}$. We arrange the vertices in $\Gamma_{\mathbf{v}}$ in any order

$$\mathbf{w}_1, \mathbf{w}_2, \ldots, \mathbf{w}_{\bar{n}}, \quad \bar{n} = |\Gamma_{\mathbf{v},0}|,$$

of satisfying that $\mathbf{w}_j$ is behind $\mathbf{w}_i$ if $\mathbf{w}_j$ is, with respect to Υ, a descendant of $\mathbf{w}_i$.

For any propagating vertex $\mathbf{w}_{\text{child}}$ of a proper tree described in Lemma 14, we assume its parent is $\mathbf{w}_{\text{parent}}$. Then, we choose

$$c^j_{4,2,2,0} \in \mathbf{x}'_{\mathbf{v},e} = (c^0_{4,2,1,1}, \ldots, c^{m-2}_{4,2,1,1}, c^j_{4,2,2,0})'$$

in (71) as $c^e_{4,2,2,0}$, where $e = [\mathbf{w}_{\text{parent}}, \mathbf{w}_{\text{child}}]$.

According to Lemma 14, it holds

$$\mathbf{x}'_{\mathbf{v},e_i} \bigcap \mathbf{x}_{\mathbf{v},e_j} = \emptyset, \qquad \text{if } i < j, \tag{76}$$

where $e_k = [\mathbf{v}, \mathbf{v}_k]$, $1 \le k \le \bar{n}$. If we denote $\mathbf{x}'_{\mathbf{v},\text{Star}}$ the vector formed by the unknowns in $\bigcup_{k=1}^{\bar{n}} \mathbf{x}''_{\mathbf{v},e_k} \setminus \bigcup_{k=1}^{\bar{n}} \mathbf{x}'_{\mathbf{v},e_k}$, then (75) is reformed as follows.

$$\tilde{W}_{\mathbf{v},\text{Star}}\tilde{\mathbf{x}}_{\mathbf{v},\text{Star}} = 0, \tag{77}$$

where

$$\tilde{\mathbf{x}}_{\mathbf{v},\text{Star}} = (\mathbf{x}'^T_{\mathbf{v},e_1}, \mathbf{x}'^T_{\mathbf{v},e_2}, \ldots, \mathbf{x}'^T_{\mathbf{v},e_{\bar{n}}}, \mathbf{x}'^T_{\mathbf{v},\text{Star}})',$$

$$\tilde{W}_{\mathbf{v},\text{Star}} = \begin{bmatrix} B_{\mathbf{v},e_1} & * & * & \cdots & * & * \\ 0 & B_{\mathbf{v},e_2} & * & \cdots & * & * \\ 0 & 0 & B_{\mathbf{v},e_3} & \cdots & * & * \\ \vdots & \vdots & \vdots & \ddots & \vdots & \vdots \\ 0 & 0 & 0 & \cdots & B_{\mathbf{v},e_{\bar{n}}} & * \end{bmatrix},$$

and $*$'s are the corresponding matrices. That is, $\tilde{W}_{\mathbf{v},\text{Star}}$ is a blocked diagonal matrix with every diagonal matrix being square and nonsingular. This means that (74) holds.

Noting that if $f = [\mathbf{v}, \mathbf{w}_1, \mathbf{w}_2] \in F_{\mathbf{v}}$, then $\mathbf{x}_{\mathbf{v},e_1} \bigcap \mathbf{x}_{\mathbf{v},e_2}$ contains only one element, where $e_1 = [\mathbf{v}, \mathbf{w}_1]$ and $e_2 = [\mathbf{v}, \mathbf{w}_2]$. We denote this element by $c^f_{4,2,2,0}$. Therefore,

$$\bigcup_{e \in E_{\mathbf{v}}} \mathbf{x}_{\mathbf{v},e} = \bigcup_{e \in E_{\mathbf{v}}} \{c^0_{e,4,2,1,1}, \ldots, c^{m-1}_{e,4,2,1,1}, c^0_{\mathbf{v},e}, c^1_{\mathbf{v},e}\} \bigcup_{f \in F_{\mathbf{v}}} \{c^f_{4,2,2,0}\}$$

and

$$|\bigcup_{e\in E_{\mathbf{v}}} \mathbf{x}_{\mathbf{v},e}| = \sum_{e\in E_{\mathbf{v}}} (degree(e) - b_e + 2) + |F_{\mathbf{v}}| = \sum_{e\in E_{\mathbf{v}}} degree(e) - |E_{\mathbf{v}}^{\partial}| + 2|E_{\mathbf{v}}| + |F_{\mathbf{v}}|.$$

Thus, together with (74), it holds that

$$\begin{aligned} \dim(\text{null}(W_{\mathbf{v},\text{Star}}) &= \dim(\text{null}(\tilde{W}_{\mathbf{v},\text{Star}}) = |\textstyle\bigcup_{e\in E_{\mathbf{v}}} \mathbf{x}_{\mathbf{v},e}| - \text{rank}(W_{\mathbf{v},\text{Star}}) \\ &= |F_{\mathbf{v}}| + 2|E_{\mathbf{v}}| + |E_{\mathbf{v}}^{\delta}| + |E_{\mathbf{v}}^{\partial}| = N_{\mathbf{v}}. \end{aligned} \tag{78}$$

According to Lemmas 4–7 and (37) and (78), it holds that

$$\begin{aligned} \dim S_8^1(\Delta) &= \dim(\text{null}(W_{V,E}))\text{+} \dim(\text{null}(W_V))\text{+} \dim(\text{null}(W_E)) \\ &\quad + \dim(\text{null}(W_\sigma)) + |E| + |T| \\ &= \textstyle\sum_{\mathbf{v}\in V} (\dim S_3^1(\text{Star}(\mathbf{v})) + N_{\mathbf{v}}) + \sum_{e\in E} (\dim S_2^1(\text{TStar}(e)) - 3) \\ &\quad + 3|F| + |E| + |T|. \end{aligned}$$

Theorem 1 is proved.

Acknowledgements Part of this author's work was finished when he visited Dalian University of Technology. This paper is partly funded by ARO (P-54412-CI-ISP) and DEPSCoR (W911NF-07-1-0422).

References

1. P. Alfeld, B. Piper and L.L. Schumaker, An explicit basis for C^1 quartic bivariate splines, SIAM J. Numer. Annl., 24(1985), 891-911.
2. P. Alfeld and L.L. Schumaker, The dimention of bivariate splines spaces of smoothness r and degree $d \geq 4r+1$, Constr. Approx., 39(1987), 189-197.
3. P. Alfeld, L. Schumaker and W. Whiteley, The generic dimension of the space of C^1 splines of degree ≥ 8 on tetrahedral decomposition, SIAM J. Numer. Anal. 3(1993), 889–920.
4. D. Hong, Spaces of bivariate spline functions over triangulation, Approximation Theory Appl., 7(1991), 56-75.
5. J. Morgan and R. Scott, A nodal basis for C^1-piecewise polynomials of degree $n \geq 5$, Math. Comp., 29(1975), 736-740.
6. Xiquan Shi, Ben Kamau, Fengshan Liu, and Baocai Yin, The Generic Dimension of the Space of Smooth Splines of Degree 7 on Tetrahedral Partitions, to appear in the Journal of Approximation Theory, Volume 157, Issue 2, Pages 89-192
7. Xiquan Shi, Dimensions of splines on 0-Star in $\mathbb{R}^3$. Approx. Th. and its Appl. 10(1994), 1-13.
8. Xiquan Shi, The dimensions of spline spaces and their singularity, J. Comp. Math., 10(1992), 224-230.
9. Xiquan Shi, Higher-Dimensional Spline Functions, Ph.D. Dissertation, Jilin University, 1988 (in Chinese).
10. R.H. Wang, X.Q. Shi, Z.X. Luo, Z.X. Su, Multivariate is Spline and its Applications, Kluwer Press, Dordrecht, 2002; Academic Press, Beijing, 1994 (in Chinese).
11. R.H. Wang and X.G. Lu, On dimension of bivariate spline spaces over triangulations, Scientia Sinica, A, No. 1(1988),585-594 (in Chinese).

On Simultaneous Approximation in Function Spaces

Eyad Abu-Sirhan

Abstract The problem of simultaneous approximation in function spaces has attracted many researchers recently. Major results on the space of vector-valued continuous functions started to appear early nineties. In 2002, results on simultaneous approximation in p-Bochner integrable function spaces were published. The objective of this paper is to give a characterization for some subspaces of Bochner integrable functions space to be simultaneously proximinal.

1 Introduction

Let X be a Banach space, G a closed subspace of X, and (Ω,Σ,μ) be a measure space.

Definition 1. Let (M,d) be a metric space. A Borel measurable function from Ω to M is called strongly measurable if it is the pointwise limit of a sequence of simple Borel measurable functions from Ω to M.

$L^1(\mu,X)$ denotes the Banach space consisting of (equivalent classes of) strongly measurable functions $f:\Omega\to X$ such that $\int_\Omega \|f(t)\|\,\mathrm{d}\mu$ is finite, with the usual norm

$$\|f\|_1=\int_\Omega \|f(s)\|\,\mathrm{d}\mu.$$

If X is the Banach space of real numbers, we simply write $L^1(\mu)$. For $A\in\Sigma$ and a strongly measurable function $f:\Omega\to X$, we write χ_A for the characteristic function of A and $\chi_A f$ denotes the function $\chi_A(s)f(s)$. In particular, for $x\in X$, $\chi_A x(s)=\chi_A(s)x$.

Eyad Abu-Sirhan
Tafila Technical University, Tafila, Jordan
e-mail: abu-sirhan@ttu.ed.jo

M. Neamtu and L. Schumaker (eds.), *Approximation Theory XIII: San Antonio 2010*, Springer Proceedings in Mathematics 13, DOI 10.1007/978-1-4614-0772-0_20,

For a finite number of elements $x_1, x_2, \ldots, x_m$ in X, we set

$$d(\{x_i : 1 \le i \le m\}, G) = \inf_{g \in G} \sum_{i=1}^{m} \|x_i - g\|.$$

G is said to be simultaneously proximinal if for any finite number of elements $x_1, x_2, \ldots, x_m$ in X there exists at least $y \in G$ such that

$$\sum_{i=1}^{m} \|x_i - y\| = d(\{x_i : 1 \le i \le m\}, G).$$

The element y is called a best simultaneous approximation of $x_1, x_2, \ldots, x_m$ in G. Of course, for $m = 1$ the preceding concepts are just best approximation and proximinality.

The theory of best simultaneous approximation has been investigated by many authors. Most of the work done has dealt with the space of continuous functions with values in a Bnach space e.g. [1,4,8]. Some recent results for best simultaneous approximation in $L^p(\mu, X)$, $1 \le p \le \infty$ have been obtained in [2]–[3], [5,7,9]. In [3], it is shown that if G is a reflexive subspace of a Banach space X, then $L^1(\mu, G)$ is simultaneously proximinal in $L^1(\mu, X)$. In [2], it is shown that if G is L^1-summand of a Banach space X, then $L^1(\mu, G)$ is simultaneously proximinal in $L^1(\mu, X)$. It is the aim of this paper to show that if G is a closed separable subspace, then $L^1(\mu, G)$ is simultaneously proximinal in $L^1(\mu, X)$ if and only if G is simultaneously proximinal in X.

2 Preliminary Results

Throughout this section X is a Banach space and G is a closed subspace of X. Let $f_1, f_2, \ldots, f_m$ be any finite number of elements in $L^1(\mu, X)$, and set

$$\phi(s) = d(\{f_i(s) : 1 \le i \le m\}, G).$$

Theorem 1. *Let (Ω, Σ, μ) be a measure space, $f_1, f_2, \ldots, f_m$ be any finite number of elements in $L^1(\mu, X)$, and $\phi(s)$ as defined above. Then, $\phi \in L^1(\mu)$ and*

$$d(\{f_i : 1 \le i \le m\},\, L^1(\mu, G)) = \int_{\Omega} |\phi(s)|\, d\mu.$$

Proof. Since $f_1, f_2, \ldots, f_m \in L^1(\mu, X)$, there exist sequences of simple functions

$$(f_{(i,n)})_{n=1}^{\infty}, i = 1, 2, \ldots, m,$$

such that

$$\lim \left\| f_{(i,n)}(s) - f_i(s) \right\| = 0,$$

for $i = 1,2,\ldots,m$, and for almost all s. We may write

$$f_{(i,n)} = \sum_{j=1}^{k(n)} \chi_{A(n,j)}(\cdot)x_{(i,n,j)}, \quad i = 1,2,\ldots,m,$$

$\sum_{j=1}^{k(n)} \chi_{A(n,j)}(\cdot) = 1$, and that $\mu(A(n,j)) > 0$. Then

$$d\left(\left\{f_{(i,n)}(s) : 1 \le i \le m\right\},\, G\right) = \sum_{j=1}^{k(n)} \chi_{A(n,j)} d\left(\left\{x_{(i,n,j)} : 1 \le i \le m\right\},\, G\right)$$

and by the continuity of d

$$\lim d\left(\left\{f_{(i,n)}(s) : 1 \le i \le m\right\},\, G\right) = d\left(\{f_i(s) : 1 \le i \le m\},\, G\right),$$

for almost all s. Thus ϕ is measurable and $\phi \in L^1(\mu)$.
Now, for any $h \in L^1(\mu, G)$,

$$\begin{aligned}\int_\Omega d\left(\{f_i(s) : 1 \le i \le m\},\, G\right) \mathrm{d}\mu &\le \int_\Omega \sum_{i=1}^{m} \|f_i(s) - h(s)\| \,\mathrm{d}\mu \\ &= \sum_{i=1}^{m} \|f_i - h\|_1\end{aligned}$$

Hence,

$$\int_\Omega d\left(\{f_i(s) : 1 \le i \le m\},\, G\right) \mathrm{d}\mu \le d\left(\{f_i : 1 \le i \le m\},\, L^1(\mu, G)\right).$$

To prove the reverse inequality, let $\varepsilon > 0$ be given and w_i, $i = 1,2,\ldots,m$, be simple functions in $L^1(\mu, X)$ such that

$$\|f_i - w_i\|_1 < \frac{\varepsilon}{3m}.$$

We may write $w_i = \sum_{k=1}^{\ell} \chi_{A_k}(\cdot)\, x_{(i,k)}$, $\sum_{k=1}^{\ell} \chi_{A_k}(\cdot) = 1$, and that $\mu(A_k) > 0$. Since $w_i \in L^1(\mu, X)$ for all i, we have $\left\|x_{(i,k)}\right\| \mu(A_k) < \infty$ for all k and i. If $\mu(A_k) < \infty$, select $h_k \in G$ so that

$$\sum_{i=1}^{m} \left\|x_{(i,k)} - h_k\right\| < d\left(\left\{x_{(i,k)} : 1 \le i \le m\right\},\, G\right) + \frac{\varepsilon}{3\mu(A_k)},$$

for all k. If $\mu(A_k) = \infty$, put $h_k = 0$.

Let $g = \sum_{k=1}^{\ell} \chi_{A_k}(\cdot)\, h_k$. It is clear that $g \in L^1(\mu, G)$. Then

$$
\begin{aligned}
d\left(\{f_i : 1 \le i \le m\},\, L^1(\mu,G)\right) &\le \sum_{i=1}^{m} \|f_i - g\|_1 \\
&= \sum_{i=1}^{m} \|f_i - w_i + w_i - g\|_1 \\
&\le \sum_{i=1}^{m} \|f_i - w_i\|_1 + \sum_{i=1}^{m} \|w_i - g\|_1 \\
&< \sum_{i=1}^{m} \left(\frac{\varepsilon}{3m}\right) + \sum_{i=1}^{m} \|w_i - g\|_1 \\
&= \frac{\varepsilon}{3} + \sum_{i=1}^{m} \sum_{k=1}^{\ell} \mu(A_k) \left\|x_{(i,k)} - h_k\right\| \\
&= \frac{\varepsilon}{3} + \sum_{k=1}^{\ell} \sum_{i=1}^{m} \mu(A_k) \left\|x_{(i,k)} - h_k\right\| \\
&= \frac{\varepsilon}{3} + \sum_{k=1}^{\ell} \mu(A_k) \sum_{i=1}^{m} \left\|x_{(i,k)} - h_k\right\| \\
&\le \frac{\varepsilon}{3} + \sum_{k=1}^{\ell} \mu(A_k)\, d\left(\left\{x_{(i,k)} : 1 \le i \le m\right\},\, G\right) + \frac{\varepsilon}{3} \\
&= \frac{2\varepsilon}{3} + \int_{\Omega} \sum_{k=1}^{\ell} \chi_{A_k}(s)\, d\left(\left\{x_{(i,k)} : 1 \le i \le m\right\},\, G\right) \mathrm{d}\mu \\
&= \frac{2\varepsilon}{3} + \int_{\Omega} d\left(\{w_i(s) : 1 \le i \le m\},\, G\right) \mathrm{d}\mu \\
&\le \frac{2\varepsilon}{3} + \int_{\Omega} \begin{bmatrix} d\left(\{f_i(s) : 1 \le i \le m\},\, G\right) \\ + \left(\sum_{i=1}^{m} \|f_i(s) - w_i(s)\|\right) \end{bmatrix} \mathrm{d}\mu \\
&\le \frac{2\varepsilon}{3} + \begin{bmatrix} \int_{\Omega} d\left(\{f_i(s) : 1 \le i \le m\},\, G\right) \mathrm{d}\mu \\ + \int_{\Omega} \sum_{i=1}^{m} \|f_i(s) - w_i(s)\| \,\mathrm{d}\mu \end{bmatrix}
\end{aligned}
$$

$$
\begin{aligned}
&\le \frac{2\varepsilon}{3} + \int_{\Omega} d\left(\{f_i(s) : 1 \le i \le m\},\, G\right) \mathrm{d}\mu + \sum_{i=1}^{m} \|f_i - w_i\|_1 \\
&\le \frac{2\varepsilon}{3} + \int_{\Omega} d\left(\{f_i(s) : 1 \le i \le m\},\, G\right) \mathrm{d}\mu + \frac{\varepsilon}{3} \\
&\le \varepsilon + \int_{\Omega} d\left(\{f_i(s) : 1 \le i \le m\},\, G\right) \mathrm{d}\mu
\end{aligned}
$$

This ends the proof.

Corollary 1. *Let* (Ω,Σ,μ) *be a measure space,* $f_1,f_2,\dots,f_m$ *be any finite number of elements in* $L^1(\mu,X)$. *Let* $g:\Omega\to G$ *be a measurable function such that* $g(s)$ *is a best simultaneous approximation of* $f_1(s),f_2(s),\dots,f_n(s)$ *for almost all* s. *Then* g *is a best simultaneous approximation of* $f_1,f_2,\dots,f_n$ *in* $L^1(\mu,G)$ *(and therefore* $g\in L^1(\mu,G)$*).*

Proof. Assume that $g(s)$ is a best simultaneous approximation of $f_1(s)$, $f_2(s)$,..., $f_m(s)$, for almost all s. Then

$$\sum_{i=1}^{m}\|f_i(s)-g(s)\|\le\sum_{i=1}^{m}\|f_i(s)-z\|,$$

for almost all s, and for all $z\in G$. Then, set $z=0$ and use triangle inequality,

$$\sum_{i=1}^{m}\|g(s)\|\le 2\sum_{i=1}^{m}\|f_i(s)\|$$

for almost all s, therefore $g\in L^1(\mu,G)$. By Theorem 2.1,

$$\begin{aligned}d\left(\{f_i:1\le i\le m\},\,L^1(\mu,G)\right)&=\int_{\Omega}d\left(\{f_i(s):1\le i\le m\},\,G\right)\mathrm{d}\mu\\&=\int_{\Omega}\sum_{i=1}^{m}\|f_i(s)-g(s)\|\,\mathrm{d}\mu\\&=\sum_{i=1}^{m}\|f_i-g\|_1.\end{aligned}$$

Therefore g is a best simultaneous approximation for $f_1,f_2,\dots,f_m$ in $L^1(\mu,G)$.

The condition in Corollary 2.1 is sufficient; $g(s)$ is a best simultaneous approximation of $f_1(s),f_2(s),\dots,f_m(s)$ for almost all s in G, implies g is a best simultaneous approximation of $f_1,f_2,\dots,f_m$ in $L^1(\mu,G)$. In fact we have the following theorem:

Theorem 2. *Let* (Ω,Σ,μ) *be a measure space. Then,* $L^1(\mu,G)$ *is simultaneously proximinal in* $L^1(\mu,X)$ *if and only if for any finite number of elements* $f_1,f_2,\dots,f_m$ *in* $L^1(\mu,X)$, *there exists* $g\in L^1(\mu,G)$ *such that* $g(s)$ *is a best simultaneous approximation of* $f_1(s),f_2(s),\dots,f_n(s)$ *for almost all* s.

Proof. Sufficiency of the condition is an immediate consequence of Corollary 2.1. We will show the necessity. Assume that $L^1(\mu,G)$ is simultaneously proximinal in $L^1(\mu,X)$ and let $f_1,f_2,\dots,f_m$ be any finite number of elements in $L^1(\mu,X)$. Then, there exists $g\in L^1(\mu,G)$ such that

$$\begin{aligned}\sum_{i=1}^{m}\|f_i-g\|_1&=d\left(\{f_i:1\le i\le m\},\,L^1(\mu,G)\right)\\&=\int_{\Omega}d\left(\{f_i(s):1\le i\le m\},\,G\right)\mathrm{d}\mu,\end{aligned}$$

hence

$$\int_{\Omega}\left(\sum_{i=1}^{m}\|f_i(s)-g(s)\|-d\left(\{f_i(s):1\le i\le m\},\,G\right)\right)\mathrm{d}\mu=0.$$

Thus

$$\sum_{i=1}^{m}\|f_i(s)-g(s)\|=d\left(\{f_i(s):1\le i\le m\},\,G\right),$$

for almost all s.

3 Main Result

Let (Ω,Σ,μ) be a measure space and X be a Banach space. We say that $f:\Omega\to X$ is measurable in the classical sense if $f^{-1}(O)$ is measurable for every open set $O\subset X$.

The following lemmas will be used to prove our main result.

Lemma 1. *([10]) Let (Ω,Σ,μ) be a complete measure space and X be a Banach space. If f is a strongly measurable function from Ω to X, then f is measurable in the classical sense.*

Lemma 2. *([10]) Let (Ω,Σ,μ) be a complete measure space and X be a Banach space. If $f:\Omega\to X$ is measurable in the classical sense and has essentially separable range, then f is strongly measurable.*

Let Φ be a set-valued mapping, taking each point of a measurable space Ω into a subset of a metric space X. We say that Φ is weakly measurable if $\Phi^{-1}(O)$ is measurable in Ω whenever O is open in X. Hence we have put, for any $A\subset X$,

$$\Phi^{-1}(A)=\{s\in\Omega:\Phi(s)\cap A\neq\phi\}.$$

The following theorem is due to Kuratowski [6], it is known as Measurable Selection Theorem.

Theorem 3. *([6]) Let Φ be a weakly measurable set-valued map which carries each point of a measurable space Ω to a closed nonvoid subset of a complete separable metric space X. Then Φ has a measurable selection; i.e., there exists a function $f:\Omega\to X$ such that $f(s)\in\Phi(s)$ for each $s\varepsilon\Omega$ and $f^{-1}(O)$ is measurable in Ω whenever O is open in X.*

The following theorem is the main result of the paper.

Theorem 4. *Let X be a Banach space and G be a closed separable subspace of X, and (Ω,Σ,μ) is σ-finite complete measure space. Then the following are equivalent:*

(i) G is simultaneously proximinal in X.
(ii) $L^1(\mu,G)$ is simultaneously proximinal in $L^1(\mu,X)$.

Proof. $(2) \Rightarrow (1)$: Let $x_1, x_2, \ldots, x_m$ be any finite number of elements in X. Since (Ω, Σ, μ) is σ-finite, we can assume that $\Omega = \bigcup_{n \in N} A_n$ such that $\mu(A_n) < \infty$ for all $n \in N$. Then there must be $k_0 \in N$ such that $0 < \mu\left(A_{k_0}\right) < \infty$. Define $f_{x_i} : \Omega \to X$, $i = 1, 2, \ldots, m$, by

$$f_{x_i}(s) = \chi_{A_{k_0}}(s) x_i,$$

for all $s \in \Omega$. Then $f_{x_i} \in L^1(\mu, X)$ for all i. By the assumption, there exists $f_0 \in L^1(\mu, G)$ such that

$$\sum_{i=1}^{m} \|f_{x_i} - f_0\|_1 = d\left(\{f_{x_i} : 1 \le i \le m\}, L^1(\mu, G)\right).$$

Then,

$$\begin{aligned}
\sum_{i=1}^{m} \|f_{x_i} - f_0\|_1 &\le \sum_{i=1}^{m} \left\|f_{x_i} - \chi_{A_{k_0}} g\right\|_1 \\
&= \sum_{i=1}^{m} \left\|\chi_{A_{k_0}} x_i - \chi_{A_{k_0}} g\right\|_1 \\
&= \int\limits_{A_{k_0}} \left(\sum_{i=1}^{m} \|x_i - g\|\right) \mathrm{d}\mu \\
&= \mu\left(A_{k_0}\right) \left(\sum_{i=1}^{m} \|x_i - g\|\right),
\end{aligned}$$

for all $g \in G$. By Theorem 2.2, $f_0(s)$ is a best simultaneous approximation of $f_{x_1}(s), f_{x_2}(s), \ldots, f_{x_m}(s)$ for almost all s. Then

$$\sum_{i=1}^{m} \|f_{x_i}(s) - f_0(s)\| \le \sum_{i=1}^{m} \|f_{x_i}(s) - h(s)\|,$$

for almost all s and for any strongly measurable function $h : \Omega \to G$, hence $f_0 = \chi_{A_{k_0}} f_0$. Put $x_0 = \int\limits_{A_{k_0}} f_0(s) \, \mathrm{d}\mu$. Then,

$$\begin{aligned}
\sum_{i=1}^{m} \left\|x_i - \frac{x_0}{\mu\left(A_{k_0}\right)}\right\| &= \mu\left(A_{k_0}\right)^{-1} \sum_{i=1}^{m} \left\| \int\limits_{A_{k_0}} f_{x_i}(s) \, \mathrm{d}\mu - \int\limits_{A_{k_0}} f_0(s) \, \mathrm{d}\mu \right\| \\
&\le \mu\left(A_{k_0}\right)^{-1} \sum_{i=1}^{m} \int\limits_{A_{k_0}} \|f_{x_i}(s) - f_0(s)\| \, \mathrm{d}\mu \\
&= \mu\left(A_{k_0}\right)^{-1} \sum_{i=1}^{m} \|f_{x_i} - f_0\|_1 \, \mathrm{d}\mu \\
&\le \sum_{i=1}^{m} \|x_i - g\|,
\end{aligned}$$

for all $g \in G$. Hence $\frac{1}{\mu\left(A_{k_0}\right)}x_0$ is a best simultaneous approximation of $x_1, x_2, \ldots, x_m$ in G.

$(1) \Rightarrow (2)$: Let $f_1, f_2, \ldots, f_m$ be any finite number of elements in $L^1(\mu, X)$. For each $s \in \Omega$ define

$$\Phi(s) = \left\{ g \in G : \sum_{i=1}^{m} \|f_i(s) - g\| = d\left(\{f_i(s) : 1 \leq i \leq m\},\ G\right) \right\}.$$

For each $s \in \Omega$, $\Phi(s)$ is closed, bounded, and nonvoid subset of G. We shall show that Φ is weakly measurable. Let O be an open set in X, the set

$$\Phi^{-1}(O) = \{s \in \Omega : \Phi(s) \cap O \neq \phi\}$$

can be also be described as

$$\Phi^{-1}(O) = \{s \in \Omega : \inf_{g \in G} \sum_{i=1}^{m} \|f_i(s) - g\| = \inf_{g \in O} \sum_{i=1}^{m} \|f_i(s) - g\|\}.$$

Since (Ω, Σ, μ) is complete, f_i is measurable in the classical since for $i = 1, 2, \ldots, m$ by Lemma 3.1. Since subtraction in G, sum, and the norm in X are continuous, then the map

$$s \to \inf_{g \in A} \sum_{i=1}^{m} \|f_i(s) - g\|$$

is measurable for any set A. It follows that $\Phi^{-1}(O)$ is measurable. By Theorem 3.3, Φ has a measurable selection; i.e., there exists a function $f : \Omega \to G$ such that $f(s) \in \Phi(s)$ for each $s \varepsilon \Omega$ and f is measurable in the classical sense. By Lemma 3.2, f is strongly measurable. Hence f is a best simultaneous approximation for $f_1, f_2, \ldots, f_m$ in $L^1(\mu, G)$ by Corollary 2.1.

References

1. Chong Li, On Best Simultaneous Approximation, J. Approx. Theory 91, (1998) 332-348.
2. E. Abu-Sirhan, Best simultaneous approximation in $L^p(I, X)$, Inter. J. Math. Analysis, vol. 3, no. 24, (2009) 1157-1168.
3. Fathi B. Saidi, Deep Hussein, and R. Khalil, Best Simultaneous Approximation in $L^p(I, E)$, J. Approx. Theory 116 (2002), 369-379.
4. G. A. Watson, A Charaterization of Best Simultaneous approximation, J. Approx. Theory 75, (1998) 175-182.
5. J. Mendoza and Tijani Pakhrou, Best Simultaneous Approximation in $L^1(\mu, X)$, J. Approx. Theory 145, (2007) 212-220.
6. K. Kuratowiski and C. Ryll-Nardzewski, A General Theorem on Selectors, Bull. Acad. Polonaise Sciences, Serie des Sciences Math. Astr. Phys. 13 (1965), 379-403 MR 32. #6421.
7. M. Khandaqji and Sh. Al-Sharif, Best simultaneous approximation in Orlicz Spaces, Inter. J. of Math. & Math. Scie. ID68017 (2007).
8. Shinji Tanimoto, On Simultaneous Approximation, Math. Japanica 48, No. 2 (2007) 275-279.

9. Tijani Pakhrou, Best Simultaneous Approximation in $L^{\infty}(\mu,X)$, Math. Nacher. 281, No. 3, (2008) 396-401.
10. W.A Light and E.W. Cheney, Approximation Theory in Tensor Product Spaces, Lecture Notes in Mathematics, 1169, Spinger-Verlag, Berlin Heidelberg New York Tokyo (1985).

Chalmers–Metcalf Operator and Uniqueness of Minimal Projections in ℓ_∞^n and ℓ_1^n Spaces

Lesław Skrzypek

Abstract We construct the Chalmers–Metcalf operator for minimal projections onto hyperplanes in ℓ_∞^n and ℓ_1^n and prove it is uniquely determined. We show how we can use Chalmers–Metcalf operator to obtain uniqueness of minimal projections. The main advantage of our approach is that it is purely algebraical and does not require consideration of the min–max problems.

1 Introduction

A projection $P : X \to V$ is taken to mean any bounded linear operator P that carries a Banach space X onto a linear subspace V in such a way that it acts as an identity on V. A projection $P : X \to V$ is called *minimal* if it has the smallest possible norm, that is if

$$\|P\| = \lambda(V,X) = \inf\{\|Q\| : Q : X \to V \text{ and } Q \text{ is a projection onto } V\}. \quad (1)$$

Observe that any projection with norm one is automatically minimal, though in general, a given subspace will not be the range of a projection of norm 1. In many cases, the existence of a minimal projection is known a priori (see [10]), which is the case when the subspace is finite-dimensional or finite-codimensional. Although typically the formula for minimal projection is difficult to find, the reader is referred to [2–4, 9–11] for more information on problems related to finding a minimal projection. The problem of the uniqueness of minimal projection is more difficult. Even in ℓ_∞^n and ℓ_1^n the situation is far from being understood. Theorem II.3.6 [11] (see also [1]) describes the minimal projections and characterizes the uniqueness of minimal

Lesław Skrzypek
Department of Mathematics and Statistics, University of South Florida, Tampa, FL 33620-5700, USA
e-mail: skrzypek@usf.edu

M. Neamtu and L. Schumaker (eds.), *Approximation Theory XIII: San Antonio 2010*, Springer Proceedings in Mathematics 13, DOI 10.1007/978-1-4614-0772-0_21,

projections of ℓ_∞^n onto hyperplanes. Theorem II.5.2 [11] (see also [1,5]) describes the minimal projections and characterizes the uniqueness of minimal projections of ℓ_1^n onto hyperplanes. The proof of the latter is lengthy and complicated. There are partial results on the description of minimal projections and its uniqueness of ℓ_∞^n onto subspaces of codimension 2 (see [6, 7]). General full characterization seems impossible.

In this paper, we explore algebraic properties of Chalmers–Metcalf operators. Without knowing the norm or actual minimal projection we are able to construct the Chalmers–Metcalf operator and use it to prove uniqueness of minimal projections. Previously, the invertibility of the Chalmers–Metcalf operator in smooth spaces has been linked to uniqueness of minimal projections [9] but ℓ_1^n and ℓ_∞^n are far from being smooth. It is interesting to see that this concept can also be used in such spaces. We will begin with introducing the notion of the Chalmers–Metcalf operator.

By $S(X)$ we denote the unit sphere of X, that is, $S(X) = \{x : ||x|| = 1\}$. Let $L: X \to Y$ be a linear operator. A functional $g \in S(Y^*)$ is called a *norming functional* for L if

$$||g \circ L|| = ||L||. \tag{2}$$

A point $x \in S(X)$ is called a *norming point* for L if

$$||L(x)|| = ||L||. \tag{3}$$

A pair $(g, f) \in S(Y^*) \times S(X)$ is called a *norming pair* for L if

$$g(Lf) = ||L||. \tag{4}$$

The set of all norming pairs for L is denoted by $\mathscr{E}(P)$. If P is a projection from X onto a finite-dimensional subspace Y, then (since P is a compact operator) it has a norming functional. If X is reflexive, then any functional attains its norm. Therefore, there is a norming pair for P. If X is not reflexive then, in general, it is not true. For example, the Fourier projection does not attain its norm in $C[0, 2\pi]$. But any functional attains its norm in X^{**}, hence we can always find a norming pair for P (extending P to P^{**}) in $S(X^*) \times S(X^{**})$.

To each extremal pair (g, f) in $S(X^*) \times S(X^{**})$ we associate the rank-one operator $g \otimes f$ from X to X^{**} given by $(g \otimes f)(z) = g(z) \cdot f$ for $z \in X$.

Theorem 1 *([2, 3]) A projection $P : X \to V$ has a minimal norm if and only if the closed convex hull of $\{g \otimes f\}_{(g,f)\in\mathscr{E}(P)}$ contains an operator E_P for which V is an invariant subspace.*

Operator E_P is called a Chalmers–Metcalf operator if it is given by the formula

$$E_P = \int_{\mathscr{E}(P)} g \otimes f \, \mathrm{d}\mu : X \to X^{**}, \tag{5}$$

where μ is a probabilistic Borel measure on $\mathscr{E}(P)$ and

$$E_P(V) \subset V \tag{6}$$

Under the assumption that X^* is separable, every operator in the closed convex hull of $\{g \otimes f\}_{(g,f)\in\mathscr{E}(P)}$ for which V is an invariant subspace is a Chalmers–Metcalf operator and vice versa (see [8]).

Here, we will list interesting properties of Chalmers–Metcalf operator (see [8] for details).

- Chalmers–Metcalf operator does not depend on a particular minimal projection, it only depends on location of V in X (that is if we construct a Chalmers–Metcalf operator for a particular minimal projection P, it will also be a Chalmers–Metcalf operator for any other minimal projection). As a result we will say Chalmers–Metcalf operator for the pair (X,V).
- The norming pairs that appears in the Chalmers–Metcalf operator are common norming pairs for any minimal projection.
- Chalmers–Metcalf operator does not have to be unique.
- The set of all Chalmers–Metcalf operators is a convex set.
- Invertibility of Chalmers–Metcalf operator (restricted to V) seems to play important role in the uniqueness of minimal projection (see [9]).

The Chalmers–Metcalf theorem has many applications especially in case of $X = L_1$, for example, it has been used to find the minimal projection onto polynomials of degree 2 (see [2]).

In this paper, we will assume that all considered spaces are real. If X is a Banach space, then denote by $Ext(X)$ the set of all extreme points of a unit ball in X. Let

$$e_i = (0,\ldots,0,1,0,\ldots,0), \tag{7}$$

where 1 is on i-th place (all other places have 0's). It is well known that

$$\mathrm{Ext}(\ell_1^n) = \{\pm e_i, i = 1,\ldots,n\}, \tag{8}$$

and

$$\mathrm{Ext}(\ell_\infty^n) = \{(\varepsilon_1,\ldots,\varepsilon_n),\ \text{where } \varepsilon_i = \pm 1\}. \tag{9}$$

Every projection $P : X \to \ker f$ has to be of the following form

$$P = Id - f \otimes z, \tag{10}$$

where $f \in S(X^*)$ and $z \in X$ such that $f(z) = 1$.

Since $X = \ell_\infty^n$ and $X = \ell_1^n$ are symmetric subspaces then there is an isometry $I : X \to X$ such that $I(\ker f) = I(\ker |f|)$. Therefore, without a loss of generality, we can assume that $f_i \geq 0$ for every $i = 1,\ldots,n$.

The following useful remark will allow us to assume that $z_i \geq 0$ for every $i = 1,\ldots,n$.

Remark 1 *Let $X = \ell_\infty^n$ or $X = \ell_1^n$. Assume that $1 \geq f_i \geq 0$ for every $i = 1,\ldots,n$ and $f \in S(X^*)$. Take a projection $P = Id - f \otimes z : X \to \ker f$ such that $||P|| > 1$. Define $x_i = 0$ if $z_i \leq 0$ and $x_i = \theta z_i$ if $z_i > 0$. Here $\theta = \frac{1}{\sum^+ f_i z_i}$, the summation symbol* *denoting the sum for $z_i > 0$. Then*

$$Q = Id - f \otimes x, \tag{11}$$

is a projection from X onto $\ker f$ *and* $||Q|| \leq ||P||$.

Proof. The proof of case $X = \ell_1^n$ can be found in Lemma 4 [1]. The case $X = \ell_\infty^n$ can be handled analogously. We are presenting it here for the sake of completeness. Using Lemma 2 [1] we have

$$||P|| = \max_{i=1,\dots,n} \{|1 - f_i z_i| + |z_i|(1 - f_i)\} \tag{12}$$

and

$$||Q|| = \max_{i=1,\dots,n} \{1 + x_i(1 - 2f_i)\}. \tag{13}$$

Denote $r_i = |1 - f_i z_i| + |z_i|(1 - f_i)$ and $s_i = 1 + x_i(1 - 2f_i)$. Since $||P|| > 1$ we have $1 - 2f_i > 0$ (see [1]). We distinguish three cases.

Case 1, $f_i z_i \leq 0$. Then $x_i = 0$ and $s_i = 1 < ||P||$.

Case 2, $0 < f_i z_i < 1$. Then $r_i = 1 + z_i(1 - 2f_i) \geq 1 + \theta z_i(1 - 2f_i) = s_i$. Therefore, $s_i \leq ||P||$.

Case 3, $f_i z_i \geq 1$. Observe that $f_i z_i \geq 1$ is equivalent to $z_i - 1 \geq 1 + z_i(1 - 2f_i)$. As a result $r_i = z_i - 1 \geq 1 + z_i(1 - 2f_i) \geq 1 + \theta z_i(1 - 2f_i) = s_i$. Therefore $s_i \leq ||P||$.

We proved that for any $i = 1, \dots, n$ $s_i \leq ||P||$. As a result $||Q|| \leq ||P||$.

2 Chalmers–Metcalf Operator for Hyperplanes in ℓ_∞^n

Minimal projections of norm 1 are a special subclass of all minimal projections (see [12]). In ℓ_∞^n we know that $\lambda(\ker f, \ell_\infty^n) = 1$ is equivalent to $||f||_\infty \geq ||f||_1/2$. (Theorem 1 in [1]). In this section, we will assume that $\lambda(\ker f, \ell_\infty^n) > 1$.

Remark 2 *(Remark 3.16 [8]) Take $X = \ell_\infty^n$ and let V be its subspace. Then any Chalmers–Metcalf operator for the pair (ℓ_∞^n, V) can be written as*

$$E_P = \sum_{i=1}^{n} \alpha_i e_i \otimes y_i, \tag{14}$$

for some $y_i \in S(\ell_\infty^n)$ and $\alpha_i \geq 0$ such that $\sum_{i=0}^{n} \alpha_i = 1$ and $e_i(Py_i) = ||P||$ for all i such that $\alpha_i > 0$ and all minimal projections $P : X \to V$.

We will prove that in most cases all α_i are strictly positive. And as a result we will see that they are uniquely determined.

Theorem 2 *Assume $f_i \neq 0$, for every $i = 1, \dots, n$, and consider $f = (f_1, \dots, f_n) \in S(\ell_1^n)$ such that $||f||_\infty < ||f||_1/2$. Then any Chalmers–Metcalf operator for the pair $(\ell_\infty^n, \ker f)$ has the form of (14), where $\alpha_i > 0$ for all $i = 1, \dots, n$.*

Proof. Let E_P be any Chalmers–Metcalf operator for the pair $(\ell_\infty^n, \ker f)$. By Remark 2

$$E_P = \sum_{i=1}^{n} \alpha_i e_i \otimes y_i, \tag{15}$$

for some $y_i \in S(\ell_\infty^n)$ and $\alpha_i \geq 0$ such that $\sum_{i=0}^n \alpha_i = 1$ and $e_i(Py_i) = ||P||$ for all i such that $\alpha_i > 0$ and all minimal projections $P : X \to V$.

We know that $E_P(\ker f) \subset \ker f$. That is for every $x \in \ker f$ we have $f(E_P(x)) = 0$. Observe that for every $x = (x_1, \dots, x_n)$

$$f(E_P(x)) = \sum_{i=1}^{n} \alpha_i x_i f(y_i) \tag{16}$$

Take $x = f_j e_i - f_i e_j$. Clearly, $x \in \ker f$ and

$$f(E_P(x)) = \alpha_i f_j f(y_i) - \alpha_j f_i f(y_j). \tag{17}$$

As a result for every $i, j = 1, \dots, n$

$$\alpha_i f_j f(y_i) = \alpha_j f_i f(y_j). \tag{18}$$

Set $\gamma_i = \alpha_i f(y_i)$. We will show that $\gamma_i \neq 0$, for all $i = 1, \dots, n$. Fix i, summing the above equations over $j = 1, \dots, n$ produces

$$\gamma_i (\sum_{k=1}^{n} f_k) = f_i (\sum_{k=1}^{n} \gamma_k) \tag{19}$$

Since $f_i \neq 0$ for all $i = 1, \dots, n$ then either $\sum_{k=1}^n \gamma_k \neq 0$ and as a result $\gamma_i \neq 0$, for all $i = 1, \dots, n$ or $\sum_{k=1}^n \gamma_k = 0$ and then $\gamma_i = 0$, for all $i = 1, \dots, n$. Therefore,

$$\alpha_i f(y_i) = 0, \quad i = 1, \dots, n. \tag{20}$$

Since $\sum_{i=0}^n \alpha_i = 1$, there is k such that $\alpha_k \neq 0$. By Remark 2

$$\lambda(\ker f, \ell_\infty^n) = e_k(Py_k) = e_k(y_k) \leq 1, \tag{21}$$

contrary to the assumption that $||f||_\infty < ||f||_1/2$ which guarantees $\lambda(\ker f, \ell_\infty^n) > 1$. The condition $\gamma_i \neq 0$, for all $i = 1, \dots, n$ immediately gives us $\alpha_i \neq 0$, for all $i = 1, \dots, n$.

Using the above theorem, we can give an easy proof of the full description of uniqueness of minimal projections (see Theorem II.3.6 [11] and also [1]). Our approach has an advantage in that it does not require a computation of the norm of minimal projection and considering min–max problems. It is completely algebraic.

Theorem 3 *Assume $f_i \neq 0$, for every $i = 1, \dots, n$, and consider $f = (f_1, \cdots, f_n) \in S(\ell_1^n)$ such that $||f||_\infty < 1/2$. Then there is only one minimal projection from ℓ_∞^1 onto $\ker f$.*

Proof. Take E_P, a Chalmers–Metcalf operator for the pair $(\ell_\infty^n, \ker f)$. By Theorem 2

$$E_P = \sum_{i=1}^{n} \alpha_i e_i \otimes y_i, \tag{22}$$

for some $y_i \in S(\ell_\infty^n)$ and $\alpha_i > 0$ such that $\sum_{i=0}^{n} \alpha_i = 1$ and $e_i(Py_i) = \lambda(\ker f, \ell_\infty^n)$ for all $i = 1, \ldots, n$ and all minimal projections $P : X \to V$.

Every projection onto $\ker f$ has to be of the form

$$Q = Id - f \otimes z, \tag{23}$$

where $f(z) = 1$. Assume Q is minimal. By (22) we have

$$e_i(Qy_i) = \lambda(\ker f, \ell_\infty^n). \tag{24}$$

That is,

$$e_i(y_i) - f(y_i)z_i = \lambda(\ker f, \ell_\infty^n). \tag{25}$$

Since $||Q|| > 1$, then $f(y_i) \neq 0$, for all $i = 1, \ldots, n$ (otherwise, if for some k, $f(y_k) = 0$ then $Q(y_k) = y_k$ and since y_k is a norming point $||Q|| = 1$). As a result

$$z_i = \frac{e_i(y_i) - \lambda(\ker f, \ell_\infty^n)}{f(y_i)}. \tag{26}$$

Therefore, z_i are uniquely determined thus Q is unique.

With a little more effort, using the above, we can find the Chalmers–Metcalf operator exactly.

Theorem 4 *Assume $f_i > 0$, for every $i = 1, \ldots, n$, and consider $f = (f_1, \ldots, f_n) \in S(\ell_1^n)$ such that $||f||_\infty < 1/2$. Then there is only one minimal projection of ℓ_∞^n onto* $\ker f$ *and the Chalmers-Metcalf operator for the pair $(\ell_\infty^n, \ker f)$ is uniquely defined by*

$$E_P = \sum_{i=1}^{n} \alpha_i e_i \otimes \eta_i, \tag{27}$$

where $\eta_i = (-1, \ldots, -1) + 2e_i$ and

$$\alpha_i = \frac{\beta_i}{\beta_1 + \cdots + \beta_n} \quad \text{where} \quad \beta_i = \frac{f_i}{1 - 2f_i}. \tag{28}$$

Moreover,

$$\lambda(\ker f, \ell_\infty^n) = 1 + \left(\sum_{i=1}^{n} \frac{f_i}{1 - 2f_i} \right)^{-1} \tag{29}$$

and projection $P = Id - f \otimes z$ is minimal for

$$z_i = \frac{\lambda(\ker f, \ell_\infty^n) - 1}{1 - 2f_i}. \tag{30}$$

Proof. Take E_P a Chalmers–Metcalf operator for the pair $(\ell_\infty^n, \ker f)$. By Theorem 2

$$E_P = \sum_{i=1}^{n} \alpha_i e_i \otimes y_i, \tag{31}$$

for some $y_i \in S(\ell_\infty^n)$ and $\alpha_i > 0$ such that $\sum_{i=0}^n \alpha_i = 1$ and $e_i(Py_i) = \lambda(\ker f, \ell_\infty^n)$ for all $i = 1,\dots,n$ and all minimal projections $P : X \to V$. Take a unique minimal projection $P = Id - f \otimes z$, where $f(z) = 1$. Using Remark 1 we may assume that $z_i \geq 0$ for every $i = 1,\dots,n$. Using $\lambda(\ker f, \ell_\infty^n) > 1$ and (26) we may further assume that $z_i > 0$ for every $i = 1,\dots,n$. Therefore

$$e_i(P(\varepsilon_1,\dots,\varepsilon_n)) = \varepsilon_i(1 - f_i z_i) - z_i \left(\sum_{j \neq i} \varepsilon_j f_j \right). \tag{32}$$

Since $\sum_i^n f_i z_i = 1$ we have $1 - f_i z_i \geq 0$, for all $i = 1,\dots,n$. In fact, $1 - f_i z_i > 0$ for all $i = 1,\dots,n$. Otherwise, $f_{i_0} z_{i_0} = 1$ and $z_i = 0$ for all $i \neq i_0$ but then for $i \neq i_0$ $e_i(P(\varepsilon_1,\dots,\varepsilon_n)) = \varepsilon_i$ and $||e_i \circ P|| = 1 < ||P||$. Therefore, for $i \neq i_0$, e_i is not a norming functional for P. That contradicts (31). As a result the norm of $e_i \circ P$ is attained only at the point $(\varepsilon_1,\dots,\varepsilon_n)$ where $\varepsilon_i = 1$ and $\varepsilon_j = -1, j \neq i$. That is, the norm is attained only at η_i. As a result

$$y_i = \eta_i \text{ for all } i = 1,\dots,n. \tag{33}$$

Since $\sum_i^n f_i = 1$ we observe that $f(\eta_i) = 2f_i - 1$. Using (26) and solving (19) gives (30) and (28). Using (30) and the fact that $\sum_{i=1}^n f_i z_i = 1$ gives (29).

3 Chalmers–Metcalf Operator for Hyperplanes in ℓ_n^1

In general, the ℓ_1^n case is more difficult. Although the full description is known, its proof is very long and complicated (Theorem II.5.2 [11], see also [1, 5]). Here we present the simple and purely algebraic proof of a special but relatively generic case.

Remark 3 *(Remark 3.16 [8]) Take $X = \ell_1^n$ and let V be its subspace. Then, any Chalmers–Metcalf operator for the pair (ℓ_1^n, V) can be written as*

$$E_P = \sum_{i=1}^{n} \alpha_i y_i \otimes e_i, \tag{34}$$

for some $y_i \in S(\ell_\infty^n)$ and $\alpha_i \geq 0$ such that $\sum_{i=1}^n \alpha_i = 1$ and $y_i(Pe_i) = ||P||$ for all i such that $\alpha_i > 0$ and all minimal projections $P : X \to V$.

Lemma 4. *Assume that $f_i \neq 0$, for every $i = 1,\dots,n$, and consider $f = (f_1,\dots,f_n) \in S(\ell_\infty^n)$. Also assume that there is a Chalmers–Metcalf operator for the pair $(\ell_1^n, \ker f)$. such that $\alpha_i > 0$, for every $i = 1,\dots,n$, and the matrix of y_i is invertible. Then the minimal projection $P : \ell_1^n \to \ker f$ is unique minimal.*

Proof. Let E_P be any Chalmers–Metcalf operator for the pair $(\ell_1^n, \ker f)$. By Remark 3

$$E_P = \sum_{i=1}^{n} \alpha_i y_i \otimes e_i, \tag{35}$$

for some $y_i \in S(\ell_\infty^n)$ and $\alpha_i > 0$ such that $\sum_{i=1}^n \alpha_i = 1$ and $y_i(Pe_i) = ||P||$ for all minimal projections $P : X \to \ker f$. Every projection onto $\ker f$ has to be of the form

$$Q = Id - f \otimes z, \tag{36}$$

where $f(z) = 1$. Assume Q is minimal. Using (35) we have

$$y_i(Qe_i) = \lambda(\ker f, \ell_1^n). \tag{37}$$

That is

$$y_i(e_i) - f_i y_i(z) = \lambda(\ker f, \ell_1^n) \tag{38}$$

and

$$y_i(z) = \frac{y_i(e_i) - \lambda(\ker f, \ell_1^n)}{f_i}. \tag{39}$$

The last equations give the system of n linear equations with n coordinates of z as unknowns. The main determinant of the system is the determinant of the matrix of y_i. Since we assume this matrix is invertible, the system has a unique solution.

The assumption that the minimal projection P in ℓ_1^n onto hyperplane attains norm on every e_i is not restrictive. One may expect the minimal projection to be equally small in every direction. Lemma I.1.1 [11] shows that every minimal projection of three dimensional space onto hyperplane have at least 6 norming points. As a result, a projection of norm greater then 1 in ℓ_1^3 has to attain norm on all $e_1, -e_1, e_2, -e_2, e_3, -e_3$. We will investigate this condition now. Remark 1 shows that there is always a minimal projection with $f_i, z_i \geq 0$, for $i = 1, \ldots, n$.

Lemma 5. *Let $n \geq 3$ and assume that $f_i, z_i \geq 0$, for $i = 1, \ldots, n$. Define*

$$P = Id - f \otimes z, \tag{40}$$

where $f(z) = 1$. Then P attains norm on every e_i, for $i = 1, \ldots, n$, if and only if every pair (η_i, e_i), for $i = 1, \ldots, n$, is a norming pair for P. Moreover if $f_i, z_i > 0$, for $i = 1, \ldots, n$ then (η_i, e_i), for $i = 1, \ldots, n$, are the only norming pairs for P.

Proof. Observe that

$$P(e_k) = (-f_k z_1, \ldots, 1 - f_k z_k, \ldots, -f_k z_n). \tag{41}$$

Since $f(z) = 1$ we can see that $0 \leq f_i z_i \leq 1$, for all $i = 1, \ldots, n$. We cannot have $f_k z_k = 1$ for any $k = 1, \ldots, n$ (otherwise for some k we have $z_k = 1/f_k$ and $z_i = 0$, for $i \neq k$ and that leads to $Pe_k = 0$). As a result, for all $i = 1, \ldots, n$ $0 \leq f_i z_i < 1$, and

$$||Pe_k|| = \eta_k(Pe_k). \tag{42}$$

If $f_i, z_i > 0$, then (41) shows that η_k is the only norming functional for Pe_k.

If we assume that this minimal projection attains norm on every e_i, for $i=1,\dots,n$ and use the above lemma to (39) we can see that z has to satisfy system of equation

$$\eta_i(z)=\frac{1-\lambda(\ker f,\ell_1^n)}{f_i},\ \text{for } i=1,\dots,n. \tag{43}$$

What we will do now is to reverse this reasoning. We will take z that satisfies the above equations and we will prove that it will give a minimal projection that attains norm on every e_i, for $i=1,\dots,n$.

Lemma 6. *Assume $n\geq 3$ and not all of b_k are zero. Then all solutions to the system of equations*

$$(b_k+b_l)x_k-(b_k+b_l)x_l+\sum_{i\neq k,l}(b_k-b_l)x_i=0,\quad k,l=1,\dots,n \tag{44}$$

are of the form

$$(x_1,\dots,x_n)=\alpha\left(\frac{\sum_{j=1}^n b_j}{n-2}-b_1,\dots,\frac{\sum_{j=1}^n b_j}{n-2}-b_n\right). \tag{45}$$

Proof. We rewrite (44) as

$$2b_lx_k-2b_kx_l+(b_k-b_l)\sum_{i=1}^n x_i=0,\quad k,l=1,\dots,n. \tag{46}$$

Fix k. Summing over l we obtain

$$2\left(\sum_{i=1}^n b_i\right)x_k-2b_k\left(\sum_{i=1}^n x_i\right)+nb_k\left(\sum_{i=1}^n x_i\right)-\left(\sum_{i=1}^n x_i\right)\left(\sum_{i=1}^n b_i\right)=0. \tag{47}$$

That is,

$$2\left(\sum_{i=1}^n b_i\right)x_k=(n-2)\left(\sum_{i=1}^n x_i\right)\left(\frac{\sum_{i=1}^n b_i}{n-2}-b_k\right), \tag{48}$$

which is (45) with $\alpha=((n-2)\sum_{i=1}^n x_i)/(2\sum_{i=1}^n b_i)$ in case $\sum_{i=1}^n b_i\neq 0$; in case $\sum_{i=1}^n b_i=0$, (48) gives $(n-2)b_k\sum_{i=1}^n x_i=0$, for all $k=1,\dots,n$. Fixing a k such that $b_k\neq 0$, we get $\sum_{i=1}^n x_i=0$, which gives from (46) $b_lx_k=b_kx_l$, i.e., $x_l=(x_k/b_k)b_l$, which is (45) with $\alpha=-x_k/b_k$.

Lemma 7. *Let $n\geq 3$ and assume $f=(f_1,\dots,f_n)\neq 0$ and $\sum_{i=1}^n\lambda_i=1$. Define*

$$E=\sum_{i=1}^n\alpha_i\eta_i\otimes e_i, \tag{49}$$

where $\eta_i=(-1,\dots,-1)+2e_i$. Then $E(\ker f)\subset\ker f$ if and only if

$$\alpha_i=\frac{\beta_i}{\beta_1+\cdots+\beta_n}\quad \textit{where}\quad \beta_i=\frac{f_1+\cdots+f_n}{(n-2)f_i}-1. \tag{50}$$

Proof. Put $x_{k,l} = f_l\, e_k - f_k\, e_l$, for $k,l = 1,\dots,n$. Observe that

$$\ker f = \operatorname{span}\{x_{k,l} : k,l = 1,\dots,n\}. \tag{51}$$

As a result $E(\ker f) \subset \ker f$ if and only if

$$f(E(x_{k,l})) = \sum_{i=1}^{n} \alpha_i\, f_i\, \eta_i(x_{k,l}) = 0, \tag{52}$$

for all $k,l = 1,\dots,n$. The last equation is equivalent to

$$\sum_{i \neq k,l} \alpha_i f_i (f_k - f_l) + \alpha_k f_k (f_k + f_l) - \alpha_l f_l (f_k + f_l) = 0, \tag{53}$$

for all $k,l = 1,\dots,n$. Putting $z_i = \alpha_i f_i$ and applying Lemma 6 we obtain

$$\alpha_i = \alpha \left(\frac{f_1 + \cdots + f_n}{(n-2) f_i} - 1 \right). \tag{54}$$

If $\alpha = 0$ then we would have $\alpha_i = 0$, for all $i = 1,\dots,n$, a contrary to $\sum_{i=1}^{n} \alpha_i = 1$. Therefore, $\alpha \neq 0$. Normalizing $\alpha_1 + \cdots + \alpha_n = 1$ gives the result.

Theorem 5 *Assume* $1 \geq f_1 \geq f_2 \geq \cdots \geq f_n > 0$, $\sum_{i=1}^{n} f_i > (n-2) f_1$ *and* $\sum_{j=1}^{n} \frac{1}{f_j} > \frac{n-2}{f_n}$. *Let* $w = (w_1,\dots,w_n)$ *be the unique solution to the system of equations*

$$\eta_i(w) = -\frac{1}{f_i}, \text{ for } i = 1,\dots,n. \tag{55}$$

Then the projection $P = Id - f \otimes z : \ell_1^n \to \ker f$, *for*

$$z_i = \frac{w_i}{\sum_{i=1}^{n} f_i w_i} \tag{56}$$

is minimal and (η_i, e_i), *for* $i = 1,\dots,n$, *are the only norming pairs for this projection.*

Proof. Let w be the unique solution to the system of equations (55). Summing these equations over $i = 1,\dots,n$ gives

$$-(n-2)\sum_{i=1}^{n} w_i = \sum_{i=1}^{n} \eta_i(w) = -\sum_{i=1}^{n} \frac{1}{f_i}. \tag{57}$$

As a result

$$\sum_{i=1}^{n} w_i = \frac{\sum_{i=1}^{n} \frac{1}{f_i}}{n-2}. \tag{58}$$

Using the above and the assumption $\sum_{j=1}^{n} \frac{1}{f_j} > \frac{n-2}{f_n} \geq \frac{n-2}{f_k}$ we get.

$$2w_k = \eta_k(w) + \sum_{i=1}^{n} w_i = -\frac{1}{f_k} + \frac{\sum_{i=1}^{n} \frac{1}{f_k}}{n-2} > 0. \tag{59}$$

Therefore,

$$z_i = \frac{w_i}{\sum_{i=1}^n f_i w_i} > 0, \tag{60}$$

for any $i = 1, \dots, n$. From the above $0 < f_i z_i < 1$. Observe that

$$f_k \eta_k(z) = \frac{f_k \eta_k(w)}{\sum_{i=1}^n f_i w_i} = \frac{f_l \eta_l(w)}{\sum_{i=1}^n f_i w_i} = f_l \eta_l(z), \tag{61}$$

for every $k, l = 1, \dots, n$. Since

$$P(e_k) = (-f_k z_1, \dots, 1 - f_k z_k, \dots, -f_k z_n) \tag{62}$$

We can see that

$$||Pe_k|| = \eta_k P(e_k) = 1 - f_k \eta_k(z) = 1 - f_l \eta_l(z) = \eta_l P(e_l) = ||Pe_l||. \tag{63}$$

To prove that P is minimal, take

$$E = \sum_{i=1}^n \alpha_i \eta_i \otimes e_i, \text{ and } \alpha_i = \frac{\beta_i}{\beta_1 + \dots + \beta_n} \quad \text{for} \quad \beta_i = \frac{f_1 + \dots + f_n}{(n-2) f_i} - 1. \tag{64}$$

From (63) $(\eta_i, e_i) \in \varepsilon(P)$, the assumption $\sum_{i=1}^n f_i > (n-2) f_1$ gives $\alpha_i > 0$. Finally, By Lemma 7, $E(\ker f) \subset \ker f$. Therefore, E is a Chalmers–Metcalf operator for P. Hence, P is minimal.

Theorem 6 *Assume* $1 \geq f_1 \geq f_2 \geq \dots \geq f_n > 0$, $\sum_{i=1}^n f_i > (n-2) f_1$ *and* $\sum_{j=1}^n \frac{1}{f_j} > \frac{n-2}{f_n}$. *Consider* $f = (f_1, \dots, f_n) \in S(\ell_\infty^n)$. *Then the minimal projection onto* $\ker f$ *is unique minimal.*

Proof. Applying Lemma 4 to Chalmers-Metcalf operator constructed in (64) gives the result.

Now we can summarize our theorems

Theorem 7 *Assume* $1 \geq f_1 \geq f_2 \geq \dots \geq f_n > 0$, $\sum_{i=1}^n f_i > (n-2) f_1$ *and* $\sum_{j=1}^n \frac{1}{f_j} > \frac{n-2}{f_n}$. *Consider* $f = (f_1, \dots, f_n) \in S(\ell_\infty^n)$. *Then, there is only one minimal projection of* ℓ_1^n *onto* $\ker f$ *and the Chalmers–Metcalf operator for the pair* $(\ell_1^n, \ker f)$ *is uniquely defined by*

$$E_P = \sum_{i=1}^n \alpha_i \eta_i \otimes e_i, \tag{65}$$

where $\eta_i = (-1, \dots, -1) + 2e_i$ *and*

$$\alpha_i = \frac{\beta_i}{\beta_1 + \dots + \beta_n} \quad \textit{where} \quad \beta_i = \frac{f_1 + \dots + f_n}{(n-2) f_i} - 1. \tag{66}$$

Moreover,

$$\lambda(\ker f, \ell_1^n) = 1 + (\sum_{i=1}^n f_i w_i)^{-1} \tag{67}$$

and projection $P = Id - f \otimes z$ *is minimal for*

$$z_i = w_i(\lambda(\ker f, \ell_1^n) - 1), \tag{68}$$

where

$$w_i = \frac{1}{2}\left(\frac{\sum_{i=1}^n \frac{1}{f_i}}{n-2} - \frac{1}{f_i}\right) \tag{69}$$

Proof. The uniqueness of minimal projection follows from the previous theorem. By the construction of a minimal projection P onto $\ker f$ in Theorem 5 we can see that (η_i, e_i) are the only norming pairs for P. Therefore, any Chalmers–Metcalf operator for the pair $(\ell_1^n, \ker f)$ has to be of the form

$$E_P = \sum_{i=1}^n \alpha_i \eta_i \otimes e_i, \tag{70}$$

where $\alpha_i \geq 0$ and $\sum_{i=1}^n \alpha_i = 1$. By Lemma 7 we see that every E_P has to be of the form (65). Also

$$\lambda(\ker f, \ell_1^n) = \eta_k(Pe_k) = 1 - f_k\eta_k(z) = 1 - \frac{f_k}{\sum_{i=1}^n f_i w_i}\eta_k(w) = 1 + \frac{1}{\sum_{i=1}^n f_i w_i}. \tag{71}$$

(68) and (69) follows from (60) and (59).

Example 1 *Assume* $f_1 \geq f_2 \geq f_3 > 0$ *and consider* $f = (f_1, f_2, f_3) \in S(\ell_\infty^3)$. *Then the minimal projection* $P : \ell_1^3 \to \ker f$ *is unique minimal and the Chalmers–Metcalf operator for* $P : \ell_1^n \to \ker f$ *is uniquely defined by*

$$E_P = \frac{\frac{f_2+f_3}{f_1}\eta_1 \otimes e_1 + \frac{f_1+f_3}{f_2}\eta_2 \otimes e_2 + \frac{f_1+f_2}{f_3}\eta_3 \otimes e_3}{\frac{f_2+f_3}{f_1} + \frac{f_1+f_3}{f_2} + \frac{f_1+f_2}{f_3}}, \tag{72}$$

where $\eta_1 = (1, -1, -1), \eta_2 = (-1, 1, -1), \eta_3 = (-1, -1, 1)$.

Proof. For $n = 3$ the assumptions of Theorem 7 are automatically satisfied.

Example 2 *The minimal projection from* ℓ_1^n *onto* $ker(1, \ldots, 1)$ *is*

$$P = Id - \frac{1}{n}(1, \ldots, 1) \otimes (1, \ldots, 1). \tag{73}$$

This projection is unique minimal.

Proof. For $f = (1, \ldots, 1)$ the assumptions of Theorem 7 are automatically satisfied.

Next, we consider the subspace $ker(1, \frac{1}{4}, \frac{1}{4}, \frac{1}{4})$. It does not satisfy the assumption of Theorem 7. It is an extreme case. Minimal projection is actually zero on one of the extreme points of ℓ_1^n and z has only one non-zero coefficient. We will show how Chalmers–Metcalf operator can be used to obtain that this projection is unique minimal. The proof is completely algebraic.

Example 3 *The minimal projection from ℓ_1^4 onto $ker(1,\frac{1}{4},\frac{1}{4},\frac{1}{4})$ is*

$$P = Id - (1,\tfrac{1}{4},\tfrac{1}{4},\tfrac{1}{4}) \otimes (1,0,0,0). \tag{74}$$

This projection is also unique minimal but does not attain the norm on every e_i. In fact $P(e_1) = 0$.

Proof. Consider projection P given by (74). Observe that $P(e_1) = 0$ and

$$P(e_k) = -1/4e_1 + e_k, \tag{75}$$

for $k = 2,3,4$. Therefore, $||Pe_k|| = 5/4$. As a result $||P|| = 5/4$ and the following pairs $((-1,1,-1,-3/4),e_2)$, $((-1,-3/4,1,-1),e_3)$, $((-1,-1,-3/4,1),e_4)$ are norming pairs for P. Consider operator

$$E = \tfrac{1}{3}(-1,1,-1,-\tfrac{3}{4}) \otimes e_2 + \tfrac{1}{3}(-1,-\tfrac{3}{4},1,-1) \otimes e_3 + \tfrac{1}{3}(-1,-1,-\tfrac{3}{4},1) \otimes e_4. \tag{76}$$

Observe that

$$\begin{aligned}(1,\tfrac{1}{4},\tfrac{1}{4},\tfrac{1}{4})(Ex) &= \tfrac{1}{3}\tfrac{1}{4}((-1,1,-1,-\tfrac{3}{4})x + (-1,-\tfrac{3}{4},1,-1)x + (-1,-1,-\tfrac{3}{4},1)x \\ &= \tfrac{1}{3}\tfrac{1}{4}(-3,-\tfrac{3}{4},-\tfrac{3}{4},-\tfrac{3}{4})x = -\tfrac{1}{4}(1,\tfrac{1}{4},\tfrac{1}{4},\tfrac{1}{4})x.\end{aligned} \tag{77}$$

Therefore, $E(ker(1,\frac{1}{4},\frac{1}{4},\frac{1}{4})) \subset ker(1,\frac{1}{4},\frac{1}{4},\frac{1}{4})$ and as a result P is minimal. We will prove uniqueness of P now. Any projection ℓ_1^4 onto $ker(1,\frac{1}{4},\frac{1}{4},\frac{1}{4})$ is given by formula

$$Q = Id - (1,\tfrac{1}{4},\tfrac{1}{4},\tfrac{1}{4}) \otimes z, \tag{78}$$

where $z_1 + (1/4)(z_2 + z_3 + z_4) = 1$. Assume Q is minimal. The norming pairs that appears in the Chalmers–Metcalf operator (76) are norming pairs for any minimal projection (see Theorem 2.18 [8]). As a result we obtain the equations

$$\begin{aligned}\tfrac{5}{4} &= (-1,1,-1,-\tfrac{3}{4})Q(e_2) = 1 + \tfrac{1}{4}z_1 - \tfrac{1}{4}z_2 + \tfrac{1}{4}z_3 + \tfrac{3}{16}z_4, \\ \tfrac{5}{4} &= (-1,-\tfrac{3}{4},1,-1)Q(e_3) = 1 + \tfrac{1}{4}z_1 + \tfrac{3}{16}z_2 - \tfrac{1}{4}z_3 + \tfrac{1}{4}z_4, \\ \tfrac{5}{4} &= (-1,-1,-\tfrac{3}{4},1)Q(e_4) = 1 + \tfrac{1}{4}z_1 + \tfrac{1}{4}z_2 + \tfrac{3}{16}z_3 - \tfrac{1}{4}z_4.\end{aligned} \tag{79}$$

Using $z_1 + (1/4)(z_2 + z_3 + z_4) = 1$ and plugging for z_1 we can reduce the above system to

$$\begin{aligned}-\tfrac{5}{16}z_2 + \tfrac{3}{16}z_3 + \tfrac{2}{16}z_4 &= 0, \\ \tfrac{2}{16}z_2 - \tfrac{5}{16}z_3 + \tfrac{3}{16}z_4 &= 0, \\ \tfrac{3}{16}z_2 + \tfrac{2}{16}z_3 - \tfrac{5}{16}z_4 &= 0.\end{aligned} \tag{80}$$

The system has the unique solution $z_2 = z_3 = z_4 = 0$. Therefore, $z_1 = 1$ and $Q = P$.

Acknowledgements The author would like to thank the referee for many helpful comments, especially for providing the easier proof of Lemma 6.

References

1. J. Blatter and E. W. Cheney, *Minimal projections on hyperplanes in sequence spaces*, Ann. Mat. Pura Appl. (4) 101 (1974), pp. 215–227. MR0358179 (50 #10644)
2. B. L. Chalmers and F. T. Metcalf, *The determination of minimal projections and extensions in* L^1, Trans. Amer. Math. Soc. 329 (1992), pp. 289–305. MR1034660 (92e:41017)
3. B. L. Chalmers and F. T. Metcalf, *A characterization and equations for minimal projections and extensions*, J. Operator Theory 32 (1994), pp. 31–46. MR1332442 (96c:46014)
4. E. W. Cheney and K. H. Price, *Minimal projections* in Approximation Theory (Proc. Sympos., Lancaster, 1969), Academic Press, London 1970, pp. 261–289. MR0265842 (42 #751)
5. E. W. Cheney and C. Franchetti, *Minimal projections of finite rank in sequence spaces* in Fourier analysis and approximation theory (Proc. Colloq., Budapest, 1976), Vol. I, North-Holland, Amsterdam 1978, pp. 241–253. MR540303 (84b:46010)
6. G. Lewicki, *Minimal projections onto subspaces of* $l_\infty^{(n)}$ *of codimension two*, Collect. Math. 44 (1993), pp. 167–179. MR1280736 (95f:46021)
7. G. Lewicki, *Minimal projections onto two-dimensional subspaces of* $l_\infty^{(4)}$, J. Approx. Theory 88 (1997), pp. 92–108. MR1426461 (97i:41047)
8. G. Lewicki and L. Skrzypek, *On the properties of Chalmers-Metcalf operator*, Banach Spaces and their Applications in Analysis (Ed. by Randrianantoanina, Beata and Randrianantoanina, Narcisse), 375-390, de Gruyter 2007
9. G. Lewicki and L. Skrzypek, *Chalmers-Metcalf operator and uniqueness of minimal projections*, J. Approx. Theory 148 (2007), pp. 71–91. MR2356576
10. P. D. Morris and E. W. Cheney, *On the existence and characterization of minimal projections*, J. Reine Angew. Math. 270 (1974), pp. 61–76. MR0358188 (50 #10653)
11. W. Odyniec and G. Lewicki, 1449, *Minimal projections in Banach spaces*. Springer-Verlag, Berlin, 1990, Problems of existence and uniqueness and their application. MR1079547 (92a:41021)
12. B. Randrianantoanina, *Norm-one projections in Banach spaces*, Taiwanese J. Math. 5 (2001), pp. 35–95.

The Polynomial Inverse Image Method

Vilmos Totik

Abstract In this survey, we discuss how to transfer results from an interval or the unit circle to more general sets. At the basis of the method is taking polynomial inverse images.

1 Introduction

In the last decade a method has been developed that (in some cases) allows one to transfer result from an interval (like $[-1,1]$) or the unit circle C_1 (which we are going to call *model cases*) to more general sets. We emphasize that the method TRANSFORMS the RESULT from the model case to the general case and is not aimed to carry over the proofs from the model cases to the general situation.

The rationale of the method is the following: on the unit circle C_1 and on $[-1,1]$ many classical and powerful tools (such as Fourier-series, classical orthogonal expansions, Poisson representation, Taylor expansions, H^p-spaces, etc.) have been developed, which are at our disposal when dealing with a problem on these model sets. When dealing with more general sets like a compact subset of the real line instead of $[-1,1]$ or a system of Jordan corves instead of C_1, either these tools are non-existent, or they are difficult to use. Therefore, if we have a method that *transforms* a model result to the general case, then

- We get the same result in many situations (as opposed to the single result in the model case).

Vilmos Totik
Bolyai Institute, Analysis Research Group of the Hungarian Academy of Sciences, University of Szeged, Szeged, Aradi v. tere 1, 6720, Hungary

Department of Mathematics and Statistics, University of South Florida 4202 E. Fowler Ave, PHY 114 Tampa, FL 33620-5700, USA
e-mail: totik@mail.usf.edu

M. Neamtu and L. Schumaker (eds.), *Approximation Theory XIII: San Antonio 2010*, Springer Proceedings in Mathematics 13, DOI 10.1007/978-1-4614-0772-0_22,

- We save the burden of finding the analogue of the model proof (which may not exist at all).

The method in question is the following: apply inverse images under polynomial mapping, i.e. if $T_N(z) = \gamma_N z^N + \cdots$ is a polynomial and E_0 is $[-1,1]$ or the unit circle C_1, then consider

$$E = T_N^{-1} E_0 = \{z \,|\, T_N(z) \in E_0\}.$$

The point is that many properties are preserved when we take polynomial inverse images, most notable, equilibrium measures and Green's functions (see the Appendix) are preserved.

Thus, in a nutshell we make the following steps:

(a) Start from a result for the model case.
(b) Apply an inverse polynomial mapping to go to a special result on the inverse images of the model sets.
(c) Approximate more general sets by inverse images as in (b).

Sometimes, (b)–(c) should be followed by an additional step:

(d) Get rid of the special properties appearing in steps (b)–(c).

Among others the polynomial inverse image method has been successful in the following situations:

1. The Bernstein-type inequality (2) below, the model case being the classical Bernstein inequality (1) on $[-1,1]$.
2. The Markoff-type inequality (16)–(17) below, the model case being the classical Markoff inequality (15).
3. Asymptotics of Christoffel functions on compact subsets of the real line, namely (25), when the model case was (23) on $[-1,1]$.
4. Asymptotics of Christoffel functions on curves, namely (26), when the model case was (22) on C_1.
5. Universality (28) on general sets, the model case being (28) on $[-1,1]$.
6. Fine zero spacing (30) of orthogonal polynomials, the model case being (29) on $[-1,1]$.
7. For a system of smooth Jordan curves the Bernstein-type inequality (19), where the model case was Bernstein's inequality (18) on the unit circle.

Before elaborating more on the method let us see how it works in a concrete case. To this we need a few things from potential theory; see the Appendix at the end of this paper for the definitions. In what follows, for a compact set $E \subset \mathsf{R}$ of positive capacity we denote by ω_E the density of the equilibrium measure with respect to the Lebesgue measure on R. This density certain exists in the (one dimensional) interior of E. On the other hand, if E is a finite family of smooth Jordan curves or arcs, then ω_E denotes the density of the equilibrium measure of E with respect to arc measure on E.

2 The Bernstein Inequality on General Sets

Let P_n denote an algebraic polynomial of degree at most n. Bernstein's inequality

$$|P_n'(x)| \le \frac{n}{\sqrt{1-x^2}} \|P_n\|_{[-1,1]}, \qquad x \in [-1,1] \tag{1}$$

relating the derivative of P_n to its supremum norm on $[-1,1]$ is of fundamental importance in approximation theory. Now with the polynomial inverse image method we can prove the following generalization of (1):

Theorem 2.1 *If $E \subset \mathrm{R}$ is compact, then*

$$|P_n'(x)| \le n\pi\omega_E(x)\|P_n\|_E, \qquad x \in \mathrm{Int}(E). \tag{2}$$

Note that for $E = [-1,1]$ we have $\omega_E(x) = 1/\pi\sqrt{1-x^2}$, so in this case (2) takes the form (1). Let us also mention that (2) is sharp: if $x_0 \in \mathrm{Int}(E)$ is arbitrary, then for every $\varepsilon > 0$ there are polynomials P_n of degree at most $n = 1,2,\ldots$ such that

$$|P_n'(x_0)| > (1-\varepsilon)n\pi\omega_E(x_0)\|P_n\|_E$$

for all large n.

Actually, more is true, namely

$$\left(\frac{|P_n'(x)|}{\pi\omega_E(x)}\right)^2 + n^2|P_n(x)|^2 \le n^2\|P_n\|_E^2, \qquad x \in \mathrm{Int}(E), \tag{3}$$

which is the analogue of the inequality

$$\left(|P_n'(x)|\sqrt{1-x^2}\right)^2 + n^2|P_n(x)|^2 \le n^2\|P_n\|_{[-1,1]}^2 \tag{4}$$

of Szegő ([6, 35]).

(2) and (3) are due to Baran [1], who actually got them also in higher dimension. Both inequalities were rediscovered in [38] with the method of the present survey. The outline of the proof of (2) using polynomial inverse images is as follows:

(a) Start from Bernstein inequality on $[-1,1]$.
(b) Next, consider the special case when $E = T_N^{-1}[-1,1]$ and $P_n = S_k(T_N)$ with some polynomial S_k. Assuming $\|P_n\|_E = 1$ we get

$$|P_n'(x)| = |S_k'(T_N(x))T_N'(x)| \le \frac{k}{\sqrt{1-T_N^2(x)}}|T_N'(x)| = kN\pi\frac{|T_N'(x)|}{\pi N\sqrt{1-T_N^2(x)}},$$

and by (6) here the right-hand side is $kN\pi\omega_E(x)$, i.e. we get (2) in this special case.
(c) Approximate a general E by $T_N^{-1}[-1,1]$ and P_n by $S_k(T_N)$ to get

$$|P_n'(x)| \le (1+o_{x,E}(1))n\pi\omega_E(x)\|P_n\|_E \tag{5}$$

where $o_{x,E}(1)$ denotes a quantity that tends to 0 as n tends to infinity. See Sect. 5 for this approximation step (the exact details for the general Bernstein inequality are in [38, Theorem 3.1]).

(d) Get rid of $o(1)$.

This very last step can be done as follows. Let P_n be any polynomial, and x_0 any point in the interior of E. We may assume $\|P_n\|_E = 1$. Let $\mathscr{T}_m(z) = \cos(m \arccos z)$ be the classical Chebyshev polynomials, and for some $0 < \alpha_m < 1$ and $0 \le \varepsilon_m < 1 - \alpha_m$ consider the polynomials

$$R_{mn}(x) = \mathscr{T}_m(\alpha_m P_n(x) + \varepsilon_m),$$

where $\alpha_m < 1$ and $0 \le \varepsilon_m < 1 - \alpha_m$ are chosen so that $\alpha_m P_n(x_0) + \varepsilon_m$ is one of the zeros of $\mathscr{T}_m$. Since the distance of neighboring zeros of $\mathscr{T}_m$ is smaller than $10/m$, we can do this with $\alpha_m = 1 - 10/m$ and with some $0 \le \varepsilon_m < 10/m$, and then $\alpha_m \to 1$ and $\varepsilon_m \to 0$ as $m \to \infty$. Now apply (5) to R_{mn}. It follows that

$$|R'_{mn}(x_0)| \le (1 + o(1))\pi\omega_E(x_0)mn\|R_{mn}\|_E,$$

where the term $o(1)$ tends to zero as $m \to \infty$. Here, on the right, $\|R_{mn}\|_E = 1$, and on the left we have

$$|R'_{mn}(x_0)| = |\mathscr{T}'_m(\alpha_m P_n(x_0) + \varepsilon_m)||P'_n(x_0)|\alpha_m.$$

Since at the zeros z of $\mathscr{T}_m$ we have $\mathscr{T}'_m(z) = m/\sqrt{1-z^2}$, it follows that

$$\frac{m}{\sqrt{1 - (\alpha_m P_n(x_0) + \varepsilon_m)^2}}|P'_n(x_0)|\alpha_m \le (1 + o(1))\pi\omega_E(x_0)mn,$$

where the term $o(1)$ tends to zero as $m \to \infty$. On dividing here by m and letting m tend to infinity we obtain

$$\frac{|P'_n(x_0)|}{\sqrt{1 - P_n^2(x_0)}} \le \pi\omega_E(x_0)n,$$

and this is the inequality (3) at the point x_0 because in our case $\|P_n\|_E = 1$. □

3 The Model case $[-1,1]$, Admissible Polynomial Maps, Approximation

As we have already mentioned, there are two model cases: the interval $[-1,1]$ and the unit circle $C_1 = \{z \mid |z| = 1\}$.

For $[-1,1]$ we allow polynomial maps with respect to real polynomials (called admissible polynomials) $T_N(z) = \gamma_n x^N + \cdots$, $\gamma_N \ne 0$ such that T_N has N zeros and $N-1$ local extremal values each of which is of size ≥ 1 in absolute value. In other

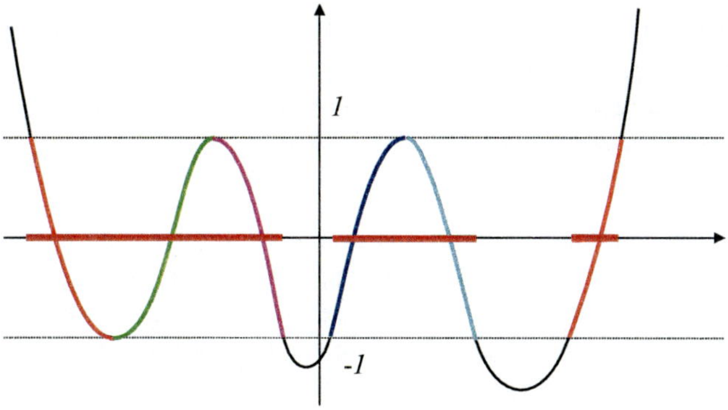

Fig. 1: The set $T_N^{-1}\,[-1,1]$

words, there are $u_1,\dots,u_N$ with $T_N'(u_j)=0$ and $|T_N(u_j)|\geq 1$. Then it easily follows that the local extremal values alternate in sign and $T_N(z)$ runs through the interval $[-1,1]$ N-times as x runs through the real line. Thus,

$$E := T_N^{-1}[-1,1] = \{x \,|\, T_N(x)\in[-1,1]\}$$

consists of N subintervals $E_{n,j}$, $1\leq j\leq N$ each of which is mapped by T_N onto $[-1,1]$ in a 1-to-1 fashion. However, some of these subintervals may be attached to one another, so $T_N^{-1}[-1,1]$ actually consists of k intervals for some $1\leq k\leq N$; see Fig. 1 where $N=6$ and $k=3$. The equilibrium measure of E is the (normalized) pull-back of the equilibrium measure on $[-1,1]$ under the mapping T_N:

$$\omega_E(x) = \frac{|T_N'(x)|}{\pi N\sqrt{1-T_N^2(x)}}, \qquad x\in E. \tag{6}$$

Polynomial inverse images of intervals, i.e. sets of the form $T_N^{-1}[-1,1]$ with admissible T_N have many interesting properties [24–27]. They are the sets $\Sigma = \cup_{j=1}^{l}[a_j,b_j]$ with the property that the equilibrium measure has rational mass on each subinterval, i.e. each $\mu_\Sigma([a_j,b_j])$, $j=1,\dots,k$ is of the form p/N. They are also the sets $\Sigma=\cup_{i=1}^{l}[a_i,b_i]$ for which the Pell-type equation

$$P^2(z)-Q(z)S^2(z)=1 \qquad \text{with} \qquad Q(x)=\prod_{i=1}^{l}(x-a_i)(x-b_i),$$

which goes back to N. H. Abel, has polynomial solutions P and Q. See [23]–[28] and the references there for many more interesting results connected with polynomial inverse images.

What we need of them is that these sets are dense among all sets consisting of finitely many intervals.

Theorem 3.1 *Given a system $\Sigma = \{[a_i,b_i]\}_{i=1}^l$ of disjoint closed intervals and an $\varepsilon > 0$, there is another system $E = \{[a_i',b_i']\}_{i=1}^l$ such that $\cup_{i=1}^l [a_i',b_i'] = T_N^{-1}[-1,1]$ for some admissible polynomial T_N, and for each $1 \le i \le l$ we have*

$$|a_i - a_i'| \le \varepsilon, \qquad |b_i - b_i'| \le \varepsilon.$$

The theorem immediately implies its strengthened form when we also prescribe if a given a_i' (or b_i') is smaller or bigger than a_i (or b_i). In particular, it is possible to require e.g. that $\Sigma \subset \Sigma'$. It is also true that in the theorem we can select $a_i' = a_i$ for all i, and even $b_l' = b_l$. Alternatively we can fix any $l+1$ of the $2l$ points a_i, b_i, $1 \le i \le l$.

Theorem 3.1 has been proven several times independently in the literature, see [7, 18, 22, 30, 38]. For a particularly simple proof see [41].

4 The Model case C_1, Sharpened form of Hilbert's Lemniscate Theorem

For the unit circle C_1 we shall take its inverse image under polynomial mappings generated by polynomials $T_N(z) = \gamma_N z^N + \cdots$ for which $T_N'(z) \neq 0$ whenever $|T_N(z)| = 1$. Then

$$\sigma := T_N^{-1} C_1 = \{z \,|\, |T_N(z)| = 1\}$$

is actually a level set of the polynomial T_N, which, from now on, we call a lemniscate. Since $T_N'(z) \neq 0$ on E, this E consists of a finite number of analytic Jordan curves (a Jordan curve is a homeomorphic image of the unit circle). Again, the equilibrium measure of E is the (normalized) pull-back of the equilibrium measure on C_1 under the mapping T_N:

$$\omega_\sigma(z) = \frac{1}{2\pi N} |T_N'(z)|, \qquad z \in E. \tag{7}$$

Hilbert's lemniscate theorem claims that if K is a compact set on the plane and U is a neighborhood of K then there is a lemniscate σ that separates K and $\mathbf{C} \setminus U$, i.e. it lies within U but encloses K. An equivalent formulation is the following. Let γ_j, Γ_j, $j = 1, \ldots, m$ be Jordan curves (i.e. homeomorphic images of the unit circle), γ_j lying interior to Γ_j and the Γ_j's lying exterior to one another, and set $\gamma^* = \cup_j \gamma_j$, $\Gamma^* = \cup_j \Gamma_j$. Then there is a lemniscate σ that is contained in the interior of Γ^* which also contains γ^* in its interior, i.e. σ separates γ^* and Γ^* in the sense that it separates each γ_j from the corresponding Γ_j. This is not enough for our purposes of approximation, what we need is the following sharpened form (see [19]).

Let γ^* and Γ^* be twice continuously differentiable in a neighborhood of P and touching each other at P. We say that they *$\mathcal{K}$-touch* each other if their (signed) curvature at P is different (signed curvature is seen from the outside of Γ^*). Equivalently we can say that in a neighborhood of P the two curves are separated by two circles one of them lying in the interior of the other one.

Theorem 4.1 *Let $\gamma^* = \cup_{j=1}^m \gamma_j$ and $\Gamma^* = \cup_{j=1}^m \Gamma_j$ be as above, and let γ^* $\mathscr{K}$-touch Γ^* in finitely many points $P_1,\ldots,P_k$ in a neighborhood of which both curves are twice continuously differentiable. Then there is a lemniscate σ that separates γ^* and Γ^* and $\mathscr{K}$-touches both γ^* and Γ^* at each P_j.*

Furthermore, σ lies strictly in between γ^* and Γ^* except for the points $P_1,\ldots,P_k$, and has precisely one connected component in between each γ_j and Γ_j, $j=1,\ldots,m$, and these m components are Jordan curves.

From our point of view the following corollary is of primary importance. Let K be the closed set enclosed by Γ^* and K_0 the closed set enclosed by γ^*. Denote by $g(K,z)$ Green's function of $\overline{\mathbb{C}}\setminus K$ with pole at infinity. Finally, let L be the closed set enclosed by σ.

Corollary 4.2 *Let Γ^*, γ^* and $P_1,\ldots,P_k \in \Gamma^*$ be as in Theorem* 4.1. *Then for every $\varepsilon > 0$ there is a lemniscate σ as in Theorem 4.1 such that for each P_j we have*

$$\frac{\partial g(L,P_j)}{\partial \mathbf{n}} \le \frac{\partial g(K,P_j)}{\partial \mathbf{n}} + \varepsilon, \tag{8}$$

where $\partial(\cdot)/\partial\mathbf{n}$ denotes (outward) normal derivative.

In a similar manner, for every $\varepsilon > 0$ there is a lemniscate σ as in Theorem 4.1 such that for each P_j we have

$$\frac{\partial g(K_0,P_j)}{\partial \mathbf{n}} \le \frac{\partial g(L,P_j)}{\partial \mathbf{n}} + \varepsilon. \tag{9}$$

Note that

$$\frac{\partial g(K,P_j)}{\partial \mathbf{n}} \le \frac{\partial g(L,P_j)}{\partial \mathbf{n}} \le \frac{\partial g(K_0,P_j)}{\partial \mathbf{n}},$$

because $K_0 \subset L \subset K$.

Now $\partial g(K,P_j)/\partial\mathbf{n}$ gives 2π-times the density of the equilibrium measure at P_j with respect to arc length on Γ^*

$$\omega_{\Gamma^*}(P_j) = \frac{1}{2\pi}\frac{\partial g(K,P_j)}{\partial \mathbf{n}};$$

hence, we can reformulate (with a different ε) (8) as

$$\omega_\sigma(P_j) \le \omega_{\Gamma^*}(P_j) + \varepsilon,$$

and similarly, (9) can be reformulated as

$$\omega_{\gamma^*}(P_j) \le \omega_\sigma(P_j) + \varepsilon.$$

5 A Critical Point in the Method

The splitting of the set appears in the step (b) when we go from the model case to its inverse image under a polynomial mapping. That is a big advance, since from then on one works with several components, and they may be sufficiently general to imitate an arbitrary set. However, there is a huge price to pay, namely in the transfer, say, from $[-1,1]$ to $E = T_N^{-1}[-1,1]$, the result is transferred into a very special statement on E, e.g. in the Bernstein inequality (1) in this step we got the extension (2) of the Bernstein inequality on E, but only for very special polynomials, namely of the form $Q_k(T_N)$. But our aim is to prove (in this case) the full analogue for ALL polynomials. Besides, in $Q_k(T_N)$ the polynomial T_N is not known, and when we approximate an arbitrary set of finitely many intervals by $T_N^{-1}[-1,1]$, it is typically of very high degree.

The idea of how to get rid of the special properties is the following. As we have already observed, $T_N^{-1}[-1,1]$ consists of N subintervals $E_i = E_{N,i}$, and we denote by $T_{N,i}^{-1}$ that branch of T_N^{-1} that maps $[-1,1]$ into E_i. Let P_n be an arbitrary polynomial of degree n, and consider the sum

$$S(x) = \sum_{i=1}^{N} P_n(T_{N,i}^{-1}(T_N(x))). \tag{10}$$

We claim that this is a polynomial of $T_N(x)$ of degree at most n/N, i.e. $S(x) = S_n(T_N(x))$ for some polynomial S_n of degree at most n/N. To this end let $x_i = T_{N,i}^{-1}(T_N(x))$, $i = 1,\ldots,N$. Then

$$S(x) = S(x_1,\ldots,x_N) = \sum_{i=1}^{N} P_n(x_i)$$

is a symmetric polynomial of the variables $x_1,\ldots,x_N$, and hence it is a polynomial of the elementary symmetric polynomials

$$S_j(x_1,\ldots,x_N) = \sum_{1\le k_1<k_2<\cdots<k_j\le N} x_{k_1}x_{k_2}\cdots x_{k_j}, \qquad 1\le j\le N.$$

However, $x_1,x_2,\ldots,x_N$ are the roots in t of the polynomial equation $T_N(t) = T_N(x)$, and so if $T_N(x) = d_N x^N + \cdots + d_0$, then it follows that

$$S_j(x_1,\ldots,x_N) = (-1)^j d_{N-j}/d_N$$

if $1 \le j < N$, while

$$S_N(x_1,\ldots,x_N) = (-1)^N(d_0 - T_N(x))/d_N,$$

from which the claim that S is a polynomial of $T_N(x)$ follows. On comparing the degree of the homogeneous parts of these polynomials, we can see that the degree of

$$S_n(u) := S(T_{N,1}^{-1}(u))$$

is at most $\deg(P_n)/N \le n/N$ in u.

There is a slight problem, namely if $x \in E_{N,i_0}$, then the sum $S(x)$ contains not only $P_n(x)$, but also the values of P_n at the conjugate points $x_i = T_{N,i}^{-1}(T_N(x))$, so $S(x)$ does not really behave like $P_n(x)$. But that is easy to correct, namely we do not form S from P_n, but rather from a P_n^*, which behaves like P_n around x and is small at conjugate points. To illustrate this crucial step, we complete the proof of (5) in the transform of the Bernstein inequality.

Let $\varepsilon > 0$ be arbitrary. Then, by Theorem 3.1, there are polynomial inverse image sets E^* consisting of the same number of intervals as E such that the corresponding endpoints of the subintervals of E and E^* are as close as we wish. Therefore, we can choose $E^* \subset \mathrm{Int}(E)$ so that

$$\omega_{E^*}(x_0) \leq (1+\varepsilon)\omega_E(x_0) \tag{11}$$

is satisfied. Let $E^* = T_N^{-1}[-1,1]$, and let $E_i^* = T_{N,i}^{-1}[-1,1]$, $i = 1,\dots,N$ be the N inverse image intervals of $[-1,1]$ under the N branches of T_N^{-1}. Since any translate of E^* is the polynomial inverse image of $[-1,1]$ via a translate of T_N, we can assume without loss of generality that x_0 is not an endpoint of any of the intervals E_i^*, i.e. x_0 is lying in the interior of $E_{i_0}^*$ for some i_0.

Let P_n be an arbitrary polynomial of degree n, and consider the polynomial

$$P_n^*(x) = (1-\alpha(x-x_0)^2)^{[\sqrt{n}]}P_n(x), \tag{12}$$

where $\alpha > 0$ is fixed so that $1-\alpha(x-x_0)^2 > 0$ on E. Clearly, P_n^* has degree at most $n+2\sqrt{n}$, $\|P_n^*\|_E \leq \|P_n\|_E$, $P_n^*(x_0) = P_n(x_0)$, $(P_n^*)'(x_0) = P_n'(x_0)$, and there is a $0<\beta<1$ such that

$$|P_n^*(x)| \leq \beta^{\sqrt{n}}\|P_n\|_E, \qquad |(P_n^*(x))'| \leq \beta^{\sqrt{n}}\|P_n\|_E \tag{13}$$

uniformly for $x \in E \setminus E_{i_0}^*$ (for the last relations just observe that the factor $1-\alpha(x-x_0)^2$ is nonnegative and strictly less than one on $E \setminus E_{i_0}^*$). For $x \in E^*$ form now

$$S(x) = \sum_{i=1}^{N} P_n^*(T_{N,i}^{-1}(T_N(x))). \tag{14}$$

As we have already observed, this is a polynomial of degree at most $(n+2\sqrt{n})/N$ of $T_N(x)$, i.e. $S(x) = S_n(T_N(x))$ for some polynomial S_n of degree at most $(n+2\sqrt{n})/N$. From the properties (13) it is also clear that

$$\|S\|_{E^*} \leq (1+N\beta^{\sqrt{n}})\|P_n\|_E, \qquad |S'(x_0) - P_n'(x_0)| \leq N\beta^{\sqrt{n}}\|P_n\|_E.$$

Now S is already of the type for which we have verified (2) above, so if we apply to S the inequality (2) at $x = x_0$, and if we use (11) and the preceding estimates we obtain (2):

$$\begin{aligned}|P_n'(x_0)| &\leq |S'(x_0)| + N\beta^{\sqrt{n}}\|P_n\|_E \\ &\leq (n+2\sqrt{n})\pi\omega_{E^*}(x_0)\|S\|_{E^*} + N\beta^{\sqrt{n}}\|P_n\|_E\end{aligned}$$

$$\leq (n+2\sqrt{n})(1+\varepsilon)\pi\omega_E(x_0)(1+N\beta^{\sqrt{n}})\|P_n\|_E + N\beta^{\sqrt{n}}\|P_n\|_E$$
$$= (1+o(1))n\pi\omega_E(x_0)\|P_n\|_E,$$

since $\varepsilon > 0$ was arbitrary.

6 The Markoff Inequality for Several Intervals

The classical Markoff inequality

$$\|P_n'\|_{[-1,1]} \leq n^2\|P_n\|_{[-1,1]} \tag{15}$$

complements Bernstein's inequality when we have to estimate the derivative of a polynomial on $[-1,1]$ close to the endpoints. What happens, if we consider more than one intervals? In [8], it was shown that if $E = [-b,-a]\cup[a,b]$, then

$$\|P_n'\|_E \leq (1+o(1))\frac{n^2 b}{b^2-a^2}\|P_n\|_E.$$

Why is $b/(b^2-a^2)$ the correct factor here? This can be answered by the transformation $x \to x^2$, but what if we have two intervals of different size, or when we have more than two intervals? With the polynomial inverse image method we proved in [38] the following extension.

Let $E = \cup_{j=1}^{l}[a_{2j-1},a_{2j}]$, $a_1 < a_2 < \cdots < a_{2l}$ consist of l intervals. When we consider the analogue of the Markoff inequality for E, actually we have to talk about one-one Markoff inequality around every endpoint of E. Let a_j be an endpoint of E, E^j part of E that lies closer to a_j than to any other endpoint. Let M_j be the best constant for which

$$\|P_n'\|_{E^j} \leq (1+o(1))M_j n^2\|P_n\|_E \tag{16}$$

holds, where $o(1)$ tends to 0 as n tends to infinity. This M_j clearly depends on what endpoint a_j we are considering. Its value is given by (see [38])

Theorem 6.1

$$M_j = 2\frac{\prod_{i=1}^{l-1}(a_j-\lambda_i)^2}{\prod_{i\neq j}|a_j-a_i|}, \tag{17}$$

where the λ_j are the numbers that appear in the equilibrium measure in (40)–(41).

Let us consider the example $E = [-b,-a]\cup[a,b]$. In this case $l=2$, $a_1 = -b$, $a_2 = -a$, $a_3 = a$, $a_4 = b$, and, by symmetry, $\lambda_1 = 0$. Hence,

$$\omega_E(t) = \frac{|t|}{\pi\sqrt{(b^2-t^2)(t^2-a^2)}},$$

$$M_1 = M_4 = \frac{2b^2}{(b-a)(b+a)(2b)} = \frac{b}{b^2-a^2}$$

$$M_2 = M_3 = \frac{2a^2}{(b-a)(b+a)(2b)} = \frac{a}{b^2-a^2}.$$

Since $M_1 = M_4 > M_2 = M_3$ we obtain that

$$\|P_n'\|_{[-b,-a]\cup[a,b]} \leq (1+o(1))n^2\frac{b}{b^2-a^2}\|P_n\|_{[-b,-a]\cup[a,b]},$$

which is the result of [8] mentioned above.

As an immediate consequence of the theorem we get the following asymptotically best possible Markoff inequality:

Corollary 6.2

$$\|P_n'\|_E \leq (1+o(1))n^2\left(\max_{1\leq j\leq 2l} M_j\right)\|P_n\|_E.$$

It is quite interesting that here the $o(1)$ term cannot be dropped. This is due to the strange fact that there are cases, where the maximum of

$$|P_n'(x)|/\|P_n\|_E$$

for all $x \in E$ and all P_n of given degree n, is attained in an inner point of E ([2]).

It seems to be a difficult problem to find on several intervals for each n the best Markoff constant for polynomials of degree at most n. The previous corollary gives the asymptotically best constant (as n tends to infinity).

7 Bernstein's Inequality on Curves

Bernstein had another inequality on the derivative of a polynomial, namely if C_1 is the unit circle, then

$$|P_n'(z)| \leq n\|P_n\|_{C_1}, \qquad z \in C_1 \tag{18}$$

for any polynomial of degree at most n. With the polynomial inverse image method in [19] we extended this to a family of C^2 Jordan curves.

Theorem 7.1 *Let E be a finite union of C^2 Jordan curves (lying exterior to one another), and ω_E the density of the equilibrium measure of E with respect to arc length. Then for any polynomial P_n of degree at most $n = 1,2,\ldots$*

$$|P_n'(z)| \leq (1+o(1))2\pi n\omega_E(z)\|P_n\|_E, \qquad z \in E. \tag{19}$$

This is sharp:

Theorem 7.2 *With the assumptions of the previous theorem for any $z_0 \in E$ there are polynomials P_n of degree at most n such that*

$$|P_n'(z_0)| > (1-o(1))2\pi n\omega_E(z_0)\|P_n\|_E.$$

for some P_n's.

We mention that the term $o(1)$ is necessary, without it the inequality is not true. Note also that, as opposed to (2), here, on the right hand side, the factor is $2\pi\omega_E(z)$ rather than $\pi\omega_E(z)$.

Corollary 7.3 *If E is a finite family of disjoint C^2 Jordan curves, then*

$$\|P_n'\|_E \le (1+o(1))n\left(2\pi \sup_{z\in E}\omega_E(z)\right)\|P_n\|_E,$$

and this is sharp for

$$\|P_n'\|_E > (1-o(1))n\left(2\pi \sup_{z\in E}\omega_E(z)\right)\|P_n\|_E$$

for some polynomials P_n, $n = 1,2,\ldots$.

8 Asymptotics for Christoffel Functions

Let μ be a finite Borel measure on the plane such that its support is compact and consists of infinitely many points. The Christoffel functions associated with μ are defined as

$$\lambda_n(\mu,z) = \inf_{P_n(z)=1}\int |P_n|^2 d\mu, \tag{20}$$

where the infimum is taken for all polynomials of degree at most n that take the value 1 at z. If $p_k(z) = p_k(\mu,z)$ denote the orthonormal polynomials with respect to μ, i.e.

$$\int p_n\overline{p_m}d\mu = \delta_{n,m},$$

then λ_n can be expressed as

$$\lambda_n^{-1}(\mu,z) = \sum_{k=0}^{n}|p_k(z)|^2.$$

In other words, $\lambda_n^{-1}(z)$ is the diagonal of the reproducing kernel

$$K_n(z,w) = \sum_{k=0}^{n} p_k(z)\overline{p_k(w)}, \tag{21}$$

which makes it an essential tool in many problems.

In the past literature, a lot of work has been devoted to Christoffel functions, e.g. the H^p theory emerged from Szegő's theorem; the density of states in statistical mechanical models of quantum physics is given by the reciprocal of the Christoffel function associated with the spectral measure (see e.g. [21]); and the recent breakthrough [16] by Lubinsky in universality connected with random matrices has also been based on them (cf. also [10,40] and particularly [31] where the importance of Christoffel functions regarding off diagonal behavior of the reproducing kernel was emphasized). See [12,14,33], and particularly [20] by Nevai and [32] by Simon for the role and various use of Christoffel functions.

In 1915 Szegő proved that if $\mathrm{d}\mu(t) = \mu'(t)\mathrm{d}t$ is an absolutely continuous measure on the unit circle (identified with $[-\pi,\pi]$) then

$$\lim_{n\to\infty} \lambda_n(z) = (1-|z|^2)\exp\left(\frac{1}{2\pi}\int_{-\pi}^{\pi}\frac{\mathrm{e}^{\mathrm{i}t}-z}{\mathrm{e}^{\mathrm{i}t}+z}\log\mu'(t)\mathrm{d}t\right), \qquad |z|<1$$

provided $\log\mu'(t)$ is integrable (otherwise the limit on the left is 0). Just to show the importance of Christoffel functions, let us mention that the $z=0$ case of this theorem immediately implies that the polynomials are dense in $L^2(\mu)$ if and only if $\int\log\mu' = -\infty$. Szegő ([36, Th. I', p. 461]) also proved that on the unit circle

$$\lim_{n\to\infty} n\lambda_n(\mu,\mathrm{e}^{\mathrm{i}\theta}) = 2\pi\mu'(\theta) \tag{22}$$

under the condition that μ is absolutely continuous and $\mu'>0$ is twice continuously differentiable. The almost everywhere result came much later, only in 1991 was it proven in [17] that (22) is true almost everywhere provided $\log\mu'$ is integrable.

All the aforestated results can be translated into theorems on $[-1,1]$, e.g.: if the support of μ is $[-1,1]$ and $\log\mu'\in L^1_{\mathrm{loc}}$, then

$$\lim_{n\to\infty} n\lambda_n(x) = \pi\sqrt{1-x^2}\mu'(x) \tag{23}$$

almost everywhere. A local result is that (23) is true on an interval I if μ is in the **Reg** class (see below), μ is absolutely continuous on I and $\log\mu'\in L^1(I)$. The measure μ is called to be in the **Reg** class (see [34, Theorem 3.2.3]) if the $L^2(\mu)$ and $L^\infty(\mu)$ norms of polynomials are asymptotically the same in n-th root sense:

$$\limsup_{n\to\infty}\frac{\|Q_n\|_{L^\infty(\mu)}^{1/n}}{\|Q_n\|_{L^2(\mu)}^{1/n}} \le 1. \tag{24}$$

An equivalent formulation is: $\lambda_n(\mu,z)^{1/n}\to 1$ uniformly on the support of μ. $\mu\in$ **Reg** is a fairly weak condition on μ; see [34] for general regularity criteria and different equivalent formulations of $\mu\in$ **Reg**. For example, $\mu'>0$ a.e. implies that $\mu\in$ **Reg**.

When the support is not $[-1,1]$, things change. Indeed, let $K=\mathrm{supp}(\mu)\subset \mathsf{R}$ be a compact set (of positive logarithmic capacity), and let ν_K denote the equilibrium measure of K. The polynomial inverse image method gives (see [37,40])

Theorem 8.1 *Let $K=\mathrm{supp}(\mu)$ be a compact set of positive capacity and suppose that $\mu\in$ **Reg** and $\log\mu'\in L^1(I)$ for some interval $I\subset K$. Then almost everywhere on I*

$$\lim_{n\to\infty} n\lambda_n(\mu,x) = \frac{\mathrm{d}\mu(x)}{d\nu_K}, \tag{25}$$

where, on the right-hand side, the expression is the Radon–Nikodym derivative of μ with respect to the equilibrium measure μ_K.

Of course, when $K = [-1,1]$, then (23) and (25) are the same.

In a similar vein, but with totally different proof (based now on the model case C_1) we have (see [42]):

Theorem 8.2 *Let $K = \mathrm{supp}(\mu)$ be a finite family of C^2 Jordan curves and suppose that $\mu \in$ **Reg** and $\log \mu' \in L^1(I)$ for some arc $I \subset K$. Then almost everywhere on I*

$$\lim_{n\to\infty} n\lambda_n(\mu,x) = \frac{\mathrm{d}\mu(x)}{d\nu_K}, \tag{26}$$

Here, $L^1(I)$ is meant with respect to arc measure on K.

We note that (26) holds at every point where the measure μ has continuous density with respect to arc length (see [42]). In this case, the support of μ can be much more general, and the result is about the asymptotics of the Christoffel function on an outer boundary arc of the support.

One can also allow a combination of Jordan arcs (homeomorphic images of $[-1,1]$) and curves for the support of μ. However, this extension does not come directly from the polynomial inverse image method, for there is a huge difference between smooth Jordan arcs and Jordan curves: the interior of Jordan curves (or family of curves) can be exhausted by lemniscates, and once an arc is in the set, this is no longer true.

Orthogonal polynomials with respect to area measures go back to Carleman [9] who gave strong asymptotics for them in the case of a Jordan domain with analytic boundary curve. For less smooth domains or for regions consisting of several components, the situation is more difficult. The polynomial inverse image method in [42] gave the asymptotics for Christoffel functions with respect to area-like measures:

Theorem 8.3 *Suppose that K is a compact set bounded by a finite number of C^2 Jordan curves and μ is a measure on K of the form $\mathrm{d}\mu = WdA$ with some continuous W such that*

$$\mathrm{cap}\big(\{z|W(z)>0\}\cap \mathrm{Int}(K)\big) = \mathrm{cap}(K).$$

Then for $z_0 \in \partial K$

$$\lim_{n\to\infty} n^2\lambda_n(\mu,z_0) = \frac{W(z_0)}{2\pi\omega_K(z_0)^2} \tag{27}$$

where ω_K is the density of the equilibrium measure with respect to arc length on ∂K (note that the equilibrium measure is supported on ∂K).

9 Lubinsky's Universality on General Sets

Let μ be a measure with compact support on the real line, and for simplicity let us assume that $\mathrm{d}\mu(x) = w(x)\mathrm{d}x$ with an L^1 function w. A form of universality in random matrix theory/statistical quantum mechanics can be expressed via orthogonal polynomials in the form (recall that K_n are the reproducing kernels from (21))

$$\lim_{n\to\infty} \frac{K_n\left(x+\frac{a}{w(x)K_n(x,x)}, x+\frac{b}{w(x)K_n(x,x)}\right)}{K_n(x,x)} = \frac{\sin\pi(a-b)}{\pi(a-b)}. \tag{28}$$

(The term "universality" comes from the fact that the right-hand side is independent of the original weight w as well as of the place x). There has been a lot of papers devoted to universality both in the mathematics and in the physics literature; the very first instance is due to E. Wigner concerning the Hermite weight. Previous approaches used rather restrictive assumptions, see [16] for references. In [16] Lubinsky recently gave a stunningly simple approach that proves (28) for measures in the **Reg** class for which $\text{supp}(\mu) = [-1,1]$ and w is continuous and positive on an interval I (then (28) holds on I uniformly in $|a|,|b| \le A$, for any $A > 0$). In [40], again with the polynomial inverse image method, universality was extended to regular measures with arbitrary support (the same result was proved by Simon in [31] using so called Jost solutions to recurrences):

Theorem 9.1 (28) *holds uniformly in* $|a|,|b| \le A$, $A > 0$ *at every continuity point of the weight* w (*lying inside the support*) *provided* $\mathrm{d}\mu(x) = w(x)\mathrm{d}x$ *is in the* **Reg** *class.*

When the support is $[-1,1]$, the almost every version of (28) under the local Szegő condition $\log w \in L^1(I)$ was proved in [10], which just pulls over to the general case (the support arbitrary) via the polynomial inverse image method (see [40]).

Theorem 9.2 (28) *holds at almost every point of an interval* I *provided* $\mathrm{d}\mu(x) = w(x)\mathrm{d}x$ *is in the* **Reg** *class and* $\log w \in L^1(I)$.

10 Fine Zero Spacing of Orthogonal Polynomials

Let μ be a measure with compact support on the real line, and let $p_n = p_n(\mu,z)$ be the n-th orthonormal polynomial with respect to μ. It is well known that classical orthogonal polynomials on $[-1,1]$ have rather uniform zero spacing: if $x_{n,j} = \cos\theta_{n,j}$ are the zeros of the n-th orthogonal polynomials, then (inside $(-1,1)$) $\theta_{n,j} - \theta_{n,j+1} \sim 1/n$. In turn, this property of zeros is of fundamental importance in quadrature and Lagrange interpolation. Several hundreds of papers have been devoted to zeros of orthogonal polynomials, still the following beautiful result has only been proven a few years ago, namely when Levin and Lubinsky [15] found that Lubinsky's universality described in Sect. 9 implies very fine zero spacing:

$$\lim_{n\to\infty} (x_{n,k+1} - x_{n,k}) \frac{n}{\pi\sqrt{1-x_{n,k}^2}} = 1. \tag{29}$$

With the polynomial inverse image method this was extended in [40] to arbitrary support (see also [31]):

Theorem 10.1 *If* $K = \mathrm{supp}(\mu) \subset \mathsf{R}$, $\mu \in \mathbf{Reg}$ *and* μ' *is continuous and positive about x, then*

$$\lim_{n\to\infty} n(x_{n,k+1} - x_{n,k})\omega_K(x) = 1, \qquad |x_{n,k} - x| \le A/n \tag{30}$$

where ω_K *is the density of the equilibrium measure of the support K.*

Furthermore, this holds locally a.e. under the local Szegő condition $\log \mu' \in L^1$:

Theorem 10.2 *If* $K = \mathrm{supp}(\mu) \subset \mathsf{R}$, $\mu \in \mathbf{Reg}$ *and* $\log \mu' \in L^1(I)$ *for some interval I, then (30) is true a.e. in I in the sense that for almost every* $x \in I$ *and for every* $A > 0$ *we have (30) for* $|x_{n,k} - x| \le A/n$.

11 Polynomial Approximation on Compact Subsets of the Real Line

The approximation of the $|x|$ function on $[-1,1]$ by polynomials is a key to many problems in approximation theory. Let $E_n(f,F)$ denote the error of best approximation to f on F by polynomials of degree at most n. Bernstein [3] proved in 1914, that the limit

$$\lim_{n\to\infty} nE_n(|x|,[-1,1]) = \sigma \tag{31}$$

exists, it is finite and positive. This is a rather difficult result (with a proof over 50 pages). For σ he showed $0.278 < \sigma < 0.286$. The exact value of σ is still unknown. Bernstein returned to the same problem some 35 years later in [4,5], and he established that for $p > 0$, p not an even integer, the finite and nonzero limit

$$\lim_{n\to\infty} n^p E_n(|x|^p,[-1,1]) = \sigma_p \tag{32}$$

exists, furthermore that for $x_0 \in (-1,1)$

$$\lim_{n\to\infty} n^p E_n(|x-x_0|^p,[-1,1]) = (1-x_0^2)^{p/2}\sigma_p \tag{33}$$

holds true, where σ_p is the same constant as in (32).

In this section, we discuss the problem that arises for more general sets. This problem was considered by Vasiliev in [43]. His approach is as follows. Let

$$F = [-1,1] \setminus \cup_{i=1}^{\infty}(\alpha_i,\beta_i),$$

and form the sets

$$F_m = [-1,1] \setminus \cup_{i=1}^{m-1}(\alpha_i,\beta_i).$$

F_m consists of m intervals

$$F_m = \cup_{j=1}^m [a_j, b_j]$$

$a_1 < b_1 < a_2 < b_2 \cdots b_{m-1} < a_m < b_m$, and for it define

$$h_{F_m}(x) = \frac{\prod_{j=1}^{m-1} |x - \lambda_j|}{\sqrt{\prod_{j=1}^{m} |x - a_j||x - b_j|}},$$

where λ_j are chosen so that

$$\int_{b_k}^{a_{k+1}} \frac{\prod_{j=1}^{m-1} (t - \lambda_j)}{\sqrt{\prod_{j=1}^{m} |t - a_j||t - b_j|}} \mathrm{d}t = 0$$

for all $k = 1, \ldots, m-1$. Set

$$h_F(x) = \lim_{m \to \infty} h_{F_m}(x) = \sup_m h_{F_m}(x),$$

where it can be shown that the limit exists (but it is not necessarily finite).

Now with these notations Vasiliev claims the following two results:

$$\lim_{n \to \infty} n^p E_n(|x - x_0|^p, F) = h_F(x_0)^{-p} \sigma_p, \tag{34}$$

$$\lim_{n \to \infty} n^p E_n(|x - x_0|^p, F) > 0 \Longleftrightarrow \int_0^1 \frac{\text{meas}\{[x_0 - t, x_0 + t] \setminus F\}^2}{t^3} \mathrm{d}t < \infty. \tag{35}$$

This second claim seems to contradict the fact (see e.g. [39, Corollary 10.4]) that there are (Cantor type) sets of measure zero for which $E_n(|x - x_0|^p, F) \geq cn^{-p}$ with some $c > 0$ (for a set F of zero measure the integral is clearly infinite). Vasiliev's paper [43] is 166 pages long, and it is dedicated solely to the proof of (34) and (35), so it is difficult to say what might be wrong in the proof. We do not know if the full (34) is correct, but we gave in [39, Theorem 10.5] a few pages proof, based on polynomial inverse images, that shows its validity provided x_0 lies in the interior of E. In fact, in this case we have transferred the original Bernstein theorem (32) into Vasiliev's theorem.

Taking into account the form (40) of the equilibrium measure for several intervals, we see that Vasiliev's function is just $h_F(x) = \pi \omega_F(x)$ if F consists of a finite number of intervals (and also if F is arbitrary compact, but x is in its interior). Hence, (34) for $x_0 \in \text{Int}(F)$ takes the following form.

Theorem 11.1 (R. K. Vasiliev) *Let $F \subseteq \mathrm{R}$ be compact and let x_0 be a point in the interior of F. Then*

$$\lim_{n \to \infty} n^p E_n(|x - x_0|^p, F) = (\pi \omega_F(x_0))^{-p} \sigma_p, \tag{36}$$

where σ_p is the constant from Bernstein's theorem (32).

For example, if $F=[-1,1]$, then

$$\pi\omega_{[-1,1]}(x)=\frac{1}{\sqrt{1-x^2}},$$

and in this special case we recapture Bernstein's result (33).

Here again, Theorem 11.1 can be obtained from Bernstein's theorems (32) via polynomial mappings and approximation.

12 Appendix: Basic Notions from Logarithmic Potential Theory

For a general reference to logarithmic potential theory see [29].

Let $E\subset \mathbf{C}$ be compact. Except for pathological cases, there is a unique probability (Borel) measure μ_E on E, called the equilibrium measure of E, that minimizes the energy integral

$$\int\int \log\frac{1}{|z-t|}\mathrm{d}\mu(z)\mathrm{d}\mu(t). \tag{37}$$

μ_E certainly exists if E has non-empty interior. One should think of μ_E as the distribution of a unit charge placed on the conductor E (in this case Coulomb's law takes the form that the repelling force between charged particles is proportional with the reciprocal of the distance).

The logarithmic capacity of E is $\mathrm{cap}(E)=\exp(-V)$, where V is the minimum of the energies (37) above. The Green's function of the unbounded component Ω of the complement $\mathbf{C}\setminus E$ with pole at infinity is denoted by $g_\Omega(z,\infty)$, and it has the form

$$g_\Omega(z,\infty)=\int\log\frac{1}{|z-t|}\mathrm{d}\mu_E(t)+\log\mathrm{cap}(E). \tag{38}$$

When $E\subset\mathbf{R}$ then we shall denote by $\omega_E(t)$ the density of μ_E with respect to Lebesgue measure wherever it exists. It certainly exists in the interior of E. For example

$$\omega_{[-1,1]}(t)=\frac{1}{\pi\sqrt{1-t^2}},\qquad t\in[-1,1]$$

is just the well known Chebyshev distribution.

If $E=T_N^{-1}[-1,1]$, $E=\cup_{i=1}^N I_i$ in such a way that T_N maps each of the intervals I_i onto $[-1,1]$ in a 1-to-1 way, then (see [13, 29])

$$\mu_E(A)=\frac{1}{N}\sum_{i=1}^{N}\mu_{[-1,1]}(T_N(A\cap I_i)),$$

which gives

$$\omega_E(t)=\frac{|T_N'(t)|}{\pi N\sqrt{1-T_N(t)^2}},\qquad t\in E. \tag{39}$$

We also know a rather explicit form for ω_E when $E = \cup_1^l [a_j, b_j]$ is a set consisting of finitely many intervals (see e.g. [38]):

$$\omega_E(x) = \frac{\prod_{j=1}^{l-1} |x-\lambda_j|}{\pi \sqrt{\prod_{j=1}^{l} |x-a_j||x-b_j|}}, \tag{40}$$

where λ_j are chosen so that

$$\int_{b_k}^{a_{k+1}} \frac{\prod_{j=1}^{l-1} (t-\lambda_j)}{\sqrt{\prod_{j=1}^{l} |t-a_j||t-b_j|}} \mathrm{d}t = 0 \tag{41}$$

for all $k = 1, \ldots, l-1$. It can be easily shown that these λ_j's are uniquely determined and there is one λ_j on any contiguous interval (b_k, a_{k+1}).

Acknowledgements This research was supported by NSF DMS0968530.

References

1. M. Baran, Complex equilibrium measure and Bernstein type theorems for compact sets in R^n, *Proc. Amer. Math. Soc.*, **123**(1995), 485–494.
2. D. Benkő and V. Totik, Sets with interior extremal points for the Markoff inequality, *J. Approx. Theory* **110** (2001), 261–265.
3. S. N. Bernstein, Sur la meilleure approximation de $|x|$ par des polynomes des degrés donnés, *Acta Math.* (Scandinavian) **37** (1914), 1–57.
4. S. N. Bernstein, On the best approximation of $|x|^p$ by means of polynomials of extremely high degree, *Izv. Akad. Nauk SSSR,* Ser. Mat. **2** (1938), 160–180. Reprinted in S. N. Bernstein "Collected Works," Vol. 2, pp. 262–272. Izdat. Nauk SSSR, Moscow, 1954. [In Russian]
5. S. N. Bernstein, On the best approximation of $|x-c|^p$, *Dokl. Akad. Nauk SSSR* **18** (1938), 379– 384. Reprinted in S. N. Bernstein "Collected Works," Vol. 2, pp. 273–260. Izdat. Nauk SSSR, Moscow, 1954. [In Russian]
6. S. N. Bernstein, Extremal properties of polynomials and best approximation of functions of a real variable, I., ONTI, 1–203. [In Russian]
7. A. B. Bogatyrev, Effective computation of Chebyshev polynomials for several intervals, *Math. USSR Sb.*, **190** (1999), 1571–1605.
8. P. Borwein, Markoff's and Bernstein inequalities on disjoint intervals, *Canad. J. Math.* **33** (1981), 201–209.
9. T. Carleman, Über die Approximation analytischer Funktionen durch lineare Aggregate von vorgegebenen Potenzen, *Ark. Mat. Astr. Fys.*, **17**(1923), 215–244.
10. M. Findley, Universality for locally Szegő measures, *J. Approx. Theory.*, **155**, 136–154.
11. M. Findley, Fine asymptotics for Christoffel functions for general measures, *Trans. Amer. Math. Soc.* **362** (2010), 2053–2087.
12. G. Freud, *Orthogonal Polynomials*, Pergamon Press, Oxford, 1971.
13. J. S. Geronimo and W. Van Assche, Orthogonal polynomials on several intervals via a polynomial mapping, *Trans. Amer. Math. Soc.* **308** (1988), 559–581.
14. U. Grenander and G. Szegő, *Toeplitz Forms and Their Applications*, University of California Press, Berkeley and Los Angeles, 1958.

15. A. L. Levin, and D. S. Lubinsky, Applications of universality limits to zeros and reproducing kernels of orthogonal polynomials, *J. Approx. Theory.* **150** (2008), 69–95.
16. D. S. Lubinsky, A new approach to universality limits involving orthogonal polynomials, *Annals of Math.* **170** (2009), 915–939.
17. A. Máté, P. Nevai and V. Totik, Szegő's extremum problem on the unit circle, *Annals of Math.*, **134**(1991), 433–453.
18. H. P. McKean and P. van Mooerbeke, Hill and Toda curves, *Comm. Pure Appl. Math.*, **33**(1980), 23–42.
19. B. Nagy and V. Totik, Sharpening of Hilbert's lemniscate theorem, *J. D'Analyse Math.*, **96**(2005), 191–223.
20. P. Nevai, Géza Freud, orthogonal polynomials and Christoffel functions. A case study, *J. Approx. Theory*, **48**(1986), 1–167.
21. L. A. Pastur, Spectral and probabilistic aspects of matrix models. *Algebraic and geometric methods in mathematical physics* (Kaciveli, 1993), 207–242, *Math. Phys. Stud.*, **19**, Kluwer Acad. Publ., Dordrecht, 1996.
22. F. Peherstorfer, Deformation of minimizing polynomials and approximation of several intervals by an inverse polynomial mapping, *J. Approx. Theory* **111** (2001), 180–195.
23. F. Peherstorfer, On Bernstein–Szegő orthogonal polynomials on several intervals, I. *SIAM J. Math. Anal.* **21** (1990), 461–482.
24. F. Peherstorfer, On Bernstein–Szegő orthogonal polynomials on several intervals, II. *J. Approx. Theory* **64** (1991), 123–161.
25. F. Peherstorfer, Orthogonal and extremal polynomials on several intervals, *J. Comp. Applied Math.* **48** (1993), 187– 205.
26. F. Peherstorfer, Elliptic orthogonal and extremal polynomials, *J. London Math. Soc.* **70** (1995), 605– 624.
27. F. Peherstorfer and K. Schiefermayr, Theoretical and numerical description of extremal polynomials on several intervals I, *Acta Math. Hungar* **83** (1999), 27–58.
28. F. Peherstorfer and R. Steinbauer, On polynomials orthogonal on several intervals, *Ann. Num. Math.* **2** (1995), 353–370.
29. T. Ransford, *Potential Theory in the Complex Plane,* Cambridge University Press, Cambridge, 1995.
30. R. M. Robinson, Conjugate algebraic integers in real point sets, *Math. Z.*, **84**(1964), 415–427.
31. B. Simon, Two extensions of Lubinsky's universality theorem, *J. D'Analyse Math.*, **105**(2008), 345–362.
32. B. Simon, The Christoffel-Darboux kernel, "Perspectives in PDE, Harmonic Analysis and Applications" in honor of V.G. Maz'ya's 70th birthday, to be published in Proceedings of Symposia in Pure Mathematics, **79**(2008), 295–335.
33. B. Simon, Weak convergence of CD kernels and applications, *Duke Math. J.* **146**(2009), 305–330.
34. H. Stahl and V. Totik, *General Orthogonal Polynomials*, Encyclopedia of Mathematics and its Applications, **43**, Cambridge University Press, Cambridge, 1992.
35. G. Szegő, Über einen Satz des Herrn Serge Bernstein, *Schriften Königsberger Gelehrten Ges. Naturwiss. Kl.*, **5** (1928/29), 59–70.
36. G. Szegő, *Collected Papers*, ed. R. Askey, Birkhaüser, Boston–Basel–Stuttgart, 1982.
37. V. Totik, Asymptotics for Christoffel functions for general measures on the real line, *J. D'Analyise Math.*, **81** (2000), 283–303.
38. V. Totik, Polynomial inverse images and polynomial inequalities, *Acta Math.*, **187**(2001), 139–160.
39. V. Totik, *Metric properties of harmonic measures*, Memoirs of the American Mathematical Society, **184**, number 867, 2006
40. V. Totik, Universality and fine zero spacing on general sets, *Arkiv för Math.*, **47**(2009), 361–391.
41. V. Totik, Chebyshev constants and the inheritance problem, *J. Approx. Theory*, **160**(2009), 187–201.

42. V. Totik, Christoffel functions on curves and domains, *Trans. Amer. Math. Soc.*, **362**(2010), 2053–2087.
43. R. K. Vasiliev, "Chebyshev Polynomials and Approximation Theory on Compact Subsets of the Real Axis," Saratov University Publishing House, 1998.

On Approximation of Periodic Analytic Functions by Linear Combinations of Complex Exponents

Vesselin Vatchev

Abstract For a 2π-periodic function f, analytic on $I = [0, 2\pi]$, we solve the minimization problem

$$E_n(f) = \min_{c_j \in R} \|f + c_1 f' + \cdots + c_n f^{(n)}\|^2_{L_2(I)},$$

and establish the convergence $\lim_{n\to\infty} E_n(f) = 0$. In case of even n i.e. $2n$ all of the zeros, $l_j, j = 1, \ldots, 2n$, of the corresponding characteristic polynomial $1 + c_1\lambda + \cdots + c_{2n}\lambda^{2n}$ are purely imaginary, $l_{-j} = -l_j$, and we prove the estimate

$$\max_{t \in I} |f(t) - \sum_{j=-n}^{n} b_j \mathrm{e}^{\mathrm{i}l_j t}| \leq \left(\frac{2\pi}{l_1^2} E_n(f)\right)^{1/2},$$

where l_1 has the smallest absolute value among all of the l's.

1 Introduction

The approximation of functions by linear combinations of exponents of the form $\sum d_j \mathrm{e}^{\lambda_j t}$ with complex λ's is well studied. The most popular and used case is the Fourier approximation when λ's are equally spaced on the imaginary axes. It is well known, see [3], that if f and its derivatives to order N, included, are 2π-periodic and continuous then the choice of coefficients $d_0^* = \frac{1}{\pi}\int_0^{2\pi} f(t)\,\mathrm{d}t$, $d_k^* = \frac{1}{2\pi}\int_0^{2\pi} f(t)\mathrm{e}^{-ikt}\,\mathrm{d}t$, $k = \pm 1, \ldots, \pm n$, where $i^2 = -1$, provides convergent

Vesselin Vatchev
University of Texas at Brownsville, 80 Fort Brown, Brownsville, TX 78520, USA
e-mail: vesselin.vatchev@utb.edu

M. Neamtu and L. Schumaker (eds.), *Approximation Theory XIII: San Antonio 2010*, Springer Proceedings in Mathematics 13, DOI 10.1007/978-1-4614-0772-0_23,

approximation of order n^{-N} in the Hilbert space $L_2(I)$ with the norm $\|f\|_2 = \left(\int_0^{2\pi} |f(t)|^2\, \mathrm{d}t\right)^{1/2}$ and in the space of bounded functions $L_\infty(I)$ with the norm $\|f\|_\infty = \sup_{t\in I} |f(t)|$. The following estimate holds true, for details see [3],

$$\|f(t) - \sum_{k=-n}^{n} d_k^* \mathrm{e}^{\mathrm{i}kt}\|_\infty < C\frac{\|f^{(N)}(t)\|_2}{n^N}. \tag{1}$$

The error estimate (1) depends on the norm of $f^{(N)}$ and the number of used exponents. The 'modes' e^{ikt} are the spanning set of the null space of the operator $F = D\prod_{j=1}^{n}(D^2 + j^2 Id)$, where $Df = f'$ and Id is the identity operator. The 'frequencies' $0, \pm 1, \ldots, \pm n$ are the zeros of the corresponding characteristic polynomial $P(\lambda) = \lambda \prod_{j=1}^{n}(\lambda^2 + j^2)$. The operator F does not depend on the particular function f although the error estimate (1) depends on it. In the current paper we consider optimization over a class of operators F with all of their λ's being purely imaginary. This particular distribution allows us to use results from the theory of orthogonal polynomials on the real line.

For an analytic 2π-periodic function f we solve the problem

$$E_n(f) = \min_{c_j \in R} \|f + c_1 f' + \cdots + c_n f^{(n)}\|_2^2. \tag{2}$$

Assuming that (2) has a solution $\{c_j^*\}_{j=1}^n$ we let $L_n(f) = f + \sum_{k=1}^{n} c_k^* f^{(k)}$. The operator L_n has n-dimensional null space spanned by the functions $\mathrm{e}^{\lambda_k t}, k = 1, \ldots, n$ where λ's are the zeros of the polynomial with real coefficients $P_n(\lambda) = 1 + \sum_{k=1}^{n} c_k^* \lambda^k$.

In [5] and [6], the problem (2) for $n = 2, 4$ and decompositions of the function f with respect to the corresponding null spaces is considered. The results in the current paper are continuation of those results.

In Sect. 2, we consider general solutions of the minimization problem (2) for 2π-periodic functions and establish a rate of convergence for analytic functions, when $n \to \infty$. In Sect. 3, we derive a three-term recurrence formula for special cases of Christoffel polynomials and relate them to the λ's. In Sect. 4, we obtain explicit formulas and error estimates for the approximation of a function f by the corresponding exponential functions generated by $L_n(f)$.

2 The Minimization Problem for Periodic Functions

We begin with the questions of existence of solutions of (2) and the convergence of $E_n(f)$ when $n \to \infty$. Since E_n contains no higher than the n-th derivative, when we deal with a fixed n we can consider f only n times continuously differentiable on the interval $I = [0, 2\pi]$, and all $f, f', \ldots, f^{(n)}$ 2π-periodic and linearly independent on I. It is clear that $E_{n+1}(f) \le E_n(f) \le \|f\|_2^2$.

Let Gram$(g_1,\dots,g_m)$ denote the determinant of the Gram matrix of the functions $g_1,\dots,g_m$. Then the following result is an exercise from the theory of the least square methods.

Proposition 1. *If f is n times differentiable function then*

$$\min_{c_j\in R}\|f+c_1f'+\dots+c_nf^{(n)}\|_2^2=\frac{\text{Gram}(f,f',\dots,f^{(n)})}{\text{Gram}(f',\dots,f^{(n)})}\le\|f\|_2^2\le\|f\|_\infty^2|I|,$$

for a unique set of real c's.

Since f is 2π-periodic analytic function, it follows that f can be represented by the Fourier series $f(t)=\sum_{j=-\infty}^{\infty}d_j\mathrm{e}^{\mathrm{i}jt}$, where for the complex coefficients d_j we have that $d_{-j}=\bar{d}_j$, $\bar{d}$ is the complex conjugated of d. It is well known, see [3], that for analytic periodic function f the sequence of its Fourier coefficients $|d_j|$ exponentially converges to zero as $|j|\to\infty$. WLOG we can assume that $d_0=\frac{1}{\pi}\int_0^{2\pi}f(t)\,\mathrm{d}t=0$, if otherwise we consider the function $f=f-d_0$. Next, we interpret the minimization problem (2) as a problem of finding a certain discrete orthogonal polynomial.

Theorem 1. *For a 2π-periodic analytic function f on I and even n the estimate $E_n(f)\le C\frac{4^n}{n!^2}$ holds true with a real constant C that depends only on f. For the optimal selection of c's all of the zeros of the corresponding characteristic polynomial $P_n(\lambda)$ are purely imaginary.*

Proof. Let

$$Q_n(\lambda)=1+\sum_{k=1}^{n/2}(-1)^kc_{2k}\lambda^{2k},$$

$$R_n(\lambda)=\sum_{k=1}^{n/2}(-1)^kc_{2k+1}\lambda^{2k+1},$$

then by using the Fourier decomposition of the derivatives, i. e. $f^{(k)}(t)=\sum_{j=-\infty}^{\infty}(ij)^kd_j\mathrm{e}^{\mathrm{i}jt}$, from the Parseval's identity we get that

$$\begin{aligned}\|f+c_1f'+\dots+c_nf^{(n)}\|_2^2&=\sum_{j=-\infty}^{\infty}|d_j|^2|P_n(ij)|^2\\&=\sum_{j=-\infty}^{\infty}|d_j|^2\left(|Q_n(j)|^2+|R_n(j)|^2\right).\end{aligned}$$

It is clear that the minimum is attained for $R_n\equiv0$, i.e. $c_{2k+1}=0$ and Q_n being the n-th discrete Christoffel polynomial (see Sect. 3 for details) with a weight $(j,|d_j|^2)$. Since the weight is even, i.e. $|d_{-j}|=|d_j|$, it follows that $Q_n(\lambda)=\frac{T'_{n/2+1}(\lambda)}{T'_{n/2+1}(0)\lambda}$, where T_n is the n-th orthogonal polynomial for the discrete weight function $(j,|d_j|^2)$. Since Q_n has only real zeros, the relation between Q_n and P_n provides that P_n is an even algebraic polynomial with purely imaginary zeros. From the fact that f is periodic and analytic it follows that the sequence $|d_j|$ is an even and exponentially decreasing when $|j| \to \infty$. Furthermore, since $d_0 = 0$, and $d_j \le Ce^{-rj}, r > 0$ we get the estimate

$$E_n(f) \le \sum_{j=-\infty}^{\infty} \frac{|d_j|^2}{((n/2)!)^2} \prod_{k=1}^{n/2} (j^2 - k^2) \le C \frac{4^n}{(n!)^2} \sum_{|j|>n/2} e^{-jr} j^{2n} \le C \frac{4^n}{(n!)^2}.$$

□

In that way, we established that if $f, f', \ldots, f^{(n)}$ are linearly independent there is a unique solution c^* of the problem (2) and E_n approaches 0 as $n \to \infty$. To the end of the paper we refer to the solution as c only. For the characterization of the solution of the problem (2) we need to consider some properties of orthogonal polynomials on the real line.

3 Related Properties of Orthogonal Polynomials

Let $w(x) > 0$ be a weight function with discrete or continuous, finite or infinite, support S on the real line. For a natural n the nth orthogonal polynomial, P_n, associated with $w(x)$ is defined as the solution of the following problem

$$\min_{\tilde{P}_n(x) = x^n + r_{n-1}(x)} \int_S w(x) \tilde{P}_n^2(x)\, \mathrm{d}x, \tag{3}$$

where r_{n-1} is an algebraic polynomial of degree less than n. In the case when S is a discrete set, the integral is replaced by a sum and the corresponding polynomials are known as discrete orthogonal polynomials. The orthogonal polynomials (3) exist and are uniquely defined for any n. The n-th *orthonormal* polynomial with leading coefficient $\gamma_n > 0$ is defined as $p_n(x) = \tilde{P}_n(x)/\|\tilde{P}_n\|_2 = \gamma_n x^n + r_{n-1}(x)$. Furthermore, p_n has exactly n simple zeros on the smallest interval containing S. We are interested only in symmetric, with respect to the origin, support set S and an even weight function $w(x)$. In that case, the orthogonal polynomials satisfy a three-term recurrence relation of the form $xP_n(x) = a_n P_{n+1}(x) + a_{n-1} P_{n-1}(x)$, where $a_n = \frac{\gamma_n}{\gamma_{n+1}} > 0$, with initialization $P_{-1}(x) = 0, P_0(x) = 1/\|w\|_2^2$, and $a_{-1} = 0$. There is an extensive literature on orthogonal polynomials, see for example [4]. An important property of any sequence of orthogonal polynomials is the Christoffel–Darboux formula

$$\sum_{k=0}^{n} p_k(t) p_k(x) = a_n \frac{p_{n+1}(x) p_n(t) - p_n(x) p_{n+1}(t)}{x - t}. \tag{4}$$

The problem $\min_{q_n(z)=1} \int_S w(x) q_n^2(x)\, \mathrm{d}x$, has a unique solution, see again [4], of the form

$$C_n(x) = \frac{\sum_{k=0}^{n} P_n(0) P_n(x)}{\sum_{k=0}^{n} P_n^2(0)},$$

with minimum equal to $1/\mu_n^2(0)$, where $\mu_n(0) = \sum_{k=0}^n P_n^2(0)$. The polynomials C_n are called Christoffel polynomials and the quantity $\mu_n(0)$ is known as one of the Christoffel numbers. By applying (4) we get that $\mu_n(0) = a_n(P'_{n+1}(0)P_n(0) - P'_n(0)P_{n+1}(0))$.

Next lemma provides a recurrence formula for the Christoffel polynomials.

Lemma 1. *For an even weight function $w(x)$ and a symmetric support S the Christoffel polynomials associated to $w(x)$ and $z = 0$ are $C_{2n+1}(x) = C_{2n}(x) = \frac{P_{2n+1}(x)}{P'_{2n+1}(0)x}$ and satisfy the three-term relation*

$$x^2 C_{2n}(x) = -k_{2n}C_{2n+2}(x) + (k_{2n} + k_{2n-1})C_{2n}(x) - k_{2n-1}C_{2n-2}(x), \tag{5}$$

with $k_{2n} = -a_{2n+1}a_{2n+2}\frac{P'_{2n+3}(0)}{P'_{2n+1}(0)}$ and $k_{2n-1} = -a_{2n}a_{2n-1}\frac{P'_{2n-1}(0)}{P'_{2n+1}(0)}$. Furthermore $k_{2n} + k_{2n-1} = a_{2n+1}^2 + a_{2n}^2$.

Proof. For any k the polynomials P_{2k+1} are odd and P_{2k} are even, hence $C_{2n+1}(x) = C_{2n}(x) = \frac{\sum_{k=0}^n P_{2k}(0)P_{2k}(x)}{\sum_{k=0}^n P_{2k}^2(0)}$ and $\mu_{2n+1}(0) = \mu_{2n}(0)$. By applying the Christoffel-Darboux formula (4) to $P_{2n}(x)$ with $t = 0$ we get

$$C_{2n}(x) = \frac{a_{2n}P_{2n}(0)}{\mu_{2n}(0)}\frac{P_{2n+1}(x)}{x} = \frac{P_{2n+1}(x)}{P'_{2n+1}(0)x}. \tag{6}$$

Since P_{2n+1} has $2n+1$ real zeros and $P_{2n+1}(0) = 0$, it follows that $P_{2n+1}(x)/x$ has $2n$ real zeros, and hence $C_{2n}(x)$ has exactly $2n$ real zeros. The recurrence relation is obtained from the recurrence for P_n and (6) in the following way

$$x^2 C_{2n}(x) = \frac{xP_{2n+1}(x)}{P'_{2n+1}(0)} = \frac{1}{P'_{2n+1}(0)}(a_{2n+1}P_{2n+2}(x) + a_{2n}P_{2n}(x))$$
$$= a_{2n+1}a_{2n+2}\frac{P_{2n+3}(x)}{P'_{2n+1}(0)x} + (a_{2n+1}^2 + a_{2n}^2)\frac{P_{2n+1}(x)}{P'_{2n+1}(0)x}) + a_{2n}a_{2n-1}\frac{P_{2n-1}(x)}{P'_{2n+1}(0)x}.$$

For any j we have that $C_{2j}(0) = 1$, and hence $k_{2n} + k_{2n-1} = a_{2n+1}^2 + a_{2n}^2$. The proof is complete. $\square$

The Christoffel polynomials are not the only ones that satisfy a recurrence of the type (5). If P_n are normalized by $P_n(0) = 1$ then they satisfy a similar relation. The matrix of recurrence coefficients is called the spectral matrix of the system P. The next remark summarizes some properties of the spectral matrices.

Remark 1. Let $\mathbf{C_{2n}}(x)$ be the vector column $(C_0(x), C_2(x), \dots, C_{2n}(x))'$ and the zeros of C_{2n} be $\pm l_j$. Then the spectral matrix, denoted by CK_n, for the system of polynomials C is defined to be the tridiagonal matrix

$$\begin{pmatrix} -(k_{2n-2}+k_{2n-1}) & k_{2n-1} & 0 & & & \\ k_{2n-4} & -(k_{2n-4}+k_{2n-3}) & k_{2n-3} & & & \\ & & \ddots & & & \\ & & k_4 & -(k_4+k_3) & k_3 & \\ & & & k_2 & -(k_2+k_1) & k_1 \\ & & & 0 & k_0 & -k_0 \end{pmatrix}.$$

From the general theory and Lemma 1 it follows that $\lambda_j = l_j^2 > 0, j = 1,\dots,n$ and $\mathbf{C_{2n}}(\lambda_j)$ are respectively the eigenvalues and the eigenvectors of CK_n.

In the next section, we consider the approximation of f by the spanning set of the null space of $L_n(f)$.

4 Approximation with Imaginary Exponents

We consider an analytic 2π-periodic function f with a Fourier series $f(t) = \sum_{j=-\infty}^{\infty} d_j e^{ijt}$, where $d_0(f) = 0$ and $d_j(f) = \frac{1}{2\pi}\int_0^{2\pi} f(t)e^{-ijt}dt$. The discrete function $w(f,k) = |d_k(f)|^2 \geq 0$ is defined for any integer k and can be considered as a discrete weight function on the real line. Furthermore, since $w(f,k)$ has an exponential decay when $|k| \to \infty$ we can define the sequence of discrete Christoffel polynomials and Christoffel numbers associated with the weight $w(f,k)$. The following result holds true

Lemma 2. *Let f be analytic and 2π-periodic on I with $d_0 = 0$, and n be fixed, then the extremal problem*

$$\min_{c_k \in R} \|f + \sum_{k=1}^{n} c_k f^{(2k)}\|_2^2 = \frac{1}{\mu_{2n}^2(0)} \tag{7}$$

has unique solution c.

Proof. Calculating the Fourier series of the even derivatives of f we get

$$f^{(2k)}(t) = \sum_{j=-\infty}^{\infty} (-1)^k j^{2k} d_j e^{ijt} \text{and } d_j(f^{(2k)}) = (-1)^k j^{2k} d_j.$$

By using the Parseval's identity, it follows that

$$\|f + \sum_{k=1}^{n} c_k f^{(2k)}\|_2^2 = \sum_{j=-\infty}^{\infty} d_j^2 \left(1 + \sum_{k=1}^{n} (-1)^k j^{2k} c_j\right)^2 = \sum_{j=-\infty}^{\infty} w(f,j) p_{2n}^2(j),$$

where $p_{2n}(\lambda)$ is an algebraic polynomial of degree at most $2n$ and $p_{2n}(0) = 1$. Taking into account that $w(f,j)$ is an even sequence we conclude that the minimum is obtained only for the Christoffel polynomial C_{2n}. This concludes the proof.

The problem (7) can be considered as a problem for finding ε-solution to a differential equation of order $2n$, for more details see [1]. It is a standard technique to relate systems of differential equations of low order to a differential equation of a higher order.

Remark 2. For f, n, and c as in Theorem 1, the differential equation

$$f(t)+\sum_{k=1}^{n} c_k f^{(2k)}(t)=\varepsilon(t) \tag{8}$$

is equivalent to the system

$$\mathbf{x}''(t)=-CK_n\mathbf{x}(t)+\mathbf{E}(t), \tag{9}$$

where $\mathbf{x}(t)=(x_n(t),\ldots,x_1(t))^{\mathrm{T}}$, $x_1(t)=f(t)$, and $\mathbf{E}(t)=(\varepsilon(t),0,\ldots,0)^{\mathrm{T}}$. From the preceding comments, it follows that the eigenvalues of CK_n are $\lambda_j=l_j^2>0$. Let $f*g(t)=\int_0^t f(t-x)g(x)\,\mathrm{d}x$ denote the one-sided convolution of f and g. If $\lambda_1=\min_j\lambda_j$, then the following theorem holds true.

Theorem 2. *Let* $\mathbf{x}$ *be the solution of (9) with prescribed initial conditions and* $\mathbf{y}$ *be the solution of the homogeneous problem* $\mathbf{u}''(t)=-CK_n\mathbf{u}(t)$ *satisfying the same initial conditions. Then*

$$\|x_j-y_j\|_\infty\le\left(\frac{2\pi}{\lambda_1}\right)^{1/2}\|\varepsilon\|_2$$

for $j=1,\ldots,n$.

Proof. Let $z_j=x_j-y_j$, then the vector function $\mathbf{z}(t)$ is the solution of the system $\mathbf{z}''(t)=-CK_n\mathbf{z}(t)+\mathbf{E}(t)$, with initial conditions $z_j(0)=z_j'(0)=0, j=1,\ldots,n$. Since the matrix CK_n is tridiagonal and symmetric it admits a Takagi factorization, see [2], in the form $CK_n=U\Lambda U^T$, where $U=(u_{i,j})$ is an unitary real matrix and Λ is diagonal with entries on the main diagonal λ_j, the eigenvalues of CK_n. Multiplying $\mathbf{z}''(t)=-CK_n\mathbf{z}(t)+\mathbf{E}(t)$ from the left by U^T and substituting $\mathbf{v}=U^T\mathbf{z}$ we obtain the following decoupled system for the v's, $v_k''=-\lambda_k v_k+\varepsilon/\sqrt{n}, k=1,\ldots,n$. Since $v_k(0)=v_k'(0)=0$, by applying the one-sided Laplace Transform to both sides of the equation for v_k we get $\mathscr{L}(v_k)=\frac{\mathscr{L}(\varepsilon)}{(\xi^2+l_k^2)\sqrt{n}}=\mathscr{L}(\varepsilon)\mathscr{L}(\cos(l_k t))/(\sqrt{n}l_k)$, and hence $v_k(t)=\varepsilon(t)*\cos(l_k t)/(\sqrt{n}l_k)$. From the inverse relation $\mathbf{z}=U\mathbf{v}$ we get that $z_j=\sum_{k=1}^n u_{k,j}v_k$. The matrix U is unitary and by using the Schwarz inequality we can get an estimate for $|z_j(t)|$, $j=1,\ldots,n$ at any $z\in I$. Since the function $\varepsilon(t)$ is 2π-periodic we have

$$\begin{aligned}|z_j(t)|&=\left|\sum_{k=1}^n u_{k,j}v_k(t)\right|\le\left(\sum_{k=1}^n u_{k,j}^2\right)^{1/2}\left(\sum_{k=1}^n v_k(t)^2\right)^{1/2}\\&=\left(\sum_{k=1}^n\frac{1}{n\lambda_k}\left|\int_0^t\varepsilon(t-x)\cos(l_k x)\,\mathrm{d}x\right|^2\right)^{1/2}\le\|\varepsilon\|_2\frac{(2\pi)^{1/2}}{|l_1|}\end{aligned}$$

The right-hand side does not depend on t and taking the maximum of the left-hand side over I we complete the proof. □

From the proof of the theorem it follows that $z_1 = f - y_1$. If the coefficients c in (8) are chosen to minimize $\|\varepsilon\|_2$ it follows that $\varepsilon(t) = L_n(f)(t)$, $L_n(y_1) = 0$, and $E_n(f) = \|\varepsilon\|_2^2$. If $y_1(t) = \sum_{|j|\geq 1}^{n} b_j e^{\mathrm{i}jt}$, then from Theorem 2 we obtain the estimate

$$\max_{t\in I} |f(t) - \sum_{j=-n}^{n} b_j e^{\mathrm{i}l_j t}| \leq \left(\frac{2\pi}{l_1^2} E_n(f)\right)^{1/2}.$$

We conclude the paper with a comment on how to obtain the explicit expression for the approximant y_1. From the equivalence stated in Remark 2, it follows that y_1 has to satisfy the initial conditions $y_1^{(k)}(0) = f^{(k)}(0), k = 0, 1, \ldots, 2n-1$. The resulting system for the unknown b's has a Van Der Monde coefficient matrix, and hence it has a unique solution.

Acknowledgements This research was supported by grant HRD-0734800.

References

1. G. Birkhoff and G.C. Rota, Differential Equations, (4-th ed) Wiley, New York, NY, 1989.
2. M. Elouafi and A. Hadj, A Takagi Factorization of a Real Symmetric Tridiagonal Matrix, Applied Math. Science **2** (2008), (46) 2289–2296.
3. Y. Katznelson, An Introduction to Harmonic Analysis, (2-nd ed.), Dover Pubns, 1976.
4. V. Totik, Orthogonal Polynomials, Surveys in Approximation Theory 1 (2005) 70 -125.
5. V. Vatchev and R. Sharpley, Decomposition of functions into pairs of intrinsic mode functions, Proc. Royal Soc. Series A, Vo 464 (2008), pp. 2265–2280.
6. V. Vatchev, Simultaneous Approximation of Intrinsic Mode Functions by Smooth Functions with Piece-Wise Linear Amplitude and Phase, " Proceedings of the 12th International Conference on Approximation Theory," eds. M. Nematu and L. Schumaker, San Antonio, (2007).

Matrix Extension with Symmetry and Its Applications

Xiaosheng Zhuang

Abstract In this paper, we are interested in the problems of matrix extension with symmetry, more precisely, the extensions of submatrices of Laurent polynomials satisfying some conditions to square matrices of Laurent polynomials with certain symmetry patterns, which are closely related to the construction of (bi)orthogonal multiwavelets in wavelet analysis and filter banks with the perfect reconstruction property in electronic engineering. We satisfactorily solve the matrix extension problems with respect to both orthogonal and biorthogonal settings. Our results show that the extension matrices do possess certain symmetry patterns and their coefficient supports can be controlled by the given submatrices in certain sense. Moreover, we provide step-by-step algorithms to derive the desired extension matrices. We show that our extension algorithms can be applied not only to the construction of (bi)orthogonal multiwavelets with symmetry, but also to the construction of tight framelets with symmetry and with high order of vanishing moments. Several examples are presented to illustrate the results in this paper.

1 Introduction and Motivation

The matrix extension problems play a fundamental role in many areas such as electronic engineering, system sciences, mathematics, etc. We mention only a few references here on this topic; see [1–3, 5, 8, 10, 12, 19–21, 23–25]. For example, matrix extension is an indispensable tool in the design of filter banks in electronic engineering (see [19,24,25]) and in the construction of multiwavelets in wavelet analysis (see [1–3, 5, 8, 10, 12, 14, 18, 20, 21]). In this section, we shall first introduce the general matrix extension problems and then discuss the connections of the general matrix extension problems to wavelet analysis and filter banks.

Xiaosheng Zhuang
Department of Mathematical and Statistical Sciences, University of Alberta, 632 CAB, Edmonton, Alberta T6G 2G1, Canada, e-mail: xzhuang@math.ualberta.ca

M. Neamtu and L. Schumaker (eds.), *Approximation Theory XIII: San Antonio 2010*, Springer Proceedings in Mathematics 13, DOI 10.1007/978-1-4614-0772-0_24,

1.1 The Matrix Extension Problems

In order to state the matrix extension problems, let us introduce some notation and definitions first. Let $\mathsf{p}(z)=\sum_{k\in\mathbb{Z}} p_k z^k, z\in\mathbb{C}\backslash\{0\}$ be a Laurent polynomial with complex coefficients $p_k\in\mathbb{C}$ for all $k\in\mathbb{Z}$. We say that p has *symmetry* if its coefficient sequence $\{p_k\}_{k\in\mathbb{Z}}$ has symmetry; more precisely, there exist $\varepsilon\in\{-1,1\}$ and $c\in\mathbb{Z}$ such that

$$p_{c-k}=\varepsilon p_k \qquad \forall\, k\in\mathbb{Z}. \tag{1}$$

If $\varepsilon=1$, then p is symmetric about the point $c/2$; if $\varepsilon=-1$, then p is antisymmetric about the point $c/2$. Symmetry of a Laurent polynomial can be conveniently expressed using a symmetry operator S defined by

$$\mathsf{Sp}(z):=\frac{\mathsf{p}(z)}{\mathsf{p}(1/z)}, \qquad z\in\mathbb{C}\backslash\{0\}. \tag{2}$$

When p is not identically zero, it is evident that (1) holds if and only if $\mathsf{Sp}(z)=\varepsilon z^c$. For the zero polynomial, it is very natural that $\mathsf{S}0$ can be assigned any symmetry pattern; i.e., for every occurrence of $\mathsf{S}0$ appearing in an identity in this paper, $\mathsf{S}0$ is understood to take an appropriate choice of εz^c for some $\varepsilon\in\{-1,1\}$ and some $c\in\mathbb{Z}$ so that the identity holds. If $\mathbb{P}$ is an $r\times s$ matrix of Laurent polynomials with symmetry, then we can apply the operator S to each entry of $\mathbb{P}$, i.e., $\mathsf{S}\mathbb{P}$ is an $r\times s$ matrix such that $[\mathsf{S}\mathbb{P}]_{j,k}:=\mathsf{S}([\mathbb{P}]_{j,k})$, where $[\mathbb{P}]_{j,k}$ is the (j,k)-entry of the matrix $\mathbb{P}$.

For two matrices $\mathbb{P}$ and Q of Laurent polynomials with symmetry, even though all the entries in $\mathbb{P}$ and Q have symmetry, their sum $\mathbb{P}+\mathsf{Q}$, difference $\mathbb{P}-\mathsf{Q}$, or product $\mathbb{P}\mathsf{Q}$, if well defined, generally may not have symmetry any more. This is one of the difficulties for matrix extension with symmetry. In order for $\mathbb{P}\pm\mathsf{Q}$ or $\mathbb{P}\mathsf{Q}$ to possess some symmetry, the symmetry patterns of $\mathbb{P}$ and Q should be compatible. For example, if $\mathsf{S}\mathbb{P}=\mathsf{SQ}$ (i.e., both $\mathbb{P}$ and Q have the same symmetry pattern), then indeed $\mathbb{P}\pm\mathsf{Q}$ has symmetry and $\mathsf{S}(\mathbb{P}\pm\mathsf{Q})=\mathsf{S}\mathbb{P}=\mathsf{SQ}$. In the following, we discuss the compatibility of symmetry patterns of matrices of Laurent polynomials.

For an $r\times s$ matrix $\mathbb{P}(z)=\sum_{k\in\mathbb{Z}} P_k z^k$, we denote

$$\mathbb{P}^*(z):=\sum_{k\in\mathbb{Z}} P_k^* z^{-k} \quad \text{with} \quad P_k^*:=\overline{P_k}^T, \qquad k\in\mathbb{Z}, \tag{3}$$

where $\overline{P_k}^{\mathrm{T}}$ denotes the transpose of the complex conjugate of the constant matrix P_k in $\mathbb{C}$. We say that *the symmetry of* $\mathbb{P}$ *is compatible* or $\mathbb{P}$ *has compatible symmetry*, if

$$\mathsf{S}\mathbb{P}(z)=(\mathsf{S}\theta_1)^*(z)\mathsf{S}\theta_2(z) \tag{4}$$

for some $1\times r$ and $1\times s$ row vectors θ_1 and θ_2 of Laurent polynomials with symmetry. For an $r\times s$ matrix $\mathbb{P}$ and an $s\times t$ matrix Q of Laurent polynomials, we say that $(\mathbb{P},\mathsf{Q})$ *has mutually compatible symmetry* if

$$\mathsf{S}\mathbb{P}(z)=(\mathsf{S}\theta_1)^*(z)\mathsf{S}\theta(z) \quad \text{and} \quad \mathsf{SQ}(z)=(\mathsf{S}\theta)^*(z)\mathsf{S}\theta_2(z) \tag{5}$$

for some $1 \times r$, $1 \times s$, $1 \times t$ row vectors $\theta_1, \theta, \theta_2$ of Laurent polynomials with symmetry. If $(\mathbb{P}, \mathsf{Q})$ has mutually compatible symmetry as in (5), then their product $\mathbb{P}\mathsf{Q}$ has compatible symmetry and in fact $\mathsf{S}(\mathbb{P}\mathsf{Q}) = (\mathsf{S}\theta_1)^*\mathsf{S}\theta_2$.

For a matrix of Laurent polynomials, another important property is the support of its coefficient sequence. For $\mathbb{P} = \sum_{k\in\mathbb{Z}} P_k z^k$ such that $P_k = 0$ for all $k \in \mathbb{Z}\backslash[m,n]$ with $P_m \neq 0$ and $P_n \neq 0$, we define its coefficient support to be $\mathrm{csupp}(\mathbb{P}) := [m,n]$ and the length of its coefficient support to be $|\mathrm{csupp}(\mathbb{P})| := n - m$. In particular, we define $\mathrm{csupp}(0) := \emptyset$, the empty set, and $|\mathrm{csupp}(0)| := -\infty$. Also, we use $\mathrm{coeff}(\mathbb{P}, k) := P_k$ to denote the coefficient matrix (vector) P_k of z^k in $\mathbb{P}$. In this paper, 0 always denotes a general zero matrix whose size can be determined in the context.

Now, we introduce the general matrix extension problems with symmetry. We shall use r and s to denote two positive integers such that $1 \leq r \leq s$. I_r denotes the $r \times r$ identity matrix.

Problem 1 (Orthogonal Matrix Extension). Let $\mathbb{F}$ be a subfield of $\mathbb{C}$. Let $\mathbb{P}$ be an $r \times s$ matrix of Laurent polynomials with coefficients in $\mathbb{F}$ such that $\mathbb{P}(z)\mathbb{P}^*(z) = I_r$ for all $z \in \mathbb{C}\backslash\{0\}$ and the symmetry of $\mathbb{P}$ is compatible. Find an $s \times s$ square matrix $\mathbb{P}_e$ of Laurent polynomials with coefficients in $\mathbb{F}$ and with symmetry such that

1. $[I_r, \mathbf{0}]\mathbb{P}_e = \mathbb{P}$ (that is, the submatrix of the first r rows of $\mathbb{P}_e$ is the given matrix $\mathbb{P}$);
2. The symmetry of $\mathbb{P}_e$ is compatible and $\mathbb{P}_e(z)\mathbb{P}_e^*(z) = I_s$ for all $z \in \mathbb{C}\backslash\{0\}$ (that is, $\mathbb{P}_e$ is *paraunitary*);
3. The length of the coefficient support of $\mathbb{P}_e$ can be controlled by that of $\mathbb{P}$ in some way.

Problem 1 is closely related to the construction of orthonormal multiwavelets in wavelet analysis and the design of filter banks with the perfect reconstruction property in electronic engineering. More generally, Problem 1 can be extended to a more general form with respect to the construction of biorthogonal multiwavelets in wavelet analysis. In a moment, we shall reveal their connections, which also serve as our motivation. The more general form of Problem 1 can be stated as follows.

Problem 2 (Biorthogonal Matrix Extension). Let $\mathbb{F}$ be a subfield of $\mathbb{C}$. Let $(\mathbb{P}, \widetilde{\mathbb{P}})$ be a pair of $r \times s$ matrices of Laurent polynomials with coefficients in $\mathbb{F}$ such that $\mathbb{P}(z)\widetilde{\mathbb{P}}^*(z) = I_r$ for all $z \in \mathbb{C}\backslash\{0\}$, the symmetry of $\mathbb{P}$ or $\widetilde{\mathbb{P}}$ is compatible, and $\mathsf{S}\mathbb{P} = \mathsf{S}\widetilde{\mathbb{P}}$. Find a pair of $s \times s$ square matrices $(\mathbb{P}_e, \widetilde{\mathbb{P}}_e)$ of Laurent polynomials with coefficients in $\mathbb{F}$ and with symmetry such that

1. $[I_r, \mathbf{0}]\mathbb{P}_e = \mathbb{P}$ and $[I_r, \mathbf{0}]\widetilde{\mathbb{P}}_e = \widetilde{\mathbb{P}}$ (that is, the submatrix of the first r rows of $\mathbb{P}_e, \widetilde{\mathbb{P}}_e$ is the given matrix $\mathbb{P}, \widetilde{\mathbb{P}}$, respectively);
2. $(\mathbb{P}_e, \widetilde{\mathbb{P}}_e)$ has mutually compatible symmetry and $\mathbb{P}_e(z)\widetilde{\mathbb{P}}_e^*(z) = I_s$ for all $z \in \mathbb{C}\backslash\{0\}$ (that is, $(\mathbb{P}_e, \widetilde{\mathbb{P}}_e)$ is *a pair of biorthogonal matrices*);
3. The lengths of the coefficient support of $\mathbb{P}_e$ and $\widetilde{\mathbb{P}}_e$ can be controlled by those of $\mathbb{P}$ and $\widetilde{\mathbb{P}}$ in some way.

1.2 Motivation

The above problems are closely connected to wavelet analysis and filter banks. The key of wavelet construction is the so-called multiresolution analysis (MRA), which contains mainly two parts. One is on the construction of refinable function vectors that satisfies certain desired conditions. Another part is on the derivation of wavelet generators from refinable function vectors obtained in first part, which should be able to inherit certain properties similar to their refinable function vectors. From the point of view of filter banks, the first part corresponds to the design of filters or filter banks with certain desired properties, while the second part can be and is formulated as some matrix extension problems stated previously. In this paper, we shall mainly focus on the second part (with symmetry) of the MRA while assume that the refinable function vectors with certain properties are given in advance (part of Sect. 3 is on the construction of refinable functions satisfying (14)).

We say that d is a *dilation factor* if d is an integer with $|\mathsf{d}|>1$. Throughout this paper, d denotes a dilation factor. For simplicity of presentation, we further assume that d is positive, while multiwavelets and filter banks with a negative dilation factor can be handled similarly by a slight modification of the statements in this paper.

We say that $\phi=[\phi_1,\ldots,\phi_r]^{\mathrm{T}}:\mathbb{R}\to\mathbb{C}^{r\times 1}$ is a d-*refinable function vector* if

$$\phi=\mathsf{d}\sum_{k\in\mathbb{Z}}a_0(k)\phi(\mathsf{d}\cdot-k), \tag{6}$$

where $a_0:\mathbb{Z}\to\mathbb{C}^{r\times r}$ is a finitely supported sequence of $r\times r$ matrices on $\mathbb{Z}$, called the *low-pass filter (or mask)* for ϕ. The *symbol* of a_0 is denoted by $\mathsf{a}_0(z):=\sum_{k\in\mathbb{Z}}a_0(k)z^k$, which is an $r\times r$ matrix of Laurent polynomials.

In the frequency domain, the refinement equation in (6) can be rewritten as

$$\widehat{\phi}(\mathsf{d}\xi)=\widehat{a_0}(\xi)\widehat{\phi}(\xi),\quad \xi\in\mathbb{R}, \tag{7}$$

where $\widehat{a_0}$ is the *Fourier series* of a_0 given by

$$\widehat{a_0}(\xi):=\sum_{k\in\mathbb{Z}}a_0(k)\mathrm{e}^{-\mathrm{i}k\xi}=\mathsf{a}_0(\mathrm{e}^{-\mathrm{i}\xi}),\quad \xi\in\mathbb{R}. \tag{8}$$

The Fourier transform $\widehat{f}$ of $f\in L_1(\mathbb{R})$ is defined to be $\widehat{f}(\xi)=\int_{\mathbb{R}}f(t)\mathrm{e}^{-\mathrm{i}t\xi}\mathrm{d}t$ and can be extended to square integrable functions and tempered distributions.

We say that a compactly supported d-refinable function vector ϕ in $L_2(\mathbb{R})$ is *orthogonal* if

$$\langle\phi,\phi(\cdot-k)\rangle=\delta(k)I_r,\quad k\in\mathbb{Z}, \tag{9}$$

where δ is the *Dirac sequence* such that $\delta(0)=1$ and $\delta(k)=0$ for all $k\neq 0$.

Usually, a wavelet system is generated by some wavelet function vectors $\psi^\ell=[\psi_1^\ell,\ldots,\psi_r^\ell]^{\mathrm{T}}$, $\ell=1,\ldots,L$, from a d-refinable function vector ϕ as follows:

$$\widehat{\psi^\ell}(\mathsf{d}\xi)=\widehat{a_\ell}(\xi)\widehat{\phi}(\xi),\quad \ell=1,\ldots,L, \tag{10}$$

where each $a_\ell:\mathbb{Z}\to\mathbb{C}^{r\times r}$ is a finitely supported sequence of $r\times r$ matrices on $\mathbb{Z}$, called the *high-pass filter (or mask)* for ψ^ℓ, $\ell=1,\ldots,L$.

We say that $\{\psi^1,\dots,\psi^L\}$ generates a d-*multiframe* in $L_2(\mathbb{R})$ if $\{\psi^\ell_{j,k} := \mathsf{d}^{j/2}\psi^\ell(\mathsf{d}^j\cdot - k) : j,k\in\mathbb{Z}, \ell=1,\dots,L\}$ is a frame in $L_2(\mathbb{R})$, that is, there exist two positive constants C_1, C_2 such that

$$C_1\|f\|^2_{L_2(\mathbb{R})} \le \sum_{\ell=1}^{L}\sum_{j\in\mathbb{Z}}\sum_{k\in\mathbb{Z}} |\langle f,\psi^\ell_{j,k}\rangle|^2 \le C_2\|f\|^2_{L_2(\mathbb{R})}, \quad \forall f\in L_2(\mathbb{R}), \qquad (11)$$

where $|\langle f,\psi^\ell_{j,k}\rangle|^2 = \langle f,\psi^\ell_{j,k}\rangle\langle\psi^\ell_{j,k},f\rangle$ and $\langle\cdot,\cdot\rangle$ is the inner product defined to be

$$\langle f,g\rangle := \int_{\mathbb{R}} f(t)\overline{g(t)}^{\mathsf{T}}\mathrm{d}t, \quad f\in (L_2(\mathbb{R}))^{s_1\times\ell}, g\in (L_2(\mathbb{R}))^{s_2\times\ell}.$$

If $C_1 = C_2 = 1$ in (11), we say that $\{\psi^1,\dots,\psi^L\}$ generates a *tight* d-*multiframe* in $L_2(\mathbb{R})$. The wavelet function vectors ψ^ℓ are called *tight multiframelets*. When $r=1$, we usually drop the prefix *multi*.

If ϕ is a compactly supported d-refinable function vector in $L_2(\mathbb{R})$ associated with a low-pass filter a_0, then it is well-known (see [6]) that $\{\psi^1,\dots,\psi^L\}$ associated with high-pass filters $\{a_1,\dots,a_L\}$ via (10) generates a tight d-multiframe *if and only if*

$$\sum_{\ell=0}^{L}\widehat{a_\ell}\overline{\widehat{a_\ell}(\cdot+2\pi k/\mathsf{d})}^{\mathsf{T}} = \delta(k)I_r, \quad k=0,\dots,\mathsf{d}-1. \qquad (12)$$

According to various requirements of problems in applications, different desired properties of a wavelet system are needed, which usually can be characterized by conditions on the low-pass filter a_0 for ϕ and the high-pass filters a_ℓ for ψ^ℓ, $\ell=1,\dots,L$. Among all properties of a wavelet system, high order of vanishing moments, (bi)orthogonality, and symmetry are highly desirable properties in wavelet and filter bank applications. High order of vanishing moments is crucial for the sparsity representation of a wavelet system, which plays an important role in image denoising and compression. (Bi)orthogonality (more general, tightness of a wavelet system) results in simple rules for guaranteeing the perfect reconstruction property. Symmetry usually produces better visual effect and less artifact in signal/image processing; not to mention the double reduction of the computational cost for a symmetric system.

A framelet ψ has *vanishing moments of order n* if

$$\int_{\mathbb{R}} t^k\psi(t)\mathrm{d}t = 0 \qquad k=0,\dots,n-1, \qquad (13)$$

which is equivalent to saying that $\frac{\mathrm{d}^k}{\mathrm{d}t^k}\widehat{\psi}(0)=0$ for all $k=0,\dots,n-1$. If (12) holds and the low-pass filter a_0 satisfies

$$1-|\widehat{a_0}(\xi)|^2 = O(|\xi|^{2n}), \quad \xi\to 0, \qquad (14)$$

which means $1-|\widehat{a_0}(\xi)|^2$ has *zero of order* $2n$ near the origin, then the framelet system generated by $\{\psi^1,\dots,\psi^L\}$ has vanishing moments of order n (see [6]). We shall see in Sect. 3 on the connection of tight frames to the orthogonal matrix extension problem and on the construction of symmetric complex tight framelets with high order of vanishing moments via the technique of matrix extension with symmetry.

Next, let us review the construction of tight d-multiframes in the point of view of filters and filter banks. Let $\mathbb{F}$ be a subfield of $\mathbb{C}$. Let $a_0 : \mathbb{Z} \to \mathbb{F}^{r\times r}$ be a low-pass filter with *multiplicity r* for a d-refinable function vector $\phi = [\phi_1, \ldots, \phi_r]^{\mathsf{T}}$. The d-*band subsymbols (polyphase components)* of a_0 are defined to be

$$\mathsf{a}_{0;\gamma}(z) := \sqrt{\mathsf{d}} \sum_{k\in\mathbb{Z}} a_0(\gamma + \mathsf{d}k) z^k, \quad \gamma \in \mathbb{Z}. \tag{15}$$

Let $a_1, \ldots, a_L : \mathbb{Z} \to \mathbb{F}^{r\times r}$ be high-pass filters for function vectors $\psi^1, \ldots, \psi^L$, respectively. The *polyphase matrix* for the filter bank $\{a_0, a_1, \ldots, a_L\}$ (or $\{\mathsf{a}_0, \mathsf{a}_1, \ldots, \mathsf{a}_L\}$) is defined to be

$$\mathbf{P}(z) = \begin{bmatrix} \mathsf{a}_{0;0}(z) & \cdots & \mathsf{a}_{0;\mathsf{d}-1}(z) \\ \mathsf{a}_{1;0}(z) & \cdots & \mathsf{a}_{1;\mathsf{d}-1}(z) \\ \vdots & \vdots & \vdots \\ \mathsf{a}_{L;0}(z) & \cdots & \mathsf{a}_{L;\mathsf{d}-1}(z) \end{bmatrix}, \tag{16}$$

where $\mathsf{a}_{\ell;\gamma}$ are subsymbols of a_ℓ similarly defined as in (15) for $\gamma = 0, \ldots, \mathsf{d}-1$ and $\ell = 1, \ldots, L$.

If ϕ is a compactly supported d-refinable function vectors in $L_2(\mathbb{R})$, then it is well-known (see [6]) that $\{\psi^1, \ldots, \psi^L\}$ associated with $\{a_1, \ldots, a_L\}$ via (10) generates a tight d-multiframe, i.e., (12) holds, *if and only if*,

$$\mathbf{P}^*(z)\mathbf{P}(z) = I_{\mathsf{d}r}, \qquad z \in \mathbb{C}\backslash\{0\}. \tag{17}$$

Note that the polyphase matrix $\mathbf{P}$ is not necessarily a square matrix (only if $L = \mathsf{d}-1$). When the d-refinable function vector ϕ associated with a low-pass filter a_0 is orthogonal, the multiframlet system generated by $\{\psi^1, \ldots, \psi^{\mathsf{d}-1}\}$ via (10) becomes an *orthonormal multiwavelet basis* for $L_2(\mathbb{R})$. In this case, the polyphase matrix $\mathbf{P}$ associated with the filter bank $\{\mathsf{a}_0, \ldots, \mathsf{a}_{\mathsf{d}-1}\}$ is indeed a square matrix. Moreover, the low-pass filter a_0 for ϕ is a d-*band orthogonal filter*:

$$\sum_{\gamma=0}^{\mathsf{d}-1} \mathsf{a}_{0;\gamma}(z)\mathsf{a}^*_{0;\gamma}(z) = I_r, \qquad z \in \mathbb{C}\backslash\{0\}. \tag{18}$$

Now, one can show that the derivation of high-pass filters $\mathsf{a}_1, \ldots, \mathsf{a}_{\mathsf{d}-1}$ from a_0 so that the filter bank $\{\mathsf{a}_0, \mathsf{a}_1, \ldots, \mathsf{a}_{\mathsf{d}-1}\}$ has the *perfect reconstruction property* as in (17) is simply a special case of Problem 1 (orthogonal matrix extension). More generally, for $L = \mathsf{d}-1$, one can consider the construction of *biorthogonal multiwavelets* (see Sect. 4), which corresponds to Problem 2. Our main focus of this paper is on matrix extension with symmetry with respect to Problems 1 and 2. We shall study in Sects. 2 and 4 on the orthogonal matrix extension problem and the biorthogonal matrix extension problem, respectively.

1.3 Prior Work and Our Contributions

Without considering symmetry issue, it is known in the engineering literature that Problem 1 or 2 can be solved by representing the given matrices in *cascade structures*; see [19, 24]. In the context of wavelet analysis, orthogonal matrix extension without symmetry was discussed by Lawton, Lee, and Shen in their paper [20]. In electronic engineering, an algorithm using the cascade structure for orthogonal matrix extension without symmetry was given in [24] for filter banks with perfect reconstruction property. The algorithms in [20,24] mainly deal with the special case that $\mathbb{P}$ is a row vector (that is, $r = 1$ in our case) without symmetry, and the coefficient support of the derived matrix $\mathbb{P}_e$ indeed can be controlled by that of $\mathbb{P}$. The algorithms in [20,24] for the special case $r = 1$ can be employed to handle a general $r \times s$ matrix $\mathbb{P}$ without symmetry; see [20, 24] for detail. However, for the general case $r > 1$, it is no longer clear whether the coefficient support of the derived matrix $\mathbb{P}_e$ obtained by the algorithms in [20, 24] can still be controlled by that of $\mathbb{P}$. For $r = 1$, Goh et al. in [9] considered the biorthogonal matrix extension problem without symmetry. They provided a step-by-step algorithm for deriving the extension matrices, yet they did not concern about the support control of the extension matrices nor the symmetry patterns of the extension matrices. For $r > 1$, there are only a few results in the literature [1,4] and most of them only consider about some very special cases. The difficulty comes from the flexibility of the biorthogonality relation between the given pair $(\mathbb{P}, \widetilde{\mathbb{P}})$ of biorthogonal matrices.

Several special cases of matrix extension with symmetry were considered in the literature. For $\mathbb{F} = \mathbb{R}$ and $r = 1$, orthogonal matrix extension with symmetry was considered in [21]. For $r = 1$, orthogonal matrix extension with symmetry was studied in [12] and a simple algorithm is given there. In the context of wavelet analysis, several particular cases of matrix extension with symmetry related to the construction of (bi)orthogonal multiwavelets were investigated in [1, 3, 10, 12, 19, 21]. However, for the general case of an $r \times s$ matrix, the approaches on orthogonal matrix extension with symmetry in [12,21] for the particular case $r = 1$ cannot be employed to handle the general case. The algorithms in [12, 21] are very difficult to be generalized to the general case $r > 1$, partially due to the complicated relations of the symmetry patterns between different rows of $\mathbb{P}$. For the general case of matrix extension with symmetry, it becomes much harder to control the coefficient support of the derived matrix $\mathbb{P}_e$, comparing with the special case $r = 1$. Extra effort is needed in any algorithm of deriving $\mathbb{P}_e$ so that its coefficient support can be controlled by that of $\mathbb{P}$.

The contributions of this paper lie in the following aspects. First, we satisfactorily solve the matrix extension problems with symmetry for any r, s such that $1 \le r \le s$. More importantly, we obtain a complete representation for any $r \times s$ paraunitary matrix $\mathbb{P}$ or pairs of biorthogonal matrices $(\mathbb{P}, \widetilde{\mathbb{P}})$ having compatible symmetry with $1 \le r \le s$. This representation leads to step-by-step algorithms for deriving a desired matrix $\mathbb{P}_e$ or the pair of extension matrices $(\mathbb{P}_e, \widetilde{\mathbb{P}}_e)$ from a given matrix $\mathbb{P}$ or a pair $(\mathbb{P}, \widetilde{\mathbb{P}})$. Second, we obtain an optimal result in the sense of (21) on controlling the coefficient support of the desired matrix $\mathbb{P}_e$ derived from a given matrix $\mathbb{P}$ by our algorithm for orthogonal matrix extension with symmetry. This is of importance in both theory and application, since short support of a filter or a multiwavelet is a highly desirable property and short support usually means a fast algorithm and simple implementation in practice. Third, we introduce the notion of compatibility of symmetry, which plays a critical role in the study of the general matrix extension problems with symmetry ($r \geq 1$). Fourth, we provide a complete analysis and a systematic construction algorithm for symmetric filter banks with the perfect reconstruction property and symmetric (bi)orthogonal multiwavelets. Finally, most of the literature on the matrix extension problem only consider Laurent polynomials with coefficients in the special field $\mathbb{C}$ (see [20]) or $\mathbb{R}$ (see [2,21]). In this paper, our setting is under a general field $\mathbb{F}$, which can be any subfield of $\mathbb{C}$ satisfies certain conditions (see (19) for the case of orthogonal matrix extension).

1.4 Outline

Here is the structure of this paper. In Sect. 2, we shall study the orthogonal matrix extension with symmetry and present a step-by-step algorithm for this problem. We shall also apply our algorithm in this section to the design of symmetric filter banks in electronic engineering and to the construction of symmetric orthonormal multiwavelets in wavelet analysis. In Sect. 3, we shall discuss the construction of symmetric complex tight framelets with high order of vanishing moments and with symmetry via our algorithm for orthogonal matrix extension with symmetry. In Sect. 4, we shall study the biorthogonal matrix extension problem corresponding to the construction of symmetric biorthogonal multiwavelets. We also provide a step-by-step algorithm for the construction of the desired pair of biorthogonal extension matrices. Examples will be provided to illustrate our algorithms and results.

2 Orthogonal Matrix Extension with Symmetry

In this section, we shall study the orthogonal matrix extension problem with symmetry. The Laurent polynomials that we shall consider in this section have their coefficients in a subfield $\mathbb{F}$ of the complex field $\mathbb{C}$ such that $\mathbb{F}$ is closed under the operations of complex conjugate of $\mathbb{F}$ and square roots of positive numbers in $\mathbb{F}$. In other words, the subfield $\mathbb{F}$ of $\mathbb{C}$ satisfies the following properties:

$$\bar{x} \in \mathbb{F} \quad \text{and} \quad \sqrt{y} \in \mathbb{F} \qquad \forall\, x, y \in \mathbb{F} \quad \text{with} \quad y > 0. \tag{19}$$

Two particular examples of such subfields $\mathbb{F}$ are $\mathbb{F} = \mathbb{R}$ (the field of real numbers) and $\mathbb{F} = \mathbb{C}$ (the field of complex numbers). A nontrivial example is the field of all algebraic number, i.e., the algebraic closure $\overline{\mathbb{Q}}$ of the rational number $\mathbb{Q}$. A subfield of $\mathbb{R}$ given by $\overline{\mathbb{Q}} \cap \mathbb{R}$ also satisfies (19).

Problem 1 is completely solved by the following theorem.

Theorem 1. *Let $\mathbb{F}$ be a subfield of $\mathbb{C}$ such that* (19) *holds. Let $\mathbb{P}$ be an $r \times s$ matrix of Laurent polynomials with coefficients in the subfield $\mathbb{F}$ such that the symmetry of $\mathbb{P}$ is compatible, i.e., $\mathsf{S}\mathbb{P} = (\mathsf{S}\theta_1)^*\mathsf{S}\theta_2$ for some $1 \times r$, $1 \times s$ vectors θ_1, θ_2 of Laurent polynomials with symmetry. Then $\mathbb{P}(z)\mathbb{P}^*(z) = I_r$ for all $z \in \mathbb{C}\backslash\{0\}$ (that is, $\mathbb{P}$ is paraunitary), if and only if, there exists an $s \times s$ square matrix $\mathbb{P}_e$ of Laurent polynomials with coefficients in $\mathbb{F}$ such that*

(1) *$[I_r, \mathbf{0}]\mathbb{P}_e = \mathbb{P}$; that is, the submatrix of the first r rows of $\mathbb{P}_e$ is $\mathbb{P}$;*
(2) *$\mathbb{P}_e$ is paraunitary: $\mathbb{P}_e(z)\mathbb{P}_e^*(z) = I_s$ for all $z \in \mathbb{C}\backslash\{0\}$;*
(3) *The symmetry of $\mathbb{P}_e$ is compatible: $\mathsf{S}\mathbb{P}_e = (\mathsf{S}\theta)^*\mathsf{S}\theta_2$ for some $1 \times s$ vector θ of Laurent polynomials with symmetry;*
(4) *$\mathbb{P}_e$ can be represented as products of some $s \times s$ matrices $\mathbb{P}_0, \mathbb{P}_1, \dots, \mathbb{P}_{J+1}$ of Laurent polynoimals with coefficient in $\mathbb{F}$:*

$$\mathbb{P}_e(z) = \mathbb{P}_{J+1}(z)\mathbb{P}_J(z)\cdots\mathbb{P}_1(z)\mathbb{P}_0(z); \tag{20}$$

(5) *$\mathbb{P}_j, 1 \le j \le J$ are elementary: $\mathbb{P}_j(z)\mathbb{P}_j^*(z) = I_s$ and* $\operatorname{csupp}(\mathbb{P}_j) \subseteq [-1, 1]$*;*
(6) *$(\mathbb{P}_{j+1}, \mathbb{P}_j)$ has mutually compatible symmetry for all $0 \le j \le J$;*
(7) *$\mathbb{P}_0 = \mathsf{U}_{\mathsf{S}\theta_2}^*$ and $\mathbb{P}_{J+1} = diag(\mathsf{U}_{\mathsf{S}\theta_1}, I_{s-r})$, where $\mathsf{U}_{\mathsf{S}\theta_1}$, $\mathsf{U}_{\mathsf{S}\theta_2}$ are products of a permutation matrix with a diagonal matrix of monomials, as defined in* (23)*;*
(8) *The coefficient support of $\mathbb{P}_e$ is controlled by that of $\mathbb{P}$ in the following sense:*

$$|\operatorname{csupp}([\mathbb{P}_e]_{j,k})| \le \max_{1 \le n \le r} |\operatorname{csupp}([\mathbb{P}]_{n,k})|, \qquad 1 \le j,k \le s. \tag{21}$$

The representation in (20) is often called the *cascade structure* in the literature of engineering, see [19, 24]. The key of Theorem 1 is to construct the elementary paraunitary matrices $\mathbb{P}_1, \dots, \mathbb{P}_J$ step by step such that $\mathbb{P}_j$'s have the properties stated as in Items (4)–(7) of the theorem. We shall provide such a step-by-step algorithm next, which not only provides a detailed construction of such $\mathbb{P}_j$'s, but also leads to a constructive proof of Theorem 1. For a complete and detailed proof of Theorem 1 using our algorithm, one may refer to [16, Sect. 4].

2.1 An Algorithm for the Orthogonal Matrix Extension with Symmetry

Now we present a step-by-step algorithm on orthogonal matrix extension with symmetry to derive the desired matrix $\mathbb{P}_e$ in Theorem 1 from a given matrix $\mathbb{P}$. Our algorithm has three steps: initialization, support reduction, and finalization. The step of initialization reduces the symmetry pattern of $\mathbb{P}$ to a standard form. The step of support reduction is the main body of the algorithm, producing a sequence of elementary matrices $\mathsf{A}_1, \dots, \mathsf{A}_J$ that reduce the length of the coefficient support of $\mathbb{P}$ to 0. The step of finalization generates the desired matrix $\mathbb{P}_e$ as in Theorem 1. More precisely, see Algorithm 1 for our algorithm written in the form of *pseudo-code*.

Algorithm 1 Orthogonal matrix extension with symmetry

(a) **Input**: $\mathbb{P}$ as in Theorem 1 with $\mathsf{S}\mathbb{P} = (\mathsf{S}\theta_1)^*\mathsf{S}\theta_2$ for some $1\times r$ and $1\times s$ row vectors θ_1 and θ_2 of Laurant polynomials with symmetry.

(b) **Initialization**: Let $\mathsf{Q} := \mathsf{U}^*_{\mathsf{S}\theta_1}\mathbb{P}\mathsf{U}_{\mathsf{S}\theta_2}$. Then the symmetry pattern of Q is

$$\mathsf{SQ} = [1_{r_1}, -1_{r_2}, z1_{r_3}, -z1_{r_4}]^{\mathrm{T}}[1_{s_1}, -1_{s_2}, z^{-1}1_{s_3}, -z^{-1}1_{s_4}], \tag{22}$$

where all nonnegative integers $r_1,\dots,r_4,s_1,\dots,s_4$ are uniquely determined by $\mathsf{S}\mathbb{P}$.

(c) **Support Reduction**: Let $\mathbb{P}_0 := \mathsf{U}^*_{\mathsf{S}\theta_2}$ and $J := 1$.

while $(|\mathrm{csupp}(\mathsf{Q})| > 0)$ **do**
  Let $\mathsf{Q}_0 := \mathsf{Q}$, $[k_1,k_2] := \mathrm{csupp}(\mathsf{Q})$, and $\mathsf{A}_J := I_s$.
  if $k_2 = -k_1$ **then**
    for $j = 1$ to r **do**
      Let $\mathsf{q} := [\mathsf{Q}_0]_{j,:}$ and $\mathsf{p} := [\mathsf{Q}]_{j,:}$ be the jth rows of Q_0 and Q, respectively. Let $[\ell_1,\ell_2] := \mathrm{csupp}(\mathsf{q})$, $\ell := \ell_2 - \ell_1$, and $\mathsf{B}_j := I_s$.
      if $\mathrm{csupp}(\mathsf{q}) = \mathrm{csupp}(\mathsf{p})$ and $\ell \ge 2$ and $(\ell_1 = k_1$ or $\ell_2 = k_2)$ **then**
        $\mathsf{B}_j := \mathsf{B}_{\mathsf{q}}$. $\mathsf{A}_J := \mathsf{A}_J\mathsf{B}_j$. $\mathsf{Q}_0 := \mathsf{Q}_0\mathsf{B}_j$.
      end if
    end for
    Q_0 takes the form in (31). Let $\mathsf{B}_{(-k_2,k_2)} := I_s$, $\mathsf{Q}_1 := \mathsf{Q}_0$, $j_1 := 1$ and $j_2 := r_3 + r_4 + 1$.
    while $j_1 \le r_1 + r_2$ and $j_2 \le r$ **do**
      Let $\mathsf{q}_1 := [\mathsf{Q}_1]_{j_1,:}$ and $\mathsf{q}_2 := [\mathsf{Q}_1]_{j_2,:}$.
      if $\mathrm{coeff}(\mathsf{q}_1,k_1) = 0$ **then** $j_1 := j_1 + 1$. **end if**
      if $\mathrm{coeff}(\mathsf{q}_2,k_2) = 0$ **then** $j_2 := j_2 + 1$. **end if**
      if $\mathrm{coeff}(\mathsf{q}_1,k_1) \ne \mathbf{0}$ and $\mathrm{coeff}(\mathsf{q}_2,k_2) \ne \mathbf{0}$ **then**
        $\mathsf{B}_{(-k_2,k_2)} := \mathsf{B}_{(-k_2,k_2)}\mathsf{B}_{(\mathsf{q}_1,\mathsf{q}_2)}$. $\mathsf{Q}_1 := \mathsf{Q}_1\mathsf{B}_{(\mathsf{q}_1,\mathsf{q}_2)}$. $\mathsf{A}_J := \mathsf{A}_J\mathsf{B}_{(\mathsf{q}_1,\mathsf{q}_2)}$. $j_1 := j_1 + 1$. $j_2 := j_2 + 1$.
      end if
    end while *// end inner while loop*
  end if
  Q_1 takes the form in (31) with either $\mathrm{coeff}(\mathsf{Q}_1,-k) = 0$ or $\mathrm{coeff}(\mathsf{Q}_1,k) = 0$. Let $\mathsf{A}_J := \mathsf{A}_J\mathsf{B}_{\mathsf{Q}_1}$ and $\mathsf{Q} := \mathsf{Q}\mathsf{A}_J$. Then

$$\mathsf{SQ} = [1_{r_1}, -1_{r_2}, z1_{r_3}, -z1_{r_4}]^{\mathrm{T}}[1_{s'_1}, -1_{s'_2}, z^{-1}1_{s'_3}, -z^{-1}1_{s'_4}].$$

  Replace $s_1,\dots,s_4$ by $s'_1,\dots,s'_4$, respectively. Let $\mathbb{P}_J := \mathsf{A}_J^*$ and $J := J+1$.
end while *// end outer while loop*

(d) **Finalization**: $\mathsf{Q} = \mathrm{diag}(F_1,F_2,F_3,F_4)$ for some $r_j \times s_j$ constant matrices F_j in $\mathbb{F}$, $j = 1,\dots,4$. Let $U := \mathrm{diag}(U_{F_1},U_{F_2},U_{F_3},U_{F_4})$ so that $QU = [I_r, 0]$. Define $\mathbb{P}_J := U^*$ and $\mathbb{P}_{J+1} := \mathrm{diag}(\mathsf{U}_{\mathsf{S}\theta_1}, I_{s-r})$.

(e) **Output**: A desired matrix $\mathbb{P}_e$ satisfying all the properties in Theorem 1

In the following subsections, we present detailed constructions of the matrices $\mathsf{U}_{\mathsf{S}\theta}$, B_{q}, $\mathsf{B}_{(\mathsf{q}_1,\mathsf{q}_2)}$, $\mathsf{B}_{\mathsf{Q}_1}$, and U_F appearing in Algorithm 1.

2.1.1 Initialization

Let θ be a $1\times n$ row vector of Laurent polynomials with symmetry such that $\mathsf{S}\theta = [\varepsilon_1 z^{c_1},\dots,\varepsilon_n z^{c_n}]$ for some $\varepsilon_1,\dots,\varepsilon_n \in \{-1,1\}$ and $c_1,\dots,c_n \in \mathbb{Z}$. Then, the symmetry of any entry in the vector $\theta \mathrm{diag}(z^{-\lceil c_1/2 \rceil}, \ldots, z^{-\lceil c_n/2 \rceil})$ belongs to $\{\pm 1, \pm z^{-1}\}$. Thus, there is a permutation matrix E_θ to regroup these four types of symmetries together so that

$$\mathsf{S}(\theta \mathsf{U}_{\mathsf{S}\theta}) = [\mathbf{1}_{n_1}, -\mathbf{1}_{n_2}, z^{-1}\mathbf{1}_{n_3}, -z^{-1}\mathbf{1}_{n_4}], \tag{23}$$

where $\mathsf{U}_{\mathsf{S}\theta} := \mathrm{diag}(z^{-\lceil c_1/2 \rceil}, \ldots, z^{-\lceil c_n/2 \rceil})E_\theta$, $\mathbf{1}_m$ denotes the $1 \times m$ row vector $[1, \ldots, 1]$, and $n_1, \ldots, n_4$ are nonnegative integers uniquely determined by $\mathsf{S}\theta$. Since $\mathbb{P}$ satisfies (4), $\mathsf{Q} := \mathsf{U}^*_{\mathsf{S}\theta_1}\mathbb{P}\mathsf{U}_{\mathsf{S}\theta_2}$ has the symmetry pattern as in (22). Note that $\mathsf{U}_{\mathsf{S}\theta_1}$ and $\mathsf{U}_{\mathsf{S}\theta_2}$ do not increase the length of the coefficient support of $\mathbb{P}$.

2.1.2 Support Reduction

For a $1 \times n$ row vector $\mathtt{f}$ in $\mathbb{F}$ such that $\|\mathtt{f}\| \neq 0$, we define $n_{\mathtt{f}}$ to be the number of nonzero entries in $\mathtt{f}$ and $\varepsilon_j := [0, \ldots, 0, 1, 0, \ldots, 0]$ to be the jth unit coordinate row vector in $\mathbb{R}^n$. Let $E_{\mathtt{f}}$ be a permutation matrix such that $\mathtt{f}E_{\mathtt{f}} = [f_1, \ldots, f_{n_{\mathtt{f}}}, 0, \ldots, 0]$ with $f_j \neq 0$ for $j = 1, \ldots, n_{\mathtt{f}}$. We define

$$V_{\mathtt{f}} := \begin{cases} \frac{\bar{f}_1}{|f_1|}, & \text{if } n_{\mathtt{f}} = 1; \\ \frac{\bar{f}_1}{|f_1|}\left(I_n - \frac{2}{\|v_{\mathtt{f}}\|^2} v_{\mathtt{f}}^* v_{\mathtt{f}}\right), & \text{if } n_{\mathtt{f}} > 1, \end{cases} \tag{24}$$

where $v_{\mathtt{f}} := \mathtt{f} - \frac{f_1}{|f_1|}\|\mathtt{f}\|\varepsilon_1$. Observing that $\|v_{\mathtt{f}}\|^2 = 2\|\mathtt{f}\|(\|\mathtt{f}\| - |f_1|)$, we can verify that $V_{\mathtt{f}}V_{\mathtt{f}}^* = I_n$ and $\mathtt{f}E_{\mathtt{f}}V_{\mathtt{f}} = \|\mathtt{f}\|\varepsilon_1$. Let $U_{\mathtt{f}} := E_{\mathtt{f}}V_{\mathtt{f}}$. Then $U_{\mathtt{f}}$ is unitary and satisfies $U_{\mathtt{f}} = [\frac{\mathtt{f}^*}{\|\mathtt{f}\|}, F^*]$ for some $(n-1) \times n$ matrix F in $\mathbb{F}$ such that $\mathtt{f}U_{\mathtt{f}} = [\|\mathtt{f}\|, 0, \ldots, 0]$. We also define $U_{\mathtt{f}} := I_n$ if $\mathtt{f} = 0$ and $U_{\mathtt{f}} := \emptyset$ if $\mathtt{f} = \emptyset$. Here, $U_{\mathtt{f}}$ plays the role of reducing the number of nonzero entries in $\mathtt{f}$. More generally, for an $r \times n$ nonzero matrix G of rank m in $\mathbb{F}$, employing the above procedure to each row of G, we can obtain an $n \times n$ unitary matrix U_G such that $GU_G = [R, 0]$ for some $r \times m$ lower triangular matrix R of rank m. If $G_1G_1^* = G_2G_2^*$, then the above procedure produces two matrices U_{G_1}, U_{G_2} such that $G_1U_{G_1} = [R, 0]$ and $G_2U_{G_2} = [R, 0]$ for some lower triangular matrix R of full rank. It is important to notice that the constructions of $U_{\mathtt{f}}$ and U_G only involve the nonzero entries of $\mathtt{f}$ and nonzero columns of G, respectively. In other words, up to a permutation, we have

$$\begin{aligned} [U_{\mathtt{f}}]_{j,:} &= ([U_{\mathtt{f}}]_{:,j})^{\mathrm{T}} = \varepsilon_j, \ \text{if } [\mathtt{f}]_j = 0, \\ [U_G]_{j,:} &= ([U_G]_{:,j})^{\mathrm{T}} = \varepsilon_j, \ \text{if } [G]_{:,j} = 0. \end{aligned} \tag{25}$$

Denote $\mathsf{Q} := \mathsf{U}^*_{\mathsf{S}\theta_1}\mathbb{P}\mathsf{U}_{\mathsf{S}\theta_2}$ as in Algorithm 1. The outer *while* loop produces a sequence of elementary paraunitary matrices $\mathsf{A}_1, \ldots, \mathsf{A}_J$ that reduce the length of the coefficient support of Q gradually to 0. The construction of each A_j has three parts: $\{\mathsf{B}_1, \ldots, \mathsf{B}_r\}$, $\mathsf{B}_{(-k,k)}$, and $\mathsf{B}_{\mathsf{Q}_1}$. The first part $\{\mathsf{B}_1, \ldots, \mathsf{B}_r\}$ (see the *for* loop) is constructed recursively for each of the r rows of Q so that $\mathsf{Q}_0 := \mathsf{Q}\mathsf{B}_1 \cdots \mathsf{B}_r$ has a special form as in (31). If both $\mathrm{coeff}(\mathsf{Q}_0, -k) \neq 0$ and $\mathrm{coeff}(\mathsf{Q}_0, k) \neq 0$, then the second part

$\mathsf{B}_{(-k,k)}$ (see the inner *while* loop) is further constructed so that $\mathsf{Q}_1 := \mathsf{Q}_0\mathsf{B}_{(-k,k)}$ takes the form in (31) with at least one of coeff$(\mathsf{Q}_1,-k)$ and coeff(Q_1,k) being 0. $\mathsf{B}_{\mathsf{Q}_1}$ is constructed to handle the case that csupp$(\mathsf{Q}_1) = [-k,k-1]$ or csupp(Q_1) $= [-k+1,k]$ so that csupp$(\mathsf{Q}_1\mathsf{B}_{\mathsf{Q}_1}) \subseteq [-k+1,k-1]$.

Let q denote an arbitrary row of Q with $|\text{csupp}(\mathsf{q})| \geq 2$. We first explain how to construct B_q for a given row q such that B_q reduces the length of the coefficient support of q by 2 and keeps its symmetry pattern. Note that in the *for* loop, B_j is simply B_q with q being the current jth row of $\mathsf{Q}\mathsf{B}_0\cdots\mathsf{B}_{j-1}$, where $\mathsf{B}_0 := I_s$.

By (22), we have $\mathsf{S}\mathsf{q} = \varepsilon z^c[\mathbf{1}_{s_1}, -\mathbf{1}_{s_2}, z^{-1}\mathbf{1}_{s_3}, -z^{-1}\mathbf{1}_{s_4}]$ for some $\varepsilon \in \{-1,1\}$ and $c \in \{0,1\}$. For $\varepsilon = -1$, there is a permutation matrix E_ε such that $\mathsf{S}(\mathsf{q}E_\varepsilon)$ $= z^c[\mathbf{1}_{s_2}, -\mathbf{1}_{s_1}, z^{-1}\mathbf{1}_{s_4}, -z^{-1}\mathbf{1}_{s_3}]$. For $\varepsilon = 1$, we let $E_\varepsilon := I_s$. Then, $\mathsf{q}E_\varepsilon$ must take the form in either (26) or (27) with $\mathtt{f}_1 \neq 0$ as follows:

$$\begin{aligned}\mathsf{q}E_\varepsilon =& [\mathtt{f}_1,-\mathtt{f}_2,\mathtt{g}_1,-\mathtt{g}_2]z^{\ell_1} + [\mathtt{f}_3,-\mathtt{f}_4,\mathtt{g}_3,-\mathtt{g}_4]z^{\ell_1+1} + \sum_{\ell=\ell_1+2}^{\ell_2-2} \text{coeff}(\mathsf{q}E_\varepsilon,\ell)z^\ell \\ &+ [\mathtt{f}_3,\mathtt{f}_4,\mathtt{g}_1,\mathtt{g}_2]z^{\ell_2-1} + [\mathtt{f}_1,\mathtt{f}_2,\mathbf{0},0]z^{\ell_2};\end{aligned} \tag{26}$$

$$\begin{aligned}\mathsf{q}E_\varepsilon =& [0,0,\mathtt{f}_1,-\mathtt{f}_2]z^{\ell_1} + [\mathtt{g}_1,-\mathtt{g}_2,\mathtt{f}_3,-\mathtt{f}_4]z^{\ell_1+1} + \sum_{\ell=\ell_1+2}^{\ell_2-2} \text{coeff}(\mathsf{q}E_\varepsilon,\ell)z^\ell \\ &+ [\mathtt{g}_3,\mathtt{g}_4,\mathtt{f}_3,\mathtt{f}_4]z^{\ell_2-1} + [\mathtt{g}_1,\mathtt{g}_2,\mathtt{f}_1,\mathtt{f}_2]z^{\ell_2}.\end{aligned} \tag{27}$$

If $\mathsf{q}E_\varepsilon$ takes the form in (27), we further construct a permutation matrix E_q such that $[\mathtt{g}_1,\mathtt{g}_2,\mathtt{f}_1,\mathtt{f}_2]E_\mathsf{q} = [\mathtt{f}_1,\mathtt{f}_2,\mathtt{g}_1,\mathtt{g}_2]$ and define $\mathsf{U}_{\mathsf{q},\varepsilon} := E_\varepsilon E_\mathsf{q}\,\text{diag}(I_{s-s_\mathsf{g}}, z^{-1}I_{s_\mathsf{g}})$, where s_g is the size of the row vector $[\mathtt{g}_1,\mathtt{g}_2]$. Then, $\mathsf{q}\mathsf{U}_{\mathsf{q},\varepsilon}$ takes the form in (26). For $\mathsf{q}E_\varepsilon$ of form (26), we simply let $\mathsf{U}_{\mathsf{q},\varepsilon} := E_\varepsilon$. In this way, $\mathsf{q}_0 := \mathsf{q}\mathsf{U}_{\mathsf{q},\varepsilon}$ always takes the form in (26) with $\mathtt{f}_1 \neq 0$.

Note that $\mathsf{U}_{\mathsf{q},\varepsilon}\mathsf{U}_{\mathsf{q},\varepsilon}^* = I_s$ and $\|\mathtt{f}_1\| = \|\mathtt{f}_2\|$ if $\mathsf{q}_0\mathsf{q}_0^* = 1$, where $\|\mathtt{f}\| := \sqrt{\mathtt{f}\mathtt{f}^*}$. Now we construct an $s\times s$ paraunitary matrix $\mathsf{B}_{\mathsf{q}_0}$ to reduce the coefficient support of q_0 as in (26) from $[\ell_1,\ell_2]$ to $[\ell_1+1,\ell_2-1]$ as follows:

$$\mathsf{B}_{\mathsf{q}_0}^* := \frac{1}{c}\left[\begin{array}{c|c|c|c}
\mathtt{f}_1(z+\frac{c_0}{c_{\mathtt{f}_1}}+\frac{1}{z}) & \mathtt{f}_2(z-\frac{1}{z}) & \mathtt{g}_1(1+\frac{1}{z}) & \mathtt{g}_2(1-\frac{1}{z}) \\
cF_1 & 0 & 0 & 0 \\ \hline
-\mathtt{f}_1(z-\frac{1}{z}) & -\mathtt{f}_2(z-\frac{c_0}{c_{\mathtt{f}_1}}+\frac{1}{z}) & -\mathtt{g}_1(1-\frac{1}{z}) & -\mathtt{g}_2(1+\frac{1}{z}) \\
0 & cF_2 & 0 & 0 \\ \hline
\frac{c_{\mathtt{g}_1}}{c_{\mathtt{f}_1}}\mathtt{f}_1(1+z) & -\frac{c_{\mathtt{g}_1}}{c_{\mathtt{f}_1}}\mathtt{f}_2(1-z) & c_{\mathtt{g}_1'}\mathtt{g}_1' & 0 \\
0 & 0 & cG_1 & 0 \\ \hline
\frac{c_{\mathtt{g}_2}}{c_{\mathtt{f}_1}}\mathtt{f}_1(1-z) & -\frac{c_{\mathtt{g}_2}}{c_{\mathtt{f}_1}}\mathtt{f}_2(1+z) & 0 & c_{\mathtt{g}_2'}\mathtt{g}_2' \\
0 & 0 & 0 & cG_2
\end{array}\right], \tag{28}$$

where $c_{\mathtt{f}_1} := \|\mathtt{f}_1\|, c_{\mathtt{g}_1} := \|\mathtt{g}_1\|, c_{\mathtt{g}_2} := \|\mathtt{g}_2\|, c_0 := \frac{1}{c_{\mathtt{f}_1}}\mathrm{coeff}(\mathsf{q}_0, \ell_1+1)\mathrm{coeff}(\mathsf{q}_0^*, -\ell_2)$,

$$c_{\mathtt{g}_1'} := \begin{cases} \frac{-2c_{\mathtt{f}_1}-\overline{c_0}}{c_{\mathtt{g}_1}} & \text{if } \mathtt{g}_1 \neq 0; \\ c & \text{otherwise}, \end{cases} \qquad c_{\mathtt{g}_2'} := \begin{cases} \frac{2c_{\mathtt{f}_1}-\overline{c_0}}{c_{\mathtt{g}_2}} & \text{if } \mathtt{g}_2 \neq 0; \\ c & \text{otherwise}, \end{cases} \tag{29}$$
$$c := (4c_{\mathtt{f}_1}^2 + 2c_{\mathtt{g}_1}^2 + 2c_{\mathtt{g}_2}^2 + |c_0|^2)^{1/2},$$

and $[\frac{\mathtt{f}_j^*}{\|\mathtt{f}_j\|}, F_j^*] = U_{\mathtt{f}_j}$, $[\mathtt{g}_j'^*, G_j^*] = U_{\mathtt{g}_j}$ for $j = 1,2$ are unitary constant extension matrices in $\mathbb{F}$ for vectors $\mathtt{f}_j, \mathtt{g}_j$ in $\mathbb{F}$, respectively. Here, the role of a unitary constant matrix $U_{\mathtt{f}}$ in $\mathbb{F}$ is to reduce the number of nonzero entries in $\mathtt{f}$ such that $\mathtt{f}U_{\mathtt{f}} = [\|\mathtt{f}\|, 0, \ldots, 0]$. The operations for the emptyset $\emptyset$ are defined by $\|\emptyset\| = \emptyset$, $\emptyset + A = A$ and $\emptyset \cdot A = \emptyset$ for any object A.

Define $\mathsf{B}_\mathsf{q} := \mathsf{U}_{\mathsf{q},\varepsilon}\mathsf{B}_{\mathsf{q}_0}\mathsf{U}_{\mathsf{q},\varepsilon}^*$. Then, B_q is paraunitary. Due to the particular form of $\mathsf{B}_{\mathsf{q}_0}$ as in (28), direct computations yield the following very important properties of the paraunitary matrix B_q:

(P1) $\mathsf{S}\mathsf{B}_\mathsf{q} = [\mathbf{1}_{s_1}, -\mathbf{1}_{s_2}, z\mathbf{1}_{s_3}, -z\mathbf{1}_{s_4}]^T[\mathbf{1}_{s_1}, -\mathbf{1}_{s_2}, z^{-1}\mathbf{1}_{s_3}, -z^{-1}\mathbf{1}_{s_4}]$, $\mathrm{csupp}(\mathsf{B}_\mathsf{q}) = [-1,1]$, and $\mathrm{csupp}(\mathsf{q}\mathsf{B}_\mathsf{q}) = [\ell_1+1, \ell_2-1]$. That is, B_q has compatible symmetry with coefficient support on $[-1,1]$ and B_q reduces the length of the coefficient support of q exactly by 2. Moreover, $\mathsf{S}(\mathsf{q}\mathsf{B}_\mathsf{q}) = \mathsf{S}\mathsf{q}$.

(P2) If $(\mathsf{p}, \mathsf{q}^*)$ has mutually compatible symmetry and $\mathsf{p}\mathsf{q}^* = 0$, then $\mathsf{S}(\mathsf{p}\mathsf{B}_\mathsf{q}) = \mathsf{S}(\mathsf{p})$ and $\mathrm{csupp}(\mathsf{p}\mathsf{B}_\mathsf{q}) \subseteq \mathrm{csupp}(\mathsf{p})$. That is, B_q keeps the symmetry pattern of p and does not increase the length of the coefficient support of p.

Next, let us explain the construction of $\mathsf{B}_{(-k,k)}$. For $\mathrm{csupp}(\mathsf{Q}) = [-k,k]$ with $k \geq 1$, Q is of the form as follows:

$$\mathsf{Q} = \left[\begin{array}{cccc} F_{11} & -F_{21} & G_{31} & -G_{41} \\ -F_{12} & F_{22} & -G_{32} & G_{42} \\ \hline 0 & 0 & F_{31} & -F_{41} \\ 0 & 0 & -F_{32} & F_{42} \end{array}\right] z^{-k} + \left[\begin{array}{cccc} F_{51} & -F_{61} & G_{71} & -G_{81} \\ -F_{52} & F_{61} & -G_{72} & G_{82} \\ \hline G_{11} & -G_{21} & F_{71} & -F_{81} \\ -G_{12} & G_{22} & -F_{72} & F_{82} \end{array}\right] z^{-k+1}$$
$$+ \sum_{n=2-k}^{k-2} \mathrm{coeff}(\mathsf{Q},n)z^n + \left[\begin{array}{cccc} F_{51} & F_{61} & G_{31} & G_{41} \\ F_{52} & F_{61} & G_{32} & G_{42} \\ \hline G_{51} & G_{61} & F_{71} & F_{81} \\ G_{52} & G_{62} & F_{72} & F_{82} \end{array}\right] z^{k-1} + \left[\begin{array}{cccc} F_{11} & F_{21} & 0 & 0 \\ F_{12} & F_{22} & 0 & 0 \\ \hline G_{11} & G_{21} & F_{31} & F_{41} \\ G_{12} & G_{22} & F_{32} & F_{42} \end{array}\right] z^k \tag{30}$$

with all F_{jk}'s and G_{jk}'s being constant matrices in $\mathbb{F}$ and $F_{11}, F_{22}, F_{31}, F_{42}$ being of size $r_1 \times s_1$, $r_2 \times s_2$, $r_3 \times s_3$, $r_4 \times s_4$, respectively. Due to Properties (P1) and (P2) of B_q, the *for* loop in Algorithm 1 reduces Q in (30) to $\mathsf{Q}_0 := \mathsf{Q}\mathsf{B}_1 \cdots \mathsf{B}_r$ as follows:

$$\left[\begin{array}{cccc} 0 & 0 & \widetilde{G}_{31} & -\widetilde{G}_{41} \\ 0 & 0 & -\widetilde{G}_{32} & \widetilde{G}_{42} \\ \hline 0 & 0 & 0 & 0 \\ 0 & 0 & 0 & 0 \end{array}\right] z^{-k} + \cdots + \left[\begin{array}{cccc} 0 & 0 & 0 & 0 \\ 0 & 0 & 0 & 0 \\ \hline \widetilde{G}_{11} & \widetilde{G}_{21} & 0 & 0 \\ \widetilde{G}_{12} & \widetilde{G}_{22} & 0 & 0 \end{array}\right] z^k. \tag{31}$$

If either $\operatorname{coeff}(\mathsf{Q}_0,-k)=0$ or $\operatorname{coeff}(\mathsf{Q}_0,k)=0$, then the inner *while* loop does nothing and $\mathsf{B}_{(-k,k)}=I_s$. If both $\operatorname{coeff}(\mathsf{Q}_0,-k)\neq 0$ and $\operatorname{coeff}(\mathsf{Q}_0,k)\neq 0$, then $\mathsf{B}_{(-k,k)}$ is constructed recursively from pairs $(\mathsf{q}_1,\mathsf{q}_2)$ with $\mathsf{q}_1,\mathsf{q}_2$ being two rows of Q_0 satisfying $\operatorname{coeff}(\mathsf{q}_1,-k)\neq 0$ and $\operatorname{coeff}(\mathsf{q}_2,k)\neq 0$. The construction of $\mathsf{B}_{(\mathsf{q}_1,\mathsf{q}_2)}$ with respect to such a pair $(\mathsf{q}_1,\mathsf{q}_2)$ in the inner *while* loop is as follows.

Similar to the discussion before (26), there is a permutation matrix $E_{(\mathsf{q}_1,\mathsf{q}_2)}$ such that $\mathsf{q}_1E_{(\mathsf{q}_1,\mathsf{q}_2)}$ and $\mathsf{q}_2E_{(\mathsf{q}_1,\mathsf{q}_2)}$ take the following form:

$$\begin{aligned}\begin{bmatrix}\widetilde{\mathsf{q}}_1\\ \widetilde{\mathsf{q}}_2\end{bmatrix} := \begin{bmatrix}\mathsf{q}_1\\ \mathsf{q}_2\end{bmatrix} E_{(\mathsf{q}_1,\mathsf{q}_2)} &= \left[\begin{array}{cccc} 0 & 0 & \widetilde{g}_3 & -\widetilde{g}_4\\ \hline 0 & 0 & 0 & 0\end{array}\right] z^{-k} + \left[\begin{array}{cccc} \widetilde{f}_5 & -\widetilde{f}_6 & \widetilde{g}_7 & -\widetilde{g}_8\\ \hline \widetilde{g}_1 & -\widetilde{g}_2 & \widetilde{f}_7 & -\widetilde{f}_8\end{array}\right] z^{-k+1}\\ &+ \sum_{n=2-k}^{k-2} \operatorname{coeff}\left(\begin{bmatrix}\widetilde{\mathsf{q}}_1\\ \widetilde{\mathsf{q}}_2\end{bmatrix}, n\right) z^n + \left[\begin{array}{cccc} \widetilde{f}_5 & \widetilde{f}_6 & \widetilde{g}_3 & \widetilde{g}_4\\ \hline \widetilde{g}_5 & \widetilde{g}_6 & \widetilde{f}_7 & \widetilde{f}_8\end{array}\right] z^{k-1} + \left[\begin{array}{cccc} 0 & 0 & 0 & 0\\ \hline \widetilde{g}_1 & \widetilde{g}_2 & 0 & 0\end{array}\right] z^{k},\end{aligned} \tag{32}$$

where $\widetilde{\mathsf{g}}_1,\widetilde{\mathsf{g}}_2,\widetilde{\mathsf{g}}_3,\widetilde{\mathsf{g}}_4$ are all nonzero row vectors. Note that $\|\widetilde{\mathsf{g}}_1\|=\|\widetilde{\mathsf{g}}_2\|=:c_{\widetilde{\mathsf{g}}_1}$ and $\|\widetilde{\mathsf{g}}_3\|=\|\widetilde{\mathsf{g}}_4\|=:c_{\widetilde{\mathsf{g}}_3}$. Construct an $s\times s$ paraunitary matrix $\mathsf{B}_{(\widetilde{\mathsf{q}}_1,\widetilde{\mathsf{q}}_2)}$ as follows:

$$\mathsf{B}^*_{(\widetilde{\mathsf{q}}_1,\widetilde{\mathsf{q}}_2)} := \frac{1}{c}\left[\begin{array}{c|c|c|c} \frac{c_0}{c_{\widetilde{\mathsf{g}}_1}}\widetilde{\mathsf{g}}_1 & 0 & \widetilde{\mathsf{g}}_3(1+\frac{1}{z}) & \widetilde{\mathsf{g}}_4(1-\frac{1}{z})\\ c\widetilde{G}_1 & 0 & 0 & 0\\ \hline 0 & \frac{c_0}{c_{\widetilde{\mathsf{g}}_1}}\widetilde{\mathsf{g}}_2 & -\widetilde{\mathsf{g}}_3(1-\frac{1}{z}) & -\widetilde{\mathsf{g}}_4(1+\frac{1}{z})\\ 0 & c\widetilde{G}_2 & 0 & 0\\ \hline \frac{c_{\widetilde{\mathsf{g}}_3}}{c_{\widetilde{\mathsf{g}}_1}}\widetilde{\mathsf{g}}_1(1+z) & -\frac{c_{\widetilde{\mathsf{g}}_3}}{c_{\widetilde{\mathsf{g}}_1}}\widetilde{\mathsf{g}}_2(1-z) & -\frac{\overline{c_0}}{c_{\widetilde{\mathsf{g}}_3}}\widetilde{\mathsf{g}}_3 & 0\\ 0 & 0 & c\widetilde{G}_3 & 0\\ \hline \frac{c_{\widetilde{\mathsf{g}}_3}}{c_{\widetilde{\mathsf{g}}_1}}\widetilde{\mathsf{g}}_1(1-z) & -\frac{c_{\widetilde{\mathsf{g}}_3}}{c_{\widetilde{\mathsf{g}}_1}}\widetilde{\mathsf{g}}_2(1+z) & 0 & -\frac{\overline{c_0}}{c_{\widetilde{\mathsf{g}}_3}}\widetilde{\mathsf{g}}_4\\ 0 & 0 & 0 & c\widetilde{G}_4 \end{array}\right], \tag{33}$$

where $c_0 := \frac{1}{c_{\widetilde{\mathsf{g}}_1}}\operatorname{coeff}(\widetilde{\mathsf{q}}_1,-k+1)\operatorname{coeff}(\widetilde{\mathsf{q}}_2^*,-k)$, $c:=(|c_0|^2+4c^2_{\widetilde{\mathsf{g}}_3})^{\frac{1}{2}}$, and $[\frac{\widetilde{\mathsf{g}}_j^*}{\|\widetilde{\mathsf{g}}_j\|},\widetilde{G}_j^*]=U_{\widetilde{\mathsf{g}}_j}$ are unitary constant extension matrices in $\mathbb{F}$ for vectors $\widetilde{\mathsf{g}}_j$ in $\mathbb{F}$, $j=1,\ldots,4$, respectively. Let $\mathsf{B}_{(\mathsf{q}_1,\mathsf{q}_2)} := E_{(\mathsf{q}_1,\mathsf{q}_2)}\mathsf{B}_{(\widetilde{\mathsf{q}}_1,\widetilde{\mathsf{q}}_2)}E^{\mathrm{T}}_{(\mathsf{q}_1,\mathsf{q}_2)}$. Similar to Properties (P1) and (P2) of B_{q}, we have the following very important properties of $\mathsf{B}_{(\mathsf{q}_1,\mathsf{q}_2)}$:

(P3) $\mathsf{S}\mathsf{B}_{(\mathsf{q}_1,\mathsf{q}_2)}=[\mathbf{1}_{s_1},-\mathbf{1}_{s_2},z\mathbf{1}_{s_3},-z\mathbf{1}_{s_4}]^{\mathrm{T}}[\mathbf{1}_{s_1},-\mathbf{1}_{s_2},z^{-1}\mathbf{1}_{s_3},-z^{-1}\mathbf{1}_{s_4}]$, $\operatorname{csupp}(\mathsf{B}_{(\mathsf{q}_1,\mathsf{q}_2)})=[-1,1]$, $\operatorname{csupp}(\mathsf{q}_1\mathsf{B}_{(\mathsf{q}_1,\mathsf{q}_2)})\subseteq[-k+1,k-1]$ and $\operatorname{csupp}(\mathsf{q}_2\mathsf{B}_{(\mathsf{q}_1,\mathsf{q}_2)})\subseteq[-k+1,k-1]$. That is, $\mathsf{B}_{(\mathsf{q}_1,\mathsf{q}_2)}$ has compatible symmetry with coefficient support on

$[-1,1]$ and $\mathsf{B}_{(\mathsf{q}_1,\mathsf{q}_2)}$ reduces the length of both the coefficient supports of q_1 and q_2 by 2. Moreover, $\mathsf{S}(\mathsf{q}_1\mathsf{B}_{(\mathsf{q}_1,\mathsf{q}_2)}) = \mathsf{S}\mathsf{q}_1$ and $\mathsf{S}(\mathsf{q}_2\mathsf{B}_{(\mathsf{q}_1,\mathsf{q}_2)}) = \mathsf{S}\mathsf{q}_2$.

(P4) If both $(\mathsf{p},\mathsf{q}_1^*)$ and $(\mathsf{p},\mathsf{q}_2^*)$ have mutually compatible symmetry and $\mathsf{p}\mathsf{q}_1^* = \mathsf{p}\mathsf{q}_2^* = 0$, then $\mathsf{S}(\mathsf{p}\mathsf{B}_{(\mathsf{q}_1,\mathsf{q}_2)}) = \mathsf{S}\mathsf{p}$ and $\operatorname{csupp}(\mathsf{p}\mathsf{B}_{(\mathsf{q}_1,\mathsf{q}_2)}) \subseteq \operatorname{csupp}(\mathsf{p})$. That is, $\mathsf{B}_{(\mathsf{q}_1,\mathsf{q}_2)}$ keeps the symmetry pattern of p and does not increase the length of the coefficient support of p.

Now, due to Properties (P3) and (P4) of $\mathsf{B}_{(\mathsf{q}_1,\mathsf{q}_2)}$, $\mathsf{B}_{(-k,k)}$ constructed in the inner *while* loop reduces Q_0 of the form in (31) with both $\operatorname{coeff}(\mathsf{Q}_0,-k) \neq 0$ and $\operatorname{coeff}(\mathsf{Q}_0,k) \neq 0$, to $\mathsf{Q}_1 := \mathsf{Q}_0\mathsf{B}_{(-k,k)}$ of the form in (31) with either $\operatorname{coeff}(\mathsf{Q}_1,-k) = \operatorname{coeff}(\mathsf{Q}_1,k) = 0$ (for this case, simply let $\mathsf{B}_{\mathsf{Q}_1} := I_s$) or one of $\operatorname{coeff}(\mathsf{Q}_1,-k)$ and $\operatorname{coeff}(\mathsf{Q}_1,k)$ is nonzero. For the latter case, $\mathsf{B}_{\mathsf{Q}_1} := \operatorname{diag}(U_1\mathsf{W}_1, I_{s_3+s_4})E$ with U_1, W_1 constructed with respect to $\operatorname{coeff}(\mathsf{Q}_1,k) \neq 0$ or $\mathsf{B}_{\mathsf{Q}_1} := \operatorname{diag}(I_{s_1+s_2}, U_3\mathsf{W}_3)E$ with U_3, W_3 constructed with respect to $\operatorname{coeff}(\mathsf{Q}_1,-k) \neq 0$, where E is a permutation matrix. $\mathsf{B}_{\mathsf{Q}_1}$ is constructed so that $\operatorname{csupp}(\mathsf{Q}_1\mathsf{B}_{\mathsf{Q}_1}) \subseteq [-k+1,k-1]$. Let Q_1 take form in (31). The matrices U_1, W_1 or U_3, W_3, and E are constructed as follows.

Let $U_1 := \operatorname{diag}(U_{\widetilde{G}_1}, U_{\widetilde{G}_2})$ and $U_3 := \operatorname{diag}(U_{\widetilde{G}_3}, U_{\widetilde{G}_4})$ with

$$\widetilde{G}_1 := \begin{bmatrix} \widetilde{G}_{11} \\ \widetilde{G}_{12} \end{bmatrix}, \widetilde{G}_2 := \begin{bmatrix} \widetilde{G}_{21} \\ \widetilde{G}_{22} \end{bmatrix}, \widetilde{G}_3 := \begin{bmatrix} \widetilde{G}_{31} \\ \widetilde{G}_{32} \end{bmatrix}, \widetilde{G}_4 := \begin{bmatrix} \widetilde{G}_{41} \\ \widetilde{G}_{42} \end{bmatrix}. \tag{34}$$

Here, for a nonzero matrix G with rank m, U_G is a unitary matrix such that $GU_G = [R,0]$ for some matrix R of rank m. For $G = 0, U_G := I$ and for $G = \emptyset, U_G := \emptyset$. When $G_1G_1^* = G_2G_2^*$, U_{G_1} and U_{G_2} can be constructed such that $G_1U_{G_1} = [R,0]$ and $G_2U_{G_2} = [R,0]$.

Let m_1, m_3 be the ranks of $\widetilde{G}_1$, $\widetilde{G}_3$, respectively ($m_1 = 0$ when $\operatorname{coeff}(\mathsf{Q}_1,k) = 0$ and $m_3 = 0$ when $\operatorname{coeff}(\mathsf{Q}_1,-k) = 0$). Note that $\widetilde{G}_1\widetilde{G}_1^* = \widetilde{G}_2\widetilde{G}_2^*$ or $\widetilde{G}_3\widetilde{G}_3^* = \widetilde{G}_4\widetilde{G}_4^*$ due to $\mathsf{Q}_1\mathsf{Q}_1^* = I_r$. The matrices $\mathsf{W}_1, \mathsf{W}_3$ are then constructed by

$$\mathsf{W}_1 := \left[\begin{array}{c|c|c|c} \mathsf{U}_1 & & \mathsf{U}_2 & \\ \hline & I_{s_1-m_1} & & \\ \hline \mathsf{U}_2 & & \mathsf{U}_1 & \\ \hline & & & I_{s_2-m_1} \end{array}\right], \mathsf{W}_3 := \left[\begin{array}{c|c|c|c} \mathsf{U}_3 & & \mathsf{U}_4 & \\ \hline & I_{s_3-m_3} & & \\ \hline \mathsf{U}_4 & & \mathsf{U}_3 & \\ \hline & & & I_{s_4-m_3} \end{array}\right], \tag{35}$$

where $\mathsf{U}_1(z) = -\mathsf{U}_2(-z) := \frac{1+z^{-1}}{2}I_{m_1}$ and $\mathsf{U}_3(z) = \mathsf{U}_4(-z) := \frac{1+z}{2}I_{m_3}$.

Let $\mathsf{W}_{\mathsf{Q}_1} := \operatorname{diag}(U_1\mathsf{W}_1, I_{s_3+s_4})$ for the case that $\operatorname{coeff}(\mathsf{Q}_1,k) \neq 0$ or $\mathsf{W}_{\mathsf{Q}_1} := \operatorname{diag}(I_{s_1+s_2}, U_3\mathsf{W}_3)$ for the case that $\operatorname{coeff}(\mathsf{Q}_1,-k) \neq 0$. Then $\mathsf{W}_{\mathsf{Q}_1}$ is paraunitary. By the symmetry pattern and orthogonality of Q_1, $\mathsf{W}_{\mathsf{Q}_1}$ reduces the coefficient support of Q_1 to $[-k+1,k-1]$, i.e., $\operatorname{csupp}(\mathsf{Q}_1\mathsf{W}_{\mathsf{Q}_1}) = [-k+1,k-1]$. Moreover, $\mathsf{W}_{\mathsf{Q}_1}$ changes the symmetry pattern of Q_1 such that

$$\mathsf{S}(\mathsf{Q}_1\mathsf{W}_{\mathsf{Q}_1}) = [\mathbf{1}_{r_1}, -\mathbf{1}_{r_2}, z\mathbf{1}_{r_3}, -z\mathbf{1}_{r_4}]^{\mathsf{T}}\mathsf{S}\theta_1,$$

with

$$\mathsf{S}\theta_1 = [z^{-1}\mathbf{1}_{m_1}, \mathbf{1}_{s_1-m_1}, -z^{-1}\mathbf{1}_{m_1}, -\mathbf{1}_{s_2-m_1}, \mathbf{1}_{m_3}, z^{-1}\mathbf{1}_{s_3-m_3}, -\mathbf{1}_{m_3}, -z^{-1}\mathbf{1}_{s_4-m_3}].$$

E is then the permutation matrix such that

$$\mathsf{S}(\mathsf{Q}_1\mathsf{W}_{\mathsf{Q}_1})E = [\mathbf{1}_{r_1}, -\mathbf{1}_{r_2}, z\mathbf{1}_{r_3}, -z\mathbf{1}_{r_4},]^{\mathsf{T}}\mathsf{S}\theta,$$

with $\mathsf{S}\theta = [\mathbf{1}_{s_1-m_1+m_3},, -\mathbf{1}_{s_2-m_1+m_3}, z^{-1}\mathbf{1}_{s_3-m_3+m_1}, -z^{-1}\mathbf{1}_{s_4-m_3+m_1}] = (\mathsf{S}\theta_1)E.$

2.2 Application to Filter Banks and Orthonormal Multiwavelets with Symmetry

In this subsection, we shall discuss the application of our results on orthogonal matrix extension with symmetry to d-band symmetric paraunitary filter banks in electronic engineering and to orthonormal multiwavelets with symmetry in wavelet analysis.

Symmetry of the filters in a filter bank is a very much desirable property in many applications. We say that the low-pass filter a_0 with multiplicity r has symmetry if

$$\mathsf{a}_0(z) = \operatorname{diag}(\varepsilon_1 z^{\mathsf{d}c_1}, \dots, \varepsilon_r z^{\mathsf{d}c_r})\mathsf{a}_0(1/z)\operatorname{diag}(\varepsilon_1 z^{-c_1}, \dots, \varepsilon_r z^{-c_r}) \qquad (36)$$

for some $\varepsilon_1, \dots, \varepsilon_r \in \{-1, 1\}$ and $c_1, \dots, c_r \in \mathbb{R}$ such that $\mathsf{d}c_\ell - c_j \in \mathbb{Z}$ for all $\ell, j = 1, \dots, r$. If a_0 has symmetry as in (36) and if 1 is a simple eigenvalue of $\mathsf{a}_0(1)$, then it is well known that the d-refinable function vector ϕ in (6) associated with the low-pass filter a_0 has the following symmetry:

$$\phi_1(c_1 - \cdot) = \varepsilon_1\phi_1, \quad \phi_2(c_2 - \cdot) = \varepsilon_2\phi_2, \quad \dots, \quad \phi_r(c_r - \cdot) = \varepsilon_r\phi_r. \qquad (37)$$

Under the symmetry condition in (36), to apply Theorem 1, we first show that there exists a suitable paraunitary matrix U acting on $\mathbb{P}_{\mathsf{a}_0} := [\mathsf{a}_{0;0}, \dots, \mathsf{a}_{0;\mathsf{d}-1}]$ so that $\mathbb{P}_{\mathsf{a}_0}\mathsf{U}$ has compatible symmetry. Note that $\mathbb{P}_{\mathsf{a}_0}$ itself may not have any symmetry.

Lemma 1. *Let $\mathbb{P}_{\mathsf{a}_0} := [\mathsf{a}_{0;0}, \dots, \mathsf{a}_{0;\mathsf{d}-1}]$, where $\mathsf{a}_{0;0}, \dots, \mathsf{a}_{0;\mathsf{d}-1}$ are* d*-band subsymbols of a* d*-band orthogonal filter a_0 satisfying* (36)*. Then there exists a $\mathsf{d}r \times \mathsf{d}r$ paraunitary matrix U such that $\mathbb{P}_{\mathsf{a}_0}\mathsf{U}$ has compatible symmetry.*

Proof. From (36), we deduce that

$$[\mathsf{a}_{0;\gamma}(z)]_{\ell,j} = \varepsilon_\ell\varepsilon_j z^{R^\gamma_{\ell,j}}[\mathsf{a}_{0;Q^\gamma_{\ell,j}}(z^{-1})]_{\ell,j}, \ \gamma = 0, \dots, \mathsf{d}-1; \ell, j = 1, \dots, r, \qquad (38)$$

where $\gamma, Q^\gamma_{\ell,j} \in \Gamma := \{0, \dots, \mathsf{d}-1\}$ and $R^\gamma_{\ell,j}$, $Q^\gamma_{\ell,j}$ are uniquely determined by

$$\mathsf{d}c_\ell - c_j - \gamma = \mathsf{d}R^\gamma_{\ell,j} + Q^\gamma_{\ell,j} \quad Q^\gamma_{\ell,j} \in \Gamma.\text{with} \quad R^\gamma_{\ell,j} \in \mathbb{Z}, \qquad (39)$$

Since $\mathsf{d}c_\ell - c_j \in \mathbb{Z}$ for all $\ell, j = 1, \dots, r$, we have $c_\ell - c_j \in \mathbb{Z}$ for all $\ell, j = 1, \dots, r$ and therefore, $Q^\gamma_{\ell,j}$ is independent of ℓ. Consequently, by (38), for every $1 \le j \le r$, the jth column of the matrix $\mathsf{a}_{0;\gamma}$ is a flipped version of the jth column of the matrix

$\mathsf{a}_{0;Q^\gamma_{\ell,j}}$. Let $\kappa_{j,\gamma} \in \mathbb{Z}$ be an integer such that $|\mathrm{csupp}([\mathsf{a}_{0;\gamma}]_{:,j} + z^{\kappa_{j,\gamma}}[\mathsf{a}_{0;Q^\gamma_{\ell,j}}]_{:,j})|$ is as small as possible. Define $\mathbb{P} := [\mathsf{b}_{0;0}, \dots, \mathsf{b}_{0;\mathsf{d}-1}]$ as follows:

$$[\mathsf{b}_{0;\gamma}]_{:,j} := \begin{cases} [\mathsf{a}_{0;\gamma}]_{:,j}, & \gamma = Q^\gamma_{\ell,j}; \\ \frac{1}{\sqrt{2}}([\mathsf{a}_{0;\gamma}]_{:,j} + z^{\kappa_{j,\gamma}}[\mathsf{a}_{0;Q^\gamma_{\ell,j}}]_{:,j}), & \gamma < Q^\gamma_{\ell,j}; \\ \frac{1}{\sqrt{2}}([\mathsf{a}_{0;\gamma}]_{:,j} - z^{\kappa_{j,\gamma}}[\mathsf{a}_{0;Q^\gamma_{\ell,j}}]_{:,j}), & \gamma > Q^\gamma_{\ell,j}, \end{cases} \tag{40}$$

where $[\mathsf{a}_{0;\gamma}]_{:,j}$ denotes the jth column of $\mathsf{a}_{0;\gamma}$. Let U denote the unique transform matrix corresponding to (40) such that $\mathbb{P} := [\mathsf{b}_{0;0}, \dots, \mathsf{b}_{0;\mathsf{d}-1}] = [\mathsf{a}_{0;0}, \dots, \mathsf{a}_{0;\mathsf{d}-1}]\mathsf{U}$. It is evident that U is paraunitary and $\mathbb{P} = \mathbb{P}_{\mathsf{a}_0}\mathsf{U}$. We now show that $\mathbb{P}$ has compatible symmetry. Indeed, by (38) and (40),

$$[\mathsf{S}\mathsf{b}_{0;\gamma}]_{\ell,j} = \mathrm{sgn}(Q^\gamma_{\ell,j} - \gamma)\varepsilon_\ell \varepsilon_j z^{R^\gamma_{\ell,j} + \kappa_{j,\gamma}}, \tag{41}$$

where $\mathrm{sgn}(x) = 1$ for $x \ge 0$ and $\mathrm{sgn}(x) = -1$ for $x < 0$. By (39) and noting that $Q^\gamma_{\ell,j}$ is independent of ℓ, we have

$$\frac{[\mathsf{S}\mathsf{b}_{0;\gamma}]_{\ell,j}}{[\mathsf{S}\mathsf{b}_{0;\gamma}]_{n,j}} = \varepsilon_\ell \varepsilon_n z^{R^\gamma_{\ell,j} - R^\gamma_{n,j}} = \varepsilon_\ell \varepsilon_n z^{c_\ell - c_n}, \quad \ell, n = 1, \dots, r,$$

which is equivalent to saying that $\mathbb{P}$ has compatible symmetry. □ □

Now, for a d-band orthogonal low-pass filter a_0 satisfying (36), we have an algorithm to construct high-pass filters $\mathsf{a}_1, \dots, \mathsf{a}_{\mathsf{d}-1}$ such that they form a symmetric paraunitary filter bank with the perfect reconstruction property. See Algorithm 2.

Algorithm 2 Construction of orthonormal multiwavelets with symmetry

(a) **Input**: An orthogonal d-band filter a_0 with symmetry in (36).

(b) **Initialization**: Construct U with respect to (40) such that $\mathbb{P} := \mathbb{P}_{\mathsf{a}_0}\mathsf{U}$ has compatible symmetry: $\mathsf{S}\mathbb{P} = [\varepsilon_1 z^{k_1}, \dots, \varepsilon_r z^{k_r}]^{\mathsf{T}}\mathsf{S}\theta$ for some $k_1, \dots, k_r \in \mathbb{Z}$ and some $1 \times \mathsf{d}r$ row vector θ of Laurent polynomials with symmetry.

(c) **Extension**: Derive $\mathbb{P}_e$ with all the properties as in Theorem 1 from $\mathbb{P}$ by Algorithm 1.

(d) **High-pass Filters**: Let $\mathbf{P} := \mathbb{P}_e\mathsf{U}^* =: (\mathsf{a}_{m;\gamma})_{0 \le m,\gamma \le \mathsf{d}-1}$ as in (16). Define high-pass filters

$$\mathsf{a}_m(z) := \frac{1}{\sqrt{\mathsf{d}}} \sum_{\gamma=0}^{\mathsf{d}-1} \mathsf{a}_{m;\gamma}(z^{\mathsf{d}}) z^\gamma, \qquad m = 1, \dots, \mathsf{d}-1. \tag{42}$$

(f) **Output**: A symmetric filter bank $\{\mathsf{a}_0, \mathsf{a}_1, \dots, \mathsf{a}_{\mathsf{d}-1}\}$ with the perfect reconstruction property, i.e., $\mathbf{P}$ in (16) is paraunitary and all filters a_m, $m = 1, \dots, \mathsf{d}-1$, have symmetry:

$$\mathsf{a}_m(z) = \mathrm{diag}(\varepsilon_1^m z^{\mathsf{d}c_1^m}, \dots, \varepsilon_r^m z^{\mathsf{d}c_r^m})\mathsf{a}_m(1/z)\,\mathrm{diag}(\varepsilon_1 z^{-c_1}, \dots, \varepsilon_r z^{-c_r}), \tag{43}$$

where $c_\ell^m := (k_\ell^m - k_\ell) + c_\ell \in \mathbb{R}$ and all $\varepsilon_\ell^m \in \{-1, 1\}$, $k_\ell^m \in \mathbb{Z}$, for $\ell, j = 1, \dots, r$ and $m = 1, \dots, \mathsf{d}-1$, are determined by the symmetry pattern of $\mathbb{P}_e$ as follows:

$$[\varepsilon_1 z^{k_1}, \dots, \varepsilon_r z^{k_r}, \varepsilon_1^1 z^{k_1^1}, \dots, \varepsilon_r^1 z^{k_r^1}, \dots, z^{k_1^{\mathsf{d}-1}}, \dots, \varepsilon_r^{\mathsf{d}-1} z^{k_r^{\mathsf{d}-1}}]^{\mathsf{T}}\mathsf{S}\theta := \mathsf{S}\mathbb{P}_e. \tag{44}$$

Proof (of Algorithm 2). Rewrite $\mathbb{P}_e = (\mathsf{b}_{m;\gamma})_{0\le m,\gamma\le \mathsf{d}-1}$ as a $\mathsf{d}\times\mathsf{d}$ block matrix with $r\times r$ blocks $\mathsf{b}_{m;\gamma}$. Since $\mathbb{P}_e$ has compatible symmetry as in (44), we have $[\mathsf{S}\mathsf{b}_{m;\gamma}]_{\ell,:} = \varepsilon_\ell^m \varepsilon_\ell z^{k_\ell^m - k_\ell}[\mathsf{S}\mathsf{b}_{0;\gamma}]_{\ell,:}$ for $\ell = 1,\dots,r$ and $m = 1,\dots,\mathsf{d}-1$. By (41), we have

$$[\mathsf{S}\mathsf{b}_{m;\gamma}]_{\ell,j} = \operatorname{sgn}(Q_{\ell,j}^\gamma - \gamma)\varepsilon_\ell^m \varepsilon_j z^{R_{\ell,j}^\gamma + k_{j,\gamma} + k_\ell^m - k_\ell}, \qquad \ell, j = 1,\dots,r. \tag{45}$$

By (45) and the definition of U^* in (40), we deduce that

$$[\mathsf{a}_{m;\gamma}]_{\ell,j} = \varepsilon_\ell^m \varepsilon_j z^{R_{\ell,j}^\gamma + k_\ell^m - k_\ell}[\mathsf{a}_{m;Q_{\ell,j}^\gamma}(z^{-1})]_{\ell,j}. \tag{46}$$

This implies that $[\mathsf{S}\mathsf{a}_m]_{\ell,j} = \varepsilon_\ell^m \varepsilon_j z^{\mathsf{d}(k_\ell^m - k_\ell + c_\ell) - c_j}$, which is equivalent to (43) with $c_\ell^m := k_\ell^m - k_\ell + c_\ell$ for $m = 1,\dots,\mathsf{d}-1$ and $\ell = 1,\dots,r$. □ □

Since the high-pass filters $\mathsf{a}_1,\dots,\mathsf{a}_{\mathsf{d}-1}$ satisfy (43), it is easy to verify that each $\psi^m = [\psi_1^m,\dots,\psi_r^m]^{\mathsf{T}}$ defined in (10) also has the following symmetry:

$$\psi_1^m(c_1^m - \cdot) = \varepsilon_1^m \psi_1^m, \quad \psi_2^m(c_2^m - \cdot) = \varepsilon_2^m \psi_2^m, \quad \dots, \quad \psi_r^m(c_r^m - \cdot) = \varepsilon_r^m \psi_r^m. \tag{47}$$

In the following, let us present an example to demonstrate our results and illustrate our algorithms (for more examples, see [16]).

Example 1. Let $\mathsf{d} = 3$ and $r = 2$. Let a_0 be the 3-band orthogonal low-pass filter with multiplicity 2 obtained in [15, Example 4]. Then

$$\mathsf{a}_0(z) = \frac{1}{540}\begin{bmatrix} a_{11}(z) + a_{11}(z^{-1}) & a_{12}(z) + z^{-1}a_{12}(z^{-1}) \\ a_{21}(z) + z^3 a_{21}(z^{-1}) & a_{22}(z) + z^2 a_{22}(z^{-1}) \end{bmatrix},$$

where

$$\begin{aligned}
a_{11}(z) &= 90 + (55 - 5\sqrt{41})z - (8 + 2\sqrt{41})z^2 + (7\sqrt{41} - 47)z^4,\\
a_{12}(z) &= 145 + 5\sqrt{41} + (1 - \sqrt{41})z^2 + (34 - 4\sqrt{41})z^3,\\
a_{21}(z) &= (111 + 9\sqrt{41})z^2 + (69 - 9\sqrt{41})z^4,\\
a_{22}(z) &= 90z + (63 - 3\sqrt{41})z^2 + (3\sqrt{41} - 63)z^3.
\end{aligned}$$

The low-pass filter a_0 satisfies (36) with $c_1 = 0, c_2 = 1$ and $\varepsilon_1 = \varepsilon_2 = 1$. From $\mathbb{P}_{\mathsf{a}_0} := [\mathsf{a}_{0;0}, \mathsf{a}_{0;1}, \mathsf{a}_{0;2}]$, the matrix U constructed by Lemma 1 is given by

$$\mathsf{U} := \frac{1}{\sqrt{2}}\begin{bmatrix} \sqrt{2} & 0 & 0 & 0 & 0 & 0 \\ 0 & 1 & 0 & 0 & 0 & 1 \\ 0 & 0 & 1 & 0 & 1 & 0 \\ 0 & 0 & 0 & \sqrt{2} & 0 & 0 \\ 0 & 0 & z & 0 & -z & 0 \\ 0 & z & 0 & 0 & 0 & -z \end{bmatrix}.$$

Let

$$c_0 = 11-\sqrt{41}, \quad t_{12} = 5(7-\sqrt{41}), \quad c_{12} = 10(29+\sqrt{41}), \quad t_{13} = -5c_0,$$
$$t_{16} = 3c_0, \quad t_{15} = 3(3\sqrt{41}-13), \quad t_{25} = 6(7+3\sqrt{41}), \quad t_{26} = 6(21-\sqrt{41}),$$
$$t_{53} = 400\sqrt{6}/c_0, \quad t_{55} = 12\sqrt{6}(\sqrt{41}-1), \quad t_{56} = 6\sqrt{6}(4+\sqrt{41}), \quad c_{66} = 3\sqrt{6}(3+7\sqrt{41}).$$

Then, $\mathbb{P} := \mathbb{P}_{\mathsf{a}_0}\mathsf{U}$ satisfies $\mathsf{S}\mathbb{P} = [1,z]^{\mathsf{T}}[1,1,1,z^{-1},-1,-1]$ and is given by

$$\mathbb{P} = \frac{\sqrt{6}}{1080}\begin{bmatrix} 180\sqrt{2} & b_{12}(z) & b_{13}(z) & 0 & t_{15}(z-z^{-1}) & t_{16}(z-z^{-1}) \\ 0 & 0 & 180(1+z) & 180\sqrt{2} & t_{25}(1-z) & t_{26}(1-z) \end{bmatrix},$$

where $b_{12}(z) = t_{12}(z+z^{-1}) + c_{12}$ and $b_{13}(z) = t_{13}(z-2+z^{-1})$. Applying Algorithm 1, we obtain a desired paraunitary matrix $\mathbb{P}_e$ as follows:

$$\mathbb{P}_e = \frac{\sqrt{6}}{1080}\left[\begin{array}{cccccc} 180\sqrt{2} & b_{12}(z) & b_{13}(z) & 0 & t_{15}(z-\frac{1}{z}) & t_{16}(z-\frac{1}{z}) \\ 0 & 0 & 180(1+z) & 180\sqrt{2} & t_{25}(1-z) & t_{26}(1-z) \\ \hline 360 & -\frac{b_{12}(z)}{\sqrt{2}} & -\frac{b_{13}(z)}{\sqrt{2}} & 0 & \frac{t_{15}}{\sqrt{2}}(\frac{1}{z}-z) & \frac{t_{16}}{\sqrt{2}}(\frac{1}{z}-z) \\ 0 & 0 & 90\sqrt{2}(1+z) & -360 & \frac{t_{25}}{\sqrt{2}}(1-z) & \frac{t_{26}}{\sqrt{2}}(1-z) \\ \hline 0 & \sqrt{6}t_{13}(1-z) & t_{53}(1-z) & 0 & t_{55}(1+z) & t_{56}(1+z) \\ 0 & \frac{\sqrt{6}t_{12}}{2}(\frac{1}{z}-z) & \frac{\sqrt{6}t_{13}}{2}(\frac{1}{z}-z) & 0 & b_{65}(z) & b_{66}(z) \end{array}\right],$$

where $b_{65}(z) = -\sqrt{6}(5t_{15}(z+z^{-1}) + 3c_{12})/10$ and $b_{66}(z) = -\sqrt{6}t_{16}(z+z^{-1})/2 + c_{66}$. Note that $\mathsf{S}\mathbb{P}_e = [1,z,1,z,-z,-1]^{\mathsf{T}}[1,1,1,z^{-1},-1,-1]$ and the coefficient support of $\mathbb{P}_e$ satisfies $\mathrm{csupp}([\mathbb{P}_e]_{:,j}) \subseteq \mathrm{csupp}([\mathbb{P}]_{:,j})$ for all $1 \le j \le 6$. From the polyphase matrix $\mathbf{P} := \mathbb{P}_e\mathsf{U}^* =: (\mathsf{a}_{m;\gamma})_{0\le m,\gamma\le 2}$, we derive two high-pass filters $\mathsf{a}_1, \mathsf{a}_2$ as follows:

$$\mathsf{a}_1(z) = \frac{\sqrt{2}}{1080}\begin{bmatrix} a^1_{11}(z)+a^1_{11}(z^{-1}) & a^1_{12}(z)+z^{-1}a^1_{12}(z^{-1}) \\ a^1_{21}(z)+z^3a^1_{21}(z^{-1}) & a^1_{22}(z)+z^2a^1_{22}(z^{-1}) \end{bmatrix},$$

$$\mathsf{a}_2(z) = \frac{\sqrt{6}}{1080}\begin{bmatrix} a^2_{11}(z)-z^3a^2_{11}(z^{-1}) & a^2_{12}(z)-z^2a^2_{12}(z^{-1}) \\ a^2_{21}(z)-a^2_{21}(z^{-1}) & a^2_{22}(z)-z^{-1}a^2_{22}(z^{-1}) \end{bmatrix},$$

where

$$a^1_{11}(z) = (47-7\sqrt{41})z^4 + 2(4+\sqrt{41})z^2 + 5(\sqrt{41}-11)z + 180,$$
$$a^1_{12}(z) = 2(2\sqrt{41}-17)z^3 + (\sqrt{41}-1)z^2 - 5(29+\sqrt{41}),$$
$$a^1_{21}(z) = 3(37+3\sqrt{41})z + 3(23-3\sqrt{41})z^{-1},$$
$$a^1_{22}(z) = -180z + 3(21-\sqrt{41}) - 3(21-\sqrt{41})z^{-1},$$
$$a^2_{11}(z) = (43+17\sqrt{41})z + (67-7\sqrt{41})z^{-1},$$
$$a^2_{12}(z) = 11\sqrt{41} - 31 - (79+\sqrt{41})z^{-1},$$
$$a^2_{21}(z) = (47-7\sqrt{41})z^4 + 2(4+\sqrt{41})z^2 - 3(29+\sqrt{41})z,$$
$$a^2_{22}(z) = 2(2\sqrt{41}-17)z^3 + (\sqrt{41}-1)z^2 + 3(3+7\sqrt{41}).$$

Then the high-pass filters a_1, a_2 satisfy (43) with $c_1^1 = 0$, $c_2^1 = 1$, $\varepsilon_1^1 = \varepsilon_2^1 = 1$ and $c_1^2 = 1$, $c_2^2 = 0$, $\varepsilon_1^2 = \varepsilon_2^2 = -1$. See Fig. 1 for graphs of the 3-refinable function vector ϕ associated with the low-pass filter a_0 and the multiwavelet function vectors ψ^1, ψ^2 associated with the high-pass filters $\mathsf{a}_1, \mathsf{a}_2$, respectively.

3 Construction of Symmetric Complex Tight Framelets

Redundant wavelet systems ($L \geq \mathsf{d}$ in (17)) have been proved to be quit useful in many applications, for examples, signal denoising, image processing, and numerical algorithm. As a redundant system, it can possess many desirable properties such as symmetry, short support, high vanishing moments, and so on, simultaneously (see [6, 7, 12, 22]). In this section, we are interested in the construction of tight framelets with such desirable properties. Due to [6], the whole picture of constructing tight framelets with high order of vanishing moments is more or less clear. Yet, when comes to symmetry, there is no general way of deriving tight framelet systems with symmetry. Especially when one requires the number of framelet generators is as less as possible. In this section, we first provide a general result on the construction of d-refinable functions with symmetry such that (14) holds. Once such a d-refinable function is obtained, we then show that using our results on orthogonal matrix extension with symmetry studied in Sect. 2, we can construct a symmetric tight framelet system with only d or $\mathsf{d}+1$ framelet generators.

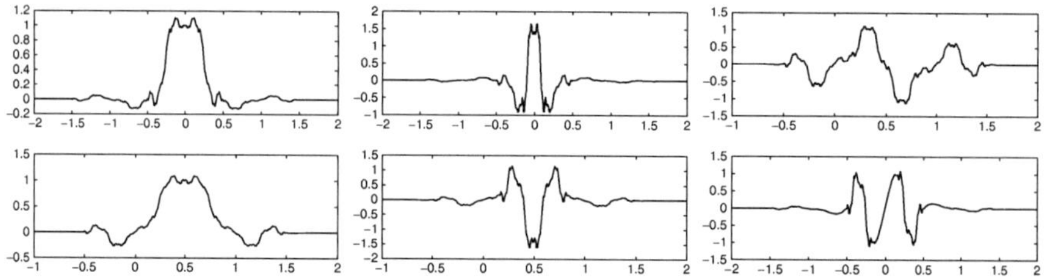

Fig. 1: Graphs of the 3-refinable function vector $\phi = [\phi_1, \phi_2]^{\mathrm{T}}$ associated with a_0 (*left column*), multiwavelet function vector $\psi^1 = [\psi_1^1, \psi_2^1]^{\mathrm{T}}$ associated with a_1 (*middle column*), and multiwavelet function vector $\psi^2 = [\psi_1^2, \psi_2^2]^{\mathrm{T}}$ associated with a_2 (*right column*) in Example 1

3.1 Symmetric Complex d*-Refinable Functions*

Let ϕ be a d-refinable functions associated with a low-pass filters a_0. To have high order of vanishing moments for a tight framelet system, we need to design a_0 such that (14) holds for some $n \in \mathbb{N}$. To guarantee that the d-refinable function ϕ associated with a_0 has certain regularity and polynomial reproducibility, usually the low-pass filter a_0 satisfies the *sum rules of order m* for some $m \in \mathbb{N}$. More precisely, $\widehat{a_0}$ is of the form:

$$\widehat{a_0}(\xi) = \left(\frac{1+e^{-i\xi}+\cdots+e^{-i(d-1)\xi}}{d}\right)^m \widehat{\mathscr{L}}(\xi), \quad \xi \in \mathbb{R} \tag{48}$$

for some 2π-periodic trigonometric polynomial $\widehat{\mathscr{L}}(\xi)$ with $\widehat{\mathscr{L}}(0)=1$. For $\widehat{\mathscr{L}}(\xi)\equiv 1$. a_0 is the low-pass filter for B-spline of order m: $\widehat{B_m}(\xi) = (1-e^{-i\xi})^m/(i\xi)^m$.

Define a function h by

$$h(y) := \prod_{k=1}^{d-1}\left(1 - \frac{y}{\sin^2(k\pi/d)}\right), \quad y \in \mathbb{R}. \tag{49}$$

One can show that

$$h(\sin^2(\xi/2)) = \frac{|1+\cdots+e^{-i(d-1)\xi}|^2}{d^2} = \frac{\sin^2(d\xi/2)}{d^2\sin^2(\xi/2)} \tag{50}$$

and

$$h(y)^{-m} = \left[\prod_{k=1}^{d-1}\left(\sum_{j_k=0}^{\infty}\frac{y^{j_k}}{\sin^{2j_k}(k\pi/d)}\right)\right]^{-m} = \sum_{j=0}^{\infty} c_{m,j}y^j, \quad |y| < \sin^2(\pi/d), \tag{51}$$

where

$$c_{m,j} = \sum_{j_1+\cdots+j_{d-1}=j}\prod_{k=1}^{d-1}\binom{m-1+j_k}{j_k}\sin(k\pi/d)^{-2j_k}, \quad j \in \mathbb{N}. \tag{52}$$

Define $P_{m,n}(y)$ a polynomial of degree $n-1$ as follows:

$$P_{m,n}(y) = \sum_{j=0}^{n-1}\left[\sum_{j_1+\cdots+j_{d-1}=j}\prod_{k=1}^{d-1}\binom{m-1+j_k}{j_k}\sin(k\pi/d)^{-2j_k}\right]y^j. \tag{53}$$

By convention, $\binom{m}{j} = 0$ if $j < 0$. Note that $P_{m,n}(y) = \sum_{j=0}^{n-1} c_{m,j}y^j$. Then, it is easy to show the following result by Taylor expansion.

Lemma 2. *Let $m,n \in \mathbb{N}$ be such that $n \le m$; let $P_{m,n}$ and h be polynomials defined as in* (53) *and* (49)*, respectively. Then $P_{m,n}(\sin^2(\xi/2))$ is the unique positive trigonometric polynomial of minimal degree such that*

$$1 - h(\sin^2(\xi/2))^m P_{m,n}(\sin^2(\xi/2)) = O(|\xi|^{2n}), \quad \xi \to 0. \tag{54}$$

For $m,n \in \mathbb{N}$ such that $1 \le n \le m$, let $\widehat{{}_{II}a_0}(\xi) := h(\sin^2(\xi/2))^m P_{m,n}(\sin^2(\xi/2))$. Then the d-refinable function ${}_{II}\phi$ associated with ${}_{II}a_0$ by (6) is called the d-*refinable pseudo spline of type II with order* (m,n). By Lemma 2, using Riesz Lemma, one can derive a low-pass filter ${}_{I}a_0$ from ${}_{II}a_0$ such that $|\widehat{{}_{I}a_0}(\xi)|^2 = \widehat{{}_{II}a_0}(\xi)$. The d-refinable function ${}_{I}\phi$ associated with such ${}_{I}a_0$ by (6) is referred as *real* d-*refinable pseudo spline of type I with order* (m,n). Interesting readers can refer to [6, 7, 22] for more details on this subject for the special case $\mathsf{d} = 2$.

Note that ${}_{I}a_0$ satisfies (14). One can construct high-pass filters $a_1,\dots,a_L$ from $a_0 := {}_{I}a_0$ such that (12) holds. Then $\psi^1,\dots,\psi^L$ defined by (10) are real-valued functions. $\{\psi^1,\dots,\psi^L\}$ has vanishing moment of order n and generates a tight d-frame. However, $\{\psi^1,\dots,\psi^L\}$ does not necessarily have symmetry since the low-pass filter ${}_{I}a_0$ from ${}_{II}a_0$ via Riesz lemma might not possess any symmetry pattern. In the following, we shall show that we can achieve symmetry for any odd integer $n \in \mathbb{N}$ if considering complex-valued wavelet generators.

For $1 \le n \le m$, we have the following lemma regarding the positiveness of $P_{m,n}(y)$, which generalizes [12, Theorem 5] and [22, Theorem 2.4]. See [26, Theorem 2] for its technical proof.

Lemma 3. *Let $m,n \in \mathbb{N}$ be such that $n \le m$. Then $P_{m,n}(y) > 0$ for all $y \in \mathbb{R}$ if and only if n is an odd number.*

Now, by $P_{m,2n-1}(y) > 0$ for all $y \in \mathbb{R}$ and $2n-1 \le m$, $P_{m,2n-1}(y)$ can only have complex roots. Hence, we must have

$$P_{m,2n-1}(y) = c_0 \prod_{j=1}^{n-1} (y - z_j)(y - \overline{z_j}), \quad z_1, \overline{z_1}, \dots, z_{n-1}, \overline{z_{n-1}} \in \mathbb{C} \setminus \mathbb{R}.$$

In view of Lemmas 2 and 3, we have the following result.

Theorem 2. *Let $\mathsf{d} > 1$ be a dilation factor. Let $m,n \in \mathbb{N}$ be positive integers such that $2n-1 \le m$. Let $P_{m,n}(y)$ be the polynomial defined in* (53). *Then,*

$$P_{m,2n-1}(y) = |Q_{m,n}(y)|^2, \tag{55}$$

where $Q_{m,n}(y) = c(y - z_1)\cdots(y - z_{n-1})$ with $c = (-1)^{n-1}(z_1 \cdots z_{n-1})^{-1}$ and $z_1, \overline{z_1}$, $\dots$, $z_{n-1}, \overline{z_{n-1}} \in \mathbb{C} \setminus \mathbb{R}$ are all the complex roots of $P_{m,2n-1}(y)$. Define a low-pass filter a_0 by

$$\widehat{a_0}(\xi) := e^{i\lfloor \frac{m(\mathsf{d}-1)}{2} \rfloor \xi} \left(\frac{1 + e^{-i\xi} + \cdots + e^{-i(\mathsf{d}-1)\xi}}{\mathsf{d}} \right)^m Q_{m,n}(\sin^2(\xi/2)), \tag{56}$$

where $\lfloor \cdot \rfloor$ is the floor operation. Then,

$$\widehat{a_0}(-\xi) = e^{i\varepsilon\xi} \widehat{a_0}(\xi) \quad \textit{with} \quad \varepsilon = m(\mathsf{d}-1) - 2\lfloor \frac{m(\mathsf{d}-1)}{2} \rfloor \tag{57}$$

and

$$\operatorname{csupp}(a_0) = \left[-\lfloor \frac{m(\mathsf{d}-1)}{2} \rfloor - n + 1, \lfloor \frac{m(\mathsf{d}-1)}{2} \rfloor + n - 1 + \varepsilon \right].$$

Let ϕ be the standard d*-refinable function associated with the low-pass filter a_0, that is, $\widehat{\phi}(\xi) := \prod_{j=1}^{\infty} \widehat{a_0}(\mathsf{d}^{-j}\xi)$. Then, ϕ is a compactly supported* d*-refinable function in $L_2(\mathbb{R})$ with symmetry satisfying $\phi(\frac{\varepsilon}{\mathsf{d}-1} - \cdot) = \phi$.*

For $m, n \in \mathbb{N}$ such that $2n - 1 \le m$, we shall refer the d-refinable function ϕ associated with the low-pass filter a_0 defined in Theorem 2 as *complex* d*-refinable pseudo spline of type I with order* $(m, 2n-1)$.

Now, we have the following result which shall play an important role in our construction of tight framelet systems in this section.

Corollary 1. *Let $\mathsf{d} > 1$ be a dilation factor. Let $m, n \in \mathbb{N}$ be such that $2n - 1 \le m$ and a_0 be the low-pass filter for the complex* d*-refinable pseudo spline of type I with order $(m, 2n-1)$. Then*

$$1 - \sum_{j=0}^{\mathsf{d}-1} |\widehat{a_0}(\xi + 2\pi j/\mathsf{d})|^2 = |\widehat{b}(\mathsf{d}\xi)|^2, \tag{58}$$

for some 2π-periodic trigonometric function $\widehat{b}(\xi)$ with real coefficients. In particular,

$$|\widehat{b}(\xi)|^2 = \begin{cases} 0 & m = 2n-1; \\ c_{2n,2n-1}[\sin^2(\xi/2)/\mathsf{d}^2]^{2n-1} & m = 2n, \end{cases}$$

where $c_{2n,2n-1}$ is the coefficient given in (52).

Proof. We first show that $1 - \sum_{j=0}^{\mathsf{d}-1} |\widehat{a_0}(\xi + 2\pi j/\mathsf{d})|^2 \ge 0$ for all $\xi \in \mathbb{R}$. Let $y_j := \sin^2(\xi/2 + \pi j/\mathsf{d})$ for $j = 0, \ldots, \mathsf{d}-1$. Noting that $|\widehat{a_0}(\xi)|^2 = h(y_0)^m P_{m,2n-1}(y_0)$, we have

$$\begin{aligned} 1 - \sum_{j=0}^{\mathsf{d}-1} |\widehat{a_0}(\xi + 2\pi j/\mathsf{d})|^2 &= 1 - \sum_{j=0}^{\mathsf{d}-1} h(y_j)^m P_{m,2n-1}(y_j) \\ &= 1 - \sum_{j=0}^{\mathsf{d}-1} h(y_j)^m P_{m,m}(y_j) + \sum_{j=0}^{\mathsf{d}-1} h(y_j)^m \sum_{k=2n-1}^{m-1} c_{m,k} y_j^k \\ &= \sum_{j=0}^{\mathsf{d}-1} h(y_j)^m \sum_{k=2n-1}^{m-1} c_{m,k} y_j^k \\ &\ge 0. \end{aligned}$$

The last equality follows from the fact that the low-pass filter a_0, which is defined by factorizing $h(y_0)^m P_{m,m}(y_0)$ such that $|\widehat{a_0}(\xi)|^2 := h(y_0)^m P_{m,m}(y_0)$, is an orthogonal low-pass filter (see [17]). Now, by that $1 - \sum_{j=0}^{\mathsf{d}-1} |\widehat{a_0}(\xi + 2\pi j/\mathsf{d})|^2$ is of period $2\pi/\mathsf{d}$, (58) follows from Riesz Lemma.

Obviously, $\widehat{b}(\xi) \equiv 0$ when $m = 2n-1$ since a_0 is then an orthogonal low-pass filter. For $m = 2n$, noting that $h(y_j)y_j = \sin^2(\mathsf{d}\xi/2)/\mathsf{d}^2$ for $j = 0,\ldots,\mathsf{d}-1$, we have

$$\begin{aligned}|\widehat{b}(\mathsf{d}\xi)|^2 &= c_{2n,2n-1}\sum_{j=0}^{\mathsf{d}-1} h(y_j)^{2n} y_j^{2n-1} = c_{2n,2n-1}\sum_{j=0}^{\mathsf{d}-1}[h(y_j)y_j]^{2n-1}h(y_j)\\ &= c_{2n,2n-1}[\sin^2(\mathsf{d}\xi/2)/\mathsf{d}^2]^{2n-1}\sum_{j=0}^{\mathsf{d}-1} h(y_j)P_{1,1}(y_j)\\ &= c_{2n,2n-1}[\sin^2(\mathsf{d}\xi/2)/\mathsf{d}^2]^{2n-1},\end{aligned}$$

which completes our proof. □

3.2 Tight Framelets via Matrix Extension

Fixed $m,n \in \mathbb{N}$ such that $1 \le 2n-1 \le m$, we next show that we can construct a vector of Laurent polynomial with symmetry from a low-pass filter a_0 for the complex d-refinable pseudo spline of type I with order $(m, 2n-1)$ to which Algorithm 1 is applicable. Indeed, by (40), we have a $1\times \mathsf{d}$ vector of Laurent polynomial $\mathsf{p}(z) := [\mathsf{b}_{0;0}(z),\ldots,\mathsf{b}_{0;\mathsf{d}-1}(z)]$ from a_0. Note $\mathsf{pp}^* = 1$ when $m = 2n-1$ while $\mathsf{pp}^* \neq 1$ when $2n-1 < m$. To apply our matrix extension algorithm, we need to append extra entries to p when $\mathsf{pp}^* < 1$. It is easy to show that

$$1-\sum_{j=0}^{\mathsf{d}-1}|\widehat{a}_0(\xi+2j\pi/\mathsf{d})|^2 = 1-\sum_{\gamma=0}^{\mathsf{d}-1}\mathsf{a}_{0;\gamma}(z^{\mathsf{d}})\mathsf{a}_{0;\gamma}^*(z^{\mathsf{d}}), \quad z = \mathsf{e}^{-\mathsf{i}\xi},$$

where $\mathsf{a}_{0;\gamma}, \gamma = 0,\ldots,\mathsf{d}-1$ are the subsymbols of a_0. By Corollary 1, we have

$$1-\sum_{j=0}^{\mathsf{d}-1}|\widehat{a}_0(\xi+2j\pi/\mathsf{d})|^2 = |\widehat{b}(\mathsf{d}\xi)|^2.$$

for some 2π-periodic trigonometric function $\widehat{b}$ with real coefficients. Hence, we can construct a Laurent polynomial $\mathsf{a}_{0;\mathsf{d}}(z)$ from $\widehat{b}$ such that $\mathsf{a}_{0;\mathsf{d}}(\mathsf{e}^{-\mathsf{i}\xi}) = \widehat{b}(\xi)$. Then, the vector of Laurent polynomials $\mathsf{q}(z) = [\mathsf{a}_{0;0}(z),\ldots,\mathsf{a}_{0;\mathsf{d}-1}(z),\mathsf{a}_{0;\mathsf{d}}(z)]$ satisfies $\mathsf{qq}^* = 1$.

For $m = 2n$, by Corollary 1, $|\widehat{b}(\xi)|^2 = c_{2n,2n-1}[\sin^2(\xi/2)/\mathsf{d}^2]^{2n-1}$. In this case, $\mathsf{a}_{0;\mathsf{d}}(z)$ can be constructed explicitly as follows:

$$\mathsf{a}_{0;\mathsf{d}}(z) = \sqrt{c_{2n,2n-1}}\left(\frac{2-z-1/z}{4\mathsf{d}^2}\right)^{n-1}\frac{1-z}{2\mathsf{d}}. \tag{59}$$

$\mathsf{a}_{0;\mathsf{d}}(z)$ has symmetry $\mathsf{Sa}_{0;\mathsf{d}} = -z$. Let $\mathsf{b}_{0;\mathsf{d}}(z) := \mathsf{a}_{0;\mathsf{d}}(z)$. Then $\mathsf{p} := [\mathsf{b}_{0;0},\ldots,\mathsf{b}_{0;\mathsf{d}}]$ is a $1\times(\mathsf{d}+1)$ vector of Laurent polynomials with symmetry satisfying $\mathsf{pp}^* = 1$.

For $m \neq 2n$, $\mathsf{a}_{0;\mathsf{d}}(z)$ does not necessary have symmetry. We can further let $\mathsf{b}_{0;\mathsf{d}}(z) := (\mathsf{a}_{0;\mathsf{d}}(z) + \mathsf{a}_{0;\mathsf{d}}(1/z))/2$ and $\mathsf{b}_{0;\mathsf{d}+1}(z) := (\mathsf{a}_{0;\mathsf{d}}(z) - \mathsf{a}_{0;\mathsf{d}}(1/z))/2$. In this way, $\mathsf{p} := [\mathsf{b}_{0;0}, \ldots, \mathsf{b}_{0;\mathsf{d}}, \mathsf{b}_{0;\mathsf{d}+1}]$ is a $1 \times (\mathsf{d}+2)$ vector of Laurent polynomials with symmetry satisfying $\mathsf{p}\mathsf{p}^* = 1$.

Consequently, we can summerize the above discussion as follows:

Theorem 3. *Let $m, n \in \mathbb{N}$ be such that $1 \le 2n-1 < m$. Let a_0 (with symbol a_0) be the low-pass filter for the complex d-refinable pseudo spline of type I with order $(m, 2n-1)$ defined in* (56). *Then one can derive Laurent polynomials $\mathsf{a}_{0;\mathsf{d}}, \ldots, \mathsf{a}_{0;L}, L \in \{\mathsf{d}, \mathsf{d}+1\}$ such that $\mathsf{p}_{\mathsf{a}_0} := [\mathsf{a}_{0;0}, \ldots, \mathsf{a}_{0;\mathsf{d}-1}, \ldots, \mathsf{a}_{0;L}]$ satisfies $\mathsf{p}_{\mathsf{a}_0}\mathsf{p}_{\mathsf{a}_0}^* = 1$, where $\mathsf{a}_{0;0}, \ldots, \mathsf{a}_{0;\mathsf{d}-1}$ are subsymbols of a_0. Moreover, one can construct an $(L+1) \times (L+1)$ paraunitary matrix U such that $\mathsf{p}_{\mathsf{a}_0}\mathsf{U}$ is a vector of Laurent polynomials with symmetry. In particular, if $m = 2n$, then $L = \mathsf{d}$ and $\mathsf{a}_{0;\mathsf{d}}$ is given by* (59).

Now, applying Theorem 3 and Algorithm 1, we have the following algorithm to construct high-pass filters $\mathsf{a}_1, \ldots, \mathsf{a}_L$ from a low-pass filter a_0 for a complex d-refinable pseudo spline of type I with order $(m, 2n-1)$ so that $\psi^1, \ldots, \psi^L$ defined by (10) generates a tight framelet system.

Algorithm 3 Construction of symmetric complex tight framelets

(a) **Input**: A low-pass filter a_0 for a complex d-refinable pseudo spline of type I with order $(m, 2n-1)$, $1 \le 2n-1 < m$. Note that a_0 satisfies (36) for $r = 1$.

(b) **Initialization**: Construct $\mathsf{p}_{\mathsf{a}_0}(z)$ and U as in Theorem 3 such that $\mathsf{p} := \mathsf{p}_{\mathsf{a}_0}\mathsf{U}$ is a $1 \times (L+1)$ row vector of Laurent polynomials with symmetry ($L = \mathsf{d}$ when $m = 2n$ while $L = \mathsf{d}+1$ when $m \neq 2n$).

(c) **Extension**: Derive $\mathbb{P}_e$ from p by Algorithm 1 with all the properties as in Theorem 1 for the case $r = 1$.

(d) **High-pass Filters**: Let $\mathbf{P} := [\mathbb{P}_e\mathsf{U}^*]_{0:L,0:\mathsf{d}-1} =: (\mathsf{a}_{m;\gamma})_{0 \le m \le L, 0 \le \gamma \le \mathsf{d}-1}$ as in (16). Define high-pass filters

$$\mathsf{a}_m(z) := \frac{1}{\sqrt{\mathsf{d}}} \sum_{\gamma=0}^{\mathsf{d}-1} \mathsf{a}_{m;\gamma}(z^{\mathsf{d}}) z^{\gamma}, \qquad m = 1, \ldots, L. \tag{60}$$

Note that we only need the first d columns of $\mathbb{P}_e\mathsf{U}^*$.

(e) **Output**: A symmetric filter bank $\{\mathsf{a}_0, \mathsf{a}_1, \ldots, \mathsf{a}_L\}$ with the perfect reconstruction property, i.e. $\mathbf{P}^*(z)\mathbf{P}(z) = I_{\mathsf{d}}$ for all $z \in \mathbb{C} \backslash \{0\}$. All filters a_m, $m = 1, \ldots, L$, have symmetry:

$$\mathsf{a}_m(z) = \varepsilon_m z^{\mathsf{d}c_m - c_0} \mathsf{a}_m(1/z), \tag{61}$$

where $c_m := k_m + c_0 \in \mathbb{R}$ and all $\varepsilon_m \in \{-1, 1\}$, $k_m \in \mathbb{Z}$ for $m = 1, \ldots, L$ are determined by the symmetry pattern of $\mathbb{P}_e$ as follows:

$$[1, \varepsilon_1 z^{k_1}, \ldots, \varepsilon_L z^{k_L}]^{\mathsf{T}} \mathsf{S}\mathsf{p} := \mathsf{S}\mathbb{P}_e. \tag{62}$$

Since the high-pass filters $\mathsf{a}_1, \ldots, \mathsf{a}_L$ satisfy (43), it is easy to verify that $\psi^1, \ldots, \psi^L$ defined in (10) also has the following symmetry:

$$\psi^1(c_1 - \cdot) = \varepsilon_1 \psi^1, \quad \psi^2(c_2 - \cdot) = \varepsilon_2 \psi^2, \quad \ldots, \quad \psi^L(c_L - \cdot) = \varepsilon_L \psi^1. \tag{63}$$

In the following, let us present an example to demonstrate our results and illustrate our algorithms. More examples can be obtained in the same way.

Example 2. Consider dilation factor $\mathsf{d}=3$. Let $m=4$ and $n=2$. Then $P_{4,3}(y)=1+\frac{32}{3}y+64y^2)$. The low-pass filter a_0 with its symbol a_0 for the complex 3-refinable pseudo spline of order $(4,3)$ is given by

$$\mathsf{a}_0(z)=\left(\frac{\frac{1}{z}+1+z}{3}\right)^4\left[-\left(\frac{4}{3}+\frac{2\sqrt{5}}{3}\mathrm{i}\right)\frac{1}{z}+\left(\frac{11}{3}+\frac{4\sqrt{5}}{3}\mathrm{i}\right)-\left(\frac{4}{3}+\frac{2\sqrt{5}}{3}\mathrm{i}\right)z\right].$$

Note that $\mathrm{csupp}(\mathsf{a}_0)=[-5,5]$ and $\mathsf{a}(z)=\mathsf{a}(z^{-1})$. In this case, $m=2n$. By Theorem 3, we can obtain $\mathsf{p}_{\mathsf{a}_0}=[\mathsf{a}_{0;0}(z),\mathsf{a}_{0;1}(z),\mathsf{a}_{0;2}(z),\mathsf{a}_{0;3}(z)]$ as follows:

$$\begin{aligned}
\mathsf{a}_{0;0}(z)&=-\frac{\sqrt{15\mathrm{i}}}{405}\left(10z+27\sqrt{5}\mathrm{i}-20+10z^{-1}\right);\\
\mathsf{a}_{0;1}(z)&=\frac{\sqrt{3}}{243}(-(4+2\sqrt{5}\mathrm{i})z^{-2}+30z^{-1}+60+6\sqrt{5}\mathrm{i}-(5+4\sqrt{5}\mathrm{i})z);\\
\mathsf{a}_{0;2}(z)&=\frac{\sqrt{3}}{243}(-(5+4\sqrt{5}\mathrm{i})z^{-2}+(60+6\sqrt{5}\mathrm{i})z^{-1}+30-(4+2\sqrt{5}\mathrm{i})z)\\
\mathsf{a}_{0;3}(z)&=-\frac{2\sqrt{10}}{81}(z-2+z^{-1})(1-z).
\end{aligned}$$

We have $\mathsf{a}_{0;1}(z)=z^{-1}\mathsf{a}_{0;2}(z^{-1})$. Let $\mathsf{p}=\mathsf{p}_{\mathsf{a}_0}\mathsf{U}$ with U being the paraunitary matrix given by

$$\mathsf{U}:=\mathrm{diag}(1,U_0,z^{-1})\quad\text{with}\quad U_0=\begin{bmatrix}\frac{1}{\sqrt{2}}&\frac{1}{\sqrt{2}}\\ \frac{1}{\sqrt{2}}&-\frac{1}{\sqrt{2}}\end{bmatrix}.$$

Then p is a 1×4 vector of Laurent polynomials with symmetry pattern satisfying $\mathsf{Sp}=[1,z^{-1},-z^{-1},-z^{-1}]$. Applying Algorithm 3, we can obtain a 4×4 extension matrix $\mathbb{P}_e^*=[\mathsf{p}_{\mathsf{a}_0}^*,\mathsf{p}_{\mathsf{a}_1}^*,\mathsf{p}_{\mathsf{a}_2}^*,\mathsf{p}_{\mathsf{a}_3}^*]$ with $\mathsf{p}_{\mathsf{a}_1}:=[\mathsf{a}_{1;0},\mathsf{a}_{1;1},\mathsf{a}_{1;2},\mathsf{a}_{1;3}]$, $\mathsf{p}_{\mathsf{a}_2}:=[\mathsf{a}_{2;0},\mathsf{a}_{2;1},\mathsf{a}_{2;2},\mathsf{a}_{2;3}]$, and $\mathsf{p}_{\mathsf{a}_3}:=[\mathsf{a}_{3;0},\mathsf{a}_{3;1},\mathsf{a}_{3;2},\mathsf{a}_{3;3}]$. The coefficient support of $\mathbb{P}_e$ satisfies $\mathrm{csupp}([\mathbb{P}_e]_{:,j})\subseteq\mathrm{csupp}([\mathsf{p}_{\mathsf{a}_0}]_j)$ for $j=1,2,3,4$. The high-pass filters a_1, $\mathsf{a}_2,\mathsf{a}_3$ constructed from $\mathsf{p}_{\mathsf{a}_1}$, $\mathsf{p}_{\mathsf{a}_2}$, and $\mathsf{p}_{\mathsf{a}_3}$ via (60) are then given by

$$\mathsf{a}_1(z)=c_1(b_1(z)+b_1(z^{-1}));\ \mathsf{a}_2(z)=c_2(b_2(z)-b_2(z^{-1}));\ \mathsf{a}_3(z)=c_3(b_3(z)-z^3b_3(z^{-1})).$$

where $c_1=\frac{\sqrt{19178}}{4660254}$, $c_2=\frac{\sqrt{218094}}{17665614}$, $c_3=\frac{2\sqrt{1338}}{54189}$, and

$$\begin{aligned}
b_1(z)&=\left(-172-86i\sqrt{5}\right)z^5+\left(-215-172i\sqrt{5}\right)z^4-258i\sqrt{5}z^3\\
&\quad+\left(1470+1224i\sqrt{5}\right)z^2+\left(1860+2328i\sqrt{5}\right)z-3036i\sqrt{5}-2943\\
b_2(z)&=\left(-652-326i\sqrt{5}\right)z^5+\left(-815-652i\sqrt{5}\right)z^4-978i\sqrt{5}z^3\\
&\quad+\left(1832i\sqrt{5}+1750\right)z^2+\left(3508i\sqrt{5}+3020\right)z\\
b_3(z)&=\left(4\sqrt{5}+10i\right)z^5+\left(5\sqrt{5}+20i\right)z^4+30iz^3+\left(-53\sqrt{5}-260i\right)z^2.
\end{aligned}$$

We have $\mathsf{a}_1(z) = \mathsf{a}_1(z^{-1})$, $\mathsf{a}_2(z) = -\mathsf{a}_2(z^{-1})$, and $\mathsf{a}_3(z) = -z^3\mathsf{a}_3(z^{-1})$. Let ϕ be the 3-refinable function associated with the low-pass filter a_0. Let ψ^1, ψ^2, ψ^3 be the wavelet functions associated with the high-pass filters $\mathsf{a}_1, \mathsf{a}_2, \mathsf{a}_3$ by (10), respectively. Then $\phi(-\cdot) = \phi$, $\psi^1(-\cdot) = \psi^1$, $\psi^2(-\cdot) = -\psi^2$, and $\psi^3(1-\cdot) = -\psi^3$. See Fig. 2 for the graphs of ϕ, ψ^1, ψ^2, and ψ^3.

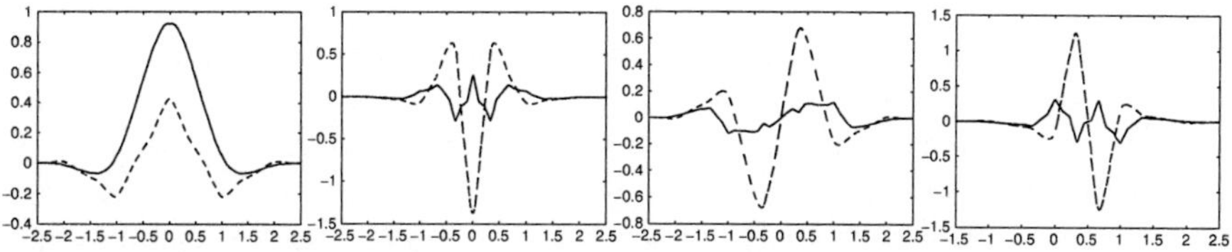

Fig. 2: The graphs of ϕ, ψ^1, ψ^2, and ψ^3 (*left to right*) in Example 2. Real part: *solid line*. Imaginary part: *dashed line*

4 Biorthogonal Matrix Extension with Symmetry

In this section, we shall discuss the construction of biorthogonal multiwavelets with symmetry, which corresponds to Problem 2. Due to the flexibility of biorthogonality relation $\mathbb{P}\widetilde{\mathbb{P}}^* = I_r$, the biorthogonal matrix extension problem becomes far more complicated than that for the orthogonal matrix extension problem we considered in Sect. 2. The difficulty here is not the symmetry patterns of the extension matrices, but the support control of the extension matrices. Without considering any issue on support control, almost all results of Theorem 1 can be transferred to the biorthogonal case without much difficulty. In Theorem 1, the length of the coefficient support of the extension matrix can never exceed the length of the coefficient support of the given matrix. Yet, for the extension matrices in the biorthogonal extension case, we can no longer expect such nice result, that is, in this case, the length of the coefficient supports of the extension matrices might not be controlled by one of the given matrices. Nevertheless, we have the following result.

Theorem 4. *Let $\mathbb{F}$ be any subfield of $\mathbb{C}$. Let $(\mathbb{P}, \widetilde{\mathbb{P}})$ be a pair of $r \times s$ matrices of Laurent polynomials with coefficients in $\mathbb{F}$ such that $\mathsf{S}\mathbb{P} = \mathsf{S}\widetilde{\mathbb{P}} = (\mathsf{S}\theta_1)^*\mathsf{S}\theta_2$ for some $1 \times r$, $1 \times s$ vectors θ_1, θ_2 of Laurent polynomials with symmetry. Moreover, $\mathbb{P}(z)\widetilde{\mathbb{P}}^*(z) = I_r$ for all $z \in \mathbb{C}\backslash\{0\}$. Then there exists a pair of $s \times s$ square matrices $(\mathbb{P}_e, \widetilde{\mathbb{P}}_e)$ of Laurent polynomials with coefficients in $\mathbb{F}$ such that*

(1) *$[I_r, 0]\mathbb{P}_e = \mathbb{P}$ and $[I_r, 0]\widetilde{\mathbb{P}}_e = \widetilde{\mathbb{P}}$; that is, the submatrices of the first r rows of $\mathbb{P}_e, \widetilde{\mathbb{P}}_e$ are $\mathbb{P}, \widetilde{\mathbb{P}}$, respectively;*
(2) *$(\mathbb{P}_e, \widetilde{\mathbb{P}}_e)$ is a pair of biorthogonal matrices: $\mathbb{P}_e(z)\widetilde{\mathbb{P}}_e^*(z) = I_s$ for all $z \in \mathbb{C}\backslash\{0\}$;*

(3) *the symmetry of each* $\mathbb{P}_e, \widetilde{\mathbb{P}}_e$ *is compatible:* $\mathsf{S}\mathbb{P}_e = \mathsf{S}\widetilde{\mathbb{P}}_e = (\mathsf{S}\theta)^*\mathsf{S}\theta_2$ *for some* $1 \times s$ *vector* θ *of Laurent polynomials with symmetry;*
(4) $\mathbb{P}_e, \widetilde{\mathbb{P}}_e$ *can be represented as:*

$$\mathbb{P}_e(z) = \mathbb{P}_J(z)\cdots\mathbb{P}_1(z), \quad \widetilde{\mathbb{P}}_e(z) = \widetilde{\mathbb{P}}_J(z)\cdots\widetilde{\mathbb{P}}_1(z), \tag{64}$$

where $(\mathbb{P}_j, \widetilde{\mathbb{P}}_j), 1 \le j \le J$ *are pairs of* $s \times s$ *biorthogonal matrices of Laurent polynomials with symmetry. Moreover, each pair of* $(\mathbb{P}_{j+1}, \mathbb{P}_j)$ *and* $(\widetilde{\mathbb{P}}_{j+1}, \widetilde{\mathbb{P}}_j)$ *has mutually compatible symmetry for all* $j = 1, \ldots, J-1$.
(5) *if* $r = 1$, *then the coefficient supports of* $\mathbb{P}_e, \widetilde{\mathbb{P}}_e$ *are controlled by those of* $\mathbb{P}, \widetilde{\mathbb{P}}$ *in the following sense:*

$$\max_{1\le j,k\le s}\{|\mathrm{csupp}([\mathbb{P}_e]_{j,k})|, |\mathrm{csupp}([\widetilde{\mathbb{P}}_e]_{j,k})|\} \le \max_{1\le \ell\le s}|\mathrm{csupp}([\mathbb{P}]_\ell)| + \max_{1\le \ell\le s}|\mathrm{csupp}([\widetilde{\mathbb{P}}]_\ell)|. \tag{65}$$

4.1 Proof of Theorem 4 and an Algorithm

In this section, we shall prove Theorem 4. Based on the proof, we shall provide a step-by-step extension algorithm for deriving the desired pair of biorthogonal extension matrices in Theorem 4.

In this section, $\mathbb{F}$ denote any subfield of $\mathbb{C}$. The next lemma shows that for a pair of constant vectors $(\mathtt{f}, \widetilde{\mathtt{f}})$ in $\mathbb{F}$, we can find a pair of constant biorthogonal matrices $(U_{(\mathtt{f},\widetilde{\mathtt{f}})}, \widetilde{U}_{(\mathtt{f},\widetilde{\mathtt{f}})})$ in $\mathbb{F}$ such that up to a constant multiplication, it normalizes $(\mathtt{f}, \widetilde{\mathtt{f}})$ to a pair of unit vectors.

Lemma 4. *Let* $(\mathtt{f}, \widetilde{\mathtt{f}})$ *be a pair of nonzero* $1 \times n$ *vectors in* $\mathbb{F}$. *Then,*

(1) *if* $\mathtt{f}\widetilde{\mathtt{f}}^* \ne 0$, *then there exists a pair of* $n \times n$ *matrices* $(U_{(\mathtt{f},\widetilde{\mathtt{f}})}, \widetilde{U}_{(\mathtt{f},\widetilde{\mathtt{f}})})$ *in* $\mathbb{F}$ *such that* $U_{(\mathtt{f},\widetilde{\mathtt{f}})} = [(\frac{\widetilde{\mathtt{f}}}{\widetilde{c}})^*, F]$, $\widetilde{U}_{(\mathtt{f},\widetilde{\mathtt{f}})} = [(\frac{\mathtt{f}}{c})^*, \widetilde{F}]$, *and* $U_{(\mathtt{f},\widetilde{\mathtt{f}})}\widetilde{U}^*_{(\mathtt{f},\widetilde{\mathtt{f}})} = I_n$, *where* $F, \widetilde{F}$ *are* $n \times (n-1)$ *constant matrices in* $\mathbb{F}$ *and* $c, \widetilde{c}$ *are two nonzero numbers in* $\mathbb{F}$ *such that* $\mathtt{f}\widetilde{\mathtt{f}}^* = c\overline{\widetilde{c}}$. *In this case,* $\mathtt{f}U_{(\mathtt{f},\widetilde{\mathtt{f}})} = c\varepsilon_1$ *and* $\widetilde{\mathtt{f}}\widetilde{U}_{(\mathtt{f},\widetilde{\mathtt{f}})} = \widetilde{c}\varepsilon_1$;
(2) *if* $\mathtt{f}\widetilde{\mathtt{f}}^* = 0$, *then there exists a pair of* $n \times n$ *matrices* $(U_{(\mathtt{f},\widetilde{\mathtt{f}})}, \widetilde{U}_{(\mathtt{f},\widetilde{\mathtt{f}})})$ *in* $\mathbb{F}$ *such that* $U_{(\mathtt{f},\widetilde{\mathtt{f}})} = [(\frac{\mathtt{f}}{\widetilde{c}_1})^*, (\frac{\widetilde{\mathtt{f}}}{c_2})^*, F]$, $\widetilde{U}_{(\mathtt{f},\widetilde{\mathtt{f}})} = [(\frac{\mathtt{f}}{c_1})^*, (\frac{\widetilde{\mathtt{f}}}{\widetilde{c}_2})^*, \widetilde{F}]$, *and* $U_{(\mathtt{f},\widetilde{\mathtt{f}})}\widetilde{U}^*_{(\mathtt{f},\widetilde{\mathtt{f}})} = I_n$, *where* $F, \widetilde{F}$ *are* $n \times (n-2)$ *constant matrices in* $\mathbb{F}$ *and* $c_1, c_2, \widetilde{c}_1, \widetilde{c}_2$ *are nonzero numbers in* $\mathbb{F}$ *such that* $\|\mathtt{f}\|^2 = c_1\overline{\widetilde{c}_1}, \|\widetilde{\mathtt{f}}\|^2 = c_2\overline{\widetilde{c}_2}$. *In this case,* $\mathtt{f}U_{(\mathtt{f},\widetilde{\mathtt{f}})} = c_1\varepsilon_1$ *and* $\widetilde{\mathtt{f}}\widetilde{U}_{(\mathtt{f},\widetilde{\mathtt{f}})} = c_2\varepsilon_2$.

Proof. If $\mathtt{f}\widetilde{\mathtt{f}}^* \ne 0$, there exists $\{\mathtt{f}_2, \ldots, \mathtt{f}_n\}$ being a basis of the orthogonal compliment of the linear span of $\{\mathtt{f}\}$ in $\mathbb{F}^n$. Let $F := [\mathtt{f}_2^*, \ldots, \mathtt{f}_n^*]$ and $U_{(\mathtt{f},\widetilde{\mathtt{f}})} := [(\frac{\widetilde{\mathtt{f}}}{\widetilde{c}})^*, F]$. Then $U_{(\mathtt{f},\widetilde{\mathtt{f}})}$ is invertible. Let $\widetilde{U}_{(\mathtt{f},\widetilde{\mathtt{f}})} := (U^{-1}_{(\mathtt{f},\widetilde{\mathtt{f}})})^*$. It is easy to show that $U_{(\mathtt{f},\widetilde{\mathtt{f}})}$ and $\widetilde{U}_{(\mathtt{f},\widetilde{\mathtt{f}})}$ are the desired matrices.

If $\mathtt{f}\widetilde{\mathtt{f}}^* = 0$, let $\{\mathtt{f}_3,\dots,\mathtt{f}_n\}$ be a basis of the orthogonal compliment of the linear span of $\{\mathtt{f},\widetilde{\mathtt{f}}\}$ in $\mathbb{F}^n$. Let $U_{(\mathtt{f},\widetilde{\mathtt{f}})} = [(\frac{\mathtt{f}}{\widetilde{c_1}})^*, (\frac{\widetilde{\mathtt{f}}}{c_2})^*, F]$ with $F := [\mathtt{f}_3^*,\dots,\mathtt{f}_n^*]$. Then $U_{(\mathtt{f},\widetilde{\mathtt{f}})}$ and $\widetilde{U}_{(\mathtt{f},\widetilde{\mathtt{f}})} := (U^{-1}_{(\mathtt{f},\widetilde{\mathtt{f}})})^*$ are the desired matrices. □ □

Thanks to Lemma 4, we can reduce the support lengths of a pair $(\mathsf{p},\widetilde{\mathsf{p}})$ of Laurent polynomials with symmetry by constructing a pair of biorthogonal matrices $(\mathsf{B},\widetilde{\mathsf{B}})$ of Laurent polynomials with symmetry as stated in the following lemma.

Lemma 5. *Let $(\mathsf{p},\widetilde{\mathsf{p}})$ be a pair of $1\times s$ vectors of Laurent polynomials with symmetry such that $\mathsf{p}\widetilde{\mathsf{p}}^* = 1$ and $\mathsf{S}\mathsf{p} = \mathsf{S}\widetilde{\mathsf{p}} = \varepsilon z^c[1_{s_1}, -1_{s_2}, z^{-1}1_{s_3}, -z^{-1}1_{s_4}] =: \mathsf{S}\theta$ for some nonnegative integers $s_1,\dots,s_4$ satisfying $s_1+\cdots+s_4 = s$ and $\varepsilon\in\{1,-1\}, c\in\{0,1\}$. Suppose $|\mathrm{csupp}(\mathsf{p})| > 0$. Then there exists a pair of $s\times s$ matrices $(\mathsf{B},\widetilde{\mathsf{B}})$ of Laurent polynomials with symmetry such that*

(1) *$(\mathsf{B},\widetilde{\mathsf{B}})$ is a pair of biorthogonal matrices: $\mathsf{B}(z)\widetilde{\mathsf{B}}^*(z) = I_n$;*
(2) *$\mathsf{S}\mathsf{B} = \mathsf{S}\widetilde{\mathsf{B}} = (\mathsf{S}\theta)^*\mathsf{S}\theta_1$ with $\mathsf{S}\theta_1 = \varepsilon z^c[1_{s_1'}, -1_{s_2'}, z^{-1}1_{s_3'}, -z^{-1}1_{s_4'}]$ for some nonnegative integers $s_1',\dots,s_4'$ such that $s_1'+\cdots+s_4' = s$;*
(3) *the length of the coefficient support of p is reduced by that of B. $\widetilde{\mathsf{B}}$ does not increase the length of the coefficient support of $\widetilde{\mathsf{p}}$. That is, $|\mathrm{csupp}(\mathsf{p}\mathsf{B})| \le |\mathrm{csupp}(\mathsf{p})| - |\mathrm{csupp}(\mathsf{B})|$ and $|\mathrm{csupp}(\widetilde{\mathsf{p}}\widetilde{\mathsf{B}})| \le |\mathrm{csupp}(\widetilde{\mathsf{p}})|$.*

Proof. We shall only prove the case that $\mathsf{S}\theta = [1_{s_1}, -1_{s_2}, z^{-1}1_{s_3}, -z^{-1}1_{s_4}]$. The proofs for other cases are similar. By their symmetry patterns, p and $\widetilde{\mathsf{p}}$ must take the forms as follows with $\ell > 0$ and $\mathrm{coeff}(\mathsf{p},-\ell)\neq 0$:

$$\begin{aligned}\mathsf{p} &= [\mathtt{f}_1,-\mathtt{f}_2,\mathtt{g}_1,-\mathtt{g}_2]z^{-\ell} + [\mathtt{f}_3,-\mathtt{f}_4,\mathtt{g}_3,-\mathtt{g}_4]z^{-\ell+1} + \sum_{k=-\ell+2}^{\ell-2}\mathrm{coeff}(\mathsf{p},k)z^k \\ &\quad + [\mathtt{f}_3,\mathtt{f}_4,\mathtt{g}_1,\mathtt{g}_2]z^{\ell-1} + [\mathtt{f}_1,\mathtt{f}_2,\mathbf{0},0]z^{\ell};\\ \widetilde{\mathsf{p}} &= [\widetilde{\mathtt{f}}_1,-\widetilde{\mathtt{f}}_2,\widetilde{\mathtt{g}}_1,-\widetilde{\mathtt{g}}_2]z^{-\widetilde{\ell}} + [\widetilde{\mathtt{f}}_3,-\widetilde{\mathtt{f}}_4,\widetilde{\mathtt{g}}_3,-\widetilde{\mathtt{g}}_4]z^{-\widetilde{\ell}+1} + \sum_{k=-\widetilde{\ell}+2}^{\widetilde{\ell}-2}\mathrm{coeff}(\widetilde{\mathsf{p}},k)z^k \\ &\quad + [\widetilde{\mathtt{f}}_3,\widetilde{\mathtt{f}}_4,\widetilde{\mathtt{g}}_1,\widetilde{\mathtt{g}}_2]z^{\widetilde{\ell}-1} + [\widetilde{\mathtt{f}}_1,\widetilde{\mathtt{f}}_2,\mathbf{0},0]z^{\widetilde{\ell}}.\end{aligned} \tag{66}$$

Then, either $\|\mathtt{f}_1\|+\|\mathtt{f}_2\|\neq 0$ or $\|\mathtt{g}_1\|+\|\mathtt{g}_2\|\neq 0$. Considering $\|\mathtt{f}_1\|+\|\mathtt{f}_2\|\neq 0$, due to $\mathsf{p}\widetilde{\mathsf{p}}^* = 1$ and $|\mathrm{csupp}(\mathsf{p})| > 0$, we have $\mathtt{f}_1\widetilde{\mathtt{f}}_1^* - \mathtt{f}_2\widetilde{\mathtt{f}}_2^* = 0$. Let $C := \mathtt{f}_1\widetilde{\mathtt{f}}_1^* = \mathtt{f}_2\widetilde{\mathtt{f}}_2^*$. There are at most three cases: (a) $C\neq 0$; (b) $C = 0$ but both $\mathtt{f}_1,\mathtt{f}_2$ are nonzero vectors; (c) $C = 0$ and one of $\mathtt{f}_1,\mathtt{f}_2$ is 0.

Case (a). In this case, we have $\mathtt{f}_1\widetilde{\mathtt{f}}_1^*\neq 0$ and $\mathtt{f}_2\widetilde{\mathtt{f}}_2^*\neq 0$. By Lemma 4, we can construct two pairs of biorthogonal matrices $(U_{(\mathtt{f}_1,\widetilde{\mathtt{f}}_1)}, \widetilde{U}_{(\mathtt{f}_1,\widetilde{\mathtt{f}}_1)})$ and $(U_{(\mathtt{f}_2,\widetilde{\mathtt{f}}_2)}, \widetilde{U}_{(\mathtt{f}_2,\widetilde{\mathtt{f}}_2)})$ with respect to the pairs $(\mathtt{f}_1,\widetilde{\mathtt{f}}_1)$ and $(\mathtt{f}_2,\widetilde{\mathtt{f}}_2)$ such that

$$U_{(\mathtt{f}_1,\widetilde{\mathtt{f}}_1)} = \left[\left(\frac{\widetilde{\mathtt{f}}_1}{\widetilde{c}_1}\right)^*, F_1\right],\quad \widetilde{U}_{(\mathtt{f}_1,\widetilde{\mathtt{f}}_1)} = \left[\left(\frac{\mathtt{f}_1}{c_1}\right)^*, \widetilde{F}_1\right],\quad \mathtt{f}_1U_{(\mathtt{f}_1,\widetilde{\mathtt{f}}_1)} = c_1\mathtt{e}_1,\quad \widetilde{\mathtt{f}}_1\widetilde{U}_{(\mathtt{f}_1,\widetilde{\mathtt{f}}_1)} = \widetilde{c}_1\mathtt{e}_1,$$

$$U_{(\mathtt{f}_2,\widetilde{\mathtt{f}}_2)} = \left[\left(\frac{\widetilde{\mathtt{f}}_2}{\widetilde{c}_1}\right)^*, F_2\right],\quad \widetilde{U}_{(\mathtt{f}_2,\widetilde{\mathtt{f}}_2)} = \left[\left(\frac{\mathtt{f}_2}{c_1}\right)^*, \widetilde{F}_2\right],\quad \mathtt{f}_2U_{(\mathtt{f}_2,\widetilde{\mathtt{f}}_2)} = c_1\mathtt{e}_1,\quad \widetilde{\mathtt{f}}_2\widetilde{U}_{(\mathtt{f}_2,\widetilde{\mathtt{f}}_2)} = \widetilde{c}_1\mathtt{e}_1,$$

where $c_1, \widetilde{c}_1$ are constants in $\mathbb{F}$ such that $C = c_1\overline{\widetilde{c}_1}$. Define $\mathsf{B}_0(z), \widetilde{\mathsf{B}}_0(z)$ as follows:

$$\begin{aligned}\mathsf{B}_0(z) &= \left[\begin{array}{cc|cc|c} \frac{1+z^{-1}}{2}(\frac{\widetilde{\mathtt{f}}_1}{\widetilde{c}_1})^* & F_1 & -\frac{1-z^{-1}}{2}(\frac{\widetilde{\mathtt{f}}_1}{\widetilde{c}_1})^* & 0 & 0 \\ -\frac{1-z^{-1}}{2}(\frac{\widetilde{\mathtt{f}}_2}{\widetilde{c}_1})^* & 0 & \frac{1+z^{-1}}{2}(\frac{\widetilde{\mathtt{f}}_2}{\widetilde{c}_1})^* & F_2 & 0 \\ \hline 0 & 0 & 0 & 0 & I_{s_3+s_4} \end{array}\right], \\ \widetilde{\mathsf{B}}_0(z) &= \left[\begin{array}{cc|cc|c} \frac{1+z^{-1}}{2}(\frac{\mathtt{f}_1}{c_1})^* & \widetilde{F}_1 & -\frac{1-z^{-1}}{2}(\frac{\mathtt{f}_1}{c_1})^* & 0 & 0 \\ -\frac{1-z^{-1}}{2}(\frac{\mathtt{f}_2}{c_1})^* & 0 & \frac{1+z^{-1}}{2}(\frac{\mathtt{f}_2}{c_1})^* & \widetilde{F}_2 & 0 \\ \hline 0 & 0 & 0 & 0 & I_{s_3+s_4} \end{array}\right].\end{aligned} \tag{67}$$

Direct computation shows that $\mathsf{B}_0(z)\widetilde{\mathsf{B}}_0(z)^* = I_s$ due to the special structures of the pairs $(U_{(\mathtt{f}_1,\widetilde{\mathtt{f}}_1)}, \widetilde{U}_{(\mathtt{f}_1,\widetilde{\mathtt{f}}_1)})$ and $(U_{(\mathtt{f}_2,\widetilde{\mathtt{f}}_2)}, \widetilde{U}_{(\mathtt{f}_2,\widetilde{\mathtt{f}}_2)})$ constructed by Lemma 4. The symmetry patterns of $\mathsf{p}\mathsf{B}_0$ and $\widetilde{\mathsf{p}}\widetilde{\mathsf{B}}_0$ satisfies

$$\mathsf{S}(\mathsf{p}\mathsf{B}_0) = \mathsf{S}(\widetilde{\mathsf{p}}\widetilde{\mathsf{B}}_0) = [z^{-1}, 1_{s_1-1}, -z^{-1}, -1_{s_2-1}, z^{-1}1_{s_3}, -z^{-1}1_{s_4}].$$

Moreover, $\mathsf{B}_0(z)$, $\widetilde{\mathsf{B}}_0(z)$ reduce the lengths of the coefficient support of p and $\widetilde{\mathsf{p}}$ by 1, respectively.

In fact, due to the above symmetry pattern and the structures of $\mathsf{B}_0, \widetilde{\mathsf{B}}_0$, we only need to show that $\mathrm{coeff}([\mathsf{p}\mathsf{B}_0]_j, \ell) = \mathrm{coeff}([\widetilde{\mathsf{p}}\widetilde{\mathsf{B}}_0]_j, \ell) = 0$ for $j = 1, s_1+1$. Note that $\mathrm{coeff}([\mathsf{p}\mathsf{B}_0]_j, \ell) = \mathrm{coeff}(\mathsf{p}, \ell)\mathrm{coeff}([\mathsf{B}_0]_{:,1}, 0) = \frac{1}{2\overline{\widetilde{c}_1}}(\mathtt{f}_1\widetilde{\mathtt{f}}_1^* - \mathtt{f}_2\widetilde{\mathtt{f}}_2^*) = 0$. Similar computations apply for other terms. Thus, $|\mathrm{csupp}(\mathsf{p}\mathsf{B}_0)| < \mathrm{csupp}(\mathsf{p})$ and $|\mathrm{csupp}(\widetilde{\mathsf{p}}\widetilde{\mathsf{B}}_0)| < |\mathrm{csupp}(\widetilde{\mathsf{p}})|$. Let E be a permutation matrix such that

$$\mathsf{S}(\mathsf{p}\mathsf{B}_0)E = \mathsf{S}(\widetilde{\mathsf{p}}\widetilde{\mathsf{B}}_0)E = [1_{s_1-1}, -1_{s_2-1}, z^{-1}1_{s_3+1}, -z^{-1}1_{s_4+1}] =: \mathsf{S}\theta_1.$$

Define $\mathsf{B}(z) = \mathsf{B}_0(z)E$ and $\widetilde{\mathsf{B}}(z) = \widetilde{\mathsf{B}}_0(z)E$. Then $\mathsf{B}(z)$ and $\widetilde{\mathsf{B}}(z)$ are the desired matrices.

Case (b). In this case, $\mathtt{f}_1\widetilde{\mathtt{f}}_1^* = \mathtt{f}_2\widetilde{\mathtt{f}}_2^* = 0$ and both $\mathtt{f}_1, \mathtt{f}_2$ are nonzero vectors. We have $\mathtt{f}_1\mathtt{f}_1^* \neq 0$ and $\mathtt{f}_2\mathtt{f}_2^* \neq 0$. Again, by Lemma 4, we can construct two pairs of biorthogonal matrices $(U_{(\mathtt{f}_1,\mathtt{f}_1)}, \widetilde{U}_{(\mathtt{f}_1,\mathtt{f}_1)})$ and $(U_{(\mathtt{f}_2,\mathtt{f}_2)}, \widetilde{U}_{(\mathtt{f}_2,\mathtt{f}_2)})$ with respect to the pairs $(\mathtt{f}_1, \mathtt{f}_1)$ and $(\mathtt{f}_2, \mathtt{f}_2)$ such that

$$\begin{aligned} U_{(\mathtt{f}_1,\mathtt{f}_1)} &= \left[\left(\frac{\mathtt{f}_1}{\widetilde{c}_1}\right)^*, F_1\right], \quad \widetilde{U}_{(\mathtt{f}_1,\mathtt{f}_1)} = \left[\left(\frac{\mathtt{f}_1}{c_0}\right)^*, F_1\right], \quad \mathtt{f}_1 U_{(\mathtt{f}_1,\mathtt{f}_1)} = c_0\varepsilon_1, \\ U_{(\mathtt{f}_2,\mathtt{f}_2)} &= \left[\left(\frac{\mathtt{f}_2}{\widetilde{c}_2}\right)^*, F_2\right], \quad \widetilde{U}_{(\mathtt{f}_2,\mathtt{f}_2)} = \left[\left(\frac{\mathtt{f}_2}{c_0}\right)^*, F_2\right], \quad \mathtt{f}_2 U_{(\mathtt{f}_2,\mathtt{f}_2)} = c_0\varepsilon_1, \end{aligned}$$

where $c_0, \widetilde{c}_1, \widetilde{c}_2$ are constants in $\mathbb{F}$ such that $\mathtt{f}_1\mathtt{f}_1^* = c_0\overline{\widetilde{c}_1}$ and $\mathtt{f}_2\mathtt{f}_2^* = c_0\overline{\widetilde{c}_2}$. Let $\mathsf{B}_0, \widetilde{\mathsf{B}}_0(z)$ be defined as follows:

$$\mathsf{B}_0(z) = \left[\begin{array}{cc|cc|c} \frac{1+z^{-1}}{2}(\frac{\mathtt{f}_1}{\widetilde{c}_1})^* & F_1 & -\frac{1-z^{-1}}{2}(\frac{\mathtt{f}_1}{\widetilde{c}_1})^* & 0 & 0 \\ -\frac{1-z^{-1}}{2}(\frac{\mathtt{f}_2}{\widetilde{c}_2})^* & 0 & \frac{1+z^{-1}}{2}(\frac{\mathtt{f}_2}{\widetilde{c}_2})^* & F_2 & 0 \\ \hline 0 & 0 & 0 & 0 & I_{s_3+s_4} \end{array}\right],$$
$$\widetilde{\mathsf{B}}_0(z) = \left[\begin{array}{cc|cc|c} \frac{1+z^{-1}}{2}(\frac{\mathtt{f}_1}{c_0})^* & F_1 & -\frac{1-z^{-1}}{2}(\frac{\mathtt{f}_1}{c_0})^* & 0 & 0 \\ -\frac{1-z^{-1}}{2}(\frac{\mathtt{f}_2}{c_0})^* & 0 & \frac{1+z^{-1}}{2}(\frac{\mathtt{f}_2}{c_0})^* & F_2 & 0 \\ \hline 0 & 0 & 0 & 0 & I_{s_3+s_4} \end{array}\right]. \qquad (68)$$

We can show that $\mathsf{B}_0(z)$ reduces the length of the coefficient support of p by 1, while $\widetilde{\mathsf{B}}_0(z)$ does not increase the support length of $\widetilde{\mathsf{p}}$. Moreover, similar to case (a), we can find a permutation matrix E such that

$$\mathsf{S}(\mathsf{p}\mathsf{B}_0)E = \mathsf{S}(\widetilde{\mathsf{p}}\widetilde{\mathsf{B}}_0)E = [\mathbf{1}_{s_1-1}, -\mathbf{1}_{s_2-1}, z^{-1}\mathbf{1}_{s_3+1}, -z^{-1}\mathbf{1}_{s_4+1}] =: \mathsf{S}\theta_1.$$

Define $\mathsf{B}(z) = \mathsf{B}_0(z)E$ and $\widetilde{\mathsf{B}}(z) = \widetilde{\mathsf{B}}_0(z)E$. Then $\mathsf{B}(z)$ and $\widetilde{\mathsf{B}}(z)$ are the desired matrices.

Case (c). In this case, $\mathtt{f}_1\widetilde{\mathtt{f}}_1^* = \mathtt{f}_2\widetilde{\mathtt{f}}_2^* = 0$ and one of $\mathtt{f}_1$ and $\mathtt{f}_2$ is nonzero. Without loss of generality, we assume that $\mathtt{f}_1 \neq 0$ and $\mathtt{f}_2 = 0$. Construct a pair of matrices $(U_{(\mathtt{f}_1,\widetilde{\mathtt{f}}_1)}, \widetilde{U}_{(\mathtt{f}_1,\widetilde{\mathtt{f}}_1)})$ by Lemma 4 such that $\mathtt{f}_1 U_{(\mathtt{f}_1,\widetilde{\mathtt{f}}_1)} = c_1\varepsilon_1$ and $\widetilde{\mathtt{f}}_1\widetilde{U}_{(\mathtt{f}_1,\widetilde{\mathtt{f}}_1)} = c_2\varepsilon_2$ (when $\widetilde{\mathtt{f}}_1 = 0$, the pair of matrices is given by $(U_{(\mathtt{f}_1,\mathtt{f}_1)}, \widetilde{U}_{(\mathtt{f}_1,\mathtt{f}_1)})$). Extend this pair to a pair of $s \times s$ matrices $(U, \widetilde{U})$ by $U := \mathrm{diag}(U_{(\mathtt{f}_1,\widetilde{\mathtt{f}}_1)}, I_{s_3+s_4})$ and $\widetilde{U} := \mathrm{diag}(\widetilde{U}_{(\mathtt{f}_1,\widetilde{\mathtt{f}}_1)}, I_{s_3+s_4})$. Then $\mathsf{p}U$ and $\widetilde{\mathsf{p}}\widetilde{U}$ must be of the form:

$$\begin{aligned}\mathsf{q} := \mathsf{p}U &= [c_1, 0, \ldots, 0, -\mathtt{f}_2, \mathtt{g}_1, -\mathtt{g}_2]z^{-\ell} + [\mathtt{f}_3, -\mathtt{f}_4, \mathtt{g}_3, -\mathtt{g}_4]z^{-\ell+1} \\ &+ \sum_{k=-\ell+2}^{\ell-2} \mathrm{coeff}(\mathsf{q}, k)z^k + [\mathtt{f}_3, \mathtt{f}_4, \mathtt{g}_1, \mathtt{g}_2]z^{\ell-1} + [c_1, 0, \ldots, 0, \mathtt{f}_2, \mathbf{0}, 0]z^{\ell};\end{aligned}$$

$$\begin{aligned}\widetilde{\mathsf{q}} := \widetilde{\mathsf{p}}\widetilde{U} &= [0, c_2, \ldots, 0, -\widetilde{\mathtt{f}}_2, \widetilde{\mathtt{g}}_1, -\widetilde{\mathtt{g}}_2]z^{-\widetilde{\ell}} + [\widetilde{\mathtt{f}}_3, -\widetilde{\mathtt{f}}_4, \widetilde{\mathtt{g}}_3, -\widetilde{\mathtt{g}}_4]z^{-\widetilde{\ell}+1} \\ &+ \sum_{k=-\widetilde{\ell}+2}^{\widetilde{\ell}-2} \mathrm{coeff}(\widetilde{\mathsf{q}}, k)z^k + [\widetilde{\mathtt{f}}_3, \widetilde{\mathtt{f}}_4, \widetilde{\mathtt{g}}_1, \widetilde{\mathtt{g}}_2]z^{\widetilde{\ell}-1} + [0, c_2, \ldots, 0, , \widetilde{\mathtt{f}}_2, \mathbf{0}, 0]z^{\widetilde{\ell}}.\end{aligned}$$

If $[\widetilde{\mathsf{q}}]_1 \equiv 0$, we choose k such that $k = \arg\min_{\ell \neq 1}\{|\mathrm{csupp}([\mathsf{q}]_1)| - |\mathrm{csupp}([\mathsf{q}]_\ell)|\}$, i.e., k is an integer such that the length of coefficient support of $|\mathrm{csupp}([\mathsf{q}]_1)| - |\mathrm{csupp}([\mathsf{q}]_k)|$ is minimal among those of all $|\mathrm{csupp}([\mathsf{q}]_1)| - |\mathrm{csupp}([\mathsf{q}]_\ell)|$, $\ell = 2, \ldots, s$; otherwise, due to $\mathsf{q}\widetilde{\mathsf{q}}^* = 0$, there must exist k such that

$$|\mathrm{csupp}([\mathsf{q}]_1)| - |\mathrm{csupp}([\mathsf{q}]_k)| \le \max_{2 \le j \le s} |\mathrm{csupp}([\widetilde{\mathsf{q}}]_j)| - |\mathrm{csupp}([\widetilde{\mathsf{q}}]_1)|,$$

(k might not be unique, we can choose one of such k so that $|\mathrm{csupp}([\mathsf{q}]_1)| - |\mathrm{csupp}([\mathsf{q}]_k)|$ is minimal among all $|\mathrm{csupp}([\mathsf{q}]_1)| - |\mathrm{csupp}([\mathsf{q}]_\ell)|, \ell = 2, \ldots, s$).

For such k (in the case of either $[\widetilde{\mathsf{q}}]_1 = 0$ or $[\widetilde{\mathsf{q}}]_1 \neq 0$), define two matrices $\mathsf{B}(z), \widetilde{\mathsf{B}}(z)$ as follows:

$$\mathsf{B}(z) = \left[\begin{array}{cccc|c} 1 & 0 & \cdots & 0 & \\ 0 & 1 & \cdots & 0 & \\ \vdots & \vdots & \ddots & \vdots & \\ -b(z) & 0 & \cdots & 1 & \\ \hline & & & & I_{s-k} \end{array}\right], \widetilde{\mathsf{B}}(z) = \left[\begin{array}{cccc|c} 1 & 0 & \cdots & b^*(z) & \\ 0 & 1 & \cdots & 0 & \\ \vdots & \vdots & \ddots & \vdots & \\ 0 & 0 & \cdots & 1 & \\ \hline & & & & I_{s-k} \end{array}\right],$$

where $b(z)$ in $\mathsf{B}(z), \widetilde{\mathsf{B}}(z)$ is a Laurent polynomial with symmetry such that $\mathsf{S}b(z) = \mathsf{S}([\mathsf{q}]_1/[\mathsf{q}]_k)$, $|\mathrm{csupp}([\mathsf{q}]_1 - b(z)[\mathsf{q}]_k)| < |\mathrm{csupp}([\mathsf{q}]_k)|$, and $|\mathrm{csupp}([\widetilde{\mathsf{q}}]_k - b^*(z)[\widetilde{\mathsf{q}}]_1)| \le \max_{1\le \ell \le s} |\mathrm{csupp}([\widetilde{\mathsf{q}}]_\ell)|$. Such $b(z)$ can be easily obtained by long division.

It is straightforward to show that $\mathsf{B}(z)\widetilde{\mathsf{B}}^*(z) = I_s$. $\mathsf{B}(z)$ reduces the length of the coefficient support of q by that of $b(z)$ due to $|\mathrm{csupp}([\mathsf{q}]_1 - b(z)[\mathsf{q}]_k)| < |\mathrm{csupp}([\mathsf{q}]_k)|$. And by our choice of k, $\widetilde{\mathsf{B}}(z)$ does not increase the length of the coefficient support of $\widetilde{\mathsf{q}}$. Moreover, the symmetry patterns of both q and $\widetilde{\mathsf{q}}$ are preserved.

In summary, for all cases (a), (b), and (c), we can always find a pair of biorthogonal matrices $(\mathsf{B}, \widetilde{\mathsf{B}})$ of Laurent polynomials such that B reduces the length of the coefficient support of p while $\widetilde{\mathsf{B}}$ does not increase the length of the coefficient support of $\widetilde{\mathsf{p}}$.

For $\|\mathtt{f}_1\| + \|\mathtt{f}_2\| = 0$, we must have $\|\mathtt{g}_1\| + \|\mathtt{g}_2\| \neq 0$. The discussion for this case is similar to above. We can find two matrices $\mathsf{B}(z), \widetilde{\mathsf{B}}(z)$ such that all items in the lemma hold. In the case that $\mathtt{g}_1\widetilde{\mathtt{g}}_1^* = \mathtt{g}_2\widetilde{\mathtt{g}}_2^* = c_1\overline{\widetilde{c}_1} \neq 0$, the pair $(\mathsf{B}_0(z), \widetilde{\mathsf{B}}_0(z))$ similar to (67) is of the form

$$\begin{aligned} \mathsf{B}_0(z) &= \left[\begin{array}{c|cc|cc} I_{s_1+s_2} & 0 & 0 & 0 & 0 \\ \hline 0 & \frac{1+z}{2}(\frac{\mathtt{g}_1}{\widetilde{c}_1})^* & G_1 & -\frac{1-z}{2}(\frac{\mathtt{g}_1}{\widetilde{c}_1})^* & 0 \\ 0 & -\frac{1-z}{2}(\frac{\mathtt{g}_2}{\widetilde{c}_1})^* & 0 & \frac{1+z}{2}(\frac{\mathtt{g}_2}{\widetilde{c}_1})^* & G_2 \end{array}\right], \\ \widetilde{\mathsf{B}}_0(z) &= \left[\begin{array}{c|cc|cc} I_{s_1+s_2} & 0 & 0 & 0 & 0 \\ \hline 0 & \frac{1+z}{2}(\frac{\mathtt{g}_1}{c_1})^* & \widetilde{G}_1 & -\frac{1-z}{2}(\frac{\mathtt{g}_1}{c_1})^* & 0 \\ 0 & -\frac{1-z}{2}(\frac{\mathtt{g}_2}{c_1})^* & 0 & \frac{1+z}{2}(\frac{\mathtt{g}_2}{c_1})^* & \widetilde{G}_2 \end{array}\right]. \end{aligned} \tag{69}$$

The pairs for other cases can be obtained similarly. We are done. □ □

Now, we can prove Theorem 4 using Lemma 5.

Proof (of Theorem 4). First, we normalize the symmetry patterns of $\mathbb{P}$ and $\widetilde{\mathbb{P}}$ to the standard form as in (22). Let $\mathsf{Q} := \mathsf{U}_{\mathsf{S}\theta_1}^* \mathbb{P} \mathsf{U}_{\mathsf{S}\theta_2}$ and $\widetilde{\mathsf{Q}} := \mathsf{U}_{\mathsf{S}\theta_1}^* \widetilde{\mathbb{P}} \mathsf{U}_{\mathsf{S}\theta_2}$ (given θ, $\mathsf{U}_{\mathsf{S}\theta}$ is obtained by (23)). Then the symmetry of each row of Q or $\widetilde{\mathsf{Q}}$ is of the form $\varepsilon z^c[1_{s_1}, -1_{s_2}, z^{-1}1_{s_3}, -z^{-1}1_{s_4}]$ for some $\varepsilon \in \{-1, 1\}$ and $c \in \{0, 1\}$.

Let $\mathsf{p} := [\mathsf{Q}]_{1,:}$ and $\widetilde{\mathsf{p}} := [\widetilde{\mathsf{Q}}]_{1,:}$ be the first row of $\mathsf{Q}, \widetilde{\mathsf{Q}}$, respectively. Applying Lemma 5 recursively, we can find pairs of biorthogonal matrices of Laurent polynomials $(\mathsf{B}_1, \widetilde{\mathsf{B}}_1)$, ..., $(\mathsf{B}_K, \widetilde{\mathsf{B}}_K)$ such that $\mathsf{p}\mathsf{B}_1 \cdots \mathsf{B}_K = [1, 0, \ldots, 0]$ and $\widetilde{\mathsf{p}}\widetilde{\mathsf{B}}_1 \cdots \widetilde{\mathsf{B}}_K = [1, \mathsf{q}(z)]$ for some $1 \times (s-1)$ vector of Laurent polynomials with symmetry. Note that by Lemma 5, all pairs $(\mathsf{B}_j, \mathsf{B}_{j+1})$ and $(\widetilde{\mathsf{B}}_j, \widetilde{\mathsf{B}}_{j+1})$ for $j = 1, \ldots, K-1$ have mutually compatible symmetry. Now construct $\mathsf{B}_{K+1}(z), \widetilde{\mathsf{B}}_{K+1}(z)$ as follows:

$$\mathsf{B}_{K+1}(z) = \begin{bmatrix} 1 & 0 \\ \mathsf{q}^*(z) & I_{s-1} \end{bmatrix}, \widetilde{\mathsf{B}}_{K+1}(z) = \begin{bmatrix} 1 & -\mathsf{q}(z) \\ 0 & I_{s-1} \end{bmatrix}.$$

B_{K+1} and $\widetilde{\mathsf{B}}_{K+1}$ are biorthogonal. Let $\mathsf{A} := \mathsf{B}_1 \cdots \mathsf{B}_K \mathsf{B}_{K+1}$ and $\widetilde{\mathsf{A}} := \widetilde{\mathsf{B}}_1 \cdots \widetilde{\mathsf{B}}_K \widetilde{\mathsf{B}}_{K+1}$. Then, $\mathsf{p}\mathsf{A} = \widetilde{\mathsf{p}}\widetilde{\mathsf{A}} = \varepsilon_1$.

Note that $\mathsf{Q}\mathsf{A}$ and $\widetilde{\mathsf{Q}}\widetilde{\mathsf{A}}$ are of the forms

$$\mathsf{Q}\mathsf{A} = \begin{bmatrix} 1 & 0 \\ 0 & \mathsf{Q}_1(z) \end{bmatrix}, \widetilde{\mathsf{Q}}\widetilde{\mathsf{A}} = \begin{bmatrix} 1 & 0 \\ 0 & \widetilde{\mathsf{Q}}_1(z) \end{bmatrix}$$

for some $(r-1) \times s$ matrices $\mathsf{Q}_1, \widetilde{\mathsf{Q}}_1$ of Laurent polynomials with symmetry. Moreover, due to Lemma 5, the symmetry patterns of Q_1 and $\widetilde{\mathsf{Q}}_1$ are compatible and satisfies $\mathsf{S}\mathsf{Q}_1 = \mathsf{S}\widetilde{\mathsf{Q}}_1$. The rest of the proof is completed by employing the standard procedure of induction. □ □

According to the proof of Theorem 4, we have an extension algorithm for Theorem 4. See Algorithm 4.

Algorithm 4 Biorthogonal matrix extension with symmetry

(a) **Input**: $\mathbb{P}, \widetilde{\mathbb{P}}$ as in Theorem 4 with $\mathsf{S}\mathbb{P} = \mathsf{S}\widetilde{\mathbb{P}} = (\mathsf{S}\theta_1)^* \mathsf{S}\theta_2$ for two $1 \times r$, $1 \times s$ row vectors θ_1, θ_2 of Laurent polynomials with symmetry.

(b) **Initialization**: Let $\mathsf{Q} := \mathsf{U}^*_{\mathsf{S}\theta_1} \mathbb{P} \mathsf{U}_{\mathsf{S}\theta_2}$ and $\widetilde{\mathsf{Q}} := \mathsf{U}^*_{\mathsf{S}\theta_1} \widetilde{\mathbb{P}} \mathsf{U}_{\mathsf{S}\theta_2}$. Then both Q and $\widetilde{\mathsf{Q}}$ have the same symmetry pattern as follows:

$$\mathsf{S}\mathsf{Q} = \mathsf{S}\widetilde{\mathsf{Q}} = [\mathbf{1}_{r_1}, -\mathbf{1}_{r_2}, z\mathbf{1}_{r_3}, -z\mathbf{1}_{r_4}]^{\mathrm{T}} [\mathbf{1}_{s_1}, -\mathbf{1}_{s_2}, z^{-1}\mathbf{1}_{s_3}, -z^{-1}\mathbf{1}_{s_4}], \tag{70}$$

where all nonnegative integers $r_1, \ldots, r_4, s_1, \ldots, s_4$ are uniquely determined by $\mathsf{S}\mathbb{P}$. Note that this step does not increase the lengths of the coefficient support of both $\mathbb{P}$ and $\widetilde{\mathbb{P}}$.

(c) **Support Reduction**:

Let $\mathsf{U}_0 := \mathsf{U}^*_{\mathsf{S}\theta_2}$ and $\mathsf{A} = \widetilde{\mathsf{A}} := I_s$.
for $k = 1$ to r **do**
  Let $\mathsf{p} := [\mathsf{Q}]_{k,k:s}$ and $\widetilde{\mathsf{p}} := [\widetilde{\mathsf{Q}}]_{k,k:s}$.
  while $|\mathrm{csupp}(\mathsf{p})| > 0$ and $|\mathrm{csupp}(\widetilde{\mathsf{p}})| > 0$ **do**
    Construct a pair of biorthogonal matrices $(\mathsf{B}, \widetilde{\mathsf{B}})$ with respect to the pair $(\mathsf{p}, \widetilde{\mathsf{p}})$ by Lemma 5 such that $|\mathrm{csupp}(\mathsf{p}\mathsf{B})| + |\mathrm{csupp}(\widetilde{\mathsf{p}}\widetilde{\mathsf{B}})| < |\mathrm{csupp}(\mathsf{p})| + |\mathrm{csupp}(\widetilde{\mathsf{p}})|$.
    Replace $\mathsf{p}, \widetilde{\mathsf{p}}$ by $\mathsf{p}\mathsf{B}$, $\widetilde{\mathsf{p}}\widetilde{\mathsf{B}}$, respectively.
    Set $\mathsf{A} := \mathsf{A}\,\mathrm{diag}(I_{k-1}, \mathsf{B})$ and $\widetilde{\mathsf{A}} := \widetilde{\mathsf{A}}\,\mathrm{diag}(I_{k-1}, \widetilde{\mathsf{B}})$.
  end while
  The pair $(\mathsf{p}, \widetilde{\mathsf{p}})$ is of the form: $([1, 0, \ldots, 0], [1, \mathsf{q}(z)])$ for some $1 \times (s-k)$ vector of Laurent polynomials $\mathsf{q}(z)$. Construct $\mathsf{B}(z), \widetilde{\mathsf{B}}(z)$ as follows:

$$\mathsf{B}(z) = \begin{bmatrix} 1 & 0 \\ \mathsf{q}^*(z) & I_{s-k} \end{bmatrix}, \widetilde{\mathsf{B}}(z) = \begin{bmatrix} 1 & -\mathsf{q}(z) \\ 0 & I_{s-k} \end{bmatrix}.$$

  Set $\mathsf{A} := \mathsf{A}\,\mathrm{diag}(I_{k-1}, \mathsf{B})$ and $\widetilde{\mathsf{A}} := \widetilde{\mathsf{A}}\,\mathrm{diag}(I_{k-1}, \widetilde{\mathsf{B}})$.
  Set $\mathsf{Q} := \mathsf{Q}\mathsf{A}$ and $\widetilde{\mathsf{Q}} := \widetilde{\mathsf{Q}}\widetilde{\mathsf{A}}$.
end for

(d) **Finalization**: Let $\mathsf{U}_1 := \mathrm{diag}(\mathsf{U}_{\mathsf{S}\theta_1}, I_{s-r})$. Set $\mathbb{P}_e := \mathsf{U}_1 \mathsf{A}^* \mathsf{U}_0$ and $\widetilde{\mathbb{P}}_e := \mathsf{U}_1 \widetilde{\mathsf{A}}^* \mathsf{U}_0$.

(e) **Output**: A pair of desired matrices $(\mathbb{P}_e, \widetilde{\mathbb{P}}_e)$ satisfying all the properties in Theorem 4.

4.2 Application to Construction of Biorthogonal Multiwavelets with Symmetry

For the construction of biorthogonal refinable function vectors (a pair of biorthogonal low-pass filters), the CBC (*coset by coset*) algorithm proposed in [11] provides a systematic way of constructing a desirable dual mask from a given primal mask that satisfies certain conditions. More precisely, given a mask (low-pass filter) satisfying the condition that a dual mask exists, following the CBC algorithm, one can construct a dual mask with any preassigned orders of sum rules, which is closely related to the regularity of the refinable function vectors. Furthermore, if the primal mask has symmetry, then the CBC algorithm also guarantees that the dual mask has symmetry. Thus, the first part of MRA corresponding to the construction of biorthogonal multiwavelets is more or less solved. However, how to derive the wavelet generators (high-pass filters) with symmetry remains open even for the scalar case ($r=1$). We shall see that using our extension algorithm for the biorthogonal case, the wavelet generators do have symmetry once the given refinable function vectors possess certain symmetry patterns.

Let $(\phi,\widetilde{\phi})$ be a pair of dual d-refinable function vectors associated with *a pair of biorthogonal low-pass filters* $(a_0,\widetilde{a}_0)$, that is, $\phi,\widetilde{\phi}$ are d-refinable function vectors associated with $a_0,\widetilde{a}_0$, respectively, and

$$\langle\phi,\widetilde{\phi}(\cdot-k)\rangle=\delta(k)I_r,\quad k\in\mathbb{Z}. \tag{71}$$

It is easy to show that the pair of biorthogonal low-pass filters $(a_0,\widetilde{a}_0)$ satisfies

$$\sum_{\gamma=0}^{\mathsf{d}-1}\mathsf{a}_{0;\gamma}(z)\widetilde{\mathsf{a}}_{0;\gamma}^{*}(z)=I_r,\qquad z\in\mathbb{C}\backslash\{0\}, \tag{72}$$

where $\mathsf{a}_{0;\gamma}$ and $\widetilde{\mathsf{a}}_{0;\gamma}$ are d-*band subsymbols (polyphase components)* of a_0 and $\widetilde{\mathsf{a}}_0$ defined similar to (15) by

$$\begin{matrix}\mathsf{a}_{0;\gamma}(z):=d_1\sum_{k\in\mathbb{Z}}a_0(k+\mathsf{d}k)z^k,\\ \widetilde{\mathsf{a}}_{0;\gamma}(z):=d_2\sum_{k\in\mathbb{Z}}\widetilde{a}_0(k+\mathsf{d}k)z^k,\end{matrix}\quad\gamma\in\mathbb{Z}. \tag{73}$$

Here, d_1,d_2 are two constants in $\mathbb{F}$ such that $\mathsf{d}=d_1d_2$.

To construct biorthogonal multiwavelets in $L_2(\mathbb{R})$, we need to design high-pass filters $a_1,\ldots,a_{\mathsf{d}-1}:\mathbb{Z}\to\mathbb{F}^{r\times r}$ and $\widetilde{a}_1,\ldots,\widetilde{a}_{\mathsf{d}-1}:\mathbb{Z}\to\mathbb{F}^{r\times r}$ such that the polyphase matrices with respect to the filter banks $\{\mathsf{a}_0,\mathsf{a}_1,\ldots,\mathsf{a}_{\mathsf{d}-1}\}$ and $\{\widetilde{\mathsf{a}}_0,\widetilde{\mathsf{a}}_1,\ldots,\widetilde{\mathsf{a}}_{\mathsf{d}-1}\}$

$$\mathbf{P}(z)=\begin{bmatrix}\mathsf{a}_{0;0}(z)&\cdots&\mathsf{a}_{0;\mathsf{d}-1}(z)\\ \mathsf{a}_{1;0}(z)&\cdots&\mathsf{a}_{1;\mathsf{d}-1}(z)\\ \vdots&\vdots&\vdots\\ \mathsf{a}_{\mathsf{d}-1;0}(z)&\cdots&\mathsf{a}_{\mathsf{d}-1;\mathsf{d}-1}(z)\end{bmatrix},\ \widetilde{\mathbf{P}}(z)=\begin{bmatrix}\widetilde{\mathsf{a}}_{0;0}(z)&\cdots&\widetilde{\mathsf{a}}_{0;\mathsf{d}-1}(z)\\ \widetilde{\mathsf{a}}_{1;0}(z)&\cdots&\widetilde{\mathsf{a}}_{1;\mathsf{d}-1}(z)\\ \vdots&\vdots&\vdots\\ \widetilde{\mathsf{a}}_{\mathsf{d}-1;0}(z)&\cdots&\widetilde{\mathsf{a}}_{\mathsf{d}-1;\mathsf{d}-1}(z)\end{bmatrix} \tag{74}$$

are biorthogonal, that is, $\mathbf{P}(z)\widetilde{\mathbf{P}}^*(z) = I_{\mathsf{d}r}$, where $\mathsf{a}_{m;\gamma}, \widetilde{\mathsf{a}}_{m;\gamma}$ are subsymbols of $\mathsf{a}_m, \widetilde{\mathsf{a}}_m$ defined similar to (73) for $m, \gamma = 0, \dots, \mathsf{d}-1$, respectively. The pair of filter banks $(\{\mathsf{a}_0, \dots, \mathsf{a}_{\mathsf{d}-1}\}, \{\widetilde{\mathsf{a}}_0, \dots, \widetilde{\mathsf{a}}_{\mathsf{d}-1}\})$ satisfying $\mathbf{P}\widetilde{\mathbf{P}}^* = I_{\mathsf{d}r}$ is called *a pair of biorthogonal filter banks with the perfect reconstruction property*.

Let $(a_0, \widetilde{a}_0)$ be a pair of biorthogonal low-pass filters such that a_0 and $\widetilde{a}_0$ have the same symmetry satisfying (36). By a slight modification of Lemma 1 (more precisely, by modifying (40)), one can easily show that there exists a suitable *invertible* matrix U, i.e., $\det(\mathsf{U})$ is a monomial, of Laurent polynomials in $\mathbb{F}$ acting on $\mathbb{P}_{\mathsf{a}_0} := [\mathsf{a}_{0;0}, \dots, \mathsf{a}_{0;\mathsf{d}-1}]$ so that $\mathbb{P}_{\mathsf{a}_0}\mathsf{U}$ and $\mathbb{P}_{\widetilde{\mathsf{a}}_0}\widetilde{\mathsf{U}}$ have compatible symmetry ($\widetilde{\mathsf{U}} = (\mathsf{U}^*)^{-1}$). Note that $\mathbb{P}_{\mathsf{a}_0}$ itself may not have compatible symmetry.

Now, for a pair of biorthogonal d-band low-pass filters $(\mathsf{a}_0, \widetilde{\mathsf{a}}_0)$ with multiplicity r satisfying (36), we have an algorithm (see Algorithm 5) to construct high-pass filters $\mathsf{a}_1, \dots, \mathsf{a}_{\mathsf{d}-1}$ and $\widetilde{\mathsf{a}}_1, \dots, \widetilde{\mathsf{a}}_{\mathsf{d}-1}$ such that the polyphase matrices $\mathbf{P}(z)$ and $\widetilde{\mathbf{P}}(z)$ defined as in (74) satisfy $\mathbf{P}(z)\widetilde{\mathbf{P}}^*(z) = I_{\mathsf{d}r}$. Here, $\mathbb{P}_{\mathsf{a}_0} := [\mathsf{a}_{0;0}, \dots, \mathsf{a}_{0;\mathsf{d}-1}]$ and $\widetilde{\mathbb{P}}_{\widetilde{\mathsf{a}}_0} := [\widetilde{\mathsf{a}}_{0;0}, \dots, \widetilde{\mathsf{a}}_{0;\mathsf{d}-1}]$ are the polyphase vectors of $\mathsf{a}_0, \widetilde{\mathsf{a}}_0$ obtained by (73), respectively.

Algorithm 5 Construction of biorthogonal multiwavelets with symmetry

---

(a) **Input**: A pair of biorthogonal d-band filters $(\mathsf{a}_0, \widetilde{\mathsf{a}}_0)$ with multiplicity r and with the same symmetry as in (36).

(b) **Initialization**: Construct a pair of biorthogonal matrices $(\mathsf{U}, \widetilde{\mathsf{U}})$ in $\mathbb{F}$ by Lemma 1 such that both $\mathbb{P} := \mathbb{P}_{\mathsf{a}_0}\mathsf{U}$ and $\widetilde{\mathbb{P}} = \widetilde{\mathbb{P}}_{\widetilde{\mathsf{a}}_0}\widetilde{\mathsf{U}}$ ($\widetilde{\mathsf{U}} = (\mathsf{U}^*)^{-1}$) are matrices of Laurent polynomials with coefficients in $\mathbb{F}$ having compatible symmetry: $\mathsf{S}\mathbb{P} = \mathsf{S}\widetilde{\mathbb{P}} = [\varepsilon_1 z^{k_1}, \dots, \varepsilon_r z^{k_r}]^{\mathsf{T}}\mathsf{S}\theta$ for some $k_1, \dots, k_r \in \mathbb{Z}$ and some $1 \times \mathsf{d}r$ row vector θ of Laurent polynomials with symmetry.

(c) **Extension**: Derive $(\mathbb{P}_e, \widetilde{\mathbb{P}}_e)$ with all the properties as in Theorem 4 from $(\mathbb{P}, \widetilde{\mathbb{P}})$ by Algorithm 4.

(d) **High-pass Filters**: Let $\mathbf{P} := \mathbb{P}_e\widetilde{\mathsf{U}}^* =: (\mathsf{a}_{m;\gamma})_{0 \le m,\gamma \le \mathsf{d}-1}$, $\widetilde{\mathbf{P}} := \widetilde{\mathbb{P}}_e\mathsf{U}^* =: (\widetilde{\mathsf{a}}_{m;\gamma})_{0 \le m,\gamma \le \mathsf{d}-1}$ as in (74). For $m = 1, \dots, \mathsf{d}-1$, define high-pass filters

$$\mathsf{a}_m(z) := \frac{1}{d_1}\sum_{\gamma=0}^{\mathsf{d}-1} \mathsf{a}_{m;\gamma}(z^{\mathsf{d}})z^{\gamma}, \quad \widetilde{\mathsf{a}}_m(z) := \frac{1}{d_2}\sum_{\gamma=0}^{\mathsf{d}-1} \widetilde{\mathsf{a}}_{m;\gamma}(z^{\mathsf{d}})z^{\gamma}. \tag{75}$$

(e) **Output**: A pair of biorthogonal filter banks $(\{\mathsf{a}_0, \mathsf{a}_1, \dots, \mathsf{a}_{\mathsf{d}-1}\}, \{\widetilde{\mathsf{a}}_0, \widetilde{\mathsf{a}}_1, \dots, \widetilde{\mathsf{a}}_{\mathsf{d}-1}\})$ with symmetry and with the perfect reconstruction property, i.e. $\mathbf{P}, \widetilde{\mathbf{P}}$ in (74) are biorthogonal and all filters $\mathsf{a}_m, \widetilde{\mathsf{a}}_m$, $m = 1, \dots, \mathsf{d}-1$, have symmetry:

$$\begin{aligned} \mathsf{a}_m(z) &= \operatorname{diag}(\varepsilon_1^m z^{\mathsf{d}c_1^m}, \dots, \varepsilon_r^m z^{\mathsf{d}c_r^m})\mathsf{a}_m(1/z)\operatorname{diag}(\varepsilon_1 z^{-c_1}, \dots, \varepsilon_r z^{-c_r}), \\ \widetilde{\mathsf{a}}_m(z) &= \operatorname{diag}(\varepsilon_1^m z^{\mathsf{d}c_1^m}, \dots, \varepsilon_r^m z^{\mathsf{d}c_r^m})\widetilde{\mathsf{a}}_m(1/z)\operatorname{diag}(\varepsilon_1 z^{-c_1}, \dots, \varepsilon_r z^{-c_r}), \end{aligned} \tag{76}$$

where $c_\ell^m := (k_\ell^m - k_\ell) + c_\ell \in \mathbb{R}$ and all $\varepsilon_\ell^m \in \{-1, 1\}$, $k_\ell^m \in \mathbb{Z}$, for $\ell = 1, \dots, r$ and $m = 1, \dots, \mathsf{d}-1$, are determined by the symmetry pattern of $\mathbb{P}_e$ as follows:

$$[\varepsilon_1 z^{k_1}, \dots, \varepsilon_r z^{k_r}, \varepsilon_1^1 z^{k_1^1}, \dots, \varepsilon_r^1 z^{k_r^1}, \dots, z^{k_1^{\mathsf{d}-1}}, \dots, \varepsilon_r^{\mathsf{d}-1} z^{k_r^{\mathsf{d}-1}}]^{\mathsf{T}}\mathsf{S}\theta := \mathsf{S}\mathbb{P}_e. \tag{77}$$

Let $(\phi, \widetilde{\phi})$ be a pair of biorthogonal d-refinable function vectors in $L_2(\mathbb{R})$ associated with a pair of biorthogonal d-band filters $(\mathsf{a}_0, \widetilde{\mathsf{a}}_0)$ and with $\phi = [\phi_1, \dots, \phi_r]^{\mathsf{T}}$,

$\widetilde{\phi} = [\widetilde{\phi}_1, \ldots, \widetilde{\phi}_r]^{\mathrm{T}}$. Define multiwavelet function vectors $\psi^m = [\psi_1^m, \ldots, \psi_r^m]^{\mathrm{T}}$, $\widetilde{\psi}^m = [\widetilde{\psi}_1^m, \ldots, \widetilde{\psi}_r^m]^{\mathrm{T}}$ associated with the high-pass filters $\mathsf{a}_m, \widetilde{\mathsf{a}}_m$, $m = 1, \ldots, \mathsf{d}-1$, by

$$\widehat{\psi^m}(\mathsf{d}\xi) := \mathsf{a}_m(\mathrm{e}^{-\mathrm{i}\xi})\widehat{\phi}(\xi), \ \widehat{\widetilde{\psi}^m}(\mathsf{d}\xi) := \widetilde{\mathsf{a}}_m(\mathrm{e}^{-\mathrm{i}\xi})\widehat{\widetilde{\phi}}(\xi), \ \xi \in \mathbb{R}. \tag{78}$$

It is well known that $\{\psi^1, \ldots, \psi^{\mathsf{d}-1}; \widetilde{\psi}^1, \ldots, \widetilde{\psi}^{\mathsf{d}-1}\}$ generates a biorthonormal multiwavelet basis in $L_2(\mathbb{R})$. Moreover, since the high-pass filters $\mathsf{a}_1, \ldots, \mathsf{a}_{\mathsf{d}-1}$, $\widetilde{\mathsf{a}}_1, \ldots, \widetilde{\mathsf{a}}_{\mathsf{d}-1}$ satisfy (76), it is easy to verify that each $\psi^m = [\psi_1^m, \ldots, \psi_r^m]^{\mathrm{T}}$, $\widetilde{\psi}^m = [\widetilde{\psi}_1^m, \ldots, \widetilde{\psi}_r^m]^{\mathrm{T}}$ defined in (78) has the following symmetry:

$$\begin{array}{llll} \psi_1^m(c_1^m - \cdot) = \varepsilon_1^m \psi_1^m, & \psi_2^m(c_2^m - \cdot) = \varepsilon_2^m \psi_2^m, & \ldots, & \psi_r^m(c_r^m - \cdot) = \varepsilon_r^m \psi_r^m, \\ \widetilde{\psi}_1^m(c_1^m - \cdot) = \varepsilon_1^m \widetilde{\psi}_1^m, & \widetilde{\psi}_2^m(c_2^m - \cdot) = \varepsilon_2^m \widetilde{\psi}_2^m, & \ldots, & \widetilde{\psi}_r^m(c_r^m - \cdot) = \varepsilon_r^m \widetilde{\psi}_r^m. \end{array} \tag{79}$$

In the following, let us present an example to demonstrate our results and illustrate our algorithms.

Example 3. Let $\mathsf{d} = 3, r = 2$, and $a_0, \widetilde{a}_0$ be a pair of dual d-filters with symbols $\mathsf{a}_0(z), \widetilde{\mathsf{a}}_0(z)$ (cf. [13]) given by

$$\mathsf{a}_0(z) = \frac{1}{243}\begin{bmatrix} a_{11}(z) & a_{12}(z) \\ a_{21}(z) & a_{22}(z) \end{bmatrix}, \quad \widetilde{\mathsf{a}}_0(z) = \frac{1}{34884}\begin{bmatrix} \widetilde{a}_{11}(z) & \widetilde{a}_{12}(z) \\ \widetilde{a}_{21}(z) & \widetilde{a}_{22}(z) \end{bmatrix}.$$

where

$$\begin{aligned} a_{11}(z) &= -21z^{-2} + 30z^{-1} + 81 + 14z - 5z^2, \\ a_{12}(z) &= 60z^{-1} + 84 - 4z^2 + 4z^3, \\ a_{21}(z) &= 4z^{-2} - 4z^{-1} + 84z + 60z^2, \\ a_{22}(z) &= -5z^{-1} + 14 + 81z + 30z^2 - 21z^3, \end{aligned}$$

and

$$\begin{aligned} \widetilde{a}_{11}(z) &= 1292z^{-2} + 2{,}844z^{-1} + 17{,}496 + 2{,}590z - 1{,}284z^2 + 1{,}866z^3, \\ \widetilde{a}_{12}(z) &= -4{,}773z^{-2} + 9{,}682z^{-1} + 8{,}715 - 2{,}961z + 386z^2 - 969z^3, \\ \widetilde{a}_{21}(z) &= -969z^{-2} + 386z^{-1} - 2{,}961 + 8{,}715z + 9{,}682z^2 - 4{,}773z^3, \\ \widetilde{a}_{22}(z) &= 1{,}866z^{-2} - 1{,}284z^{-1} + 2{,}590 + 17{,}496z + 2{,}844z^2 + 1{,}292z^3. \end{aligned}$$

The low-pass filters a_0 and $\widetilde{a}_0$ do not satisfy (36). However, we can employ a very simple orthogonal transform $E := \left[\begin{smallmatrix} 1 & 1 \\ 1 & -1 \end{smallmatrix}\right]$ to $\mathsf{a}_0, \widetilde{\mathsf{a}}_0$ so that the symmetry in (36) holds. That is, for $\mathsf{b}_0(z) := E\mathsf{a}_0(z)E^{-1}$ and $\widetilde{\mathsf{b}}_0(z) := E^{-1}\widetilde{\mathsf{a}}_0(z)E$, it is easy to verify that b_0 and $\widetilde{\mathsf{b}}_0$ satisfy (36) with $c_1 = c_2 = 1/2$ and $\varepsilon_1 = 1, \varepsilon_2 = -1$. Let $\mathsf{d} = d_1 d_2$ with $d_1 = 1$ and $d_2 = 3$. Construct $\mathbb{P}_{\mathsf{b}_0} := [\mathsf{b}_{0;0}, \mathsf{b}_{0;1}, \mathsf{b}_{0;2}]$ and $\widetilde{\mathbb{P}}_{\widetilde{\mathsf{b}}_0} := [\widetilde{\mathsf{b}}_{0;0}, \widetilde{\mathsf{b}}_{0;1}, \widetilde{\mathsf{b}}_{0;2}]$ from b_0 and $\widetilde{\mathsf{b}}_0$. Let U be given by

$$\mathsf{U} = \begin{bmatrix} z^{-1} & 0 & z^{-1} & 0 & 0 & 0 \\ 0 & z^{-1} & 0 & z^{-1} & 0 & 0 \\ 1 & 0 & -1 & 0 & 0 & 0 \\ 0 & 1 & 0 & -1 & 0 & 0 \\ 0 & 0 & 0 & 0 & 1 & 0 \\ 0 & 0 & 0 & 0 & 0 & 1 \end{bmatrix}$$

and define $\widetilde{\mathsf{U}} := (\mathsf{U}^*)^{-1}$. Let $\mathbb{P} := \mathbb{P}_{\mathsf{b}_0}\mathsf{U}$ and $\widetilde{\mathbb{P}} := \widetilde{\mathbb{P}}_{\widetilde{\mathsf{b}}_0}\widetilde{\mathsf{U}}$. Then we have $\mathsf{S}\mathbb{P} = \mathsf{S}\widetilde{\mathbb{P}} = [z^{-1}, -z^{-1}]^{\mathsf{T}}[1,-1,-1,1,1,-1]$ and $\mathbb{P}, \widetilde{\mathbb{P}}$ are given by

$$\mathbb{P} = c\begin{bmatrix} t_{11}(1+\frac{1}{z}) & t_{12}(1-\frac{1}{z}) & t_{13}(1-\frac{1}{z}) & t_{14} & t_{15}(1+\frac{1}{z}) & t_{16}(1-\frac{1}{z}) \\ t_{21}(1-\frac{1}{z}) & t_{22}(1+\frac{1}{z}) & t_{23}(1+\frac{1}{z}) & t_{24}(1-\frac{1}{z}) & t_{25}(1-\frac{1}{z}) & t_{26}(1+\frac{1}{z}) \end{bmatrix},$$

$$\widetilde{\mathbb{P}} = \widetilde{c}\begin{bmatrix} \widetilde{t}_{11}(1+\frac{1}{z}) & \widetilde{t}_{12}(1-\frac{1}{z}) & \widetilde{t}_{13}(1-\frac{1}{z}) & \widetilde{t}_{14} & \widetilde{t}_{15}(1+\frac{1}{z}) & \widetilde{t}_{16}(1-\frac{1}{z}) \\ \widetilde{t}_{21}(1-\frac{1}{z}) & \widetilde{t}_{22}(1+\frac{1}{z}) & \widetilde{t}_{23}(1+\frac{1}{z}) & \widetilde{t}_{24}(1-\frac{1}{z}) & \widetilde{t}_{25}(1-\frac{1}{z}) & \widetilde{t}_{26}(1+\frac{1}{z}) \end{bmatrix},$$

where $c = \frac{1}{486}, \widetilde{c} = \frac{3}{34,884}$ and t_{jk}'s, $\widetilde{t}_{jk}$'s are constants defined as follows:

$$\begin{array}{llllll} t_{11} = 162, & t_{12} = 34, & t_{13} = -196, & t_{14} = 0, & t_{15} = 81, & t_{16} = 29, \\ t_{21} = -126, & t_{22} = -14, & t_{13} = 176, & t_{24} = -36, & t_{15} = -99, & t_{16} = -31, \\ \widetilde{t}_{11} = 5,814, & \widetilde{t}_{12} = -1,615, & \widetilde{t}_{13} = -7,160, & \widetilde{t}_{14} = 0, & \widetilde{t}_{15} = 5,814, & \widetilde{t}_{16} = 2,584, \\ \widetilde{t}_{21} = -5,551, & \widetilde{t}_{22} = 5,808, & \widetilde{t}_{13} = 7,740, & \widetilde{t}_{24} = -1,358, & \widetilde{t}_{15} = -6,712, & \widetilde{t}_{16} = -4,254. \end{array}$$

Applying Algorithm 2, we obtain $\mathbb{P}_e$ and $\widetilde{\mathbb{P}}_e$ as follows:

$$\mathbb{P}_e = c\left[\begin{array}{cccccc} t_{11}(1+\frac{1}{z}) & t_{12}(1-\frac{1}{z}) & t_{13}(1-\frac{1}{z}) & t_{14} & t_{15}(1+\frac{1}{z}) & t_{16}(1-\frac{1}{z}) \\ t_{21}(1-\frac{1}{z}) & t_{22}(1+\frac{1}{z}) & t_{23}(1+\frac{1}{z}) & t_{24}(1-\frac{1}{z}) & t_{25}(1-\frac{1}{z}) & t_{26}(1+\frac{1}{z}) \\ \hline t_{31}(1+\frac{1}{z}) & t_{32}(1-\frac{1}{z}) & t_{33}(1-\frac{1}{z}) & t_{34}(1+\frac{1}{z}) & t_{35}(1+\frac{1}{z}) & t_{36}(1-\frac{1}{z}) \\ t_{41} & 0 & 0 & t_{44} & t_{45} & 0 \\ \hline 0 & t_{52} & t_{53} & 0 & 0 & t_{56} \\ t_{61}(1-\frac{1}{z}) & t_{62}(1+\frac{1}{z}) & t_{63}(1+\frac{1}{z}) & t_{64}(1-\frac{1}{z}) & t_{65}(1-\frac{1}{z}) & t_{66}(1+\frac{1}{z}) \end{array}\right],$$

where all t_{jk}'s are constants given by

$$\begin{array}{llll} t_{31} = 24, & t_{32} = \frac{472}{27}, & t_{33} = -\frac{148}{27}, & \\ t_{34} = -36, & t_{35} = -24, & t_{36} = -\frac{112}{27}, & \\ t_{41} = \frac{1,09,998}{533}, & t_{44} = \frac{94,041}{533}, & t_{45} = -\frac{1,09,989}{533}, & \\ t_{52} = 406c_0, & t_{53} = 323c_0, & t_{56} = 1,142c_0, & c_0 = \frac{16,09,537}{13,122}, \\ t_{61} = 24,210c_1, & t_{62} = 14,318c_1, & t_{63} = -11,807c_1, & t_{64} = -26,721c_1, \\ t_{65} = -14,616c_1, & t_{66} = -1,934c_1, & c_1 = 200/26,163. & \end{array}$$

And

$$\widetilde{\mathbb{P}}_e = \widetilde{c}\left[\begin{array}{cccccc} \widetilde{t}_{11}(1+\frac{1}{z}) & \widetilde{t}_{12}(1-\frac{1}{z}) & \widetilde{t}_{13}(1-\frac{1}{z}) & \widetilde{t}_{14} & \widetilde{t}_{15}(1+\frac{1}{z}) & \widetilde{t}_{16}(1-\frac{1}{z}) \\ \widetilde{t}_{21}(1-\frac{1}{z}) & \widetilde{t}_{22}(1+\frac{1}{z}) & \widetilde{t}_{23}(1+\frac{1}{z}) & \widetilde{t}_{24}(1-\frac{1}{z}) & \widetilde{t}_{25}(1-\frac{1}{z}) & \widetilde{t}_{26}(1+\frac{1}{z}) \\ \hline \widetilde{t}_{31}(1+\frac{1}{z}) & \widetilde{t}_{32}(1-\frac{1}{z}) & \widetilde{t}_{33}(1-\frac{1}{z}) & \widetilde{t}_{34}(1+\frac{1}{z}) & \widetilde{t}_{35}(1+\frac{1}{z}) & \widetilde{t}_{36}(1-\frac{1}{z}) \\ \widetilde{t}_{41} & 0 & 0 & \widetilde{t}_{44} & \widetilde{t}_{45} & 0 \\ \hline 0 & \widetilde{t}_{52} & \widetilde{t}_{53} & 0 & 0 & \widetilde{t}_{56} \\ \widetilde{t}_{61}(1-\frac{1}{z}) & \widetilde{t}_{62}(1+\frac{1}{z}) & \widetilde{t}_{63}(1+\frac{1}{z}) & \widetilde{t}_{64}(1-\frac{1}{z}) & \widetilde{t}_{65}(1-\frac{1}{z}) & \widetilde{t}_{66}(1+\frac{1}{z}) \end{array}\right],$$

where all $\widetilde{t}_{jk}$'s are constants given by

$$\widetilde{t}_{31} = 3,483\widetilde{c}_0, \quad \widetilde{t}_{32} = 37,427\widetilde{c}_0, \quad \widetilde{t}_{33} = 4,342\widetilde{c}_0, \quad \widetilde{t}_{34} = -12,222\widetilde{c}_0,$$
$$\widetilde{t}_{35} = -3,483\widetilde{c}_0, \quad \widetilde{t}_{36} = -7,267, \quad \widetilde{c}_0 = \frac{8,721}{4,264},$$
$$\widetilde{t}_{41} = 5,814, \quad \widetilde{t}_{44} = 1,1628, \quad \widetilde{t}_{45} = -1,1628,$$
$$\widetilde{t}_{52} = 3\widetilde{c}_1, \quad \widetilde{t}_{53} = 2\widetilde{c}_1, \quad \widetilde{t}_{56} = 10\widetilde{c}_1, \quad \widetilde{c}_1 = \frac{12,680,011}{243};$$
$$\widetilde{t}_{61} = 18,203\widetilde{c}_2, \quad \widetilde{t}_{62} = 1,01,595\widetilde{c}_2, \quad \widetilde{t}_{63} = 1,638\widetilde{c}_2, \quad \widetilde{t}_{64} = -33,950\widetilde{c}_2,$$
$$\widetilde{t}_{65} = -10,822\widetilde{c}_2, \quad \widetilde{t}_{66} = -36,582\widetilde{c}_2, \quad \widetilde{c}_2 = \frac{26,163}{2,13,200}.$$

Note that $\mathbb{P}_e$ and $\widetilde{\mathbb{P}}_e$ satisfy

$$\mathsf{S}\mathbb{P}_e = \mathsf{S}\mathbb{P}_e = [z^{-1}, -z^{-1}, z^{-1}, 1, -1, -z^{-1}]^{\mathrm{T}}[1, -1, -1, 1, 1, -1].$$

From the polyphase matrices $\mathbf{P} := \mathbb{P}_e\widetilde{\mathsf{U}}^*$ and $\widetilde{\mathbf{P}} := \widetilde{\mathbb{P}}_e\mathsf{U}^*$, we derive high-pass filters $\mathsf{b}_1, \mathsf{b}_2$ and $\widetilde{\mathsf{b}}_1, \widetilde{\mathsf{b}}_2$ as follows:

$$\mathsf{b}_1(z) = \begin{bmatrix} b^1_{11}(z) & b^1_{12}(z) \\ b^1_{21}(z) & b^1_{22}(z) \end{bmatrix}, \mathsf{b}_2(z) = \begin{bmatrix} b^2_{11}(z) & b^2_{12}(z) \\ b^2_{21}(z) & b^2_{22}(z) \end{bmatrix},$$

where

$$b^1_{11}(z) = \frac{199}{6,561} + \frac{125}{6,561}z^3 - \frac{4}{81}z^2 + \frac{199}{6,561}z - \frac{4}{81}z^{-1} + \frac{125}{6,561}z^{-2},$$
$$b^1_{12}(z) = -\frac{361}{6,561} - \frac{125}{6,561}z^3 - \frac{56}{6,561}z^2 + \frac{361}{6,561}z + \frac{56}{6,561}z^{-1} + \frac{125}{6,561}z^{-2},$$
$$b^1_{21}(z) = \frac{679}{3,198}z^3 + \frac{679}{3,198}z - \frac{679}{1,599}z^2, \quad b^1_{22}(z) = \frac{387}{2,132}z^3 - \frac{387}{2,132}z,$$
$$b^2_{11}(z) = c_3(323z^3 - 323z),$$
$$b^2_{12}(z) = c_3(406z^3 + 2,284z^2 + 406z),$$
$$b^2_{21}(z) = c_4(-36,017 + 12,403z^3 - 29,232z^2 + 36,017z + 29,232z^{-1} - 12,403z^{-2}),$$
$$b^2_{22}(z) = c_4(41,039 - 12,403z^3 - 3,868z^2 + 41,039z - 3,868z^{-1} - 12,403z^{-2}),$$
$$c_3 = \frac{27}{32,19,074}, \quad c_4 = \frac{50}{63,57,609}.$$

And

$$\widetilde{\mathsf{b}}_1(z) = \begin{bmatrix} \widetilde{b}^1_{11}(z) & \widetilde{b}^1_{12}(z) \\ \widetilde{b}^1_{21}(z) & \widetilde{b}^1_{22}(z) \end{bmatrix}, \widetilde{\mathsf{b}}_2(z) = \begin{bmatrix} \widetilde{b}^2_{11}(z) & \widetilde{b}^2_{12}(z) \\ \widetilde{b}^2_{21}(z) & \widetilde{b}^2_{22}(z) \end{bmatrix},$$

where

$$\widetilde{b}^1_{11}(z) = -\frac{859}{17,056} + \frac{7,825}{17,056}z^3 - \frac{3,483}{8,528}z^2 - \frac{859}{17,056}z - \frac{3,483}{8,528}z^{-1} + \frac{7,825}{17,056}z^{-2},$$

$$\widetilde{b}^1_{12}(z) = -\frac{49,649}{17,056} + \frac{25,205}{17,056}z^3 - \frac{559}{656}z^2 + \frac{49,649}{17,056}z + \frac{559}{656}z^{-1} - \frac{25,205}{17,056}z^{-2},$$

$$\widetilde{b}^1_{21}(z) = \frac{1}{6}(z^3 + z - 2z^2), \quad \widetilde{b}^1_{22}(z) = \frac{1}{3}(z^3 - z),$$

$$\widetilde{b}^2_{11}(z) = 2\widetilde{c}_3(z^3 - z),$$

$$\widetilde{b}^2_{12}(z) = \widetilde{c}_3(3z^3 + 10z^2 + 3z), \ \widetilde{c}_3 = \frac{39,257}{26,244};$$

$$\widetilde{b}^2_{21}(z) = -\frac{9,939}{1,70,560} + \frac{59,523}{8,52,800}z^3 - \frac{16,233}{4,26,400}z^2 + \frac{9,939}{1,70,560}z + \frac{16,233}{4,26,400}z^{-1} - \frac{59,523}{8,52,800}z^{-2},$$

$$\widetilde{b}^2_{22}(z) = \frac{81,327}{1,70,560} + \frac{40,587}{1,70,560}z^3 - \frac{4,221}{32,800}z^2 + \frac{81,327}{1,70,560}z - \frac{4,221}{32,800}z^{-1} + \frac{40,587}{1,70,560}z^{-2}.$$

Then the high-pass filters $\mathrm{b}_1, \mathrm{b}_2$ and $\widetilde{\mathrm{b}}_1, \widetilde{\mathrm{b}}_2$ satisfy (76) with $c^1_1 = c^1_2 = 1/2$, $\varepsilon^1_1 = 1, \varepsilon^1_2 = 1$ and $c^2_1 = c^2_2 = 3/2$, $\varepsilon^1_1 = -1, \varepsilon^1_2 = -1$, respectively. Using E, we can define $\mathrm{a}_1, \mathrm{a}_2$ and $\widetilde{\mathrm{a}}_1, \widetilde{\mathrm{a}}_2$ to be the high-pass filters constructed from $\mathrm{b}_1, \mathrm{b}_2$ and $\widetilde{\mathrm{b}}_1, \widetilde{\mathrm{b}}_2$ by $\mathrm{a}_1(z) := E^{-1}\mathrm{b}_1(z)E, \mathrm{a}_2 := E^{-1}\mathrm{b}_2E$ and $\widetilde{\mathrm{a}}_1(z) := E\widetilde{\mathrm{b}}_1(z)E^{-1}, \widetilde{\mathrm{a}}_2 := E\widetilde{\mathrm{b}}_2E^{-1}$.

See Fig. 4 for graphs of the 3-refinable function vectors $\phi, \widetilde{\phi}$ associated with the low-pass filters $\mathrm{a}_0, \widetilde{\mathrm{a}}_0$, respectively, and the biorthogonal multiwavelet function vectors ψ^1, ψ^2 and $\widetilde{\psi}^1, \widetilde{\psi}^2$ associated with the high-pass filters $\mathrm{a}_1, \mathrm{a}_2$ and $\widetilde{\mathrm{a}}_1, \widetilde{\mathrm{a}}_2$, respectively. Also, see Fig. 3 for graphs of the 3-refinable function vectors $\eta, \widetilde{\eta}$ associated with the low-pass filters $\mathrm{b}_0, \widetilde{\mathrm{b}}_0$, respectively, and the biorthogonal multiwavelet function vectors ζ^1, ζ^2 and $\widetilde{\zeta}^1, \widetilde{\zeta}^2$ associated with the high-pass filters $\mathrm{b}_1, \mathrm{b}_2$ and $\widetilde{\mathrm{b}}_1, \widetilde{\mathrm{b}}_2$, respectively.

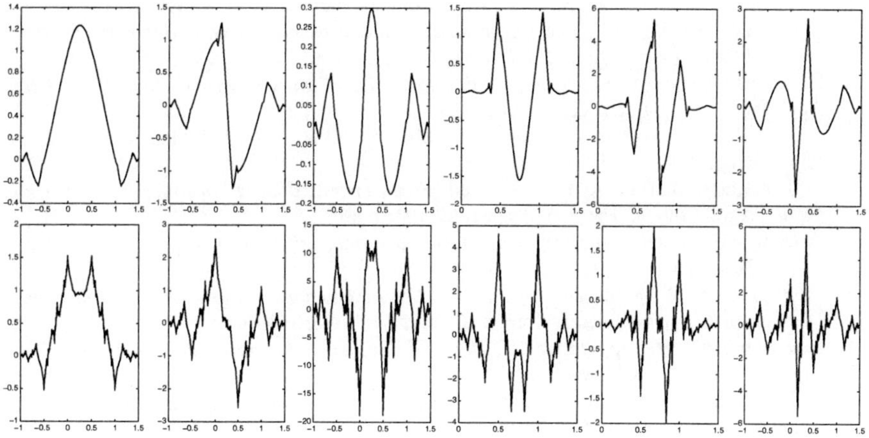

Fig. 3: Graphs of $\phi = [\phi_1, \phi_2]^{\mathrm{T}}$, $\psi^1 = [\psi^1_1, \psi^1_2]^{\mathrm{T}}$, and $\psi^2 = [\psi^2_1, \psi^2_2]^{\mathrm{T}}$ *(top, left to right)*, and $\widetilde{\phi} = [\widetilde{\phi}_1, \widetilde{\phi}_2]^{\mathrm{T}}$, $\widetilde{\psi}^1 = [\widetilde{\psi}^1_1, \widetilde{\psi}^1_2]^{\mathrm{T}}$, and $\widetilde{\psi}^2 = [\widetilde{\psi}^2_1, \widetilde{\psi}^2_2]^{\mathrm{T}}$ *(bottom, left to right)*

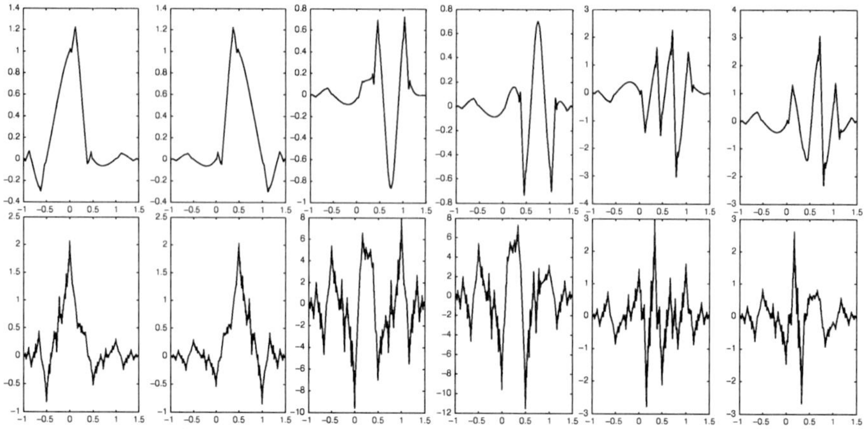

Fig. 4: Graphs of $\eta = [\eta_1, \eta_2]^{\mathrm{T}}$, $\zeta^1 = [\zeta_1^1, \zeta_2^1]^{\mathrm{T}}$, and $\zeta^2 = [\zeta_1^2, \zeta_2^2]^{\mathrm{T}}$ *(top, left to right)*, and $\widetilde{\eta} = [\widetilde{\eta}_1, \widetilde{\eta}_2]^{\mathrm{T}}$, $\widetilde{\zeta}^1 = [\widetilde{\zeta}_1^1, \widetilde{\zeta}_2^1]^{\mathrm{T}}$, and $\widetilde{\zeta}^2 = [\widetilde{\zeta}_1^2, \widetilde{\zeta}_2^2]^{\mathrm{T}}$ *(bottom, left to right)*

References

1. Y. G. Cen and L. H. Cen, *Explicit construction of compactly supported biorthogonal multiwavelets based on the matrix extension*, IEEE Int. Conference Neural Networks & Signal Processing, Zhenjiang, China, June 8∼10, 2008.
2. C. K. Chui and J. A. Lian, *Construction of compactly supported symmetric and antisymmetric orthonormal wavelets with scale* = 3, Appl. Comput. Harmon. Anal., 2 (1995), 21–51.
3. L. H. Cui, *Some properties and construction of multiwavelets related to different symmetric centers*, Math. Comput. Simul., 70 (2005), 69–89.
4. L. H. Cui, *A method of construction for biorthogonal multiwavelets system with* $2r$ *multiplicity*, Appl. Math. Comput., 167 (2005), 901–918.
5. I. Daubechies, *Ten lectures on wavelets*, CBMS-NSF Series in Applied Mathematics, SIAM, Philadelphia, 1992.
6. I. Daubechies, B. Han, A. Ron, Z. Shen, *Framelets:MRA-based constructions of wavelet frames*, Appl. Comput. Harmon. Anal, 14 (2003), 1–46.
7. B. Dong and Z. Shen, *Pseudo-splines, wavelets and framelets*, Appl. Comput. Harmon. Anal., 22 (2007), 78–104.
8. J. Geronimo, D. P. Hardin, and P. Massopust, *Fractal functions and wavelet expansions based on several scaling functions*, J. Approx. Theory, 78 (1994), 373–401.
9. S. S. Goh and V. B. Yap, *Matrix extension and biorthogonal multiwavelet construction*, Lin. Alg. Appl., 269 (1998), 139–157.
10. B. Han, *Symmetric orthonormal scaling functions and wavelets with dilation factor 4*, Adv. Comput. Math., 8 (1998), 221–247.
11. B. Han, *Approximation properties and construction of Hermite interpolants and biorthogonal multiwavelets*, J. Approx. Theory, 110 (2001), 18–53.
12. B. Han, *Matrix extension with symmetry and applications to symmetric orthonormal complex* M*-wavelets*, J. Fourier Anal. Appl., 15 (2009), 684–705.
13. B. Han, S. Kwon and X. Zhuang, *Generalized interpolating refinable function vectors*, J. Comput. Appl. Math., 227 (2009), 254–270.
14. B. Han and Q. Mo, *Splitting a matrix of Laurent polynomials with symmetry and its application to symmetric framelet filter banks*, SIAM J. Matrix Anal. Appl., 26 (2004), 97–124.

15. B. Han and X. Zhuang, *Analysis and construction of multivariate interpoalting refinable function vectors*, Acta Appl. Math., 107 (2009), 143–171.
16. B. Han and X. Zhuang, *Matrix extension with symmetry and its application to filter banks*, SIAM J. Math. Anal., 42 (5) (2010), 2297–2317.
17. P. N. .Heller, Rank M wavelets with N vanishing moments. *SIAM J. Matrix Anal. Appl.* **16**(1995), 502–519.
18. H. Ji and Z. Shen, *Compactly supported (bi)orthogonal wavelets generated by interpolatory refinable functions*, Adv. Comput. Math., 11 (1999), 81–104.
19. Q. T. Jiang, *Symmetric paraunitary matrix extension and parameterization of symmetric orthogonal multifilter banks*, SIAM J. Matrix Anal. Appl., 22 (2001), 138–166.
20. W. Lawton, S. L. Lee, and Z. Shen, *An algorithm for matrix extension and wavelet construction*, Math. Comp., 65 (1996), 723–737.
21. A. Petukhov, *Construction of symmetric orthogonal bases of wavelets and tight wavelet frames with integer dilation factor*, Appl. Comput. Harmon. Anal., 17 (2004), 198–210.
22. Y. Shen, S. Li, and Q. Mo, *Complex wavelets and framelets from pseudo splines*, J. Fourier Anal. Appl., 16 (6) (2010), 885–900.
23. Z. Shen, *Refinable function vectors*, SIAM J. Math. Anal., 29 (1998), 235–250.
24. P. P. Vaidyanathan, *Multirate systems and filter banks*, Prentice Hall, New Jersey, 1992. (1984), 513–518.
25. D. C. Youla and P. F. Pickel, *The Quillen-Suslin theorem and the structure of n-dimesional elementary polynomial matrices*, IEEE Trans. Circ. Syst., 31 (1984), 513–518.
26. X. Zhuang, *Construction of symmetric complex tight wavelet frames from pseudo splines via matrix extension with symmetry*, Preprint, `http://arxiv.org/abs/1003.3500`.